S0-CFF-687

Introductory Engineering Statistics

Second Edition

IRWIN GUTTMAN
University of Montréal and the
University of Wisconsin

The Late S. S. WILKS
Princeton University

J. STUART HUNTER
Princeton University

JOHN WILEY & SONS, INC.,
New York . London . Sydney . Toronto

131842

519
G985-1

Copyright © 1965, 1971, by John Wiley & Sons, Inc.

All rights reserved. Published simultaneously in Canada.

No part of this book may be reproduced by any means, nor transmitted, nor translated into a machine language without the written permission of the publisher.

Library of Congress Catalogue Card Number: 72-160214

ISBN O-471-33770-6

Printed in the United States of America.

10 9 8 7 6 5 4 3 2

To Mary Alice and Tady

Preface

The previous edition was planned by S. S. Wilks and Irwin Guttman in the summer of 1959, and much statistical water has flowed under statistical bridges since that time. In preparing this second edition we have been guided by two significant currents of the decade of the 1960s, which are especially important to engineering sciences (and, indeed, other physical sciences).

One of these currents was generated by the attention of the statistical fraternity to *Bayesian statistics*. We feel that engineers are constantly in situations where prior information about a process under study is available. Since Bayesian statistics gives a formal method of incorporating prior information into statistical analyses, an up-to-date introductory account of engineering statistics should give some idea of this important area. Accordingly we have introduced new material in Chapter 2 (Elements of Probability) and have added a complete new chapter (Chapter 9A) to give the background and some essential ideas necessary for the use and understanding of Bayesian statistics. (Indeed, because we feel that this subject is very important, we were tempted to recast the entire text in a Bayesian light. But since this is an introductory book, we resisted this temptation, promising ourselves that the third edition will include "still more" material on Bayes.)

The second significant current, alluded to above, is the one generated by the study of experimental designs suitable for use in this modern age of technology—factorial designs and their use in "response surface" studies. This area of statistics has seen enormous development in the past decade and should be an important instrument in the engineering statisticians' tool chest. With this in mind, we have introduced a new chapter (Chapter 17), which surveys the elements of this area.

Many other changes and revisions of the text have been made, hopefully to improve the exposition and to clarify ideas. Also, we have added many more problems, and answers to selected problems are given at the end of the text. Mrs. Stella Yu contributed importantly to this aspect.

This second edition is suitable for a two-semester sequence of introductory statistics to engineers. A one-semester sequence also could be drawn from the text by covering Chapters 2–7, 9, 9A, and 15.

We happily acknowledge the advice and critical comments of D. G. Watts and N. R. Draper, and thank them for their constructive help.

We are indebted to the Literary Executor of the late Sir Ronald A. Fisher, F.R.S., to Dr. Frank Yates, F.R.S., and to Oliver and Boyd, Edinburgh, for permission to reprint Tables V1 and V2 from their book *Statistical Tables for use in Biological, Agricultural and Medical Research.*

Finally, we again give special thanks to Mrs. R. C. Werkman for her careful typing, good humor, and encouragement of this revision.

IRWIN GUTTMAN
J. STUART HUNTER

Preface to First Edition

The purpose of this book is to give undergraduate engineers a facility in, and understanding, of some elementary applied statistical techniques. No previous background in probability and statistics is assumed. It is assumed, however, that the reader has had at least one introductory course in Calculus.

The present version of the book is an outgrowth of notes written by the authors for courses at Princeton University, McGill University, and the University of Wisconsin. At both McGill and Princeton, the courses consisted of one semester and comprised topics from Chapters 2 through 10, and Chapter 15, plus statements and discussion of the relevant theorems from the Appendices (this latter material is presented after Section 4 of Chapter 8). The course at Wisconsin carries on two semesters and topics from all the chapters are covered.

Naturally, the choice of the topics, their order of presentation, and extent of coverage are subjective at the best of times. What we have included reflects our feelings on what initially a beginning student in statistics should be exposed to, when indeed said student is more likely than not one day to be applying statistical techniques to real data. About 240 problems are provided to give the student practice in the techniques discussed.

The authors have benefited from the advice and counsel of Professor J. Stuart Hunter, Princeton University. To David Stoneman, for his help with the problem sections, and Charles Palit, for his enthusiastic aid with all stages of the manuscript, we give our sincere thanks. The authors also would like to express their appreciation to Mrs. R. C. Werkman for her careful typing of the various stages of the notes, and especially, for the wonderful job on the final version of the manuscript.

IRWIN GUTTMAN

Madison, Wisconsin
February 1965

ix

Dedication of First Edition

One imagines that when a book such as this finally emerges into the light of day, all who had a hand in its birth would have a certain satisfaction and joy. There is, however, but sadness to this writer.

For unbelievably, Sam Wilks has suddenly departed from this life. This warm, vital, and multitalented man is lost.

It is true that Sam Wilks was an eminently successful statistician, but more importantly, he was as eminently successful as a human being, friend, colleague, citizen. It was a rare privilege to know and work with him, and we are richer for it in many, many ways.

In memory, then, of a wonderful man, this book is humbly and respectfully dedicated.

IRWIN GUTTMAN

Madison, Wisconsin
April 23, 1965

Contents

xiii

Introductory
Engineering Statistics

CHAPTER 1

Introduction

In this book we shall be concerned with an introductory discussion of the application of the mathematical theory of probability to certain kinds of problems in engineering, particularly in mass production and in the analysis of experimental data. For some kinds of problems the application of probability theory is fairly simple and direct. For others it leads into rather sophisticated methods of statistical inference. In this Chapter we shall give a general indication of the material to be covered in this book.

Since probability is fundamental to the techniques of statistical inference, we present the basic principles of elementary probability theory in Chapter 2. First, however, there are some fundamental concepts underlying the mathematical theory of probability which can be discussed informally and intuitively here before making a mathematically more formal presentation in Chapter 2.

The first basic concept underlying the theory of probability is that of a *repetitive operation* or *repetitive experiment*, that is, the notion of being able to repeat an operation over and over again "under essentially the same conditions." Some examples of repetitive operations are: (1) Tossing a coin; (2) throwing a die; (3) shuffling a pack of playing cards and then cutting the pack; (4) taking a can of beer "at random" from a lot of 12,000 cans; (5) taking 2 screws "at random" from a box of 100 screws.

The second basic concept underlying probability theory is the set or collection of all possible outcomes of a repetitive operation. This set of outcomes is called the *sample space* of the operation. Thus, in example 1 the sample space is a set of two outcomes $\{H, T\}$, where H denotes "head" and T denotes "tail." In example 2, the sample space is a set of six outcomes $\{1, 2, 3, 4, 5, 6\}$ corresponding to the six faces of the die. The sample space of example 3 is a set of 52 outcomes corresponding to the 52

1

different cards in the deck. The sample space of example 4 is a set of 12,000 outcomes corresponding to the different cans which can be drawn. The sample space of example 5 is a set of 4950 outcomes corresponding to all possible pairs of screws which can be drawn.

The third basic concept underlying probability theory is that of an *event*, which is some subset of the sample space of particular interest. For instance, in example 3, "cutting a spade" is an event which consists of 13 of the 52 possible outcomes in the sample space. In example 5, suppose the box contains 10 defective screws; the event of "getting no defective screws" in the pair drawn is the subset of 4005 of the 4950 possible outcomes in the sample space in which the pair of screws drawn contains no defectives.

The fourth basic concept is that of the *probability of an event*. If we conceive of performing a repetitive operation over and over again indefinitely many times under essentially the same conditions and consider the fraction of trials in which a specified event occurs, this fraction is what is meant by the probability of the occurrence of that event. It is obviously impossible to determine exactly the value of the probability of a specified event, since it is physically impossible to perform indefinitely many trials of any operation. There are two possibilities, however, of making some progress. The first is to set up *randomization* procedures which will make it safe to assign specified numerical values of probabilities. Such randomization procedures can often be implemented by one device or another, the most satisfactory and widely used one being *random numbers*. For instance, by the proper use of random numbers, one can select a particular card from a deck of playing cards so that it is safe to assign the value 1/52 to the probability of that particular card being drawn. This would, of course, be equivalent to "cutting a thoroughly shuffled deck" of cards.

The second possibility, in case randomization devices are inapplicable, is to repeat the operation (or allow it to be repeated) under the given conditions a large number of times and use the fraction of times the given event occurs as an estimate of the probability. The probabilities of death at various ages in mortality tables, for instance, are estimated in this way.

The fifth basic concept is that of *conditional probability*. If we are given the information that an event F has occurred, what then is the probability of the event E, where both E and F are events in our sample space? Knowledge of the occurrence of the event F means that we are considering a new sample space. For example, if a card is drawn at random from an ordinary deck, and we are given that it is a heart, and then wish to find the probability that the card is a three or a four, then the new sample space we must consider is the set of outcomes {2 of hearts, 3 of hearts, . . . , 10 of hearts, jack of hearts, queen of hearts, king of hearts, ace of hearts}.

The sixth basic concept underlying the theory of probability is that of a *random variable*, which is simply some function (real and single-valued) of the outcome of a "repetitive operation" in which we have some particular interest. For instance, if a pair of dice is thrown the total number of dots which appears is a random variable. If a sample of 50 light bulbs is drawn "at random" from a lot of 1000, and tested, the number of defective bulbs in the sample is a random variable. If a light bulb is taken at random from a large lot and burned, the time required for it to "expire" is a random variable.

The seventh basic concept in probability theory is that of the *distribution function* or *probability law* of a random variable, which provides the probability that the random variable has any specified value (or is less than any specified number) when the basic "repetitive operation" is performed once.

Distribution functions or probability laws play very important roles in probability theory and statistical inference. Accordingly, Chapters 3, 6, and 7 are devoted to some of the most important distribution functions.

Chapter 4 is concerned with what is usually called *acceptance sampling* or *sampling inspection* methods. These are procedures based directly on some fairly simple probability laws for deciding whether to accept or reject a *lot* of articles, with specified degrees of risk of accepting a substandard lot and of rejecting an acceptable lot, on the basis of the number of defective articles found in one or more samples drawn at random from the lot.

Chapter 8 deals with a similar problem, and uses the information contained in the mean of observations or measurements made on the articles in a random sample of articles from a lot as the basis for deciding whether to accept or reject the lot.

Chapter 5 is devoted to procedures for numerically describing and summarizing information contained in measurements on the items in a random sample from a lot or a population of items.

In some problems in probability and statistics, the functional form of the underlying probability laws often can be determined or safely assumed, except for numerical values of one or more *parameters*. Chapters 9 and 9A deal with *statistical estimation*, that is, with problems of estimating parameters from samples of observations, together with methods of ascribing the degree of uncertainty inherent in such estimators. Two approaches to this problem are given—Chapter 9 gives the classical approach, while Chapter 9A gives an alternative approach, called the *Bayesian approach*.

Chapter 10 is devoted to *statistical tests*, that is, methods of using information in one or more samples of observations to decide whether a sample could "reasonably" have come from a population with prescribed characteristics, or whether two samples could have "reasonably" come from the same population or from populations which are identical with respect to certain characteristics.

Chapter 11 deals with *control charts*, which are devices for determining whether a repetitive operation such as a mass production process is "under statistical control" on the basis of information contained in a sequence of small samples of items which have been subjected to the given mass production process.

Chapter 12 is concerned with *goodness of fit tests*, that is, methods for testing whether the observed frequencies with which several mutually exclusive events have occurred are consistent or not "beyond a reasonable doubt" with probabilities prescribed for these events according to some specified hypothesis.

Chapter 13 is devoted to *order statistics*, that is, with the probability theory and some applications of the theory, of the smallest, largest, and other quantities obtained when measurements on the items in a random sample are ordered from smallest to largest.

Chapter 14 gives some of the more important applications of the results of Chapter 13 to problems of statistical inference.

Chapter 15 is concerned with *regression analysis*, that is, methods of estimating values of one variable from observed values of one or more additional variables, where the relationships among the variables are estimated from observations on all variables in a random sample.

Chapter 16 is concerned with *analysis of variance*, that is, with principles and methods of designing certain kinds of experiments and statistically analyzing the results of such experiments.

In Chapter 17, we continue our discussion of designing experiments, and introduce some concepts necessary for such a discussion. We then describe a class of designs (2-level factorial designs) and their use in response-surface methodology. Other designs useful in response-surface methods are also introduced.

CHAPTER 2

Elements of Probability

In Chapter 1 we have indicated the kinds of problems with which modern statistical methods are concerned. It is evident throughout that discussion that the theory of probability plays a fundamental role in dealing with these problems. We have also discussed the nature of probability from an intuitive point of view. Before we proceed further, we shall need a more formal treatment of the mathematical elements of probability theory. The purpose of the present chapter is to present such a treatment.

2.1 REPETITIVE OPERATIONS AND SAMPLE SPACES

Inherent in any situation to which the theory of probability is applicable is the notion of *performing a repetitive operation*, that is, repeating a trial or experiment over and over again "under essentially the same conditions." A few examples of performing a repetitive operation are:

1. Rolling a die.
2. Tossing two coins.
3. Drawing 5 screws "at random" from a box of 100 screws.
4. Dealing 13 cards from a thoroughly shuffled deck of playing cards.
5. Filling a 12-ounce can with beer by an automatic filling machine.
6. Drawing a piece of steel rod by a testing machine until it breaks.
7. Firing a rifle at a target 100 yards away.
8. Burning ten 60-watt bulbs with filament of type Z continuously until they all "expire."

One of the basic features to be found in repetitive operations under specified conditions is that the *outcome* varies from trial to trial. This leads us to analyze the possible outcomes which could arise if a trial or experiment were performed once. [The set of all possible outcomes which could arise if the experiment were performed once is called the *sample space* of the experiment, which will be denoted by S. It is convenient to call each outcome e in a sample space S a *sample space element* or simply element. We sometimes say that the sample space S of such elements or points is *generated* by the operation or trial or experiment.

    Example 2.1(a). If a die is rolled once, the sample space S thus generated consists of six possible outcomes; the die can turn up faces numbered 1, 2, 3, 4, 5, or 6. Thus, in this case we may write

$$S = \{1, 2, 3, 4, 5, 6\}.$$

    Example 2.1(b). If two coins are tossed, let us say a nickel and a dime, and if we designate head and tail on a nickel by H and T respectively, and head and tail on a dime by h and t respectively, the sample space S generated by tossing the two coins consists of four possible outcomes. That is, we may write

$$S = \{Hh, Ht, Th, Tt\}.$$

    Example 2.1(c). The sample space S for drawing 5 screws "at random" from a box of 100 consists of all possible sets of 5 screws which could be drawn from 100; S contains 75,287,520 elements.

    Example 2.1(d). In dealing 13 cards from a thoroughly shuffled deck of playing cards, the sample space S consists of the 635,013,559,600 possible hands of 13 cards which could be drawn.

The sample spaces for the preceding examples are all *finite sample spaces;* they contain only a finite number of sample points.

Many problems in probability involve *infinite sample spaces*, that is, sample spaces containing an indefinitely large number of elements.

    Example 2.1(e). If a coin is tossed indefinitely many times, the sample space thus generated is infinite: it consists of all possible unending sequences which can be formed by using the letters H and T.

    Example 2.1(f). The sample space S for filling a 12-ounce can with beer with the use of an automatic filling machine under factory conditions would contain an indefinitely large number of elements.

The sample spaces for examples 6, 7, 8 of repetitive operations listed above are also infinite.

2.2 EVENTS

Suppose S is the sample space of some repetitive operation which contains a finite number of outcomes or elements $e_1, \ldots, e_m$. In most probability problems we are more interested in whether an outcome belongs to some set E of outcomes than in individual outcomes. For instance, if playing craps, one is usually more interested in the total number of dots which appear when the two dice are thrown than in any particular outcome obtained from throwing a pair of dice. An inspector who examines 5 screws taken "at random" from a box of 100 is not really interested in any particular one of the 75,287,520 different sets of 5 screws he could have drawn; he is interested in the number of defective screws he gets in the 5 he draws. In other words, he is interested in whether the outcome he obtains belongs to the set of outcomes with 0 defectives, or the set with 1 defective, or the set with 2 defectives, and so on.

[A set of outcomes (in which there might be some particular interest) is called an *event.*]

>*Example 2.2(a).* The event E of getting exactly one head in throwing the 2 coins of Example 2.1(b) consists of the set of two elements $\{Ht, Th\}$ from the sample space $S = \{Hh, Ht, Th, Tt\}$.

>*Example 2.2(b).* The event E of getting a hand of cards with 12 spades consists of the set of 507 elements out of the 635,013,559,600 elements in the sample space S for dealing a hand of 13 cards which contain 12 spades.

Schematically, if we let the set of points inside the rectangle in Figure 2.2.1 represent a sample space S, we may represent an event E by the set of points inside the circle, and $\bar{E}$ (in S) by the region outside the circle. Such a representation is called a *Venn diagram.*

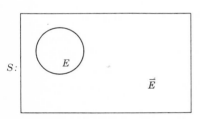

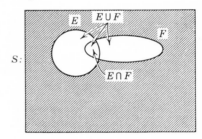

Figure 2.2.1. Venn diagram representing events E and $\bar{E}$.

Figure 2.2.2. Venn diagram representing events E, F, $E \cup F$ and $E \cap F$.

It is evident that we can describe events in the language of sets, in which the words *set* and *event* can be used interchangeably.

[If E contains no elements, it is called the *empty, impossible,* or *null* event, and is denoted by ϕ.]

The *complement* $\bar{E}$ of an event E consists of all elements in S which are not in E. $\bar{E}$, of course, is an event. Note that $\bar{S} = \phi$.

Now suppose we have two events E and F. [The event consisting of all elements contained in E or F, or both, is called the *union* of E and F; it is written as]

(2.2.1) $E \cup F.$

The event consisting of all elements in a sample space S contained in both E and F is called the *intersection* of E and F: it is written

(2.2.2) $$E \cap F.$$

Referring to the Venn diagram in Figure 2.2.2, if S is represented by the points inside the rectangle, E by the points inside the circle, and F by the points inside the ellipse, then $E \cup F$ is represented by the points in the region not cross-hatched in the rectangle, and $E \cap F$ is represented by the points in the region in which the circle and ellipse overlap.

> **Example 2.2(c).** Suppose S is the set of all possible hands of 13 cards, E is the set of all hands containing 5 spades, and F is the set of all hands containing 6 honor cards. Then $E \cup F$ is the set of all hands containing 5 spades *or* 6 honor cards, *or both*. $E \cap F$ is the set of all hands containing 5 spades *and* 6 honor cards. If there are no elements which belong to both E and F, then
>
> (2.2.3) $$E \cap F = \phi,$$
>
> and the sets E and F are said to be *disjoint*.

If all elements in E are also contained in F, then we say that E is a *subevent* of F, and we write

(2.2.4) $$E \subset F, \quad \text{or} \quad F \supset E.$$

This means that, if E occurs, then F necessarily occurs.

> **Example 2.2(d).** Let S be the sample space obtained if 5 screws are drawn from a box of 100 screws of which 10 are defective. If E is the event consisting of all possible sets of 5 screws containing one defective screw and F is the event consisting of all possible sets of five screws containing at least one defective, then $E \subset F$.

If $E \subset F$ and $F \subset E$, then every element of E is an element of F and vice versa. In this case, we say that E and F are *equal* or *equivalent* events; we write

(2.2.5) $$E = F.$$

The set of elements in E which are not contained in F is called the *difference* between E and F: it is written

(2.2.6) $$E - F.$$

If F is contained in E, then $E - F$ is the *proper difference* between E and F. Notice that in either case

(2.2.7) $$E - F = E \cap \bar{F}.$$

> **Example 2.2(e).** If E is the set of all possible bridge hands with exactly 5 spades and if F is the set of all possible hands with exactly 6 honor cards, then $E - F$ is the set of all hands with exactly 5 spades but not containing exactly 6 honor cards.

number of dots which appear when the two dice are thrown than in any particular outcome obtained from throwing a pair of dice. An inspector who examines 5 screws taken "at random" from a box of 100 is not really interested in any particular one of the 75,287,520 different sets of 5 screws he could have drawn; he is interested in the number of defective screws he gets in the 5 he draws. In other words, he is interested in whether the outcome he obtains belongs to the set of outcomes with 0 defectives, or the set with 1 defective, or the set with 2 defectives, and so on.

A set of outcomes (in which there might be some particular interest) is called an *event.*

> **Example 2.2(a).** The event E of getting exactly one head in throwing the 2 coins of Example 2.1(*b*) consists of the set of two elements $\{Ht, Th\}$ from the sample space $S = \{Hh, Ht, Th, Tt\}$.

> **Example 2.2(b).** The event E of getting a hand of cards with 12 spades consists of the set of 507 elements out of the 635, 013, 559, 600 elements in the sample space S for dealing a hand of 13 cards which contain 12 spades.

Schematically, if we let the set of points inside the rectangle in Figure 2.2.1 represent a sample space S, we may represent an event E by the set of points inside the circle, and $\bar{E}$ (in S) by the region outside the circle. Such a representation is called a *Venn diagram.*

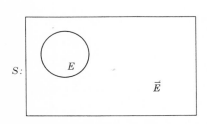

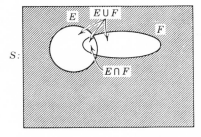

Figure 2.2.1. Venn diagram representing events E and $\bar{E}$.

Figure 2.2.2. Venn diagram representing events $E, F, E \cup F$ and $E \cap F$.

It is evident that we can describe events in the language of sets, in which the words *set* and *event* can be used interchangeably.

If E contains no elements, it is called the *empty, impossible,* or *null* event, and is denoted by ϕ.

The *complement* $\bar{E}$ of an event E consists of all elements in S which are not in E. $\bar{E}$, of course, is an event. Note that $\bar{S} = \phi$.

Now suppose we have two events E and F. The event consisting of all elements contained in E or F, or both, is called the *union* of E and F; it is written as

(2.2.1) $$E \cup F.$$

The event consisting of all elements in a sample space S contained in both E and F is called the *intersection* of E and F: it is written

(2.2.2) $E \cap F.$

Referring to the Venn diagram in Figure 2.2.2, if S is represented by the points inside the rectangle, E by the points inside the circle, and F by the points inside the ellipse, then $E \cup F$ is represented by the points in the region not cross-hatched in the rectangle, and $E \cap F$ is represented by the points in the region in which the circle and ellipse overlap.

> **Example 2.2(c).** Suppose S is the set of all possible hands of 13 cards, E is the set of all hands containing 5 spades, and F is the set of all hands containing 6 honor cards. Then $E \cup F$ is the set of all hands containing 5 spades *or* 6 honor cards, *or both*. $E \cap F$ is the set of all hands containing 5 spades *and* 6 honor cards. If there are no elements which belong to both E and F, then
>
> (2.2.3) $E \cap F = \phi,$
>
> and the sets E and F are said to be *disjoint.*

If all elements in E are also contained in F, then we say that E is a *subevent* of F, and we write

(2.2.4) $E \subset F,$ or $F \supset E.$

This means that, if E occurs, then F necessarily occurs.

> **Example 2.2(d).** Let S be the sample space obtained if 5 screws are drawn from a box of 100 screws of which 10 are defective. If E is the event consisting of all possible sets of 5 screws containing one defective screw and F is the event consisting of all possible sets of five screws containing at least one defective, then $E \subset F$.

If $E \subset F$ and $F \subset E$, then every element of E is an element of F and vice versa. In this case, we say that E and F are *equal* or *equivalent* events; we write

(2.2.5) $E = F.$

The set of elements in E which are not contained in F is called the *difference* between E and F: it is written

(2.2.6) $E - F.$

If F is contained in E, then $E - F$ is the *proper difference* between E and F. Notice that in either case

(2.2.7) $E - F = E \cap \bar{F}.$

> **Example 2.2(e).** If E is the set of all possible bridge hands with exactly 5 spades and if F is the set of all possible hands with exactly 6 honor cards, then $E - F$ is the set of all hands with exactly 5 spades but not containing exactly 6 honor cards.

If $E_1, \ldots, E_k$ are several events in a sample space S, the event consisting of all elements contained in one or more of the E's is the union of $E_1, \ldots, E_k$ and is written

(2.2.8) $$E_1 \cup \cdots \cup E_k$$

or, more briefly,

(2.2.8a) $$\bigcup_{i=1}^{k} E_i.$$

Similarly, the event consisting of all elements contained in all E's is the intersection of $E_1, \ldots, E_k$ and is written as

(2.2.9) $$E_1 \cap \cdots \cap E_k$$

or, more briefly, as

(2.2.9a) $$\bigcap_{i=1}^{k} E_i.$$

If for every pair of events (E_i, E_j) in $E_1, \ldots, E_k$ we have $E_i \cap E_j = \phi$, we say that $E_1, \ldots, E_k$ are *disjoint* and *mutually* exclusive events.

An important result concerning several events is the following:

THEOREM 2.2.1. *If $E_1, \ldots, E_k$ are events in a sample space S, then $\bigcup_{i=1}^{k} E_i$ and $\bigcap_{i=1}^{k} \bar{E}_i$ are disjoint events whose union is S.*

This can be verified by the reader by reference to a Venn diagram.

2.3 PROBABILITY

Suppose we have a sample space S which consists of a finite number of elements $e_1, \ldots, e_m$. If we think of repeating the operation whose sample space is S a large number of times, some of these repetitions will result in e_1, some in e_2, and so on. Let $f_1, \ldots, f_m$ be the fractions of the total number of trials resulting in $e_1, \ldots, e_m$ respectively. Then $f_1, \ldots, f_m$ are all non-negative and their sum is 1. We may think of $f_1, \ldots, f_m$ as observed weights or measures of occurrence of $e_1, \ldots, e_m$ obtained on the basis of an experiment consisting of a large number of repeated trails. If we were to repeat the entire experiment, we would obtain another set of f's with slightly different values, and so on for further repetitions. If we think of indefinitely many repetitions, we can conceive of idealized values being obtained for the f's. It is impossible, of course, to show that in a physical experiment the f's converge to limiting values, in a strict mathematical sense, as the number of trials increase indefinitely. So we postulate values $p(e_1), \ldots, p(e_m)$ corresponding to the idealized values of

$f_1, \ldots, f_m$ respectively, for an indefinitely large number of trials. It is assumed that $p(e_1), \ldots, p(e_m)$ are all positive numbers and that

(2.3.1) $$p(e_1) + \cdots + p(e_m) = 1.$$

The quantities $p(e_1), \ldots, p(e_m)$ are called *probabilities* of occurrence of $e_1, \ldots, e_m$ respectively.

Now suppose E is any event in S. It will consist of a set of one or more e's, say $e_{i_1}, \ldots, e_{i_r}$. Thus $E = \{e_{i_1}, \ldots, e_{i_r}\}$. The probability of the occurrence of E is denoted by $P(E)$, and is defined as follows:

$$P(E) = p(e_{i_1}) + \cdots + p(e_{i_r})$$

or, more briefly,

$$P(E) = \sum_{\alpha=1}^{r} p(e_{i_\alpha}).$$

If E contains only one element, say e_j, we may write

$$E = \{e_j\}, \quad \text{and} \quad P(E) = p(e_j).$$

It is evident that probabilities of events in a finite sample space S are values of an *additive set function* $P(E)$ defined on sets E in S, satisfying the following conditions:

1. *If E is any event in S, then*
$$P(E) \geq 0.$$

2. $$P(S) = 1.$$

3. *If E and F are two disjoint events, then*

(2.3.2) $$P(E \cup F) = P(E) + P(F).$$

In the case of an infinite sample space S, we replace condition 3 by 3a, as follows:

3a. *If $E_1, E_2, \ldots$ is an infinite sequence of disjoint events, we assume that*

(2.3.2a) $$P(E_1 \cup E_2 \cup \cdots) = P(E_1) + P(E_2) + \cdots.$$

Note that E and $\bar{E}$ are disjoint events, and from condition 3, we obtain

(2.3.3) $$P(E \cup \bar{E}) = P(E) + P(\bar{E}).$$

But $E \cup \bar{E} = S$ and, since $P(S) = 1$, we have the following:

THEOREM 2.3.1 (*Rule of complementation*). *If E is an event in a sample space S, then*

(2.3.4) $$P(\bar{E}) = 1 - P(E).$$

We sometimes say that the *odds* in favor of E are

(2.3.4a)   Odds $(E) = P(E)/P(\bar{E})$,

which from (2.3.4) takes the form $P(E)/[1 - P(E)]$. The reader will note that $0 \leq$ Odds $(E) \leq \infty$.

The law of complementation provides a simple method of finding the probability of an event $\bar{E}$ if E is an event whose probability is easy to find.

> **Example 2.3(a).** Suppose 10 coins are tossed and we ask for the probability of getting at least 1 head. In this example, the sample space S has $2^{10} = 1024$ sample points, and if the coins are "true," we assign equal probabilities to these 1024 sample points, that is, the probability of $1/1024$ at each point. If we denote by E the event of getting no heads, then E contains only one sample point (and $\bar{E}$, of course, has 1023 sample points). Thus,
>
> $$P(\bar{E}) = 1 - \tfrac{1}{1024} = \tfrac{1023}{1024}.$$
>
> The odds on E are easily seen to be: Odds $(E) = 1/1023$.

Referring to the statement in Theorem 2.2.1 that

$$\bigcup_{i=1}^{k} E_i \quad \text{and} \quad \bigcap_{i=1}^{k} \bar{E}_i$$

are disjoint events whose union is S, we have the following:

THEOREM 2.3.2 (*General rule of complementation*). If $E_1, \ldots, E_k$ are *events in a sample space S, then*

(2.3.5) $$P\left(\bigcap_{i=1}^{k} \bar{E}_i\right) = 1 - P\left(\bigcup_{i=1}^{k} E_i\right).$$

Another useful result, which follows readily from condition 3, by mathematical induction is

THEOREM 2.3.3 (*Rule of addition of probabilities for mutually exclusive events*). If $E_1, \ldots, E_k$ are *disjoint events in a sample space S, then*

(2.3.6) $$P(E_1 \cup \cdots \cup E_k) = P(E_1) + \cdots + P(E_k).$$

> **Example 2.3(b).** Suppose a nickel and a dime are tossed with H and T denoting head and tail for the nickel and h and t denoting head and tail for the dime. The sample space S consists of the four elements Hh, Ht, Th, Tt. If these four elements are all assigned equal probabilities and if E is the event of getting exactly one head, then $E = \{Ht\} \cup \{Th\}$ and
>
> $$P(E) = P(\{Ht\} \cup \{Th\}) = P(\{Ht\}) + P(\{Th\}) = \tfrac{1}{4} + \tfrac{1}{4} = \tfrac{1}{2}.$$

Now suppose E_1 and E_2 are arbitrary events in S. Then $E_1 \cap E_2$, $E_1 \cap \bar{E}_2$, $\bar{E}_1 \cap E_2$ are three disjoint events whose union is $E_1 \cup E_2$. Therefore,

(2.3.7) $$P(E_1 \cup E_2) = P(E_1 \cap E_2) + P(E_1 \cap \bar{E}_2) + P(\bar{E}_1 \cap E_2).$$

Also $E_1 \cap \bar{E}_2$ and $E_1 \cap E_2$ are disjoint sets whose union is E_1. Hence,

(2.3.8) $$P(E_1) = P(E_1 \cap \bar{E}_2) + P(E_1 \cap E_2).$$

Similarly,

(2.3.9) $$P(E_2) = P(\bar{E}_1 \cap E_2) + P(E_1 \cap E_2).$$

Solving (2.3.8) for $P(E_1 \cap E_2)$, and (2.3.9) for $P(\bar{E}_1 \cap E_2)$ and substituting in (2.3.7), we obtain the following:

THEOREM 2.3.4 (*Rule for addition of probabilities for two arbitrary events*). *If E_1 and E_2 are any two events in a sample space S, then*

(2.3.10) $$P(E_1 \cup E_2) = P(E_1) + P(E_2) - P(E_1 \cap E_2).$$

The rule for three events E_1, E_2, E_3 is given by

$$P(E_1 \cup E_2 \cup E_3) = P(E_1) + P(E_2) + P(E_3) - P(E_1 \cap E_2)$$
(2.3.11) $$- P(E_1 \cap E_3) - P(E_2 \cap E_3) + P(E_1 \cap E_2 \cap E_3).$$

Rules (2.3.10) and (2.3.11) can be verified by the reader by drawing Venn diagrams.

More generally, for n events $E_1, \ldots, E_n$, we have

(2.3.12) $$P(E_1 \cup \cdots \cup E_n) = \sum_{i=1}^{n} P(E_i) - \sum_{j>i=1}^{n} P(E_i \cap E_j)$$
$$+ \sum_{k>j>i=1}^{n} P(E_i \cap E_j \cap E_k) + \cdots$$
$$+ (-1)^{n-1} P(E_1 \cap \cdots \cap E_n).$$

Note that, if E_1 and E_2 are disjoint, $P(E_1 \cap E_2) = 0$, and (2.3.10) reduces to (2.3.6) for $k = 2$, that is,

$$P(E_1 \cup E_2) = P(E_1) + P(E_2).$$

Similarly, if E_1, E_2, E_3 are disjoint, (2.3.11) reduces to (2.3.6) with $k = 3$.

Example 2.3(c). A box of 1000 rivets contains:
50 rivets with type A defects;
32 rivets with type B defects;
18 rivets with type C defects;
7 rivets with type A and type B defects;
5 rivets with type A and type C defects;
4 rivets with type B and type C defects;
2 rivets with all three types of defects.

Suppose a rivet is picked at random from the box. Denote the event of getting a rivet with a type A defect by E_A, with similar meanings for E_B, E_C. The sample space S for this operation contains 1000 elements to each of which we assign the same probability, namely, 1/1000. The numbers of elements in various events with which we are concerned are as follows:

Event	Number of Points
E_A	50
E_B	32
E_C	18
$E_A \cap E_B$	7
$E_A \cap E_C$	5
$E_B \cap E_C$	4
$E_A \cap E_B \cap E_C$	2

The probability that the rivet will have a type A defect or type B defect, or both, is

$$P(E_A \cup E_B) = P(E_A) + P(E_B) - P(E_A \cap E_B)$$
$$= \tfrac{50}{1000} + \tfrac{32}{1000} - \tfrac{7}{1000} = \tfrac{75}{1000}.$$

The probability that the rivet will have at least one of the three types of defect is given by

$$P(E_A \cup E_B \cup E_C) = P(E_A) + P(E_B) + P(E_C) - P(E_A \cap E_B)$$
$$- P(E_A \cap E_C) - P(E_B \cap E_C) + P(E_A \cap E_B \cap E_C)$$
$$= \tfrac{1}{1000}(50 + 32 + 18 - 7 - 5 - 4 + 2) = \tfrac{86}{1000}.$$

The probability that the rivet will be free of these defects is given by

$$P(\bar{E}_A \cap \bar{E}_B \cap \bar{E}_C) = 1 - P(E_A \cup E_B \cup E_C)$$
$$= 1 - \tfrac{86}{1000} = \tfrac{914}{1000}.$$

It is to be noticed that the evaluation of the probability of an event implies two different types of operation. For example, if $A = E_A \cup E_B \cup E_C$ and we wish to determine $P(A)$, we first must interpret the event A using the *algebra of sets* developed in Section 2.2. This is the first important step. Having done so, then we must "operate" on A with the function P in accordance with the "axioms of probability" listed on page 10 and their consequences (Theorems 2.3.1–2.3.4), if necessary. This dichotomy is extremely important and should be borne in mind by the reader at all times.

2.4 PERMUTATIONS AND COMBINATIONS

The problem of computing probabilities of events in finite sample spaces in which equal probabilities are assigned to the elements in any given sample space reduces to that of counting the elements which make up the events. The counting of such elements is often greatly simplified by the use of rules for permutations and combinations.

(a) Permutations

Suppose we have n distinct objects, $O_1, O_2, \ldots, O_n$. Let us consider how many different sequences of x objects we can form by choosing x objects in succession from the n objects where $1 \leq x \leq n$. For convenience we may think of a sequence of x places which are to be filled with x objects. We have n choices of objects to fill the first place. After the first place is filled, we have $n - 1$ objects left and, hence, $n - 1$ choices to fill the second place. Each of the n choices for filling the first place can be combined with each of the $n - 1$ choices for filling the second place, thus yielding $n(n - 1)$ ways of filling the first two places. By continuing this argument, we see that there are $n(n - 1) \cdots (n - x + 1)$ ways of filling the x places by choosing x objects from the set of n objects. Each of these sequences or arrangements of x objects is called a *permutation* of x objects from n. The total number of permutations of x objects from n, denoted by $P(n, x)$, is therefore given by

$$(2.4.1) \qquad P(n, x) = n(n - 1) \cdots (n - x + 1).$$

Note that the number of ways of permuting n objects is given by

$$(2.4.2) \qquad P(n, n) = n(n - 1) \cdots (2)(1) = n!,$$

where $n!$ is read "n factorial."

Expressed in terms of factorials, we have

$$(2.4.3) \qquad P(n, x) = \frac{n!}{(n - x)!}.$$

(b) Combinations

It will be noted that, if we select any permutation of x objects from n, there are $x!$ ways this particular set of x objects can be permuted. Hence there are $x!$ permutations which contain any set of x objects taken from the n. Any set of x objects from n distinct objects is called a *combination* of x objects from n objects. The number of such combinations, usually denoted by $\binom{n}{x}$ is therefore given by

$$(2.4.4) \qquad \binom{n}{x} = \frac{1}{x!} P(n, x) = \frac{n!}{x! \, (n - x)!}.$$

Example 2.4(a). The number of different possible hands of 13 cards in a pack of 52 ordinary playing cards (combinations of 13 cards from 52 cards) is

$$(2.4.5) \qquad \binom{52}{13} = \frac{52!}{13! \, 39!} = 635{,}013{,}559{,}600.$$

Example 2.4(b). The number of samples of 10 objects which can be selected from a lot of 100 objects is

$$(2.4.6) \qquad \binom{100}{10} = \frac{100!}{10!\,90!} = 17,310,309,456,440.$$

Example 2.4(c). The number of distinguishable arrangements of x A's and $n - x$ B's written in n places is $\binom{n}{x}$. For we can think of all n places filled with B's, and then select x of these places and replace the B's in them by A's. The number of such selections is $\binom{n}{x}$. This is equivalent to the number of ways we can arrange x A's and $n - x$ B's in n places.

Example 2.4(d). The coefficient of $a^x b^{n-x}$ in the expansion of $(a + b)^n$ is $\binom{n}{x}$. For we can write $(a + b)^n$ as

$$(2.4.7) \qquad (a + b)(a + b) \cdots (a + b) \qquad (n \text{ factors}),$$

and the coefficient of $a^x b^{n-x}$ is the number of ways we can pick x of these factors, and then choose a from each factor while choosing b from the remaining $n - x$ factors.

(c) Arrangements of n Objects Involving Several Kinds of Identical Objects

Suppose we have x_1 A_1's, x_2 A_2's, $\ldots$, x_k A_k's, where $x_1 + \cdots + x_k = n$. The total number of distinguishable arrangements of these several kinds of A's is given by

$$(2.4.8) \qquad \binom{n}{x_1, \ldots, x_k} = \frac{n!}{x_1! \cdots x_k!}.$$

For if we think of each of the n places being originally filled with A, there are $\binom{n}{x_1}$ ways of choosing x_1 A's to be replaced by A_1's. In each of these $\binom{n}{x_1}$ ways there are $\binom{n - x_1}{x_2}$ ways of choosing more x_2 A's to be replaced by A_2's. Hence, the number of ways of choosing x_1 A's and replacing them by A_1's and choosing x_2 of the remaining $n - x_1$ A's and replacing them by A_2's is $\binom{n}{x_1}\binom{n - x_1}{x_2}$. Continuing this argument, it is seen that the number of ways of choosing x_1 A's and replacing them by A_1's, x_2 A's and replacing them by A_2's, and so on until the last x_k A's replaced by A_k's, is

$$\binom{n}{x_1}\binom{n - x_1}{x_2} \cdots \binom{n - x_1 - \cdots - x_{k-1}}{x_k} = \frac{n!}{x_1!\, x_2! \cdots x_k!},$$

which we denote by $\binom{n}{x_1, \ldots, x_k}$.

Example 2.4(e). If n ordinary dice are thrown, the number of ways x_1 aces, x_2 two's, $\ldots$, x_6 sixes can appear, where $x_1 + \cdots + x_6 = n$, is $\binom{n}{x_1, \ldots, x_6}$.

Example 2.4(f). The coefficient of $a_1^{x_1} \cdots a_k^{x_k}$ in $(a_1 + \cdots + a_k)^n$ is $\binom{n}{x_1, \ldots, x_k}$. The argument is a direct extension of that used in Example 2.4(d).

(d) Applications of Combinations to Probability Problems

To illustrate the application of combinations to probability problems involving finite sample spaces, we consider several examples.

Example 2.4(g). If 13 cards are dealt from a "shuffled" deck of 52 ordinary playing cards, the probability of getting 5 spades is

$$\frac{\binom{13}{5}\binom{39}{8}}{\binom{52}{13}}.$$

For the numbers of ways of getting 5 spades from the 13 spades in the deck is $\binom{13}{5}$ and the number of ways of getting 8 nonspades from the 39 non-spades in the deck is $\binom{39}{8}$, and hence the number of ways of getting 5 spades and 8 nonspades in a hand of 13 cards is the product $\binom{13}{5} \cdot \binom{39}{8}$. This is the number of elements in the sample space constituting the event "getting 5 spades in dealing 13 cards from a shuffled deck." Each sample point is assigned the same probability, namely, $1 \big/ \binom{52}{13}$. Hence the probability of getting 5 spades in dealing 13 cards is

$$\binom{13}{5}\binom{39}{8} \cdot \frac{1}{\binom{52}{13}} = \frac{\binom{13}{5}\binom{39}{8}}{\binom{52}{13}}.$$

Example 2.4(h). A lot of 1000 articles contain 100 defectives. If a sample of 50 articles is drawn at random, the probability that the sample contains exactly 4 defectives is given by

$$\frac{\binom{100}{4}\binom{900}{46}}{\binom{1000}{50}}.$$

The reasoning is similar to that used in Example 2.4(*g*).

The actual numerical answers to the problems above are, of course, tedious to compute by hand, for they involve factorials of large numbers. A useful approximation for $n!$, where n is large, is the so-called Stirling approximation, given by

(2.4.9) $n! \sim (2\pi)^{1/2} n^{n+1/2} e^{-n}$

where the symbol $\sim$ denotes the fact that the ratio of the two sides of (2.4.9) tends to one as n tends to infinity.

Some example of how good the approximation is are as follows: the right hand side of (2.4.9) approximates

$1! =$	1	as	.9221
$2! =$	2	as	1.919
$5! =$	120	as	118.019
$10! =$	3,628,000	as	3,598,600

and the percentage errors are 8, 4, 2, and .8 % respectively.

2.5   CONDITIONAL PROBABILITY

In some probability problems we wish to determine the probability that an event F occurs, if it is known or given that an event E has occurred. This probability, denoted by $P(F \mid E)$, and called the *conditional probability of F given E*, is obtained essentially by letting E be a new (reduced) sample space and then computing the fraction of probability on E which lies on $E \cap F$, that is

(2.5.1) $P(F \mid E) = \dfrac{P(E \cap F)}{P(E)},$

where, of course, $P(E) \neq 0$.

Example 2.5(a). In Example 2.3(c) a rivet drawn from the box is known to have a type A defect. What is the probability it also has a type B defect? We have

$$P(E_B \mid E_A) = \frac{P(E_A \cap E_B)}{P(E_A)}$$

$$= (\tfrac{7}{1000})/(\tfrac{50}{1000}) = \tfrac{7}{50}.$$

If we rewrite (2.5.1) we obtain the following:

THEOREM 2.5.1 (*Rule of multiplication of probabilities*). *If E and F are events in a sample space S, such that $P(E) \neq 0$, then*

(2.5.2) $$P(E \cap F) = P(E) \cdot P(F \mid E).$$

It should be noted that (2.5.2) provides a two-step rule for determining the probability of the occurrence of E and F by first determining the probability of E and multiplying by the conditional probability of F given E.

Example 2.5(b). Two of the light bulbs in a box of six have broken filaments. If the bulbs are tested at random one at a time, what is the probability that the second defective bulb is found as soon as the third bulb is tested? Let E be the event of getting one good and one defective bulb in the first two bulbs tested, and let F be the event of getting a defective one on the drawing of the third bulb. Then E $\cap$ F is the event whose probability we are seeking. Hence,

$$P(E \cap F) = P(E) \cdot P(F \mid E).$$

The sample space S in which E lies consists of all possible selections of 2 bulbs out of 6, the number of elements in S being 15. The event E consists of all selections of one good and one defective bulb out of four good and two defective bulbs, and the number of such selections is eight. Therefore $P(E) = 8/15$. We now compute $P(F \mid E)$, the probability that F occurs given that E occurs. If E has occurred, there are three good and one defective bulb left in the box, and F is the event of drawing the defective one on the next draw. Thus $P(F \mid E) = \tfrac{1}{4}$. Therefore we have as our required probability

$$P(E \cap F) = P(E) \cdot P(F \mid E) = \tfrac{8}{15} \cdot \tfrac{1}{4} = \tfrac{2}{15}.$$

For the case of three events, E_1, E_2, E_3, the extended form of (2.5.2) is

(2.5.3) $$P(E_1 \cap E_2 \cap E_3) = P(E_1)P(E_2 \mid E_1) \cdot P(E_3 \mid E_1 \cap E_2),$$

provided $P(E_1)$ and $P(E_1 \cap E_2)$ are both $\neq 0$. Formula (2.5.3) extends to any finite number of events $E_1, \ldots, E_k$.

If the conditional probability of F given E has the same value as the probability of F, that is, if

$$P(F \mid E) = P(F),$$

we say that E and F are *independent* events. Thus we have the following important theorem:

THEOREM 2.5.2 (*Rule of multiplication of probabilities of two independent events*). *E and F are independent events if and only if*

$$(2.5.4) \qquad P(E \cap F) = P(E) \cdot P(F).$$

Theorem 2.5.2 can be extended to the case of several independent events $E_1, \ldots, E_n$, in which case (2.5.4) generalizes to

$$(2.5.5) \qquad P(E_1 \cap \cdots \cap E_n) = P(E_1) \cdots P(E_n).$$

Example 2.5(c). If a "true" die is thrown n times, what is the probability of never getting an ace? Let E_1 be the event of not getting an ace on the first throw, E_2 the event of not getting an ace on the second throw and so on. Assuming independence of the events $E_1, \ldots, E_n$ and a "true" die, we have $P(E_1) = \cdots = P(E_n) = \frac{5}{6}$. Hence the required probability from (2.5.5) is

$$P(E_1 \cap \cdots \cap E_n) = P(E_1) \cdots P(E_n) = (\tfrac{5}{6})^n.$$

It should be noted that the basic sample space S in which the events $E_1 \cap \cdots \cap E_n$ as well as $E_1, \ldots, E_n$ lie has 6^n elements. The number of elements in $E_1 \cap \cdots \cap E_n$ is 5^n, the number of elements in each of the events $E_1, \ldots, E_n$ is $5 \cdot 6^{n-1}$. Thus considered in S we have $P(E_1) = \cdots = P(E_n) = 5 \cdot 6^{n-1}/6^n = 5/6$. But if we consider a sample space S_1 for the first roll, S_2 for the second, and so on, then $S_1, \ldots, S_n$ each contains six elements, and each of $E_1, \ldots, E_n$ in $S_1, \ldots, S_n$ respectively, consists of five elements. Thus in S_1, we find $P(E_1) = 5/6$, in S_2, $P(E_2) = 5/6$, and so on. We say that the sample space S for all n throws is the *cartesian product* of $S_1, \ldots, S_n$, and we write $S = S_1 \times \cdots \times S_n$.

Remark. The reader should note that if $E_1, E_2, \ldots, E_k$ are events whose relative frequencies of occurrence in n trials are $F(E_1), F(E_2), \ldots, F(E_k)$, then the relative frequency function of an event, say $F(E)$, is an additive set function that obeys all the rules of probability stated in the present section, and of course, Section 2.3.

Example 2.5(d). Referring to Example 2.5(c) above, suppose the "true" die is rolled until the occurrence of the rth ace. What is the probability that this event occurs on the xth roll? ($x = r, r + 1, \ldots; r = 1, 2, \ldots$).

Suppose we let E be the event that in the first $(x - 1)$ rolls of the die, that there are exactly $(r - 1)$ aces, and thus $(x - r)$ "nonaces." Further, let F be the event that the xth roll results in an ace. Then $E \cap F$ is the event whose probability we are seeking and we have

$$(2.5.6) \qquad P(E \cap F) = P(E)P(F \mid E).$$

Since successive rolls of the die are independent, any particular sequence of $(x - 1)$ rolls that has $(r - 1)$ aces and $(x - r)$ nonaces has probability

$$(2.5.7) \qquad (\tfrac{1}{6})^{r-1}(\tfrac{5}{6})^{x-r}.$$

The number of different sequences of $(x - 1)$ rolls having $(r - 1)$ aces and $(x - r)$ nonaces is clearly the number of ways we may choose $(r - 1)$ places from $(x - 1)$ available slots for the occurrences of the aces, so that applying (2.3.6) yields

$$(2.5.8) \qquad P(E) = \binom{x-1}{r-1}(\tfrac{1}{6})^{r-1}(\tfrac{5}{6})^{x-r}.$$

Further, we note that $P(F \mid E) = P(F) = 1/6$, so that (2.5.6) and (2.5.8) yields

$$(2.5.9) \qquad P(E \cap F) = \binom{x-1}{r-1}(\tfrac{1}{6})^{r}(\tfrac{5}{6})^{x-r} = \binom{x-1}{r-1}\frac{5^{x-r}}{6^{x}}$$

for $x \geq r$, $r = 1, 2, \ldots$.

2.6 BAYES' THEOREM

We discuss now an interesting version of the conditional probability formula (2.5.1), due to the Reverend Thomas Bayes. Bayes' result was published posthumously in 1763.

Suppose E and F are two events in a sample space S and such that $E \cap F \neq \phi$. The reader will recall (see for example the Venn Diagram in Figure 2.2.2) that the events $(E \cap F)$ and $(E \cap \bar{F})$ are disjoint and that their union is E, so that

$$(2.6.1) \qquad P(E) = P(E \cap F) + P(E \cap \bar{F}).$$

Using the rule given by (2.5.2), we have

$$(2.6.2) \qquad P(E) = P(F)\,P(E \mid F) + P(\bar{F})\,P(E \mid \bar{F}).$$

Thus we may write (2.5.1) in the form

$$(2.6.3) \qquad P(F \mid E) = \frac{P(F)P(E \mid F)}{P(F)P(E \mid F) + P(\bar{F})P(E \mid \bar{F})}.$$

The rule provided by (2.6.3) is known as Bayes' Theorem for two events E and F; the probabilities $P(F)$ and $P(\bar{F})$ are sometimes referred to as the *prior* probabilities of events F and $\bar{F}$ respectively (note that $F \cup \bar{F} = S$, $F \cap \bar{F} = \phi$), and $P(F \mid E)$ as given by Bayes' Theorem (2.6.3) is referred to as the *posterior* probability of F, given the event E has occurred. An interpretation of (2.6.3) is that *posterior* to observing that the event E has occurred, the probability of F changes from $P(F)$, the prior probability, to $P(F \mid E)$, the posterior probability.

    Example 2.6(a). The Gimmick TV Model A uses a printed circuit, and the company has a routine method for diagnosing defects in the circuitry, if a set fails. Over the years, the experience with this routine diagnostic method yields the following pertinent information:

the probability that a set, which fails due to printed circuit defects (PCD), is correctly diagnosed as failing because of PCD, is 80%;

the probability that a set, which fails due to causes other than PCD, is diagnosed incorrectly as failing because of PCD, is 30%.

Further, experience with the printed circuit shows that about 25% of all Model A failures are due to PCD. Find the probability that a Model A set's failure is due to PCD, given that it has been diagnosed as being due to PCD.

To answer this question, we use Bayes' Theorem (2.6.3) to find the posterior probability of a set's failure being due to PCD, after observing that the failure is diagnosed as being due to PCD. We let

$$F = \text{event, set fails due to PCD}$$
$$E = \text{event, set failure is diagnosed as being due to PCD}$$

and we wish to determine the posterior probability $P(F \mid E)$.

We have that $P(F) = .25$, so that $P(\bar{F}) = .75$, and that $P(E \mid F) = .80$, $P(E \mid \bar{F}) = .30$. Hence, applying (2.6.3) gives

(2.6.4)
$$P(F \mid E) = \frac{P(F)P(E \mid F)}{P(F)P(E \mid F) + P(\bar{F})P(E \mid \bar{F})}$$
$$= \frac{(.25)(.80)}{(.25)(.80) + (.75)(.30)}$$
$$= .465.$$

Notice that in light of the event E having occurred, the probability of F has changed from the *prior* probability of 25%, to the *posterior* probability of 46.5%.

The formula (2.6.3) may be generalized to more complicated situations, and indeed Bayes stated his theorem for the more general situation that we now consider.

THEOREM 2.6.1. *Suppose* $F_1, F_2, \ldots, F_k$ *are mutually exclusive events, and such that* $\sum_{i=1}^{k} P(F_i) = 1$. *Then*

(2.6.5)
$$P(F_i \mid E) = \frac{P(F_i)P(E \mid F_i)}{\sum_{i=1}^{k} P(F_i)P(E \mid F_i)} \qquad \text{(Bayes' Theorem)}.$$

We note that (2.6.3) is a special case of (2.6.5), with $k = 2$, $F_1 = F$ and $F_2 = \bar{F}$. Bayes' Theorem (for k events F_i) has aroused much controversy over its use. The reason for this is that in many situations the *prior* probabilities $P(F_i)$ are unknown, and in practice when not much is known a priori, these have often been set equal to $1/k$, as advocated by Bayes himself. The setting of $P(F_i) = 1/k$ in the so-called "in-ignorance" situation is the source of the controversy. Of course, when the $P(F_i)$

are known, or may be estimated on the basis of much past experience, (2.6.5) provides a way of incorporating prior knowledge about the F_i to determine the conditional probabilities $P(F_i \mid E)$, as given by (2.6.5). We shall return to Bayes' Theorem and its use in statistical inference in Chapter 9A. At this point, we illustrate (2.6.5) with the following example.

Example 2.6(b). Miss Speed, Secretary of the Year, uses various model typewriters T_1, T_2, T_3, and T_4 in 40%, 30%, 20%, and 10%, respectively, of all speed-typewriting contests she enters. In the last few years, she has won 20%, $33\frac{1}{3}$%, 20%, and 10% of all the contests in which she has used models T_1, T_2, T_3, and T_4 respectively. This year she entered one contest and won. What is the probability that she used T_1? Let

$$F_i = \text{event, typist uses model } T_i$$
$$E = \text{event, typist wins speed contest.}$$

We have

(2.6.6)
$$P(F_1 \mid E) = \frac{P(F_1)P(E \mid F_1)}{\sum_{i=1}^{4} P(F_i)P(E \mid F_i)}$$

$$= \frac{(.4)(.2)}{(.4)(.2) + (.3)(.33) + (.2)(.2) + (.1)(.1)} = .349.$$

We note that the posterior probability of F_1, given E, is .349, while the prior probability of F_1 was $P(F_1) = .4$.

2.7   RANDOM VARIABLES

Suppose we have a finite sample space S listing m elements $e_1, \ldots, e_m$. There are 2^m possible events which can be formed from these elements, provided we count the empty event ϕ and the entire sample space S as two of the events. This is seen by the fact that we have the choice of selecting or not selecting each of the m elements in making up an event. We are rarely, if ever, really interested in all of these 2^m events and their probabilities. We are usually interested in a relatively small number of them produced by specified values of some function defined on the elements. For instance, in the sample space S of the $\binom{52}{13}$ possible hands of 13 bridge cards, we are usually interested in such events as getting 2 aces, 8 spades, 10 honor cards, and so on.

A (real and single-valued) function $x(e)$ defined on each element e in the sample space S is called a *random variable*. Suppose $x(e)$ can take on only the values $x_1, x_2, \ldots, x_k$, and $E_1, \ldots, E_k$ be the events in the sample space S for which $x(e) = x_1, \ldots, x(e) = x_k$ respectively. Let $P(E_1) = p(x_1), \ldots, P(E_k) = p(x_k)$. Then we say that the random variable $x(e)$

defined on S is a *discrete random variable* which takes the values $x_1, \ldots, x_k$ with probabilities $p(x_1), \ldots, p(x_k)$ respectively. Since $E_1, \ldots, E_k$ are disjoint and their union is equal to the entire sample space S, we have

(2.7.1) $$p(x_1) + \cdots + p(x_k) = 1.$$

We can arrange these values of x and the corresponding probabilities in table form as follows:

(2.7.2)

x	x_1	x_2	$\cdots$	x_k
$p(x)$	$p(x_1)$	$p(x_2)$	$\cdots$	$p(x_k)$

The values of the discrete random variable $x(e)$ together with their associated probabilities are called the *probability function* (p.f.) of $x(e)$. Ordinarily we drop the e and refer to the random variable as x. The set of values $x_1, \ldots, x_k$ is called the *sample space of the random variable x.*

Example 2.7(a). Let x be a random variable denoting the number of dots which appear when two dice are thrown. If all 36 elements in the sample space are assigned the same probability, namely 1/36, then the probability function of x is as follows:

x	2	3	4	5	6	7	8	9	10	11	12
$p(x)$	1/36	2/36	3/36	4/36	5/36	6/36	5/36	4/36	3/36	2/36	1/36

If the sample space S has an infinite number of elements and if the random variable x can take on a countably infinite set of values $x_1, x_2, \ldots,$ we have a *discrete* random variable with sample space $x_1, x_2, \ldots.$

Example 2.7(b). Let x be a random variable denoting the number of times a die is thrown until an ace appears. The sample space of x is $1, 2, \ldots,$ and the probability function $p(x)$ is given by the table:

x	1	2	$\cdots$	x	$\cdots$
$p(x)$	$\dfrac{1}{6}$	$\dfrac{5}{6^2}$	$\cdots$	$\dfrac{5^{x-1}}{6^x}$	$\cdots$

or more briefly by

$$p(x) = \frac{5^{x-1}}{6^x}, \qquad x = 1, 2, \ldots.$$

2.8   RANDOMIZATION AND RANDOM NUMBERS

One of the most important problems in the application of probability and statistics is that of setting up events, the probabilities of which can be conveniently physically implemented or realized. For example, if there are n balls marked $1, \ldots, n$ in an urn and if a ball is drawn at random from the urn, we talk about the probability being $1/n$ that a ball having a specified number will be drawn, and we use this number in any mathematical formulas we may wish to derive about probabilities of events involved in drawing balls from the urn. But if we consider physically drawing a ball from the urn, returning it, stirring the balls, and repeating this many, many times, it is at best a clumsy procedure. Furthermore, it is hazardous to assume that the fraction of times the ball numbered i is drawn would be close to $1/n$, if n is a very large number. This situation becomes more awkward and hazardous if two balls are drawn simultaneously and replaced, and if this operation is repeated over and over again for drawing larger numbers of balls. The kinds of physical factors which can bias results of such experimental procedures are many, varied, and subtle.

There are statistically more interesting problems: suppose, for instance, that a manufacturer has a lot of 1000 items and that an interested purchaser wishes to draw a sample of 100 of these items at random and inspect them before deciding whether to purchase the lot. The question arises how to select physically 100 items from the lot so that there is reasonable assurance that all possible $\binom{1000}{100}$ selections of 100 items would be equally probable.

It is risky, at best, to attempt to simulate faithfully the drawing of 100 objects out of 1000 by physically drawing 100 balls from an urn containing 1000 balls, or by drawing 100 tags from a hat containing 1000 numbered tags, or by any other such procedure. What is needed is some mechanism for simulating the drawing which requires little effort and which has a high degree of validity for realizing the desired probabilities.

Such a mechanism exists in the form of *random numbers* made up of sets of *random digits*. Thus, suppose we consider a "perfect" icosohedral die (a 20-sided polyhedron, each side being an equilateral triangle) with two sides marked 0, two marked 1, . . . , two marked 9. If this die were thrown once, the sample space S would consist of the digits (1, 2, 3, 4, 5, 6, 7, 8, 9, 0) each with probability $1/10$. If the die were thrown twice, the sample space S would contain the 100 pairs of digits $(01, \ldots, 98, 99, 00)$ each with probability $1/100$. In the two cases we would identify 0 and 00 with 10 and 100 respectively. In general, if the die were thrown n times, the sample space S would contain 10^n outcomes, each outcome being a set of n digits which are automatically numbered from 1 to 10^n.

Although an icosohedral die may be satisfactory for many practical purposes for generating random digits, there are other methods that are faster and superior as far as being free from bias is concerned. Scientists at the RAND Corporation devised an electronic procedure for generating random digits which seems to have been remarkably free from bias. With the use of this device, a table of random digits was generated and published by the RAND Corporation (1955). The RAND table contains 400 pages of random digits, each page containing 50 lines, each line containing 10 sets of 5 digits each—1,000,000 random digits in all. Table I, Appendix VII, shows the first two pages from the RAND tables.

To illustrate the use of Table I, suppose we have a lot of 1532 articles, numbered serially from 1 to 1532, from which we wish to draw 10 at random so that all $\binom{1532}{10}$ possible selections which could be made would be equally likely. We would pick a line number from 1 to 100 and a column number from 1 to 10 by some simple randomization procedure. A simple one is to toss a coin seven times denoting a head by 1 and tail by 0, thus generating a 7-place binary number which will represent 0, 1, . . . , 127 with equal probabilities. Use the number so generated if it is 0, 1, . . . , 99, otherwise ignore it and repeat the process. By similarly generating and using a 4-place binary number, we can pick one of the ten columns of 5-digit numbers at random. This pair of random numbers (one for line number and one for column number) provides an entry position into the table. After performing the experiment and flipping a coin seven times, suppose we find 0010111 ($= 23$) as the first binary number and 0101 ($= 5$) as the second. Then, we enter Table I at line 23 and column 5 and find the random number 12860. We keep the first four digits, namely, 1286. We now move down the column of 5-digit numbers beginning with 12860, keeping only the first four digits until we find ten numbers between 0001 and 1532 (inclusive) which are all different. The numbers thus found are 1286, 1686, 0556, 0533, 0229 (and going to the top of column 6), 0256, 0892, 0021, 0700, 0889. These, then, are the serial numbers of the sample of 10 articles to be drawn from the lot of 1532. This process provides as nearly equal probabilities as we can physically achieve for all possible $\binom{1532}{10}$ samples of 10 articles which could be drawn from the lot of 1532 articles.

It should be noted that it is not really necessary to give serial numbers to the articles in the lot, but it is necessary that we have a rule for numbering them without ambiguity if we wish. It should be further noted that, if there are rs objects in the lot or population which are arrayed in r sets, each with s objects, we can draw an object at random by using a pair of random numbers (a, b), say, where a is a random number between 1 and r (inclusive) and b is a random number between 1 and s (inclusive). Thus (a, b) would mean the bth object in the ath set.

Example 2.8(a). A directory has 512 pages and 70 lines per page, some persons requiring 2 or even 3 lines for listing. The problem is to select a name at random from the directory so that every name in the directory has the same probability of being drawn. We shall pick a pair of random numbers (a, b), where $001 \leq a \leq 512$ and $01 \leq b \leq 70$. By tossing a coin 7 times and then four times, suppose we obtained the binary number 0110100 ($=52$) and 1001 ($=9$). Entering Table I at line 52 and column 9, we find 41361 and retain the first three digits. Bearing in mind that $01 \leq b \leq 70$, we proceed down the column until 07706. Retaining the first two digits, we now use (413, 07) as our pair of random numbers, that is, we select the seventh line on page 413 of the directory if this line starts with the name of a person (and not his address). Otherwise we pass down the column to the next pair of 5-digit numbers and repeat the process. This procedure insures that each name listed in the table is as likely to be drawn as any other. If we want n names from the directory, we repeat the process enough times to obtain n different names.

PROBLEMS

2.1. Certain pieces made by an automatic lathe are subject to three kinds of defects, namely, X, Y, Z. A sample of 1000 pieces was inspected with the following results:

2.1 % had type X defect;
2.4 % had type Y defect;
2.8 % had type Z defect;
.3 % had both type X and type Y defects;
.4 % had both type X and type Z defects;
.6 % had both type Y and type Z defects;
.1 % had type X, type Y, and type Z defects.

(a) What percent had none of these defects?
(b) What percent had at least one of these defects?
(c) What percent were free of type X and type Y defects?
(d) What percent had not more than one of these defects?

2.2. Two inspectors A and B independently inspected the same lot of items. Four percent of the items are actually defective. It turns out that:

5 % of the items are called defective by A;
6 % of the items are called defective by B;
2 % of the items are correctly called defective by A;
3 % of the items are correctly called defective by B;
4 % of the items are called defective by both A and B;
1 % of the items are correctly called defective by both A and B.

(a) Make a Venn diagram showing percentages of items in the eight possible disjoint classes generated by the classification of the two inspectors and the true classification of the items.
(b) What percent of the truly defective items are missed by inspectors?

2.3. In testing an automatic inspection device for detecting defective items, an experiment was set up as follows: 1000 items, of which 100 were true defectives, were run through the device. The device indicated 105 of the items defective of which 90 were found to be true defectives. The items were then run through the device again. This time the device indicated 103 to be defective including 3 of the true defectives it missed before, 85 of the true defectives it caught before, and 4 of the items it falsely indicated to be defectives before.

(a) Make a Venn diagram of the eight classes of items obtained in this experiment.

(b) How many items were indicated as defectives at least once by the device?

(c) How many items were either truly defective or were indicated as defective in at least one of the runs?

(d) If an item is truly defective, what would you estimate the probability to be that, if it is run through the device once, it will be caught? If run through the machine twice, it will be caught at least once?

(e) What would you estimate the probability to be that a good item will be indicated as defective if it is run through the device once?

2.4. (a) A box of 100 items contains 90 nondefective items, 7 with type A defects, 5 with type B defects and 2 with both types of defects. Let S be the sample space generated by the operation of drawing one item blindly from the box. Let E_A be the event of getting a type A defective, E_B the event of getting a type B defective. Set up a Venn diagram and show the following events: $E_A \cap E_B$, $E_A \cap \bar{E}_B$, $\bar{E}_A \cap E_B$, $\bar{E}_A \cap \bar{E}_B$, indicating the number of elements in each. How many elements are in $E_A \cup E_B$?

(b) In the box described in part (a) of this problem let S^* be the sample space generated by blindly drawing two items (simultaneously) from the box. Let G be the event that corresponds to pairs of nondefective items. Let G_A be the event that corresponds to pairs of items containing at least one type A defective while G_B is the event which corresponds to pairs of items containing at least one type B defective. Set up a Venn diagram and show the eight basic events in S^* obtained by placing or omitting a bar over each G in $G \cap G_A \cap G_B$, and write in the number of elements in each of the eight basic events.

2.5 A "true" icosahedral (Japanese) die has 20 sides, 2 sides marked with 0, 2 sides with 1, 2 sides with 2, . . . , 2 sides with 9; the probabilities assigned to the 20 faces are all equal. Suppose three such dice are thrown. Find the following probabilities:

(a) That no two top faces will be alike.

(b) That at least two top faces will be alike.

(c) That all three top faces will be different even numbers (0 is considered an even number).

(d) Generalize parts (a) and (b) to the case of a "true" 2 n-sided die with 2 sides marked 1, 2 sides marked 2, . . . , 2 sides marked n.

2.6. Ten defective items are known to be in a box of 100 items.

(a) If they are located by testing the items one at a time until all defectives are found, what is the probability that the 10th (last) defective item is located as soon as the 50th item is tested?

(b) What is the probability that, if 50 items are drawn at random from the box and tested, all 10 defectives will be found?

(c) If 20 are tested and found to be nondefective, what is the probability that all defectives will be found among the next 30 tested?

2.7. If a lot of 1000 articles has 100 defectives and if a sample of 10 articles is selected at random from the lot, what is the probability that the sample will contain:

(a) No defectives?

(b) At least 1 defective?

2.8. Assume that a given type of aircraft motor will operate eight hours without failure with probability .99. Assume that a 2-motor plane can fly with at least one motor, and that a 4-motor plane can fly with at least 2 motors, and that failure of one motor is independent of the failure of another.

(a) If a 2-motor plane and a 4-motor plane take off for an 8-hour flight, show that the 2-motor plane is more than 25 times more likely to be forced down by motor failure than the 4-motor plane.

(b) Compute the respective probabilities that the planes will not be forced down by motor failure.

(c) What is the answer to (b) if the probability of failure of a motor during an 8-hour period is p rather than .01?

2.9. Suppose 10 chips are marked 1, 2, . . . , 10 respectively and put in a hat. If two chips are drawn at random, what is the probability that:

(a) Their difference will be exactly 1?

(b) Neither number will exceed 5?

(c) Both numbers will be even?

(d) At least one of the numbers will be 1 or 10?

2.10. If the probability is 0.001 that a type-X 20-watt bulb will fail in a 10-hour test, what is the probability that a sign constructed from 1000 such bulbs will burn 10 hours:

(a) With no bulb failures?

(b) With one bulb failure?

(c) With k bulb failures?

2.11. The game of craps is played with two ordinary 6-sided dice as follows:

If the shooter throws 7 or 11 he wins without further throwing; if he throws 2, 3, or 12, he loses without further throwing. If he throws 4, 5, 6, 8, 9, or 10, he must continue throwing until a 7 or the "point" he initially threw appears. If 7 appears first he loses, if the "point" he initially threw appears first, he wins. Show that the probability is approximately 0.4929 that the shooter wins (assuming "true" dice).

2.12. In a group of 11 persons, no two persons are the same age. We are to choose 5 people at random from this group of 11.

(a) What is the probability that the oldest and the youngest persons of the 11 will be among those chosen?

(b) What is the probability that the third youngest of the 5 chosen will be the 6th youngest of the 11?

(c) What is the probability that at least 3 of the 4 youngest of the 11 will be chosen?

2.13. Suppose the probability is 1/365 that a person selected at random was born on any specified day of the year (ignoring persons born on February 29). What is the probability that, if r people are thus selected, no two will have the same birthday? (The smallest value of r for which the probability that at least two will have a common birthday exceeds .5 is 23.)

2.14. Suppose six "true" dice are rolled simultaneously. What is the probability of getting:

(a) All faces alike?
(b) No two faces alike?
(c) Only five different faces?

2.15. If A, B, C, and D are four events such that

$$P(A) = P(B) = P(C) = P(D) = p_1,$$

$$P(A \cap B) = P(A \cap C) = \cdots = P(C \cap D) = p_2,$$

$$P(A \cap B \cap C) = P(A \cap B \cap D) = P(A \cap C \cap D) = P(B \cap C \cap D) = p_3,$$

$$P(A \cap B \cap C \cap D) = p_4,$$

express the values of the following probabilities in terms of p_1, p_2, p_3, p_4:

(a) $P(A \cup B \cup C)$;
(b) $P(A \cup B \cup C \cup D)$;
(c) $P(A \cap B \mid C \cap D)$.
(d) Probability of the occurrence of exactly 1, exactly 2, exactly 3, of the events of A, B, C, D.

2.16. If four addressed letters are inserted into four addressed envelopes at random, what is the probability that:

(a) At least one letter is inserted into its own envelope?
(b) No letter is inserted into its own envelope?
(c) Find the expressions for the probabilities for (a) and (b) if there are n letters and n envelopes.

2.17. Five segments of a 1000-foot wire, each 1/4 inch long are to be selected at random as sites for diameter measurements. Describe how you would select these segments by random numbers.

2.18. Five squares, each 1/4 by 1/4 inches, are to be selected at random on a sheet of tin 4 feet by 8 feet in size for sites of thickness measurements. Describe how you would select these squares by random numbers.

2.19. By the use of random numbers, explain how you could select a bridge hand so that every hand is just as likely to be drawn as any other hand.

2.20. A local television station covers an area served by telephone listed in 5 directories, having 132, 87, 286, 164, and 130 pages. Each page has four columns, and each column has at most 100 names. The master of ceremonies (MC) of a TV program wants to call the number of a private residence in a contest so that all private residences are equally likely to be called. Explain precisely how you would use triplets of random numbers to select a telephone number for the MC.

2.21. In each day of a 5-day week each of 10 machines in a plant makes between 80 and 100 articles of a certain kind. The machines are numbered serially and the articles from each machine are numbered serially each day. By using random numbers, explain how you would design a sample of 100 articles from the week's production so that each article could be designated for the sample as soon as it is made, and so that each article manufactured is as likely to be drawn as any other.

2.22. There are 3016 counties in the United States (mainland). The area of each county in square miles is given. Explain precisely how you would select five counties by random numbers so that the probability of drawing a county is proportional to its area.

2.23. Three machines A, B, and C produce, 40%, 45%, and 15%, respectively, of the total number of ball bearings produced by a certain factory. The percentages of defective output of these machines are 3%, 6%, and 9%. If a ball bearing is selected at random, find the probability that the item is defective.

2.24. In Problem 2.23, suppose a ball bearing is selected at random, and is found to be defective. Find the probability that the item was produced by machine A.

2.25. Enrollment data at a certain college shows that 30% of the men and 10% of the women are studying statistics, and that the men form 45% of the student body. If a student is selected at random and is found to be studying statistics, determine the probability that the student is a woman.

2.26. A certain cancer diagnostic test is 95% accurate on those that do have cancer, and 90% accurate on those that do not have cancer. If 1/2% of the population actually does have cancer, compute the probability that a particular individual has cancer if the test finds that he has cancer.

2.27. An urn contains 3 blue and 7 white chips. A chip is selected at random, and if the color of the chip so selected is white, it is replaced and two more white chips are added to the urn. If the chip drawn is blue, it is not replaced and no additional chips are put in the urn. A chip is then drawn from the urn a second time. What is the probability that it is white?

2.28. Referring to Problem 2.27, suppose we are given that the chip selected for the second time is white. What is the probability that the chip selected at the first stage is blue?

2.29. Referring to Problem 2.6, suppose that it takes eleven tests to find all the 10 defectives (i.e., the eleventh test produces the last defective). What is the probability that the first item is nondefective?

2.30. A bag contains a nickel, quarter, and "dime," with the dime being a fake coin and having two heads. A coin is chosen at random from the bag and tossed four times in succession. If the result is 4 heads, what is the probability that the fake dime was used?

CHAPTER 3

Some Important Discrete Distributions

3.1 THE HYPERGEOMETRIC DISTRIBUTION

The hypergeometric probability function provides probabilities of certain events when a sample of n objects is drawn at random from a finite population of N objects, where the sampling is done *without replacement* and where each element of the population may be dichotomized in some simple fashion as belonging to one of two disjoint classes.

As an example of sampling from a dichotomized population, suppose we consider first a hand of bridge and ask for the probability of getting x spades. Here the dichotomy is "spades" and "nonspades;" the population consists of 52 cards; the sample consists of 13 cards. Or consider a lot of N electronic tubes, of which $100\theta\%$ are "defective" and $100(1 - \theta)\%$ are "nondefective," where $0 \leq \theta \leq 1$. If these are packaged in boxes of n tubes, what is the probability that x tubes in a given box are defective? Here, the population is the lot of N tubes, and the sample size is n. The dichotomy in this example is, of course, "defective" and "nondefective."

In general, suppose that a random sample of n observations is taken from a population of size N without replacement, and the observations are examined as to their having an attribute A or not having this attribute. Objects of the former kind will be called type A and the latter called type $\bar{A}$. Suppose N_1 objects of the population are of type A and $N_2 = N - N_1$ are of type $\bar{A}$. We ask for the probability that x of the objects in the sample of size n are of type A, $0 \leq x \leq n$ (and thus $n - x$ are of type $\bar{A}$).

We know that the total number of ways of choosing a sample of n from N is $\binom{N}{n}$, which is the number of elements in the sample space S for this problem. Each of these elements will be assigned the probability $1 \Big/ \binom{N}{n}$. The number of ways of choosing x objects of type A from N_1 is

$\binom{N_1}{x}$, $0 \leq x \leq N_1$, and the number of ways of obtaining $n - x$ objects of type $\bar{A}$ from N_2 is$\binom{N_2}{n-x}$, $0 \leq n - x \leq N_2 = N - N_1$. Each of these latter ways may be combined with each of the former, so that the number of elements in S comprising the event of getting x objects of type A is $\binom{N_1}{x} \cdot \binom{N - N_1}{n - x}$. Note that x is a random variable. Hence the probability we seek is given by

(3.1.1)
$$h(x) = \frac{\binom{N_1}{x}\binom{N - N_1}{n - x}}{\binom{N}{n}},$$

where the sample space of x is the set of integers x that satisfies the inequalities $\max [0, n - (N - N_1)] \leq x \leq \min (n, N_1)$.

The expression (3.1.1) is the probability function of the *hypergeometric distribution*. As shown in Section 3.6 the sum of $h(x)$ over all values of x in the sample space of x is 1. We note that $h(x)$ is the probability of obtaining x objects of type A in a random sample of size n, when sampling *without* replacement from a population of size N which contains N_1 objects of type A and $N_2 = N - N_1$ objects of type $\bar{A}$.

Example 3.1(a). Let us consider the probability of getting x spades in a hand of bridge. In the above framework, A denotes a spade, and $\bar{A}$ denotes any card other than a spade. Hence $N_1 = 13$, $N - N_1 = 39$, and if x denotes the number of spades in a hand of $n = 13$ cards, x has the sample space $0 \leq x \leq 13$, and

$$h(x) = \frac{\binom{13}{x}\binom{39}{13 - x}}{\binom{52}{13}}$$

For instance, if $x = 0$, then the probability of getting 0 spades in a hand is given by

$$h(0) = \frac{\binom{13}{0}\binom{39}{13}}{\binom{52}{13}} = 0.0128.$$

In engineering applications, it is more convenient to consider the hypergeometric distribution from the following point of view. Consider a dichotomized finite population or lot of size N consisting of $N\theta$ objects of type A, and $N(1 - \theta)$ objects of type $\bar{A}$, where $0 \leq \theta \leq 1$. Thus, we have replaced N_1 and N_2 by $N\theta$ and $N(1 - \theta)$. If a sample of size n is drawn without replacement, we have as the probability of obtaining x articles of type A,

(3.1.1a)
$$h(x) = \frac{\binom{N\theta}{x}\binom{N(1 - \theta)}{n - x}}{\binom{N}{n}},$$

where the sample space of x consists of the integers x that satisfy the inequalities max $[0, n - N(1 - \theta)] \le x \le$ min $(n, N\theta)$.

Example 3.1(b).   A carton contains 24 light bulbs, $12\frac{1}{2}\%$ of which are defective. What is the probability that, if a sample of six is chosen at random from the carton of the bulbs, x will be defective? Using (3.1.1a), we have, since $12\frac{1}{2}\%$ of 24 is 3, that

$$h(x) = \frac{\binom{3}{x}\binom{21}{6 - x}}{\binom{24}{6}}, \qquad 0 \le x \le 3.$$

This gives
$$h(0) = \frac{\binom{3}{0}\binom{21}{6}}{\binom{24}{6}} = .40316,$$

$$h(1) = \frac{\binom{3}{1}\binom{21}{5}}{\binom{24}{6}} = .45356,$$

$$h(2) = \frac{\binom{3}{2}\binom{21}{4}}{\binom{24}{6}} = .13340,$$

$$h(3) = \frac{\binom{3}{3}\binom{21}{3}}{\binom{24}{6}} = .00988.$$

Note that $\sum_{x=0}^{3} h(x) = 1$.

3.2 THE BINOMIAL DISTRIBUTION

The probabilities given by the binomial distribution may arise in the following ways: (1) When sampling from a finite population with replacement; (2) when sampling from an infinite population (often referred to as an indefinitely large population) with or without replacement.

Suppose we wish to draw a sample of size n from a lot of N objects of which $N\theta$ are defectives, and $N(1 - \theta)$ are nondefectives, and that sampling is done with replacement. In other words, we draw at random a member of the lot, examine it, record the result, and replace it in the lot, "mix thoroughly," and repeat the procedure a further $n - 1$ times. We ask for the probability of obtaining x defectives in the sample of n trials.

For example, if we let D denote the event "defective" and $\bar{D}$ denote the event "nondefective," then our sample space consists of the 2^n possible sequences of n letters, each letter being a D or a $\bar{D}$. Thus, x defectives would occur in n trials in some sequence of x D's and $(n - x)$ $\bar{D}$'s such as

$$D\bar{D}\ DD\bar{D}\ \bar{D}D \cdots D\bar{D}.$$

Since the trials are independent, the probability associated with such a sequence is

$$\theta(1 - \theta)\theta\theta(1 - \theta)(1 - \theta)\theta \cdots \theta(1 - \theta),$$

where, of course, there are x factors having the value θ, and $n - x$ factors having the value $1 - \theta$; that is, the probability of such a sequence is

$$(3.2.1) \qquad\qquad \theta^x(1 - \theta)^{n-x}.$$

Now there are $\binom{n}{x}$ different possible sequences of x D's and $(n - x)$ $\bar{D}$'s in our sample space. The probability that any one of these occurs is $\theta^x(1 - \theta)^{n-x}$. Since these $\binom{n}{x}$ different sequences are mutually exclusive elements in the sample space, the probability of obtaining x defectives in n trials is therefore given by

$$(3.2.2) \qquad\qquad b(x) = \binom{n}{x}\theta^x(1 - \theta)^{n-x},$$

where the sample space of x is $0, 1, \ldots, n$. Note that x, the number of D's in an element of the sample space, is a random variable.

The expression given by (3.2.2) is the probability function of the *binomial distribution* and derives its name from the fact that the probabilities are the successive terms in the expansion of the binomial

$$[(1 - \theta) + \theta]^n,$$

the $(x + 1)$st term being the expression for $b(x)$. Thus it is seen that
$$\sum_{x=0}^{n} b(x) = \sum_{x=0}^{n} \binom{n}{x} \theta^x (1 - \theta)^{n-x} = [(1 - \theta) + \theta]^n = 1;$$ that is, the sum of
$b(x)$ over all points in the sample space of x is unity.

We can also obtain the probability function of the binomial distribution by applying a limiting process to that of the hypergeometric distribution. We first give the method and then interpret the result.

It is convenient to use $(3.1.1a)$ for this purpose. There we obtained

$$h(x) = \frac{\binom{N\theta}{x}\binom{N(1 - \theta)}{n - x}}{\binom{N}{n}}$$

which may be written as

$$\frac{(N\theta)!}{x!\,(N\theta - x)!} \cdot \frac{[N(1 - \theta)]!}{(n - x)!\,[N(1 - \theta) - n + x]!} \cdot \frac{n!\,(N - n)!}{N!}$$

After expanding the factorials and performing some algebra, this may be expressed as

$$\binom{n}{x} \frac{\theta\left(\theta - \dfrac{1}{N}\right) \cdots \left(\theta - \dfrac{x - 1}{N}\right)(1 - \theta)\left(1 - \theta - \dfrac{1}{N}\right) \cdots \left(1 - \theta - \dfrac{n - x - 1}{N}\right)}{1\left(1 - \dfrac{1}{N}\right) \cdots \left(1 - \dfrac{x - 1}{N}\right) \cdots \left(1 - \dfrac{n - 1}{N}\right)}$$

If we let N increase indefinitely while θ, the proportion of articles of type A, remains constant, we find

$$\lim_{N \to \infty} h(x) = \binom{n}{x} \theta^x (1 - \theta)^{n-x} = b(x).$$

This result shows that, if N—the population size—is large, the probability function $h(x)$ of the hypergeometric distribution can be approximated by the probability function $b(x)$ of the binomial distribution, for any value of x which can occur, the approximation error tending to zero as $N \to \infty$. This implies that, for very large N, sampling with replacement gives approximately the same probabilities as sampling without replacement.

To give some idea of the closeness of the approximation, suppose we have a lot of 10,000 articles, of which 500 are defective. If a sample of size 10 is taken at random, the probabilities of obtaining x defectives for $x = 0$,

1, . . . , 10 are given in Table 3.2.1 for the cases of sampling with replacement (binomial) and without replacement (hypergeometric). The table shows that, if we use the probability function of the binomial distribution to approximate that of the hypergeometric distribution for this case, the error is very small indeed.

TABLE 3.2.1

VALUES OF $b(x)$ $(n = 10, \theta = .05)$ AND $h(x)$ $(N = 10,000, N_1 = 500, n = 10)$

x	$b(x)$	$h(x)$
0	.5987	.5986
1	.3151	.3153
2	.0746	.0746
3 .	.0105	.0104
4	.0009	.0007
5	.0001	.0000

Example 3.2(a). Two dice are thrown 100 times and the number of "nines" is recorded. What is the probability that x "nines" occur? That at least three "nines" occur?

It is apparent that we are examining each roll of the two dice for the events "nine" or "not-nine." The probability of obtaining a nine by throwing two dice is $4/36 = 1/9$, that is, $\theta = 1/9$. Hence

$$b(x) = \binom{100}{x}\left(\frac{1}{9}\right)^x\left(\frac{8}{9}\right)^{100-x}, \qquad x = 0, 1, 2, \ldots, 100.$$

In answer to the second question, we have

$$P(x \geq 3) = 1 - P(x < 3) = 1 - P(x \leq 2)$$

$$= 1 - \left[\sum_{x=0}^{2}\binom{100}{x}\left(\frac{1}{9}\right)^x\left(\frac{8}{9}\right)^{100-x}\right]$$

$$= 1 - (.000078 + .000097 + .000603)$$

$$= .9993.$$

3.3   THE POISSON DISTRIBUTION

We shall now consider an important probability distribution obtained as a limiting form of the probability function $b(x)$ of the binomial distribution as $n \to \infty$ and $\theta \to 0$ so that $n\theta$ remains constant.

Referring to (3.2.2), we can write

$$b(x) = \frac{n!}{x!\,(n-x)!}\,\theta^x \frac{(1-\theta)^n}{(1-\theta)^x}.$$

If we let $n \to \infty$ and $\theta \to 0$ in such a way that $n\theta$ remains fixed at a value μ, we obtain

$$\lim b(x) = \lim \frac{n!}{x!\,(n-x)!\,n^x}\,\frac{\mu^x(1-\mu/n)^n}{(1-\mu/n)^x}$$

$$= \lim \frac{n(n-1)\cdots(n-x+1)}{x!}\,\frac{\mu^x(1-\mu/n)^n}{n^x(1-\mu/n)^x}$$

$$= \lim 1\left(1-\frac{1}{n}\right)\cdots\left(1-\frac{x-1}{n}\right)\frac{\mu^x(1-\mu/n)^n}{x!(1-\mu/n)^x}$$

$$= \frac{\mu^x}{x!}\,e^{-\mu},$$

where the sample space of x is $0, 1, 2, \ldots$, and, of course, e is the base of natural logarithms having the value 2.71828. Denoting this limit by $p(x)$, we may write

$$(3.3.1) \qquad\qquad p(x) = \frac{\mu^x}{x!}\,e^{-\mu},$$

which is the probability function of the *Poisson distribution*. As the proof suggests, we may use it to approximate $b(x)$ for large n and small θ as follows:

$$(3.3.2) \qquad\qquad b(x) \cong \frac{(n\theta)^x e^{-n\theta}}{x!}$$

Note that

$$\sum_{x=0}^{\infty} p(x) = \sum_{x=0}^{\infty} \frac{\mu^x}{x!}\,e^{-\mu} = e^{-\mu}\left(1 + \mu + \frac{\mu^2}{2!} + \cdots\right) = e^{-\mu} \cdot e^{\mu} = 1,$$

that is, the sum of $p(x)$ over all points in the sample space of x is 1.

Example 3.3(a). Two percent of the screws made by a machine are defective, the defectives occurring at random during production. If the screws are packaged 100 per box, what is the probability that a given box will contain x defectives?

The probability that the box contains x defectives is given by the binomial distribution

$$b(x) = \binom{100}{x}\left(\frac{2}{100}\right)^x\left(1-\frac{2}{100}\right)^{100-x}, \qquad x = 0, 1, \ldots, 100.$$

Since $n = 100$, $\theta = .02$, and $n\theta = 2$, the Poisson approximation to $b(x)$ is given by

$$b(x) \simeq \frac{2^x e^{-2}}{x!}$$

A comparison of $b(x)$ and its Poisson approximation is given in Table 3.3.1.

TABLE 3.3.1

VALUES OF $b(x)$ ($n = 100$, $\theta = .02$) AND $p(x)$ ($\mu = 2$)

x	$b(x)$	$p(x)$
0	.1326	.1353
1	.2707	.2707
2	.2734	.2707
3	.1823	.1804
4	.0902	.0902
5	.0353	.0361
6	.0114	.0120
7	.0031	.0034
8	.0007	.0009
9	.0002	.0002
10	.0000	.0000

We have already noted that under certain conditions the binomial distribution provides an approximation for the hypergeometric distribution, and that under certain conditions the Poisson distribution provides an approximation to the binomial distribution. It is evident that, under both sets of conditions, the Poisson distribution provides an approximation for the hypergeometric distribution.

The Poisson distribution is a distribution in its own right which arises in many different situations. For instance, it provides probabilities of specified numbers of telephone calls per unit interval of time, of given numbers of defects per unit length of wire or thread, of specified numbers of defects per unit area of glass or textiles or papers, of various numbers of bacterial colonies per unit volume of a dilution, and so on.

For example, a certain type of wire is insulated by an enameling process, and it is found that the occurrence of an insulation break in a piece of wire ΔL units long is proportional to ΔL, say $\lambda \Delta L$, and independent of both the position on the wire and the number of insulation breaks that occur elsewhere on the wire. We show in Section 3.6b that the probability that x insulation breaks will occur in a length L of wire, say $p_x(L)$, is given by

$$(3.3.3) \qquad p_x(L) = \frac{(\lambda L)^x}{x!} e^{-\lambda L},$$

where λ is the mean number of insulation breaks per unit of length.

Example 3.3(b). It is known that, in a certain enameling process, the number of insulation breaks per yard is .07. What is the probability of finding x such breaks in a piece of wire 16 yards long? Here

$$\lambda = .07, \quad \text{and} \quad \mu = \lambda L = (.07)(16) = 1.12.$$

Hence,

$$p_x(L) = \frac{(1.12)^x}{x!} e^{-1.12}.$$

3.4 THE PASCAL DISTRIBUTION

Suppose trials are to be performed successively, independently of one another, and that the outcome of each trial is either an A or an $\bar{A}$ (see Section 3.1). Let $P(A) = \theta$ so that $P(\bar{A}) = 1 - \theta = \tau$. We now ask for the probability that the number of trials required, in order to obtain exactly k A's, is x.

To answer this question, suppose we let E be the event of obtaining $k - 1$ A's in the first $x - 1$ trials and F be the event of getting an A on the xth trial. If C is the event of having to make exactly x trials to get exactly k A's, then $C = E \cap F$ so that

(3.4.1) $$P(C) = P(E)\, P(F \mid E).$$

But $P(F \mid E) = \theta$ and it is easy to see that

(3.4.2) $$P(E) = \binom{x-1}{k-1} \theta^{k-1}(1-\theta)^{x-k}.$$

Hence, denoting $P(C)$ by $p_A(x)$, we have that

(3.4.3) $$p_A(x) = \binom{x-1}{k-1} \theta^k(1-\theta)^{x-k}, \qquad x = k, \quad k+1, \ldots.$$

This is the *Pascal distribution*, and is sometimes called the negative binomial distribution. The latter name derives from the fact that $p_A(x)$ is the $(x - k + 1)$th term obtained in the expansion of $\theta^k (1 - \tau)^{-k}$ if $(1 - \tau)^{-k}$ is expanded into a series in powers of τ, where of course $\tau = 1 - \theta$. It is interesting also to note that (3.4.3) is sometimes referred to as the *binomial waiting-time distribution*, for the probability function $p_A(x)$ is simply the probability that one must wait through x independent trials in order to obtain k A's. It can also be shown (see problem 6.24) that the "average wait" is k/θ.

Example 3.4(a). Two dice are thrown and the sum of the dots obtained on the uppermost faces recorded. What is the probability that a "7" occurs for the third time on the third throw? on the fourth throw? on the xth throw $(x \geq 3)$?

Consulting Example 2.7(a), we have here that $\theta = 6/36 = 1/6$ and that the number of throws x, needed to get $k = 3$ "sevens" (event A is the event "a seven") has probabilities given by the Pascal distribution, that is

$$(3.4.4) \qquad p_A(x) = \binom{x-1}{2} \theta^3 (1 - \theta)^{x-3} = \binom{x-1}{2} \frac{5^{x-3}}{6^x}.$$

Hence

$$p_A(3) = 1/216 = .00463$$
$$p_A(4) = 15/1296 = .01157$$
$$p_A(5) = 150/7776 = 0.1929, \text{ etc.}$$

If this experiment were repeated a large number of times, the average wait (k/θ) needed to get 3 "sevens" would be $3/[1/6] = 18$ throws.

3.5 THE MULTINOMIAL DISTRIBUTION

Suppose a trial can result in one and only one of k mutually exclusive events $E_1, E_2, \ldots, E_k$ with probabilities $\theta_1, \theta_2, \ldots, \theta_k$ respectively, where $\theta_1 + \cdots + \theta_k = 1$. If n independent trials are made, then it is seen by argument similar to that by which the probability function (3.2.2) of the binomial distribution was obtained that the probability of getting x_1 E_1's, x_2 E_2's, $\ldots, x_k$ E_k's is given by

$$(3.5.1) \qquad m(x_1, \ldots, x_k) = \frac{n!}{x_1! \cdots x_k!} \theta_1^{x_1} \cdots \theta_k^{x_k},$$

where $0 \leq x_i \leq n$, $i = 1, \ldots, k$, and $x_1 + \cdots + x_k = n$. This is the probability function of the *multinomial distribution*. The names derives from the fact that the probabilities are the terms in the expansion of $(\theta_1 + \cdots + \theta_k)^n$.

Note that, if $k = 2$, we have the binomial distribution, and hence the multinomial distribution is essentially an extension of the binomial distribution.

Example 3.5(a). Consider the production of ball bearings of a certain type, the diameters of which should be .2500 inch. Because of the inherent variability in the manufacturing process, and because of consumer demands, the bearings are classified as *undersize*, *oversize*, and *acceptable* if they measure less than .2495 inch, more than .2505 inch, and between .2495 and .2505 inch, respectively.

Suppose the production process for these bearings is such that 4% of the bearings are undersize, 6% are oversize, and 90% are acceptable. If 100 of these bearings are picked at random, the probability of getting x_1 undersize, x_2 oversize, and x_3 acceptable bearings is given by

$$(3.5.2) \qquad m(x_1, x_2, x_3) = \frac{100!}{x_1! \, x_2! \, x_3!} (.04)^{x_1}(.06)^{x_2}(.90)^{x_3},$$

where $0 \leq x_1 \leq 100, 0 \leq x_2 \leq 100, 0 \leq x_3 \leq 100$, and $\sum_{i=1}^{3} x_i = 100$.

3.6 SOME DERIVATIONS AND PROOFS

(a) Proof that the Probability Function of the Hypergeometric Distribution Sums to 1

We consider the identity

(3.6.1) $$(1 + a)^{N_1}(1 + a)^{N-N_1} = (1 + a)^N.$$

The coefficient of a^n in the expansion of the right-hand side is $\binom{N}{n}$. The left-hand side of (3.6.1) can be written as

(3.6.2) $$\left[1 + \binom{N_1}{1}a + \cdots + \binom{N_1}{x}a^x + \cdots + a^{N_1}\right]$$
$$\times \left[1 + \cdots + \binom{N - N_1}{n - x}a^{n-x} + \cdots + a^{N-N_1}\right],$$

and, therefore, the coefficient of a^n in the expansion of the left-hand side of (3.6.1) is seen to be

$$\sum_x \binom{N_1}{x}\binom{N - N_1}{n - x},$$

where the summation is over values of x satisfying

$$\max\,[0, n - (N - N_1)] \le x \le \min\,[n, N_1].$$

But the coefficients of a^n of the left and right sides of (3.6.1) must be equal, and so we have

(3.6.3) $$\sum_x \binom{N_1}{x}\binom{N - N_1}{n - x} = \binom{N}{n},$$

and division by $\binom{N}{n}$ yields

$$\sum_x h(x) = 1.$$

(b) Derivation of (3.3.3)

We have assumed that the probability of occurrence of an insulation break in a piece of wire ΔL units long is $\lambda\,\Delta L$, and is independent of both the position on the wire (of length L units) and the number of insulation breaks that occur elsewhere on the wire.

Under these assumptions, and dividing up a piece of wire of length L units into n small pieces of length $\Delta L = L/n$ units each, we further assume that:

1. The probability that exactly one insulation break occurs in a specified piece of length ΔL units (ΔL small) is $\lambda\,\Delta L$.

2. The probability that no breaks occur in a piece of length ΔL is $1 - \lambda\, \Delta L$.

3. The probability that two or more breaks occur in a piece of wire of length ΔL, which we express by $o(\Delta L)$, is such that $o(\Delta L)/\Delta L \to 0$ as $\Delta L \to 0$.

The event that x breaks occur in a piece of wire of length $L + \Delta L$ is the union of the following mutually exclusive events:

(a) Exactly x breaks occur in the piece of wire of length L and none in the "bordering" piece of length ΔL.

(b) Exactly $x - 1$ breaks occur in the piece of wire of length L and exactly one in the "bordering" piece of length ΔL.

(c) Exactly $x - i$ breaks occur in the piece of length L and exactly i in the "bordering" piece of length ΔL, where $i = 2, \ldots, x$.

The probabilities of the events described in assumptions (1), (2), and (3) are respectively $p_x(L)(1 - \lambda\, \Delta L)$, $p_{x-1}(L)\lambda\, \Delta L$, and a quantity which we denote by $o(\Delta L)$, which goes to zero faster than ΔL goes to zero. Hence,

$$p_x(L + \Delta L) = p_x(L)(1 - \lambda\, \Delta L) + p_{x-1}(L)\lambda\, \Delta L + o(\Delta L)$$

if $x = 1, 2, \ldots$, and

$$p_0(L + \Delta L) = p_0(L)(1 - \lambda \Delta L) + o(\Delta L).$$

These equations may be rewritten as follows:

$$\frac{p_x(L + \Delta L) - p_x(L)}{\Delta L} = \lambda[p_{x-1}(L) - p_x(L)] + \frac{o(\Delta L)}{\Delta L}, \qquad x = 1, 2, \ldots,$$

$$\frac{p_0(L + \Delta L) - p_0(L)}{\Delta L} = -\lambda p_0(L) + \frac{o(\Delta L)}{\Delta L}.$$

As $\Delta L \to 0$, that is, as we allow the length of $L + \Delta L$ to shrink to L, we obtain the differential equations

$$\frac{dp_x(L)}{dL} = \lambda[p_{x-1}(L) - p_x(L)]$$

and

$$\frac{dp_0(L)}{dL} = -\lambda p_0(L).$$

If we define $p_0(0) = 1$ and let $p_x(0) = 0$ for $x > 1$, we obtain the solution

$$p_x(L) = \frac{(\lambda L)^x e^{-\lambda L}}{x!}, \qquad x = 0, 1, \ldots,$$

which is the Poisson distribution $p(x)$ given in (3.3.1) with $\mu = \lambda L$.

PROBLEMS

3.1. Suppose a lot of 50 fuses, of which 7 are known to be defective, are available for sampling. It is proposed to draw ten fuses at random (without replacement) and test them. What is the probability that such random samples of ten will contain 0, 1, 2, . . . , 7 defective fuses?

3.2. A lot contains 30 items, 6 of which are defective. What is the probability that a random sample of five items from the lot will contain no defective items? Not more than one defective? More than two defectives?

3.3. In rolling five true dice, find the probability of obtaining at least one ace? Exactly one ace? Exactly two aces?

3.4. If the probability of hitting a target is .2 and ten shots are fired independently, what is the probability that the target will be hit at least once? At least twice?

3.5. What is the probability of drawing a 13-card hand containing no aces, kings, queens, or jacks?

3.6. Suppose 5% of the aspirin tablets pressed by a certain type of machine are chipped. The tablets are boxed 12 per box. What percent of the boxes would you estimate:
 (a) to be free of chipped tablets;
 (b) to have not more than 1 chipped tablet;
 (c) to have exactly x chipped tablets?

3.7. Suppose 13 cards are dealt from a thoroughly shuffled deck of ordinary playing cards.
 (a) What is the probability of getting x spades?
 (b) What is the probability of getting y hearts? Describe the sample space of y.
 (c) What is the probability of getting x spades and y hearts?
Describe the sample space of (x, y).

3.8. What is the probability of throwing two heads three times in four throws of five coins?

3.9. It is known that .0005% of the insured males die from a certain kind of accident each year. What is the probability that an insurance company must pay off on more than 3 of 10,000 insured risks against such accidents in a given year?

3.10. A bag of grass seed is known to contain 1% weed seeds. A sample of 100 seeds is drawn. Find the probabilities of 0, 1, 2, 3, . . . , 7 weed seeds being in the sample.

3.11. A process for making plate glass produces an average of four "seeds" (small bubbles) scattered at random in the glass per 100 square feet. With the use of the Poisson distribution, what is the probability that:
 (a) A piece of plate glass 5 feet by 10 feet will contain more than two seeds?
 (b) Six pieces of plate glass 5 feet by 5 feet will all be free of seeds?

3.12. Small brass pins are made by company ABC. Two percent of these pins are "undersized," 6% are "oversized," and 92% are "satisfactory." The pins are boxed 100 per box. A box is taken at random. Write down the expression for the probabilities of the following events:

(a) The box contains only satisfactory pins.

(b) The box contains x satisfactory pins, y undersized pins, the remaining pins being oversized.

(c) The box contains no undersized pins.

3.13. Suppose ten people each throw two coins. What is the probability that:

(a) Three people throw two heads, three people throw two tails, and four people throw one head and one tail?

(b) No one threw a head and a tail?

3.14. Two coins are tossed n times. Find the probability of x, the number of times no heads appear, y, the number of times one head appears, and z, the number of times two heads appear.

3.15. An urn contains 10 white and 20 black balls. Balls are drawn one by one, without replacement, until five white ones have appeared. Find an expression for the probability that the total number drawn is x.

3.16. Suppose a lot of 10,000 articles has 200 defectives, and that a random sample of 100 articles is drawn from the lot.

(a) What is the probability of getting exactly x defectives in the sample?

(b) Determine the binomial approximation for the probability in (a).

(c) Determine the Poisson approximation for the probability in (a).

3.17. The fraction of articles turned out by a machine which are defective is equal to θ. The defectives occur "at random" during production. The articles are boxed m per box and cartoned n boxes per carton.

(a) If a box of articles is taken at random, what is the probability it contains exactly x defectives?

(b) If a carton is taken at random, what is the probability that exactly y of its boxes will be free of defectives?

3.18. In Problem 3.17, what is the probability that:

(a) The machine will produce k good (nondefective) articles before it turns out a defective?

(b) The machine has to turn out a total of x articles in order to produce exactly k good ones?

3.19. A certain process turns out articles of which $100\theta\%$ are defective. It is inspected by randomly choosing a large number of samples of four articles and inspecting them. The following information is known: in 60% of the samples none of the four was defective, while in 30% of the samples one of the four was defective. What fraction of the articles produced by this process would you estimate to be defective, that is, what estimate would you make for θ?

3.20. A lot of N articles has d defectives. If articles are taken at random from the lot one at a time, what is the probability that:

(a) Exactly x articles have to be examined to find the first defective?

(b) Exactly y articles have to be examined to find the dth (last) defective?

3.21. In testing a relay, suppose the probability is θ that it fails to make satisfactory contact in a single trial and that θ remains unchanged over a very large number of trials. By assuming the outcomes of successive trials to be independent, what is the probability that:

(a) x trials have to be made to obtain the first failure?

(b) x trials have to be made to obtain the kth failure?

3.22. In the past decade, the proportion of successful launches at a test missile site has been .85. Suppose that an experiment is planned which requires 3 successful launches. What is the probability that exactly 5 attempts will be necessary? exactly 7? exactly x? Fewer than 6?

3.23. In Problem 3.22, suppose another experiment requires 4 consecutive successful launches. What is the probability that six attempts will be required? What is the probability that x attempts will be required?

CHAPTER 4

Acceptance Sampling

4.1 INTRODUCTION

In the procurement of *lots* of mass-produced articles the decision to accept or reject an individual lot is often based on the results of inspecting or testing a random *sample* of articles from the lot. More precisely, suppose we have a lot of N articles and that a random sample of n articles is drawn from the lot. The lot is accepted if not more than an *allowable number c* of articles in the sample is defective. Otherwise the lot is rejected. Thus, for a given lot of size N, a *sampling plan* depends on the two numbers (n, c). A sampling plan based on one sample is sometimes called a *single sampling plan* as contrasted with a *double sampling* or *multiple sampling plan* to be discussed in Section 4.5.

If articles cannot be tested without destroying them, rejection usually means scrapping the lot. For instance, if testing a fuse is to determine its blowing time, the fuse is destroyed in the process of testing. Hence, to reject a lot of fuses as a result of blowing-time tests on a sample of fuses from the lot is tantamount to scrapping the lot, since tests on the remaining fuses in the lot would destroy them in any case.

On the other hand, if articles can be tested or inspected without destroying or damaging the good ones, rejection of the lot means consignment of the lot to 100% inspection if it is economical to pick out the defectives in the lot by inspection. A lot in which defectives are removed and replaced by nondefectives is sometimes called a *rectified* lot. If the entire lot is not worth the cost involved in removing defectives, the lot would be scrapped or possibly sold at cut-rate prices under certain conditions.

The probability of accepting a lot is obtained by adding the probabilities of getting 0 or 1 or 2, or . . . , or c defectives in the sample. If $p(x)$ is the probability of getting x defectives in the sample, and if we denote the event of acceptance by A and let $P(A)$ denote the probability of acceptance of the lot, then

$$(4.1.1) \qquad\qquad P(A) = \sum_{x=0}^{c} p(x).$$

The value of $P(A)$ as well as $p(x)$ depends not only on the lot size N, the sample size n, and the allowable number of defectives c but also on the fraction θ of defectives in the lot. Hence $P(A)$ and $p(x)$ are functions of θ which may be written as $P(A \mid \theta)$ and $p(x \mid \theta)$, where

$$(4.1.2) \qquad\qquad P(A \mid \theta) = \sum_{x=0}^{c} p(x \mid \theta).$$

It is evident that $P(A \mid 0) = 1$, that is, the probability of accepting a lot with no defectives is 1. Also, it is evident that $P(A \mid \theta)$ decreases as θ

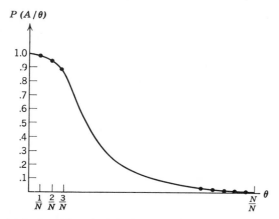

Figure 4.1.1. Operating characteristic curve for a typical sampling plan.

increases. In other words, if we use the same sample size n and the same allowable defective number c on two lots L_1 and L_2 of the same size N, but having fractions defective θ_1 and θ_2, where $\theta_1 < \theta_2$, then we have $P(A \mid \theta_1) > P(A \mid \theta_2)$. Note that $P(A \mid \theta)$ is defined only for $\theta = 0$, $1/N$, $2/N, \ldots, N/N$, and hence the graph of $P(A \mid \theta)$ is simply the sequence of points

$$(0, P(A \mid 0)), \left(\frac{1}{N}, P\left(A \mid \frac{1}{N}\right)\right), \ldots, \left(\frac{N}{N}, P\left(A \mid \frac{N}{N}\right)\right).$$

If we draw a smooth curve through these points, then we obtain the *operating characteristic* (OC) curve of the sampling plan, which is illustrated in Figure 4.1.1.

Example 4.1(a). To illustrate the preceding ideas, let us take a simple example of a single sampling plan. Suppose articles are packaged 25 to a box and that the following acceptance sampling plan is used for accepting (or rejecting) boxes of these articles:

1. A random sample of two articles is drawn from a box and the articles are tested.

2. The box is accepted if both articles in the sample are good; otherwise the box of articles is rejected.

In this case, $N = 25$, $n = 2$, $c = 0$, and $P(A \mid \theta)$ is simply the probability of getting no defectives in the sample. Written out explicitly,

$$P(A \mid \theta) = \frac{\binom{25\theta}{0}\binom{25 - 25\theta}{2}}{\binom{25}{2}},$$

where the possible values of θ are $0, 1/25, \ldots, 25/25$. Writing out the values of $P(A \mid \theta)$ in table form, we find (to two decimal places):

θ	$P(A \mid \theta)$	θ	$P(A \mid \theta)$
0	1.00	.52	.22
.04	.92	.56	.18
.08	.84	.60	.15
.12	.77	.64	.12
.16	.70	.68	.10
.20	.64	.72	.07
.24	.57	.76	.05
.28	.51	.80	.03
.32	.45	.84	.02
.36	.40	.88	.01
.40	.35	.92	.00
.44	.30	.96	.00
.48	.26	1.00	.00

To illustrate the meaning of the entries of this table, let us take the entry (.16, .70). This means that, if the box has four defectives (fraction defectives $= 4/25 = .16$), the probability of accepting the box on the basis of the sampling plan is .70.

The OC curve shown in Figure 4.1.2 is the smooth curve drawn through the points plotted from the preceding table.

Remark. It should be noted that the concept of acceptance sampling can be applied to any binomial or dichotomized population of objects where each object is either defective or nondefective. For example, the population of light bulbs where a bulb with a broken filament is a defective; screws where a slotless screw is a defective; bills where an incorrect bill is a defective, and so on.

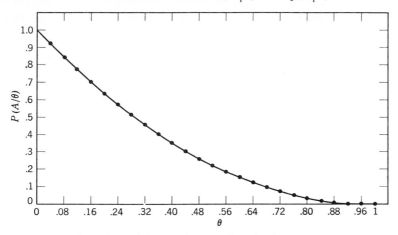

Figure 4.1.2. OC curve for sampling plan in Example 4.1(*a*).

4.2 APPROXIMATION FOR $P(A \mid \theta)$ AND $p(x \mid \theta)$ FOR LARGE LOTS AND SMALL VALUES OF θ

(a) The Binomial Approximation for $p(x \mid \theta)$ and $P(A \mid \theta)$

In many situations, acceptance sampling plans are applied to large lots, that is, situations where N runs into hundreds, thousands, or even larger. Furthermore, they are usually applied to situations in which θ is quite small, say, not more than .10 or .15. (A successful manufacturer of mass-produced articles is one who has usually managed to keep his fraction of defectives down to few percentage points!) In such cases we have seen in Section 3.1 that the probability of getting exactly x defectives in a random sample of size n from a lot of size N containing $N\theta$ defectives is given by (3.1.1*a*), that is,

$$(4.2.1) \qquad p(x \mid \theta) = \frac{\binom{N\theta}{x}\binom{N - N\theta}{n - x}}{\binom{N}{n}}.$$

But we also know from Section 3.2 that, as $N \to \infty$, we may approximate $p(x \mid \theta)$ by the binomial probability function, that is,

$$(4.2.2) \qquad p(x \mid \theta) \cong \binom{n}{x}\theta^x(1 - \theta)^{n-x}.$$

Now the binomial approximation to $p(x \mid \theta)$ given in (4.2.2) for large lots is simpler to compute than the exact expression for $p(x \mid \theta)$ given in

TABLE 4.2.1

VALUES OF $P(A \mid \theta)$ FOR EXAMPLE 4.2(a)

θ	$P(A \mid \theta)$
0	1.00
.01	.91
.02	.74
.03	.56
.04	.41
.05	.29
.06	.20
.07	.14
.08	.09
.09	.06
.10	.04
.11	.03
.12	.02
.13	.01
.14	.01
.15	.00

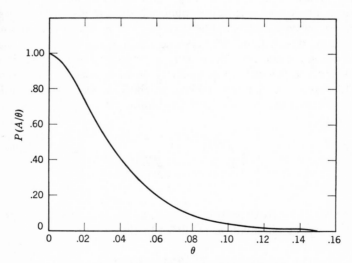

Figure 4.2.1. OC curve for Example 4.2(a).

(4.2.1). The values of $P(A \mid \theta)$ from this approximation is given by

$$P(A \mid \theta) = \sum_{x=0}^{c} \binom{n}{x} \theta^x (1 - \theta)^{n-x}.$$

(b) The Poisson Approximation for $p(x \mid \theta)$ and $P(A \mid \theta)$

We have seen in Section 3.3 that, if $n \to \infty$ and $\theta \to 0$, so that $n\theta = \mu$, the binomial distribution has as its limit the Poisson distribution. Hence, if n is "large" and θ is "small" so that $n\theta$ is "moderate" (in practice less than about 25), the Poisson provides a good approximation for $p(x \mid \theta)$. That is,

$$(4.2.3) \qquad p(x \mid \theta) \simeq \frac{\mu^x e^{-\mu}}{x!} ,$$

where $\mu = n\theta$.

The Poisson approximation to $P(A \mid \theta)$ is given by

$$(4.2.4) \qquad P(A \mid \theta) \simeq \sum_{x=0}^{c} \frac{\mu^x}{x!} e^{-\mu}.$$

> *Example 4.2(a).* Suppose the following single sampling plan is used for accepting (or rejecting) large lots of mass-produced articles.
>
> 1. Draw a sample of size 50 from the lot and inspect the 50 items.
> 2. Accept the lot if the sample contains not more than one defective; otherwise reject the lot.
>
> Here we have a sampling plan with $n = 50$, $c = 1$. If θ is the fraction of defectives in the lot, then $\mu = 50\theta$. The Poisson approximation to $p(x \mid \theta)$ is
>
> $$(4.2.5) \qquad p(x \mid \theta) = \frac{(50\theta)^x e^{-50\theta}}{x!}$$
>
> and the Poisson approximation to $P(A \mid \theta)$ is
>
> $$(4.2.6) \qquad P(A \mid \theta) = \sum_{x=0}^{1} p(x \mid \theta) = e^{-50\theta}(1 + 50\theta).$$
>
> A tabulation of $P(A \mid \theta)$ for a set of conveniently chosen values of θ is given in Table 4.2.1. The OC curve for the sampling plan given in Example 4.2(a) is shown in Figure 4.2.1.

4.3   AVERAGE OUTGOING QUALITY

Suppose we have a given sampling plan with the probability $P(A \mid \theta)$ that this sampling plan will accept a large lot containing a fraction of defectives equal to θ. The probability that this plan will reject the lot is $1 - P(A \mid \theta)$. Let us assume that inspection is nondestructive and that rejection of a lot means rectifying the lot, that is, consignment of the lot

TABLE 4.3.1

AOQ FOR EXAMPLE 4.2(*a*)

θ	$P(A \mid \theta)$	$AOQ(\theta) = \theta P(A \mid \theta)$
0	1.00	0
.01	.91	.009
.02	.74	.015
.03	.56	.017
.04	.41	.016
.05	.29	.015
.06	.20	.012
.07	.14	.010
.08	.09	.007
.09	.06	.005
.10	.04	.004
.11	.03	.003
.12	.02	.002
.13	.01	.001
.14	.01	.001
.15	.00	.000

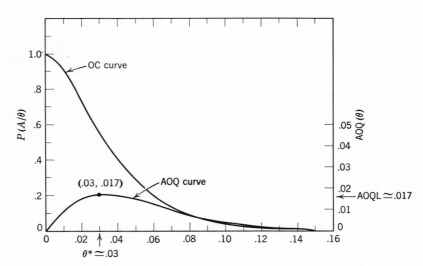

Figure 4.3.1. The OC and AOQ curves for the sampling plan in Example 4.2(*a*).

to 100% inspection, in which case all defectives are removed and replaced by nondefectives. If the allowable number of defectives c exceeds 0, we assume that any defectives found in a sample from a lot subsequently accepted, are replaced by nondefectives.

Now if we consider lot after lot from a plant producing 100θ% defectives being subjected to this sampling plan, then a certain fraction of these lots, namely $P(A \mid \theta)$, will each contain a fraction of defectives equal to θ. A certain fraction of the lots, namely, $1 - P(A \mid \theta)$ will be rectified, that is, will contain no defectives since all defectives in these lots will have been removed and replaced by nondefective articles.

If we now consider the pool of all lots (all out-going material) from the plant which are thus subjected to the same sampling plan, the fraction of this pool of articles which will be defective is called the *Average Outgoing Quality* (AOQ). This fraction of defectives is simply the fraction of lots with defectives times the fraction of defectives per lot, that is, we have

(4.3.1) $$\text{AOQ}(\theta) = \theta P(A \mid \theta),$$

where, since AOQ is a function of θ, we write it $\text{AOQ}(\theta)$.

Remembering that $P(A \mid \theta)$ is a decreasing function from $P(A \mid 0) = 1$ to $P(A \mid 1) = 0$, and noting that θ increases from 0 at $\theta = 0$ to 1 at $\theta = 1$, it is evident that the graph of $\text{AOQ}(\theta)$ will increase from 0 to a maximum value $\theta^* P(A \mid \theta^*)$ at $\theta = \theta^*$ as θ increases from 0 to θ^*, and then decreases from $\theta^* P(A \mid \theta^*)$ to 0 as θ increases from θ^* to 1. The maximum value of $\theta P(A \mid \theta)$, namely $\theta^* P(A \mid \theta^*)$, is called the *Average Outgoing Quality Limit* (AOQL). It is the largest value of the fraction defective which can possibly occur in the pool of outgoing material, after being subjected to the sampling plan, no matter what fraction defective θ the plant is actually producing. Note that, if the plant happens to be producing fraction defective equal to θ^*, then the fraction of defectives in the pool of out-going material will attain its maximum value, namely the AOQL. If the value of θ fluctuates from lot to lot, then the AOQ for a set of lots will be an average of AOQ's for the individual lots and, hence will not exceed the AOQL.

Let us illustrate this situation graphically by considering Example 4.2(a). Values of $\text{AOQ}(\theta)$ for various values of θ are shown in Table 4.3.1.

The graphs of $P(A \mid \theta)$ and $\theta P(A \mid \theta)$, that is, the OC curve and the AOQ curve are shown in Figure 4.3.1. The scale for graphing $\text{AOQ}(\theta)$ is placed at the right of the graph.

Note that $\theta^* \simeq .03$ and $\text{AOQL} \simeq .017$. Thus no matter what fraction defective exists in the material being produced, the average outgoing fraction defective in the pool of material subjected to the sampling plan of Example 4.2(a) cannot exceed (approximately) 1.7%. This would occur if the plant were producing approximately 3% of defective articles. If the plant produces a fraction defective higher or lower than 3%, the fraction defective in the outgoing material, after being subjected to the sampling, will be less than 1.7%. If the fraction defective fluctuates from lot to lot the AOQ for a collection of lots will not exceed 1.7%.

A set of tables has been developed by Dodge and Romig (1959) for determining values of n and c for sampling plans with various values of AOQL. These tables are widely used in military and other large procurement programs. Other widely used tables have been prepared by the U.S. Department of Defense (1950), and by the Chemical Corps Engineering Agency of the U.S. Army (1953).

4.4 PRODUCER'S AND CONSUMER'S RISKS

The AOQL of a sampling plan provides protection to a buyer of a quantity of lots against having a fraction of defectives in his pool of material greater than a certain amount, provided the lots are rectifiable, that is, provided defectives can be removed without destructive testing.

If lots are not rectifiable, the protection provided by a sampling plan is furnished by the information contained in the OC curve of the sampling plan. To make the concept of protection more precise both for the producer (seller) and consumer (buyer) the concepts of *producer's risk* and *consumer's risk* of a sampling plan have been introduced. In general, the producer's (seller's) risk of a sampling plan is the risk that the producer takes in having a *good* or *acceptable* lot rejected by the sampling plan, whereas the consumer's (buyer's) risk is the risk the buyer takes in having a *bad* or *unacceptable* lot accepted by the sampling plan.

To describe these concepts numerically, suppose a producer and a buyer agree that a lot containing fraction defective $\leq \theta_1$ is an acceptable lot and one having fraction defective $\geq \theta_2$ (where $\theta_1 < \theta_2$) is an unacceptable lot. Lots having $\theta_1 < \theta < \theta_2$ might be called lots of *indifferent* quality. Suppose α is the probability of the sampling plan rejecting a lot which has fraction defective θ_1 and β is the probability of the sampling plan accepting a lot which has fraction defective θ_2. Then it follows from the fact that the OC curve decreases as θ increases that the probability is at most α that the sampling plan will reject an acceptable lot and the probability is at most β that the plan will accept an unacceptable lot. For given θ_1 and θ_2, α and β are called the *producer's risk* and the *consumer's risk*, respectively: These concepts are represented graphically in Figure 4.4.1.

If the producer and consumer agree on values of α and β and of θ_1 and θ_2, this is equivalent to their agreeing that any sampling plan they use should have an OC curve passing through the two points $(\theta_1, 1 - \alpha)$ and

(θ_2, β). For large lots it is possible to find an n and c (that is, a sampling plan) so that the OC curve does approximately pass through these two points, although we shall not go into the techniques for doing this. In practice α and β are usually chosen as .05 or .10. Use of these values in plans which have their OC curve passing through the points $(\theta_1, 1 - \alpha)$ and (θ_2, β), will ensure that the average percent of defective items that will reach the consumer is less than $100\theta_2\%$, and in some instances, may be even less than $100\theta_1\%$.

Referring to Figure 4.4.1, suppose the producer's risk α is .05, and the consumer's risk β is .10. Then acceptable lots would be defined as those

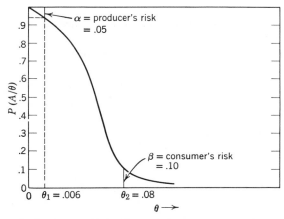

Figure 4.4.1. Graph showing producer's risk and consumer's risk for a sampling plan. Acceptable lots have $\theta \leq \theta_1$, unacceptable lots have $\theta \geq \theta_2$, and indifferent lots have values of θ between θ_1 and θ_2.

with fraction defective $\leq$ .006 (approximately) and unacceptable lots as those with fraction defective $\geq$ .08 (approximately). This, of course, leaves a range of values of θ for lots of indifferent quality. The only way to shorten the range of θ for lots of indifferent quality for fixed values of α and β is to use a larger sample and an appropriate value of c. Sampling plans for specified values of α, β, θ_1, and θ_2 have been published by Dodge and Romig (1959) and by the Statistical Research Group of Columbia University (1948). These plans have been widely applied in military and other large procurement programs.

4.5 DOUBLE SAMPLING

Some saving of inspection costs can often be achieved if acceptance is based on double sampling, that is, on a sampling plan involving two successively drawn samples rather than one. There is also a psychological advantage under some conditions in using a double sampling plan rather than a single sampling plan. A double sampling plan gives a doubtful lot a "second chance," so to speak, although the risks involved for both producer and consumer are contained in the OC function of the plan. As a matter of fact, a single sampling plan can usually be found for which producer and consumer risks are approximately the same as those for any given double sampling plan.

In general, a double sampling plan is defined as follows:

1. A sample of size n_1 is drawn from a lot of size N and the items are inspected.

2. If the number of defective items in the sample does not exceed c_1, the lot is accepted.

3. If the number of defective items in the sample exceeds c_2, the lot is rejected, where $c_1 < c_2$.

4. If the number of defective items in the sample exceeds c_1 but not c_2, a second sample of size n_2 is drawn from the remainder of the lot.

5. If the number of defectives in the two samples combined does not exceed c_2, the lot is accepted.

6. If the number of defectives in the two samples combined exceeds c_2, the lot is rejected.

If we let θ be the fraction of defectives in the lot, and let $p(x_1, x_2 \mid \theta)$ be the probability of getting x_1 defectives in the first sample and x_2 in the second, then we have for all *possible* combinations of values (x_1, x_2):

(4.5.1)
$$p(x_1, x_2 \mid \theta) = \frac{\binom{N\theta}{x_1}\binom{N(1-\theta)}{n_1 - x_1}}{\binom{N}{n_1}} \cdot \frac{\binom{N\theta - x_1}{x_2}\binom{N(1-\theta) - n_1 + x_1}{n_2 - x_2}}{\binom{N - n_1}{n_2}}.$$

If we let $p_1(x_1 \mid \theta)$ be the probability distribution of x_1 and let $P(A \mid \theta)$ be the probability of accepting the lot, then we have

(4.5.2)
$$p_1(x_1 \mid \theta) = \frac{\binom{N\theta}{x_1}\binom{N(1-\theta)}{n_1 - x_1}}{\binom{N}{n_1}},$$

(4.5.3)
$$P(A \mid \theta) = \sum_{x_1=0}^{c_1} p_1(x_1 \mid \theta) + \sum{}^* p(x_1, x_2 \mid \theta),$$

where $\sum^*$ denotes summation over all points (x_1, x_2) in the x_1, x_2-plane, where x_1 and x_2 is each 0 or a positive integer such that $x_1 + x_2 \le c_2$ and $c_1 < x_1 \le c_2$.

$P(A \mid \theta)$ is a function of $N, n_1, n_2, c_1, c_2, \theta$ and, in general, its computation is tedious. However, under the conditions in Section 4.2a we have for large N the binomial approximation

$$(4.5.4) \quad P(A \mid \theta) \cong \sum_{x_1=0}^{c_1} \binom{n_1}{x_1} \theta^{x_1} (1 - \theta)^{n_1 - x_1}$$

$$+ \sum^* \binom{n_1}{x_1}\binom{n_2}{x_2} \theta^{x_1 + x_2}(1 - \theta)^{n_1 + n_2 - x_1 - x_2}$$

where $\sum^*$ has the same meaning as in (4.5.3). Under the further conditions of Section 4.2b, which, as we have pointed out, commonly prevail in industry, we may use the Poisson approximation, thus obtaining

$$(4.5.5) \quad P(A \mid \theta) \cong \sum_{x_1=0}^{c_1} \frac{(n_1\theta)^{x_1} e^{-n_1\theta}}{x_1!} + \sum^* \frac{(n_1\theta)^{x_1}(n_2\theta)^{x_2} e^{-(n_1\theta) - (n_2\theta)}}{x_1! \, x_2!}.$$

$P(A \mid \theta)$, under either the binomial or Poisson approximations, depends on $n_1, n_2, c_1, c_2,$ and θ. Thus, when either of these approximations is used, we may characterize the double sampling plan set forth at the beginning of this section by $(n_1, n_2; c_1, c_2)$.

In the case of rectifiable inspection and large lots, the average outgoing quality AOQ of the double sampling plan $(n_1, n_2; c_1, c_2)$ is given approximately by

$$(4.5.6) \qquad\qquad \text{AOQ}(\theta) = \theta P(A \mid \theta),$$

where $P(A \mid \theta)$ is given by (4.5.3) and, under appropriate conditions, can be approximated by (4.5.4) or (4.5.5).

Example 4.5(a). A large lot is accepted (or rejected) by the double sampling plan $(50, 100; 0, 2)$, that is, the following plan is used:

1. A sample of 50 items is drawn from a large lot and tested.
2. The lot is accepted if the sample contains no defectives.
3. The lot is rejected if the sample contains three or more defectives.
4. A second sample of 100 items is drawn from the lot if 1 or 2 defectives are contained in the sample of 50.
5. If the number of defectives in the two samples combined does not exceed two, the lot is accepted.
6. Otherwise the lot is rejected.

As before, allowing x_1 and x_2 to denote the numbers of defectives in the first and second lots respectively, the lot is accepted under the following

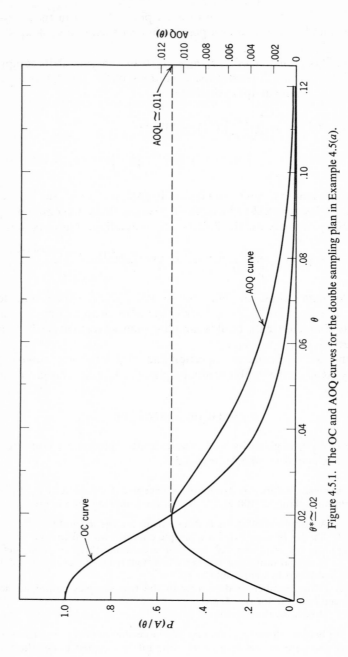

Figure 4.5.1. The OC and AOQ curves for the double sampling plan in Example 4.5(a).

four mutually exclusive conditions:

$$x_1 = 0;$$
$$x_1 = 1, \quad x_2 = 0;$$
$$x_1 = 1, \quad x_2 = 1;$$
$$x_1 = 2, \quad x_2 = 0.$$

With the case of the Poisson approximation, the expression for $P(A \mid \theta)$ is given by the following approximation:

(4.5.7) $P(A \mid \theta) \simeq e^{-50\theta} + 50\theta e^{-150\theta}(1 + 125\theta).$

Since we are assuming that lots are large, we have

(4.5.8) $\text{AOQ}(\theta) = \theta P(A \mid \theta) \simeq \theta e^{-50\theta} + 50\theta^2 e^{-150\theta}(1 + 125\theta).$

Values of $P(A \mid \theta)$ and $\text{AOQ}(\theta)$ are given in Table 4.5.1. The OC and AOQ curves for Example 4.5(a) are given in Figure 4.5.1.

TABLE 4.5.1

VALUES OF $P(A \mid \theta)$ AND $\text{AOQ}(\theta)$ FOR EXAMPLE 4.5(a)

θ	$P(A \mid \theta)$	$\text{AOQ}(\theta)$
0	1.000	0
.005	.970	.0048
.010	.857	.0083
.015	.699	.0105
.020	.542	.0108
.025	.409	.0102
.030	.312	.0089
.035	.223	.0078
.040	.165	.0066
.045	.122	.0055
.050	.092	.0046
.055	.069	.0038
.060	.052	.0032
.065	.041	.0027
.070	.031	.0022
.075	.024	.0018
.080	.019	.0015
.085	.015	.0012
.090	.011	.0010
.095	.009	.0008
.100	.007	.0007
.105	.005	.0005
.110	.004	.0004
.115	.003	.0003
.120	.002	.0003

PROBLEMS

4.1. A purchaser buys boxes of 12 electronic tubes in accordance with the following sampling plan for each box:
   (i) Three tubes are drawn at random from a box and tested.
   (ii) If the three tubes are good, they are returned to the box and the box is accepted without further testing.
   (iii) If the three tubes are not all good, the remaining nine are tested and all defectives are replaced by good tubes.
   (a) Tabulate exact values of $P(A \mid \theta)$, the probability of accepting a box of tubes having fraction of defectives equal to θ, and graph the OC curve of this sampling plan.
   (b) Tabulate exact values of AOQ(θ) and graph the AOQ curve of the sampling plan.
   (c) Determine the AOQL of this sampling plan and give the value of θ which yields the AOQL.

4.2. A purchaser buys ten articles at a time according to the following plan:
   (i) He takes three at random from the ten and tests them.
   (ii) If none of the three are defective, he accepts the entire ten articles.
   (iii) If one or more of the three are defective, he rejects the ten.
   If θ is the fraction of defectives in a batch of ten articles, let $P(A \mid \theta)$ be the probability of accepting the batch of ten.
   (a) Graph $P(A \mid \theta)$ as a function of θ.
   (b) If not more than two of the batch of ten are defective, the probability is at least what value of accepting the batch of ten?
   (c) If at least six of the batch of ten are defective, the probability is at most what value of accepting the batch?

4.3. A wholesaler buys cartons of 25 articles according to the following sampling plan for each carton:
   (i) Three articles are drawn at random from a carton and inspected.
   (ii) If none of the three is defective, he accepts the entire carton.
   (iii) If one or more are defective, he rejects the carton.
   (a) Graph $P(A \mid \theta)$ for this sampling plan.
   (b) If lots are capable of rectification, what would the AOQL be?

4.4. A buyer purchases spark plugs in boxes of 24 according to the following plan:
   (i) Twelve are drawn at random from a box and tested.
   (ii) If none or one is defective, he accepts the lot.
   (iii) If more than one are defective, all items of the lot are tested and all the defective items are replaced by good ones.
   (a) Graph the OC curve for this sampling plan.
   (b) Graph the AOQ curve for the plan.
   (c) Estimate the AOQL for the plan.
   (d) Estimate θ_1 if the producer's risk is .05.
   (e) Estimate θ_2 if the consumer's risk is .10.

4.5. A company purchases large lots of items by using the sampling plan $n = 5, c = 0$.

   (a) Graph the OC curve for this plan by using the binomial approximation.
   (b) Graph the AOQ curve for this plan.
   (c) Estimate the AOQL for the plan.
   (d) Estimate θ_1 if the producer's risk is .05.
   (e) Estimate θ_2 if the consumer's risk is .10.

4.6. Do the work required in Problem 4.5 for the single sampling plan (6, 1).

4.7. Do the work required in Problem 4.5 for the single sampling plan (10, 2).

4.8. An accountant checks large batches of bills for errors by using the single sampling plan for which $n = 50, c = 2$. (Rejection here means checking all bills in a batch.)
   (a) Graph the OC function $P(A \mid \theta)$ for the sampling plan by using the Poisson approximation.
   (b) Graph the AOQ curve for the plan.
   (c) Estimate the AOQL for the plan.
   (d) Estimate θ_1 for which the producer's risk is .05.
   (e) Estimate θ_2 for which the consumer's risk is .10.

4.9. A military procurement agency uses the following single sampling plan for purchasing large lots of items of a certain type: $n = 100, c = 2$.
   (a) Graph the OC function $P(A \mid \theta)$ for the plan by using the Poisson approximation for the probability of getting x defectives in the sample if $\theta = $ fraction defective in the lot.
   (b) Graph the AOQ curve for this sampling plan.
   (c) What is the value of the AOQL of the plan and for what value of θ does it occur?
   (d) Estimate θ_1 if the producer's risk is .05.
   (e) Estimate θ_2 if the consumer's risk is .10.

4.10. Carry out the work of Problem 4.9. for the single sampling plan (75, 1).

4.11. Carry out the work of Problem 4.9. for the single sampling plan (50, 0).

4.12. Large lots of items are purchased by using the double sampling plan $n_1 = 5, n_2 = 5, c_1 = 0, c_2 = 1$.
   (a) Graph the OC function $P(A \mid \theta)$ for the plan by using the binomial approximation.
   (b) Graph the AOQ curve for this sampling plan.
   (c) What is the value of the AOQL of the plan and for what value of θ does it occur?
   (d) Estimate θ_1 if the producer's risk is .05.
   (e) Estimate θ_2 if the consumer's risk is .10.

4.13. Do the work required in Problem 4.12 for the double sampling plan (10, 10; 0, 2).

4.14. Large lots of items are purchased according to the double sampling plan for which $n_1 = 50, n_2 = 50, c_1 = 0, c_2 = 1$.
   (a) Graph the OC function $P(A \mid \theta)$ for the plan using the Poisson approximation.
   (b) Graph the AOQ curve for this sampling plan.

(c) What is the value of the AOQL of the plan and for what value of θ does it occur.

(d) Estimate θ_1 if the producer's risk is .05.

(e) Estimate θ_2 if the consumer's risk is .05.

4.15. Carry out the work of Problem 4.14 for the double sampling plan (50, 100; 1, 2).

4.16. Carry out the work of Problem 4.14 for the double sampling plan (75, 100; 1, 3).

CHAPTER 5

Description of a Sample of Measurements

In Chapter 4 we discussed methods of acceptance sampling in which a sample of objects taken from a lot is inspected as to whether they are defective or nondefective with respect to some characteristic. The measurement or test applied to each object is thus essentially used to classify the object as defective or nondefective. In many studies of samples of objects, more refined analyses are performed by using the measurements which are made on objects in the sample. It is convenient to refer to such a set of measurements as a *sample of measurements*. The number of measurements is called the *sample size*. Even before analyses are made on such a sample of measurements, it is often necessary to summarize statistically the measurements. The purpose of this chapter is to present some of these summarization methods. These methods will be used in various succeeding chapters.

5.1 GRAPHICAL DESCRIPTION OF A SAMPLE OF MEASUREMENTS

First we shall discuss some simple, but widely used, graphical procedures for describing a sample of measurements. It is sufficient to discuss these procedures in terms of an example.

As an example of a sample of measurements, Table 5.1.1 lists the weights (to the nearest .01 ounce) of zinc coatings of 75 small galvanized iron sheets of a given size taken from an American Society of Testing Materials Manual (1947).

The 75 iron sheets, the coating weights of which are given in Table 5.1.1, may be considered as a *sample* from a large *population* of iron sheets that might have been taken. Thus, if we had taken a further sample of 75 iron sheets, we would have obtained another set of 75 numbers, and so on, for any number of samples of 75 we can imagine as having been taken.

TABLE 5.1.1

WEIGHTS (IN OUNCES) OF ZINC COATINGS OF
75 GALVANIZED IRON SHEETS

1.47	1.60	1.58	1.56	1.44
1.62	1.60	1.58	1.39	1.35
1.52	1.38	1.32	1.65	1.53
1.77	1.73	1.62	1.62	1.38
1.55	1.70	1.47	1.53	1.46
1.53	1.60	1.42	1.47	1.44
1.38	1.60	1.45	1.34	1.47
1.37	1.48	1.34	1.58	1.43
1.64	1.51	1.44	1.49	1.64
1.46	1.53	1.56	1.56	1.50
1.63	1.59	1.48	1.54	1.61
1.54	1.50	1.48	1.57	1.42
1.53	1.60	1.55	1.67	1.57
1.34	1.54	1.64	1.47	1.75
1.60	1.57	1.57	1.63	1.47

We will consider the mathematical theory of sampling in Chapter 6. Our immediate task is to summarize the information furnished by the sample of 75 numbers in Table 5.1.1.

The 75 numbers may be regarded as 75 values of a variable x. Thus a simple graphical representation of the 75 measurements in Table 5.1.1 is given by the *dot frequency diagram* shown in Figure 5.1.1, in which each of the 75 numbers is indicated by a dot placed directly above the point on the x-axis corresponding to that number. The display in Figure 5.1.1

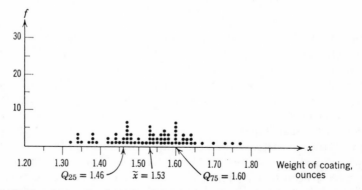

Figure 5.1.1. Dot frequency diagram of the measurements in Table 5.1.1, showing the lower quartile Q_{25}, the median $\tilde{x}$, and the upper quartile Q_{75}.

gives a quick graphical picture of the data. It is also useful to indicate the *lower quartile* Q_{25}, the median $\tilde{x}$ (or *middle quartile* Q_{50}), and *upper quartile* Q_{75} of the dot frequency diagram. It is convenient to define these quantities in terms of percentiles of the dot frequency diagram. The *p*th percentile is a number Q_p such that at least $p\%$ of the dots lie on or to the left of the line $x = Q_p$ in the x, f-plane and *also* at least $(100 - p)\%$ of the dots lie on or to the right of the line $x = Q_p$. If there is more than one such number (in which case there will be an interval of them), the *p*th percentile is defined as the average of the numbers (mid-point of the interval). Thus the lower quartile Q_{25} is the 25th percentile, the median $\tilde{x}$ (or Q_{50}) is the 50th percentile, and the upper quartile Q_{75} is the 75th percentile. We call $Q_{75} - Q_{25}$ the *interquartile range*.

5.2 FREQUENCY TABLES, FREQUENCY DISTRIBUTIONS, AND CUMULATIVE POLYGONS

If there are more than about 25 measurements in the sample, the construction of dot frequency diagrams involves more detail than is usually needed for practical purposes. One does not distort the information provided by the sample of measurements very much from a practical point of view if each measurement is *approximated* by the nearest of a set of conveniently chosen and equally spaced numbers. To define this approximating process we first construct a *frequency table*. Returning to Table 5.1.1 (or looking at Figure 5.1.1), we find the smallest measurement in the table to be 1.32 and the largest to be 1.77. The difference between these two numbers, which is called the *range*, is .45. We now divide the range into a number of equal *intervals* of convenient length. This means that the interval length should be some "simple number." The number of intervals is usually taken to be between 10 and 25. A convenient interval length for our example is .05, and we will use ten intervals. We might also have used .04, .03, or .02, but we would usually avoid inconvenient interval lengths such as .0333, .035, etc.

We now choose a set of ten equally spaced "simple numbers" at which to locate the mid-points of the ten equal intervals. In particular, we shall choose these interval mid-points as 1.30, 1.35, . . . , 1.75. The end points of the intervals will then be the 11 numbers 1.275, 1.325, . . . , 1.775. It is convenient to call these located intervals *cells*. Thus, the *cell mid-points* are 1.30, 1.35, . . . , 1.75, and the corresponding cell boundaries are the cell end points, namely, (1.275, 1.325), (1.325, 1.375), . . . , (1.725, 1.775).

Note that each of the 75 measurements in Table 5.1.1 will lie inside one of the ten cells we have defined. Each measurement which falls inside a cell is *approximated* by the mid-point of that cell. For instance, the measurement 1.47 is approximated by 1.45, which is the mid-point of the cell into which 1.47 falls. In case the measurements are such that a measurement could fall on a cell boundary, such a measurement is assigned, as a matter of convention, to the cell for which it falls on the upper boundary. For instance, if the measurement in Table 5.1.1 had been made to three decimal places and if 1.325 had been a measurement in the table, this measurement would be assigned to the cell (1.275, 1.325) and not (1.325, 1.375).

TABLE 5.2.1

FREQUENCY TABLE OF THE WEIGHTS OF 75 ZINC COATINGS

(a)	(b)	(c)	(d)	(e)	(f)	(g)
Cell Boundaries	Cell Mid-points	Tallied Frequency	Frequency	Relative Frequency	Cumulative Frequency	Cumulative Relative Frequency
1.275–1.325	1.30	1	1	.013	1	.013
1.325–1.375	1.35	⊬⊬	5	.067	6	.080
1.375–1.425	1.40	⊬⊬ 1	6	.080	12	.160
1.425–1.475	1.45	⊬⊬ ⊬⊬ 111	13	.173	25	.333
1.475–1.525	1.50	⊬⊬ 111	8	.107	33	.440
1.525–1.575	1.55	⊬⊬ ⊬⊬ ⊬⊬ 11	17	.227	50	.667
1.575–1.625	1.60	⊬⊬ ⊬⊬ ⊬⊬	14	.187	64	.854
1.625–1.675	1.65	⊬⊬ 11	7	.093	71	.947
1.675–1.725	1.70	1	1	.013	72	.960
1.725–1.775	1.75	111	3	.040	75	1.000
			75	1.000		

It is convenient to arrange the cell boundaries and cell mid-points as shown in columns (a) and (b) of Table 5.2.1. Then, as each measurement in Table 5.1.1 is approximated by a cell mid-point, it can be tallied, as shown in column (c). The counted tallies are the *frequencies* shown in column (d). The *relative frequencies* in column (e) are the frequencies divided by the sample size 75. Columns (f) and (g) are cumulated forms of the frequencies and relative frequencies respectively.

The frequencies [columns (d) and (e)] in Table 5.2.1 can be represented graphically as a *frequency histogram*, as shown in Figure 5.2.1. Note that two scales are provided for the ordinates—one scale refers to frequency and the other to relative frequency expressed in terms of percent.

A more useful graphical representation of the material in Table 5.2.1 is given by a *cumulative polygon*, as shown by the heavy graph in Figure 5.2.2. This graph together with the two scales of ordinates is a graphical representation of columns (f) and (g) of Table 5.2.1. In this graph, note that the points are plotted above the upper cell boundaries and *not* above the cell mid-points.

The cumulative polygon provides a simple and quick graphical procedure for approximately determining percentiles, fractiles, quartiles, and

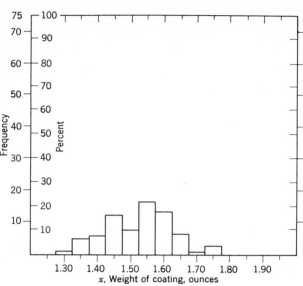

Figure 5.2.1. Frequency histogram for the frequencies of Table 5.2.1.

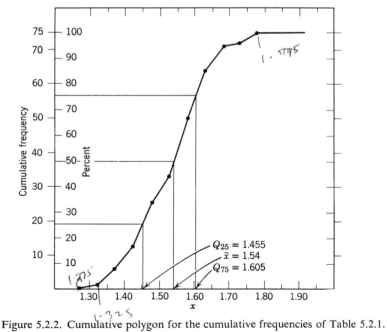

Figure 5.2.2. Cumulative polygon for the cumulative frequencies of Table 5.2.1.

the median, without going to the degree of detail involved in using the dot frequency diagram of Figure 5.1.1. The approximate values of the lower quartile, median, and upper quartile, are 1.455, 1.54, and 1.605 respectively, as shown in the figure.

We can determine from a cumulative polygon *approximately* the number of cases in the sample lying between two values of x. For example, suppose we are interested in the number of cases in the sample having zinc-coating weight *between* 1.42 and 1.68 ounces. As seen from Figure 5.2.2, 1.68 is the 94.8 percentile, and 1.42 is the 14.7 percentile. The difference 94.8 − 14.7 = 80.1 % is approximately the percentage of cases having zinc coatings between 1.42 and 1.68 ounces. The actual number of cases is 59 (or 78.7 %).

Although the procedures described in this section for constructing frequency tables, frequency histograms, and cumulative polygons have been exhibited in terms of a specific example, it is fairly evident how they can be generalized for an arbitrary sample of n measurements. We shall therefore not attempt to retrace the procedures for a general sample of n measurements, but will leave this as an exercise for the reader who may be interested in doing so.

5.3 MEAN AND STANDARD DEVIATION OF A SAMPLE OF MEASUREMENTS

In Section 5.1 we have seen how a sample of measurements can be condensed into a dot frequency diagram and how information pertaining to percentiles can be obtained from it. In Section 5.2 we have seen how similar, but simpler, procedures can be carried out on a sample of grouped measurements. In both situations we have seen how the 50th percentile or median is the "middle" of the distribution of measurements in a certain well-defined graphical sense. The *interquartile* range is an indicator of the amount of "scatter" or "spread" of the distribution of measurements in a well-defined graphical sense.

There are other important *statistics* for describing the middle of the distribution and the spread of the distribution. In this section we shall discuss the *arithmetic mean* or simply the *mean* of the distribution of sample measurements as another statistic for indicating the middle of the distribution, and the *standard deviation* of the measurements as another indicator for the amount of spread of the distribution.

(a) Definition of the Mean of a Sample

As a simple example, suppose the weights (in pounds) of five students are 141, 136, 157, 143 and 138. The *mean* of this sample of five weights is the sum of the weights divided by five, that is,

$$\text{mean} = \frac{141 + 136 + 157 + 143 + 138}{5} = \frac{715}{5} = 143 \text{ pounds.}$$

In general, if $x_1, x_2, \ldots, x_n$ is a sample of n measurements, the *sample mean* $\bar{x}$ is defined by the following relation:

$$(5.3.1) \qquad n\bar{x} = x_1 + x_2 + \cdots + x_n.$$

We can write the *sample sum* $x_1 + x_2 + \cdots + x_n$ (sometimes denoted by T) more compactly as $\sum_{j=1}^{n} x_j$. Hence (5.3.1) can be written more compactly as

$$(5.3.2) \qquad n\bar{x} = \sum_{j=1}^{n} x_j,$$

from which the formula for the mean $\bar{x}$ is written explicitly as

$$(5.3.3) \qquad \bar{x} = \frac{1}{n} \sum_{j=1}^{n} x_j.$$

If we refer to our example of five weights, we would have $n = 5$, $x_1 = 141$, $x_2 = 136$, $x_3 = 157$, $x_4 = 143$, $x_5 = 138$, $\sum_{j=1}^{n} x_j = 715$, and applying (5.3.3) to the five weights would result in $\bar{x} = 715/5 = 143$ pounds. The mean of the sample of 75 measurements in Table 5.1.1 is given by

$$75\bar{x} = 1.47 + 1.62 + \cdots + 1.47$$

$$= 114.51,$$

that is,

$$\bar{x} = 1.527 \text{ ounces.}$$

Note from Figure 5.1.1 that the value of $\bar{x}$ is very close to the value of the median $\tilde{x}$, which when we round off is 1.53 ounces.

Suppose we take the difference between each x and the mean $\bar{x}$. We have $x_1 - \bar{x}, x_2 - \bar{x}, \ldots, x_n - \bar{x}$. If we add these differences, we get

$$(x_1 - \bar{x}) + (x_2 - \bar{x}) + \cdots + (x_n - \bar{x}) = (x_1 + x_2 + \cdots + x_n) - n\bar{x} = 0$$

because of (5.3.1). Hence, by using summation notation, we have

$$(5.3.4) \qquad \sum_{j=1}^{n} (x_j - \bar{x}) = 0.$$

That is, the sum of the differences between each measurement in a sample and the mean of all measurements in the sample is equal to zero.

(b) Definition of the Standard Deviation of a Sample

Considering the example of the five weights again, suppose we square the difference between each measurement and the mean and add them. The *standard deviation s* of the five weights is given by the following relation:

$$(5 - 1)s^2 = (141 - 143)^2 + (136 - 143)^2$$
$$+ (157 - 143)^2 + (143 - 143)^2 + (138 - 143)^2.$$

Performing the indicated arithmetic, we find

$$s^2 = 68.50$$

or

$$s = 8.28.$$

More generally, if $x_1, x_2, \ldots, x_n$ is a sample of n measurements, the standard deviation s of the sample is defined by

$$(5.3.5) \quad (n - 1)s^2 = (x_1 - \bar{x})^2 + (x_2 - \bar{x})^2 + \cdots + (x_n - \bar{x})^2.$$

Using the summation notation, this can be written more briefly as

$$(5.3.6) \quad (n - 1)s^2 = \sum_{j=1}^{n} (x_j - \bar{x})^2.$$

The quantity s^2 (the *square* of the standard deviation s) is called the *variance* of the sample. We shall not rewrite (5.3.5), (5.3.6), or any similar formula so as to give an explicit formula for the standard deviation s. For we can perfectly well talk about the standard deviation s given by (5.3.6) or the variance s^2 given by (5.3.6) without having to write down two formulas.

From the point of view of computation, a formula which is often more convenient when a calculating machine is available can be found from (5.3.5). For, by squaring each term on the right-hand side of (5.3.5), we have

$$(5.3.7) \quad (n - 1)s^2 = (x_1^2 - 2x_1\bar{x} + \bar{x}^2)$$
$$+ (x_2^2 - 2x_2\bar{x} + \bar{x}^2) + \cdots + (x_n^2 - 2x_n\bar{x} + \bar{x}^2)$$

or, collecting terms,

$$(5.3.8) \quad (n - 1)s^2 = (x_1^2 + x_2^2 + \cdots + x_n^2) - 2\bar{x}(x_1 + x_2 + \cdots + x_n) + n\bar{x}^2.$$

But substituting the value of $\bar{x}$ from (5.3.1) into (5.3.8) and simplifying, we obtain

$$(5.3.9) \quad (n - 1)s^2 = \sum_{j=1}^{n} x_j^2 - n\bar{x}^2,$$

which is the desired formula. In practice, it is convenient to delay the division by n in calculating $\bar{x}$ and to calculate s^2 from the formula

$$(5.3.10) \qquad (n-1)s^2 = \sum_{j=1}^{n} x_j^2 - \frac{1}{n}\left(\sum_{j=1}^{n} x_j\right)^2$$

The quantity $\frac{1}{n}\left(\sum_{j=1}^{n} x_j\right)^2$ is sometimes called the "correction factor" or "correction term."

As an example, if (5.3.10) is applied to the 75 measurements of Table 5.1.1, we find

$$74s^2 = (1.47)^2 + (1.62)^2 + \cdots + (1.47)^2 - \tfrac{1}{75}(1.47 + 1.62 + \cdots + 1.47)^2$$

$$= (175.5849) - \tfrac{1}{75}(114.51)^2 = .7510,$$

$$s^2 = .01015, \qquad s = .101.$$

It should be noticed that $n - 1$ appears in (5.3.6) for the variance where one might have expected n. Justification of the use of $n - 1$ depends on concepts introduced in subsequent chapters and is treated in Chapter 9.

(c) Further Remarks on the Interpretation of the Mean and Standard Deviation of a Sample

It was found that the mean of the sample of 75 measurements in Table 5.1.1 is 1.527. In Figure 5.1.1 it will be seen that the mean falls near the "middle" of the distribution of dots. Actually the mean is at the *center of gravity* of the distribution. By this we mean that, if the dots as arranged in Figure 5.1.1 were all of equal weight bearing down on the x-axis, this arrangement would be balanced by a knife edge held under the x-axis at the mean, 1.527 ounces. The mean has another important property. If we take each measurement minus the mean, we obtain 75 "discrepancies" or differences; some of these are positive and some are negative, but, as we have seen from (5.3.4), the algebraic sum of all the differences is equal to zero.

The mean of a set of measurements gives us some information about the location of the "middle" or "center of gravity" of the set of measurements, but it gives no information about the "scatter" (or "amount of concentration") of the measurements. For example, the five measurements 14.0, 24.5, 25, 25.5, and 36.0 have the same mean as the five measurements 24.0, 24.5, 25, 25.5, and 26.0, but the two sets of measurements have different amounts of scatter. One simple indication of the scatter of a set of measurements is the *range*, that is, the largest measurement minus the smallest. In the two sets of measurements mentioned, the ranges are 22 and 2 respectively. If we always worked with fairly small samples of the same size n (as is the case in *industrial quality control*, where $n = 4$ and $n = 5$ are widely used sample sizes), then we would find the range very convenient. It is difficult, however, to compare a range for one sample size with that for a different sample size. For this and other reasons, the range, in spite of its simplicity, convenience, and importance, is used only in rather restricted situations, such as industrial quality control charts which are discussed in Chapter 11.

The *interquartile range* defined in Section 5.1 is only useful when the samples are large enough to establish the quartiles fairly well. For n less than about 25, the quartiles are of doubtful value.

We clearly need a measure of scatter which can be used in samples of any size and in some sense makes use of all the measurements in the sample. There are several measures of scatter that can be used for this purpose, and the most common of these is the *standard deviation*. For normal (or Gaussian) distributions, discussed in detail in Chapter 7, the standard deviation is the "natural" measure of scatter, as will be seen there.

Samples of measurements in a wide variety of practical situations yield frequency histograms which are fairly symmetrical and bell-shaped such that:

1. *About 95% of the measurements fall within the interval*
$$(\bar{x} - 2s, \bar{x} + 2s).$$
2. *About 68% of the measurements fall within the interval*
$$(\bar{x} - s, \bar{x} + s).$$
3. *About 50% of the measurements fall within the interval*
$$(\bar{x} - \tfrac{2}{3}s, \bar{x} + \tfrac{2}{3}s).$$

The larger the samples and the closer the frequency histograms are to normal distributions the more accurate statements 1, 2, and 3 become.

In the example of the zinc-coating measurements of Table 5.1.1, we found $\bar{x} = 1.527$ and $s = .101$; thus within the interval $(1.527 \pm 2(.101))$ (that is, between 1.325 and 1.729), there are 71 out of 75 measurements or 94.7% instead of about 95% of the measurements.

5.4 WEIGHTED MEANS AND VARIANCES

Suppose we consider first the following sample of 12 measurements,

(5.4.1) $3, 7, 6, 3, 2, 7, 3, 6, 2, 7, 2, 3.$

The mean of this sample is

(5.4.2) $\text{mean} = \dfrac{3 + 7 + 6 + 3 + 2 + 7 + 3 + 6 + 2 + 7 + 2 + 3}{12}$

$= \dfrac{51}{12} = 4.25$

and we note that this may be written as

$$(5.4.3) \quad \text{mean} = \frac{(4)[3] + (3)[7] + (2)[6] + (3)[2]}{4 + 3 + 2 + 3} = \frac{51}{12} = 4.25.$$

We may describe (5.4.3) in the following way: as the sample contains the different values 3, 7, 6, and 2, the sample mean may be found by "weighting," i.e., multiplying, these values by the frequency with which they occurred in the sample, in this case, 4, 3, 2, and 3, respectively, and on finding the weighted sum, dividing by the sum of the weights, $4 + 3 + 2 + 3 = 12$. Such a mean is called a *weighted mean*.

In general, suppose we have a sample of n measurements in which there are k different values $y_1, \ldots, y_k$ and in which the value y_1 occurs f_1 times, the value y_2 occurs f_2 times, $\ldots$, the value y_k occurs f_k times, where

$$(5.4.4) \qquad f_1 + f_2 + \cdots + f_k = n.$$

We then call

$$(5.4.5) \qquad \bar{y} = \frac{f_1 y_1 + \cdots + f_k y_k}{\sum f_i} = \frac{\sum_{i=1}^{k} f_i y_i}{n}$$

the weighted mean of the sample, or the weighted mean of $y_1, \ldots, y_k$, where it is to be understood that the weight of y_i is f_i and $n = \sum_{i=1}^{k} f_i$.

Returning to the sample data (5.4.1), it is clear that the sample variance is

$$(5.4.6) \quad [(3 - 4.25)^2 + (7 - 4.25)^2 + (6 - 4.25)^2 + \cdots +$$
$$(2 - 4.25)^2 + (3 - 4.25)^2]/(12 - 1)$$
$$= [4(3 - 4.25)^2 + 3(7 - 4.25)^2 + 2(6 - 4.25)^2 + 3(2 - 4.25)^2]/11$$

so that in general, the sample variance of a sample in which $y_1, \ldots, y_k$ occur $f_1, \ldots, f_k$ times respectively, with $\sum_{i=1}^{k} f_i = n$ is

$$(5.4.7) \qquad s^2 = (n - 1)^{-1} \sum_{i=1}^{k} f_i(y_i - \bar{y})^2,$$

where $\bar{y}$ is given by (5.4.5). We note that from (5.4.5) we may prove the analogous property to (5.3.4), viz,

$$(5.4.8) \qquad \sum_{i=1}^{k} f_i(y_i - \bar{y}) = 0$$

and, similar to (5.3.9), that

$$(5.4.9) \qquad s^2 = (n-1)^{-1}\left[\sum_{i=1}^{k} f_i y_i^2 - n\bar{y}^2\right]$$

$$= (n-1)^{-1}\left[\sum_{i=1}^{n} f_i y_i^2 - \frac{1}{n}\left(\sum_{i=1}^{k} f_i y_i\right)^2\right].$$

(See Problem 5.17.)

5.5 THE MEAN AND STANDARD DEVIATION OF A SAMPLE OF APPROXIMATED MEASUREMENTS

Suppose we have a sample of measurements $x_1, x_2, \ldots, x_n$. Suppose these measurements are approximated by a procedure such as described in Section 5.2, where k cells are chosen, each of cell length δ, the cell midpoints being $m_1, m_2, \ldots, m_k$. Let $f_1, f_2, \ldots, f_k$ be the cell frequencies, that is, f_1 is the number of measurements approximated by m_1, f_2 is the number approximated by m_2, and so on. The frequency table for this setup is shown in Table 5.5.1.

TABLE 5.5.1
GENERAL FREQUENCY TABLE

(a) Cell Mid-Points, m_i	(b) Cell Frequencies, f_i	(c) $f_i m_i$	(d) $f_i m_i^2$	(e) u_i	(f) $f_i u_i$	(g) $f_i u_i^2$
m_1	f_1	$f_1 m_1$	$f_1 m_1^2$	u_1	$f_1 u_1$	$f_1 u_1^2$
m_2	f_2	$f_2 m_2$	$f_2 m_2^2$	u_2	$f_2 u_2$	$f_2 u_2^2$
.	.	.	.	.	.	.
.	.	.	.	.	.	.
.	.	.	.	.	.	.
m_k	f_k	$f_k m_k$	$f_k m_k^2$	u_k	$f_k u_k$	$f_k u_k^2$
Total	n	$\sum_{i=1}^{k} f_i m_i$	$\sum_{i=1}^{k} f_i m_i^2$		$\sum_{i=1}^{k} f_i u_i$	$\sum_{i=1}^{k} f_i u_i^2$

Since f_1 of the x's have been approximated by m_1, f_2 of the x's to m_2, and so on, it is evident that the expression $\sum_{j=1}^{n} x_j$ in (5.3.2) is *approximated* by $\sum_{i=1}^{k} f_i m_i$, and hence by substituting this approximation into (5.3.3), we obtain the following approximation for $\bar{x}$, say,

$$(5.5.1) \qquad \bar{x} \cong \frac{1}{n}\sum_{i=1}^{k} f_i m_i = \bar{m},$$

that is to say, $\bar{x}$ is approximated by the weighted mean of the cell mid-points, m_i, where the weights are the cell frequencies f_i.

Referring to Table 5.5.1, we see that $\bar{x}$ is approximated by dividing the total of column (c) by the total of column (b).

Referring to (5.3.10), we note that, by using the approximate values of the x's, $\sum_{j=1}^{n} x_j^2$ is approximated by $\sum_{i=1}^{k} f_i m_i^2$, and $\sum_{i=1}^{n} x_i$ is approximated by $\sum_{i=1}^{k} f_i m_i$. Hence we have from (5.3.10) the following approximation to s^2:

$$(5.5.2) \qquad s^2 \simeq \frac{1}{n-1}\left[\sum_{i=1}^{k} f_i m_i^2 - \frac{1}{n}\left(\sum_{i=1}^{k} f_i m_i\right)^2\right],$$

which is to say that s^2 is approximated by the weighted variance of the m_i, the cell mid-points [see (5.4.9)].

Referring to Table 5.5.1, it will be seen that the constituents n, $\sum_{i=1}^{k} f_i m_i$, and $\sum_{i=1}^{k} f_i m_i^2$, required for computing the right-hand side of (5.5.2), are available as the totals of columns (b), (c), and (d) respectively.

If a computing machine is not available, an easy way to compute the value of the right-hand sides of (5.5.1) and of (5.5.2) by hand is as follows:

First note that

$$(5.5.3) \qquad \frac{1}{n-1}\left[\sum_{i=1}^{k} f_i m_i^2 - \frac{1}{n}\left(\sum_{i=1}^{k} f_i m_i\right)^2\right] = \frac{1}{n-1}\sum_{i=1}^{k} f_i(m_i - \bar{m})^2,$$

where [referring to (5.5.1)],

$$\bar{m} = \frac{1}{n}\sum_{i=1}^{k} f_i m_i,$$

that is, $\bar{m}$ is the mean of the approximated values of the x's.

Now let us introduce some new quantities $u_1, u_2, \ldots, u_k$ defined linearly in terms of $m_1, m_2, \ldots, m_k$ as follows:

$$m_1 = \alpha + \delta u_1,$$
$$m_2 = \alpha + \delta u_2,$$

(5.5.4)

$$m_k = \alpha + \delta u_k,$$

where α denotes any one of the cell mid-points we please, and δ is the cell length used in approximating the x's. With α and δ thus defined, it will be noted that $u_1, u_2, \ldots, u_k$ will all be positive or negative integers or zero.

Then, if we multiply the first equation in (5.5.4) by f_1, the second by f_2, and so on, add the resulting equations and divide by n, we obtain

$$(5.5.5) \qquad \bar{m} = \alpha + \delta \bar{u},$$

where

$$\bar{u} = \frac{1}{n} \sum_{i=1}^{k} f_i u_i$$

and, hence, the approximation

$$(5.5.6) \qquad \bar{x} \cong \alpha + \delta \bar{u}.$$

Remembering that α is some cell mid-point, δ is the cell length, and $u_1, u_2, \ldots, u_k$ are positive or negative integers or zero, we can set up columns (e) and (f) in Table 5.5.1, and readily compute the value of $\bar{u}$ by dividing the total of column (f) by the total of column (b). Then $\bar{m}$ is computed from (5.5.5), which is our approximate value of $\bar{x}$.

Hence by substituting from (5.5.4) and (5.5.5) into the right-hand side of (5.5.3), we obtain

$$(5.5.7) \qquad \frac{1}{n-1} \sum_{i=1}^{k} f_i (m_i - \bar{m})^2 = \frac{\delta^2}{n-1} \sum_{i=1}^{k} f_i (u_i - \bar{u})^2,$$

that is,

$$s_m^2 = \delta^2 s_u^2.$$

But

$$(5.5.8) \qquad \frac{\delta^2}{n-1} \sum_{i=1}^{k} f_i (u_i - \bar{u})^2 = \frac{\delta^2}{n-1} \left[\sum_{i=1}^{k} f_i u_i^2 - \frac{1}{n} \left(\sum_{i=1}^{k} f_i u_i \right)^2 \right].$$

Hence we have the approximation

$$(5.5.9) \qquad s^2 \cong \frac{\delta^2}{n-1} \left[\sum_{i=1}^{k} f_i u_i^2 - \frac{1}{n} \left(\sum_{i=1}^{k} f_i u_i \right)^2 \right].$$

By inserting column (g) in Table 5.5.1, we note that the constituents n, $\sum_{i=1}^{k} f_i u_i$, and $\sum_{i=1}^{k} f_i u_i^2$ of the right-hand side of (5.5.9) are the totals of columns (b), (f), and (g) of Table 5.5.1 and δ is the cell length. Thus the short-cut method we have described for computing approximate values of $\bar{x}$ and s^2 from (5.5.6) and (5.5.9) does not require columns (c) and (d) of Table 5.5.1.

The short-cut method involved in using (5.5.6) and 5.5.9) is sometimes called the *fully coded* method. If we choose $\delta = 1$, the method is sometimes called the *working origin* method.

Let us illustrate for the example of zinc coatings the two methods of approximating $\bar{x}$ and s^2, first by the *long* method using (5.5.1) and (5.5.2) and then by the *fully coded* method using (5.5.6) and (5.5.9). The basic cell and frequency information from Table 5.2.1 together with the required columns for computing are given in Table 5.5.2.

TABLE 5.5.2

TABLE SHOWING CALCULATIONS FOR APPROXIMATING MEAN AND STANDARD DEVIATION OF MEASUREMENTS IN TABLE 5.1.1

(a)	(b)	(c)	(d)	(e)	(f)	(g)
Cell Mid-Points, m_i	Cell Frequencies, f_i	$f_i m_i$	$f_i m_i^2$	u_i	$f_i u_i$	$f_i u_i^2$
1.30	1	1.30	1.6900	−4	−4	16
1.35	5	6.75	9.1125	−3	−15	45
1.40	6	8.40	11.7600	−2	−12	24
1.45	13	18.85	27.3325	−1	−13	13
1.50	8	12.00	18.0000	0	0	0
1.55	17	26.35	40.8425	1	17	17
1.60	14	22.40	35.8400	2	28	56
1.65	7	11.55	19.0575	3	21	63
1.70	1	1.70	2.8900	4	4	16
1.75	3	5.25	9.1875	5	15	75
Total	75	114.55	175.7125		41	325

Approximating $\bar{x}$ and s^2 by the long method using (5.5.1) and (5.5.2), we obtain

$$\bar{x} \cong \frac{114.55}{75} = 1.527$$

and

$$s^2 \cong \frac{1}{74}\left[175.7125 - 75\left(\frac{114.55}{75}\right)^2\right] = .0102,$$

or

$$s \cong .101.$$

Approximating $\bar{x}$ and s^2 by the fully coded method using (5.5.6) and (5.5.9), and noting that α was chosen to be the cell mid-point 1.50, and $\delta = .05$, we obtain

$$\bar{x} \cong 1.50 + (.05)(\tfrac{41}{75}) = 1.527$$

and

$$s^2 \cong \frac{(.05)^2}{74}\left[325 - \frac{1}{75}(41)^2\right] = .0102,$$

or

$$s \cong .101.$$

It should be noted that it is very often convenient to use a short-cut method for computing $\bar{x}$ and s^2 for a sample $(x_1, \ldots, x_n)$ of ungrouped measurements. Letting $f_i \equiv 1$, $i = 1, \ldots, n$, the reader should verify that by replacing the m_i's by x_i (the ith observation) in (5.5.4) we again obtain

$$\bar{x} = \alpha + \delta\bar{u}$$

and

$$s_x^2 = \delta^2 s_u^2.$$

This method is sometimes referred to as using a fully coded scheme for ungrouped samples.

PROBLEMS

5.1. "On" temperatures at which a certain thermostatic switch operated in 25 trials were as follows (Grant data):

55	54	55	51	53
55	55	54	51	56
55	55	54	53	55
54	53	50	52	56
55	55	50	56	55

(a) Make a dot frequency diagram of these measurements and indicate the median, lower quartile, and upper quartile.

(b) Compute $\bar{x}$ and s by using a coded method.

5.2. Twenty-five repeated measurements of a spectral line gave the following results (in millimicrons):

65.171	65.174	65.173	65.184	65.177
65.178	65.185	65.180	65.181	65.180
65.175	65.169	65.172	65.183	65.179
65.175	65.180	65.179	65.174	65.178
65.170	65.174	65.183	65.172	65.179

(a) Make a dot frequency diagram of these measurements, and indicate the median, the upper quartile, and the lower quartile on the diagram.

(b) Compute $\bar{x}$ and s of the sample of 25 measurements by using a fully coded scheme.

5.3. Experiments to study the variability of coke yield by a certain coke oven resulted in the following observations (percent of coal fed into the oven on 20 successive days).

69.4	70.0	71.1	71.5
70.5	71.6	71.1	72.2
73.3	67.5	72.3	67.1
70.3	72.1	72.2	70.1
70.8	70.9	69.7	73.0

(a) Draw up a dot frequency diagram of this sample and indicate the median, lower quartile, and upper quartile.

(b) Compute $\bar{x}$ and s by using a coded scheme.

5.4. Two hundred rivet heads are drawn at random from a mass-production process, and their diameter in millimeters are listed below. [Data taken from Hald (1952)]:

13.39	13.43	13.54	13.64	13.40	13.55	13.40	13.26
13.42	13.50	13.32	13.31	13.28	13.52	13.46	13.63
13.38	13.44	13.52	13.53	13.37	13.33	13.24	13.13
13.53	13.53	13.39	13.57	13.51	13.34	13.39	13.47
13.51	13.48	13.62	13.58	13.57	13.33	13.51	13.40
13.30	13.48	13.40	13.57	13.51	13.40	13.52	13.56
13.40	13.34	13.23	13.37	13.48	13.48	13.62	13.35
13.40	13.36	13.45	13.48	13.29	13.58	13.44	13.56
13.28	13.59	13.47	13.46	13.62	13.54	13.20	13.38
13.43	13.35	13.56	13.51	13.47	13.40	13.29	13.20
13.46	13.44	13.42	13.29	13.41	13.39	13.50	13.48
13.53	13.34	13.45	13.42	13.29	13.38	13.45	13.50
13.55	13.33	13.32	13.69	13.46	13.32	13.32	13.48
13.29	13.25	13.44	13.60	13.43	13.51	13.43	13.38
13.24	13.28	13.58	13.31	13.31	13.45	13.45	13.44
13.34	13.49	13.50	13.38	13.48	13.43	13.37	13.29
13.54	13.33	13.36	13.46	13.23	13.44	13.38	13.27
13.66	13.26	13.40	13.52	13.59	13.48	13.46	13.40
13.43	13.26	13.50	13.38	13.43	13.34	13.41	13.24
13.42	13.55	13.37	13.41	13.38	13.14	13.42	13.52
13.38	13.54	13.30	13.18	13.32	13.46	13.39	13.35
13.34	13.37	13.50	13.61	13.42	13.32	13.35	13.40
13.57	13.31	13.40	13.36	13.28	13.58	13.58	13.38
13.26	13.37	13.28	13.39	13.32	13.20	13.43	13.34
13.33	13.33	13.31	13.45	13.39	13.45	13.41	13.45

(a) Make a frequency table for these data.

(b) Construct a histogram and a cumulative polygon (on the same sheet of graph paper) and indicate the median and two quartiles on the graph.

(c) By "approximating" the measurements appropriately and using the short-cut scheme of Section 5.5, approximate the value of the mean $\bar{x}$ and of the standard deviation s of the original measurements.

5.5. In a study of the variation of a critical dimension of electric contacts, the distribution of measurements of this dimension in a sample of 1000 electric contacts taken from the production line gave the frequencies shown in Table P5.5 [the tally mark column (c) has been omitted].

TABLE P5.5

(a)	(b)	(d)	(e)	(f)	(g)
All Boundaries (inches)	All Mid-Points (inches)	Number of Contacts (frequency)	Relative Frequency	Cumulative Frequency	Cumulative Relative Frequency
	.412	6			
	.417	34			
	.422	132			
	.427	179			
	.432	218			
	.437	183			
	.442	146			
	.447	69			
	.452	30			
	.457	3			
	Total	1000			

(a) Provide the required information of columns (a), (e), (f), and (g).

(b) Approximate the value of the mean $\bar{x}$ and of the standard deviation s of the original 1000 measurements using the short-cut scheme of Section 5.5.

5.6. Tests of stiffness of a number of aluminum-alloy channels resulted in the frequency of measurements of stiffness shown in Table P5.6 (effective EI in psi).

TABLE P5.6

(a)	(b)	(d)	(e)	(f)	(g)
Cell Boundaries (psi)	Cell Mid-Points (psi)	Frequency	Relative Frequency	Cumulative Frequency	Cumulative Relative Frequency
	2160	1			
	2200	3			
	2240	5			
	2280	14			
	2320	22			
	2360	35			
	2400	41			
	2440	33			
	2480	25			
	2520	11			
	2560	7			
	2600	2			
	2640	1			
	Total	200			

Perform the same operations for this set of measurements as requested in (*a*) and (*b*) of Problem 5.5.

5.7. The grouped frequency distribution of thickness measurements (in inches) determined from 50 places on a coil of sheet metal were found to be the data given in Table P5.7 (Westman data).

TABLE P5.7

Cell Mid-Point, m_i	Frequency, f_i
.147	1
.148	1
.149	4
.150	6
.151	10
.152	6
.153	7
.154	6
.155	6
.156	3
	50

(*a*) Construct a cumulative polygon of these data and estimate the median and the two quartiles.

(*b*) Find the mean, variance, and standard deviation of this distribution (by using a fully coded method).

5.8. The distribution of carbon content (percent) obtained from 178 determinations on a certain mixed powder is:

TABLE P5.8

Percent Carbon (Cell Mid-Points), m_i	Frequency, f_i
4.145	1
4.245	2
4.345	7
4.445	20
4.545	24
4.645	31
4.745	38
4.845	24
4.945	21
5.045	7
5.145	3
	178

(*a*) Construct a cumulative polygon of these data and estimate the median and two quartiles.

(*b*) Find the mean, variance, and standard deviation of the carbon content.

5.9. The grouped sample distribution of 500 rivets according to the diameter of the heads, in millimeters, is shown in Table P5.9:

TABLE P5.9

Diameter (Cell Mid-Points), m_i	Frequency, f_i
13.07	1
13.12	4
13.17	4
13.22	18
13.27	38
13.32	56
13.37	69
13.42	96
13.47	72
13.52	68
13.57	41
13.62	18
13.67	12
13.72	2
13.77	1
	500

(a) Construct a cumulative polygon of these data and estimate the median and the two quartiles.

(b) With the use of a fully coded method, find the mean and the variance of this distribution.

TABLE P5.10

Capacity (Cell Mid-Point), m_i	Frequency, f_i
1.86	2
1.89	3
1.92	13
1.95	16
1.98	45
2.01	68
2.04	77
2.07	103
2.10	99
2.13	81
2.16	61
2.19	35
2.22	24
2.25	14
2.28	8
2.31	1
	650

5.10. Table P5.10 shows the distribution of the capacities of 650 capacitors in microfarads.

(*a*) Construct the cumulative polygon of this frequency distribution and estimate the median and the two quartiles.

(*b*) Find the mean and variance of this distribution by using a fully coded method.

5.11. In a germination experiment, 80 rows of cabbage seed with 10 seeds per row were incubated. The distribution of number of cabbage seeds which germinated per row is shown in Table P5.11 (Tippett data):

TABLE P5.11

No. Seeds Germinated per Row (Cell Mid-Point), m_i	Frequency of Rows f_i
0	6
1	20
2	28
3	12
4	8
5	6
	80

Find the mean and variance of the numbers of seeds germinating per row.

5.12. Suppose 1000 pieces of enameled ware are inspected and the number of surface defects on each piece is recorded. The distribution of number of defects is shown in Table P5.12:

TABLE P5.12

No. of Defects (Cell Mid-Point), m_i	Frequency, f_i
0	600
1	310
2	75
3	13
4	2
	1000

Find the mean and variance of the numbers of defects.

5.13. A pair of cheap dice were thrown 100 times. The distribution of the total number of dots obtained is shown in Table P5.13.

TABLE P5.13

No. of Dots per Throw (Cell Mid-Point), m_i	Frequency, f_i
2	0
3	7
4	9
5	19
6	16
7	13
8	11
9	4
10	6
11	11
12	4
	100

Find the mean and variance of the numbers of dots obtained (using a working origin).

5.14. The frequency distribution shown in Table P5.14 was obtained for the numbers of heads that came up in 100 tosses of 20 coins (Allendoerfer):

TABLE P5.14

No. of Heads in One Toss of 20 Coins, x_i	Frequency, f_i
3	1
4	1
5	0
6	3
7	5
8	11
9	15
10	23
11	11
12	12
13	11
14	5
15	2
	100

Using a working origin, find the mean, variance, and standard deviation of this distribution.

5.15. Thirty-two dice are thrown 100 times. The distribution of the total number of dots per throw is shown in Table P5.15.

TABLE P5.15

Total Number of Dots (Cell Mid-Points), m_i	Frequency, f_i
90	3
95	3
100	4
105	15
110	17
115	15
120	16
125	12
130	8
135	3
140	3
145	0
150	1
	100

Find the mean, variance, and standard deviation of the total number of dots per throw.

5.16. A pair of dice can fall in 36 different "ways." One of these ways will yield a total of two dots, two will yield three dots, and so on. The frequencies of ways which will yield various total numbers of dots are shown in Table P5.16

TABLE P5.16

Total No. of Dots x_i	Frequency, f_i
2	1
3	2
4	3
5	4
6	5
7	6
8	5
9	4
10	3
11	2
12	1

Find the mean, variance, and standard deviation of total number of dots. (Use a working origin.)

5.17. Suppose a sample of n observations contains the values $y_1, \ldots, y_k$ with frequency of occurrence $f_1, \ldots, f_k$, respectively, where $\sum_{i=1}^{k} f_i = n$. Denote the "weighted" mean by $\bar{y}$, i.e., $\bar{y} = \sum_{i=1}^{k} f_i y_i / n$. Prove that $\sum_{i=1}^{k} f_i(y_i - \bar{y}) = 0$ and that

$$\sum_{i=1}^{k} f_i(y_i - \bar{y})^2 = \sum_{i=1}^{k} f_i y_i^2 - n\bar{y}^2 = \sum_{i=1}^{k} f_i y_i^2 - \left(\sum_{i=1}^{k} f_i y_i\right)^2 / n.$$

CHAPTER 6

Probability Distributions

6.1 GRAPHICAL DESCRIPTION OF DISCRETE RANDOM VARIABLES

In Chapter 5 we have discussed methods of describing *empirical distributions*, that is, distributions of the numerical values of the measurements obtained in a sample. Similar methods can be used for describing *theoretical distributions*, that is, distributions of random variables. The simplest type of a random variable, namely, a discrete random variable, was defined in Section 2.7. The set of possible values of a discrete random variable x together with their associated probabilities, as we have pointed out in Section 2.7, is called the *probabilty function* (p.f.) of the discrete random variable or, more briefly, a *discrete distribution*. Special cases of important discrete distributions are given in Chapter 3.

Discrete distributions can be conveniently described graphically by *probability bar charts*. Thus, in general, suppose the probability function of a discrete random variable x is:

(6.1.1)

x	x_1	x_2	$\cdots$	x_k
$p(x)$	$p(x_1)$	$p(x_2)$	$\cdots$	$p(x_k)$

as originally stated in (2.7.2). The probability bar chart for this probability function (or discrete distribution) is shown in Figure 6.1.1 where the lengths of the vertical bars represent the magnitudes of the probabilities.

The *cumulative distribution function* (c.d.f.) $F(x)$ provides an alternative way of describing a discrete random variable x, where for any real number x', $F(x')$ is defined as follows:

(6.1.2)
$$F(x') = P(x \leq x') = \Sigma' p(x_i)$$

where Σ' denotes summation for all values of i for which $x_i \leq x'$.

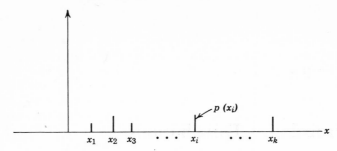

Figure 6.1.1. A probability bar chart.

It will be observed that

$$F(x_1) = p(x_1),$$
$$F(x_2) = p(x_1) + p(x_2),$$

.

.

.

(6.1.3) $$F(x_i) = p(x_1) + \cdots + p(x_i),$$

.

.

.

$$F(x_k) = p(x_1) + \cdots + p(x_k) = 1.$$

Dropping the prime it will be further seen that $F(x)$ is a "step function" defined for *all* (real) values of x having a graph as shown in Figure 6.1.2.

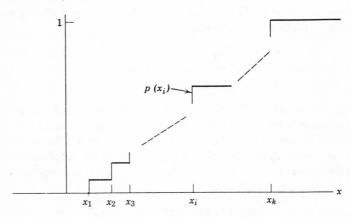

Figure 6.1.2. Graph of a cumulative distribution function $F(x)$ of a discrete random variable x.

Note that the steps in the graph occur at the points for which

$$x = x_1, x_2, \ldots, x_k.$$

Actually, the cumulative distribution function $F(x)$ is a more convenient device for describing the probability distributions of what are called *continuous random variables* to be discussed in Section 6.3. But the introduction of $F(x)$ for the case of discrete random variables at this point will simplify the interpretation of $F(x)$ for a continuous random variable in Section 6.3.

6.2   MEAN AND VARIANCE OF A DISCRETE RANDOM VARIABLE

In Chapter 5 we defined the mean of a sample of measurements. In an entirely similar manner we define the (*population*) *mean $E(x)$* (denoted by μ) of the discrete random variable x having probability function given in (6.1.1) as follows:

$$(6.2.1) \qquad E(x) = \sum_{i=1}^{k} x_i p(x_i) = \mu,$$

which is a weighted average of the possible values $x_1, x_2, \ldots, x_k$ of the random variable x, the weights being the associated probabilities $p(x_1)$, $p(x_2), \ldots, p(x_k)$. $E(x)$ is often called the *expected value* or *expectation* of the random variable x.

Corresponding to the procedure for defining the variance s^2 of a sample of measurements, we define the *variance* of the discrete random variable x, Var (x), (denoted by σ^2) as follows:

$$(6.2.2) \qquad \text{Var } (x) = \sum_{i=1}^{k} (x_i - \mu)^2 p(x_i) = \sigma^2,$$

which, it should be noted, is a weighted average of the squared deviations of $x_1, x_2, \ldots, x_k$ from the mean μ, the weights again being the probabilities.

A useful alternative expression for σ^2 is given by

$$(6.2.2a) \qquad \sigma^2 = \sum_{i=1}^{k} x_i^2 p(x_i) - \mu^2,$$

which is obtained by expanding the squared term $(x_i - \mu)^2$ in (6.2.2), performing the summation, and simplifying the results.

Example 6.2(a). The mean μ and variance σ^2 of the *hypergeometric distribution*, the p.f. of which is given by

$$h(x) = \frac{\dbinom{N\theta}{x} \dbinom{N(1-\theta)}{n-x}}{\dbinom{N}{n}}$$

where the sample space of x is the set of nonnegative integers that fall between max $[0, n - N(1 - \theta)]$ and min $(n, N\theta)$, inclusive, are given by

$$\mu = n\theta,$$
$$\sigma^2 = \frac{N - n}{N - 1}\, n\theta(1 - \theta).$$

These formulas are established in Section 6.11a.

Example 6.2(b). The mean μ and the variance σ^2 of the *binomial* distribution, the p.f. of which is given by

$$b(x) = \binom{n}{x} \theta^x (1 - \theta)^{n-x},$$

$x = 0, 1, \ldots, n$, are given by

$$\mu = n\theta,$$
$$\sigma^2 = n\theta(1 - \theta).$$

These formulas are established in Section 6.11b.

Example 6.2(c). The mean and variance of the *Poisson distribution* the p.f. of which is given by

$$p(x) = e^{-\mu} \frac{\mu^x}{x!},$$

$x = 0, 1, 2, \ldots$, both have the same value μ (the parameter appearing in the distribution itself). The proof is given in Section 6.11c.

6.3   CONTINUOUS RANDOM VARIABLES

Many sample spaces contain an infinite number of elements which cannot be put into one-one correspondence with the positive integers. Such sample spaces are said to have the property of the continuum, and random variables defined on them may vary over a continuum of values. In such a case we are not able to associate probabilities with the different possible values which the random variable can take on in the same way as for a discrete random variable.

More precisely, suppose S is a sample space with a continuum of values, and let $E_{x'}$ be the event consisting of all elements e for which $x(e) \leq x'$, or expressed more briefly by $\{e: x(e) \leq x'\}$. We assign a probability, say $F(x')$ to $E_{x'}$ for every value of x'. That is, we have

$$P[x(e) \leq x'] = P(E_{x'}) = F(x').$$

If we drop the prime and consider $F(x)$, we see that

$$F(-\infty) = 0,$$
(6.3.1)
$$F(+\infty) = 1,$$
$$F(x) \text{ is nondecreasing in } x.$$

If $F(x)$ is continuous in x, we say that x is a *continuous random variable* and that $F(x)$ is the *cumulative distribution function* (c.d.f.) of x. We have

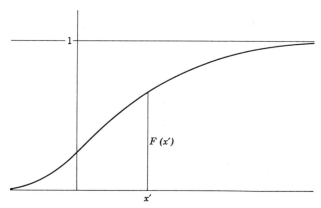

Figure 6.3.1. Graph of the cumulative distribution for a continuous random variable.

already seen in Section 6.1, that, if x is a discrete random variable, then the c.d.f. $F(x)$ is a step function as shown in Figure 6.1.2.

In the case of a continuous random variable, the graph of $F(x)$ is illustrated in Figure 6.3.1, that is, it has no vertical jumps. For any point x', $F(x')$ represents the total amount of probability "smeared" along the x-axis to the left of x'.

If $F(x)$ has a derivative $f(x)$, then $f(x)$ is nonnegative and is called the *probability density function* (p.d.f.) of x. The relationship between $F(x)$ and $f(x)$ is as follows:

(6.3.2) $$f(x) = \frac{dF(x)}{dx},$$

(6.3.3) $$F(x') = \int_{-\infty}^{x'} f(x)\,dx.$$

For a given c.d.f. $F(x)$ as illustrated in Figure 6.3.1, the graph of its p.d.f. $f(x)$ is illustrated in Figure 6.3.2.

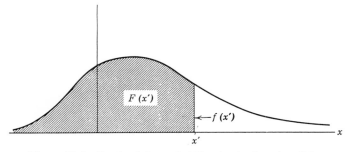

Figure 6.3.2. Graph of the probability density function $f(x)$.

If one wishes to calculate the probability that the random variable x has a value between a and b $(a < b)$, this can be done either from $F(x)$ or from $f(x)$ as follows:

(6.3.4) $$P(a < x < b) = F(b) - F(a) = \int_a^b f(x)\, dx.$$

Note that

(6.3.5) $$\int_{-\infty}^{\infty} f(x)\, dx = 1.$$

Example 6.3(a). Suppose a point e is picked "at random" inside the circle in the u,v-plane having equation

$$u^2 + v^2 = R^2,$$

so that the probability that e falls into any region (inside the circle) of area A is given by $A/\pi R^2$. The sample space here consists of all points inside the circle. Let x be a random variable denoting the distance from the point e to the center of the circle. Then the c.d.f. $F(x)$ of the random variable x is given by

$$F(x) = \frac{\pi x^2}{\pi R^2} = \frac{x^2}{R^2}$$

and the p.d.f. $f(x)$ is given by

$$f(x) = \frac{2x}{R^2}.$$

The sample space of x is the interval $(0, R)$.

Example 6.3(b). Suppose the c.d.f. of a continuous random variable x is given by

$$F(x) = 0, \qquad x \le 0,$$
$$F(x) = x, \qquad 0 < x \le 1,$$
$$F(x) = 1, \qquad x > 1.$$

Then the p.d.f. $f(x)$ is given by

$$f(x) = 1, 0 < x < 1,$$
$$= 0, \text{ otherwise.}$$

The random variable x is said to have a *rectangular distribution* on the interval $(0, 1)$.

6.4 MEAN AND VARIANCE OF CONTINUOUS RANDOM VARIABLES

Suppose a continuous random variable x has a probability density function $f(x)$. The expectation $E(x)$ of the random variable x is defined as

(6.4.1) $$E(x) = \int_{-\infty}^{\infty} xf(x)\, dx = \mu,$$

which may be interpreted as the center of gravity of the probability density function $f(x)$.

The variance of a continuous random variable x is defined by the expression

$$(6.4.2) \qquad \text{Var}\,(x) = \int_{-\infty}^{\infty} (x - \mu)^2 f(x)\, dx = \sigma^2,$$

which reduces to the alternative, and often more convenient, expression

$$(6.4.3) \qquad \sigma^2 = \int_{-\infty}^{\infty} x^2 f(x)\, dx - \mu^2.$$

6.5 MEAN VALUES OF MORE GENERAL FUNCTIONS OF RANDOM VARIABLES

In Section 6.2 we have defined the mean and variance of discrete random variables. Similar definitions have been given in Section 6.4 for the mean and variance of continuous random variables.

More generally if x is a discrete random variable as defined in Section 6.2, and $g(x)$ is a function (real and single-valued) of x, the mean value, or expectation of $g(x)$, which we write as $E(g(x))$, is defined by

$$(6.5.1) \qquad E(g(x)) = \sum_{i=1}^{k} g(x_i) p(x_i).$$

In the case of a continuous random variable x, the mean value of $g(x)$ [again denoted by $E(g(x))$] is defined by

$$(6.5.2) \qquad E(g(x)) = \int_{-\infty}^{\infty} g(x) f(x)\, dx$$

(whenever the integral exists).

If we put $g(x) = x^k$, then (6.5.1) and (6.5.2) take the forms

$$(6.5.3) \qquad E(x^k) = \sum x_i^k p(x_i),$$

$$(6.5.4) \qquad E(x^k) = \int_{-\infty}^{\infty} x^k f(x)\, dx.$$

In either case, the quantity defined by (6.5.3) or (6.5.4) is called the kth *moment* of the random variable x about the origin. Note that, if $k = 0$, we have

$$E(x^0) = E(1) = 1,$$

and if $k = 1$, we have

$$E(x) = \text{''}.$$

We usually denote $E(x^k)$ by μ_k'.

Now if we put $g(x) = (x - \mu)^k$, then (6.5.1) and (6.5.2) take the form

$$(6.5.5) \qquad E[(x - \mu)^k] = \sum (x_i - \mu)^k p(x_i),$$

$$(6.5.6) \qquad E[(x - \mu)^k] = \int_{-\infty}^{\infty} (x - \mu)^k f(x)\, dx.$$

In either case the quantity defined by (6.5.5) or (6.5.6) is called the kth *moment* of x about its mean and denoted by μ_k. Note that $k = 0$ gives $E[(x - \mu)^0] = 1$ and if $k = 1$, we have $\mu_1 = 0$. We also note that $\mu_2 = E[(x - \mu)^2]$ is the variance of x which we denote by σ^2.

The reader should verify that, if c is a constant, then

(6.5.7)
$$E(cx) = cE(x),$$

$$\text{Var}(cx) = c^2 \text{Var}(x).$$

6.6 CHEBYCHEV'S INEQUALITY

Sometimes it is desirable to have some notion as to how much probability there is in the tails of a probability function or a probability density function. A theorem which gives an upper bound for such probabilities, useful for some purposes is the following:

THEOREM 6.6.1. *If x is a random variable having finite mean μ and variance σ^2, then*

(6.6.1)
$$P(|x - \mu| > \lambda \sigma) \leq \frac{1}{\lambda^2}.$$

This inequality is sometimes called the Chebychev inequality. Note that the theorem states that, no matter what distribution a random variable x may have, if its mean μ and variance σ^2 are finite, the total amount of probability lying in the two tails $(-\infty, \mu - \lambda\sigma)$ and $(\mu + \lambda\sigma, +\infty)$ is not greater than $1/\lambda^2$. This, of course, implies that the amount of probability lying on the closed interval $[\mu - \lambda\sigma, \mu + \lambda\sigma]$ is greater than $1 - 1/\lambda^2$.

We shall prove Theorem 6.6.1 for the case in which x has a probability density function $f(x)$. Let I_1, I_2, I_3 be the three intervals $(-\infty, \mu - \lambda\sigma)$, $[\mu - \lambda\sigma, \mu + \lambda\sigma]$, $(\mu + \lambda\sigma, +\infty)$ respectively. Then, since

(6.6.2)
$$\sigma^2 = \int_{-\infty}^{\infty} (x - \mu)^2 f(x)\, dx,$$

we have

(6.6.3)
$$\sigma^2 = \int_{I_1} (x - \mu)^2 f(x)\, dx + \int_{I_2} (x - \mu)^2 f(x)\, dx + \int_{I_3} (x - \mu)^2 f(x)\, dx.$$

Dropping the middle integral completely, we have

(6.6.4)
$$\sigma^2 \geq \int_{I_1} (x - \mu)^2 f(x)\, dx + \int_{I_3} (x - \mu)^2 f(x)\, dx,$$

since the middle integral is nonnegative. If we replace x by $\mu - \lambda\sigma$ in the first integral on the right and x by $\mu + \lambda\sigma$ in the second integral, we do not increase the value of either integral. Hence,

(6.6.5) $$\sigma^2 \geq \lambda^2 \sigma^2 \int_{I_1} f(x)\, dx + \lambda^2 \sigma^2 \int_{I_3} f(x)\, dx,$$

that is,

(6.6.6) $$\sigma^2 \geq \lambda^2 \sigma^2 [P(X \in I_1) + P(x \in I_3)].$$

But the quantity in [ ] is $P(|x - \mu| > \lambda\sigma^2)$. Hence we have

$$\frac{1}{\lambda^2} \geq P(|x - \mu| > \lambda\sigma^2),$$

thus completing the proof of Theorem 6.6.1.

To prove Theorem 6.6.1 for the case of a discrete random variable x with probability function $p(x)$, we replace $f(x)$ by $p(x)$ and the integrals by sums in (6.6.2), (6.6.3), (6.6.4), and (6.6.5). The details are left to the reader as an exercise.

The reader should note that the Chebychev inequality for the case of a discrete random variable can be extended to the case of a sample of measurements $(x_1, \ldots, x_n)$ having mean $\bar{x}$ and standard deviation s, and states that the fraction of measurements in the sample $x_1, \ldots, x_n$ lying in the intervals $(-\infty, \bar{x} - \lambda s)$ and $(\bar{x} + \lambda s, +\infty)$ cannot exceed $1/\lambda^2$. This, of course, implies that the fraction of measurements in the sample lying in the closed interval $[\bar{x} - \lambda s, \bar{x} + \lambda s]$ exceeds $1 - 1/\lambda^2$.

Example 6.6(a). If a sample of size n is taken from a lot of N items containing 10% defectives, show by using the Chebychev inequality that the probability exceeds .99 that the number of defectives in the sample differs from $n/10$ by not more than $3\sqrt{n}\sqrt{(N-n)/(N-1)}$.

We know from Example 6.2(a) that, if d is the number of defectives in the sample and θ is the fraction of defectives in the lot, and noticing that $\theta = 1/10$ in the present example,

$$E(d) = \frac{n}{10},$$

$$\mathrm{Var}\,(d) = \frac{9n}{100}\frac{N-n}{N-1}.$$

Thus

$$\mu_d = n/10, \quad \sigma_d = \tfrac{3}{10}\sqrt{n}\sqrt{(N-n)/(N-1)},$$

and

$$\lambda\sigma_d = 3\sqrt{n}\sqrt{(N-n)/(N-1)},$$

from which we see that $\lambda = 10$. Hence, it follows from (6.6.1) that

$$P\left(\left|d - \frac{n}{10}\right| \leq 3\sqrt{n}\sqrt{\frac{N-n}{N-1}}\right) > 1 - \frac{1}{100} = .99,$$

thus establishing the desired result.

6.7 DISTRIBUTION FUNCTIONS OF TWO RANDOM VARIABLES

(a) The Case of Two Discrete Random Variables

If, for each element e in a finite sample space S, we make two measurements on e, say $(x_1(e), x_2(e))$, if (x_1', x_2') is a possible value of $(x_1(e), x_2(e))$, and if we let

$$(6.7.1) \qquad p(x_1', x_2') = P[x_1(e) = x_1', x_2(e) = x_2'],$$

then the set of all possible values $\{(x_1^{(i)}, x_2^{(i)})\}$ of $(x_1(e), x_2(e))$ together with the associated probabilities $p(x_1^{(i)}, x_2^{(i)})$ is the probability function of the pair of discrete random variables (x_1, x_2) (dropping e).

Thus we may think of k points $(x_1^{(1)}, x_2^{(1)}), \ldots, (x_1^{(k)}, x_2^{(k)})$ in the x_1, x_2-plane at which are located probabilities

$$p(x_1^{(1)}, x_2^{(1)}), \ldots, p(x_1^{(k)}, x_2^{(k)})$$

respectively, which are all positive and whose sum is 1.

If we sum $p(x_1^{(i)}, x_2^{(i)})$ for all i such that $x_1^{(i)} = x_{1\alpha}$ (a possible value of x_1), we obtain a probability which we shall denote by $p_1(x_{1\alpha})$. That is, we have

$$(6.7.2) \qquad p_1(x_{1\alpha}) = \sum{}' p(x_1^{(i)}, x_2^{(i)})$$

where the operation $\sum'$ stands for the summation over all points $(x_1^{(i)}, x_2^{(i)})$ in the sample space for which the first coordinate has the value $x_{1\alpha}$. If the possible values of $x_{1\alpha}$ are $x_{11}, \ldots, x_{1k_1}$, then these values together with the probabilities $p_1(x_{11}), \ldots, p_1(x_{1k_1})$, constitute the *marginal distribution* $p_1(x_1)$ of the random variable x_1. It gives the probability function of x_1, *ignoring* x_2, and is therefore merely the probability function of x_1. In a similar manner, if $x_{21}, \ldots, x_{2k_2}$ are the possible values of x_2, these values and their associated probabilities $p_2(x_{21}), \ldots, p_2(x_{2k_2})$ constitute the *marginal distribution* $p_2(x_2)$ of the random variable x_2.

Geometrically, if x_l is the usual horizontal axis and x_2 the vertical axis and if we project the probabilities $p(x_1^{(1)}, x_2^{(1)}), \ldots, p(x_1^{(k)}, x_2^{(k)})$ [located at the points $(x_1^{(1)}, x_2^{(1)}), \ldots, (x_1^{(k)}, x_2^{(k)})$] vertically onto the x_1-axis, we obtain the marginal distribution $p_1(x_1)$ of the random variable x_1. If, on the other hand, we project these probabilities horizontally onto the x_2-axis, we obtain the marginal distribution $p_2(x_2)$ of the random variable x_2.

The mean μ_1 and variance σ_1^2 of x_1 are defined by applying (6.2.1) and (6.2.2) to the probability function $p_1(x_1)$. Similarly μ_2 and σ_2^2 are defined by applying those formulas to $p_2(x_2)$.

If the probability function $p(x_1, x_2)$ factors into the product of the two marginal probability functions, that is,

(6.7.3) $$p(x_1, x_2) = p_1(x_1)p_2(x_2),$$

then x_1 and x_2 are said to be *independent random variables*.

Example 6.7(a). In dealing a hand of 13 cards from a deck of ordinary playing cards, let x_1 and x_2 be random variables denoting the numbers of spades and of hearts respectively. Obviously, $0 \leq x_1 \leq 13$, $0 \leq x_2 \leq 13$, and $0 \leq x_1 + x_2 \leq 13$. Then we see that $p(x_1, x_2)$, the p.f. of (x_1, x_2), is given by

$$p(x_1, x_2) = \frac{\binom{13}{x_1}\binom{13}{x_2}\binom{26}{13 - x_1 - x_2}}{\binom{52}{13}},$$

where the sample space of (x_1, x_2) is all pairs of nonnegative integers (x_1, x_2) for which $0 \leq x_1, x_2 \leq 13$ and $0 \leq x_1 + x_2 \leq 13$, that is, the sample space (x_1, x_2) consists of the 105 points

$$[(0, 0), (0, 1), (1, 0), \ldots, (12, 1), (13, 0)].$$

Now it is possible by a direct probability argument to find the marginal distribution of x_1, for, the probability of x_1 spades in a hand of 13 is clearly given by

$$p_1(x_1) = \binom{13}{x_1}\binom{39}{13 - x_1} \Big/ \binom{52}{13}$$

where $0 \leq x_1 \leq 13$.

But we may also arrive at this result by following the prescription of this section, namely

$$p_1(x_{1\alpha}) = \sum' p(x_1^{(i)}, x_2^{(i)}) = \sum_{x_2^{(i)}}' \frac{\binom{13}{x_{1\alpha}}\binom{13}{x_2^{(i)}}\binom{26}{13 - x_{1\alpha} - x_2^{(i)}}}{\binom{52}{13}}$$

where $\sum'_{x_2^{(i)}}$ denotes the operation of summing over all points $(x_1^{(i)}, x_2^{(i)})$ in the sample space of (x_1, x_2) for which the first coordinate is $x_{1\alpha}$. But clearly this implies that we sum $x_2^{(i)}$, the number of hearts in the hand of 13 when there is $x_{1\alpha}$ spades, from 0 to $13 - x_{1\alpha}$.

Hence

$$p_1(x_{1\alpha}) = \frac{\binom{13}{x_{1\alpha}}}{\binom{52}{13}} \sum_{x_2^{(i)}=0}^{13-x_{1\alpha}} \binom{13}{x_2^{(i)}}\binom{26}{13 - x_{1\alpha} - x_2^{(i)}}.$$

Now using the identity (3.6.3), which restated here for our purposes is

$$\sum_{x=0}^{m} \binom{N_1}{x} \binom{N_2}{m-x} = \binom{N_1 + N_2}{m} \quad \text{where} \quad m \leq N_1 < N_2$$

we see that

$$p_1(x_{1\alpha}) = \binom{13}{x_{1\alpha}} \binom{39}{13 - x_{1\alpha}} \bigg/ \binom{52}{13},$$

and $x_{1\alpha}$ can take on the values $0, 1, 2, \ldots, 13$. Dropping the subscript α we have

$$p_1(x_1) = \binom{13}{x_1} \binom{39}{13 - x_1} \bigg/ \binom{52}{13}$$

for $x_1 = 0, 1, \ldots, 13$.

(b) The Case of Two Continuous Random Variables

If the sample space S consists of a continuum of elements and if for any point (x_1', x_2') in the x_1, x_2-plane we let

(6.7.4) $$F(x_1', x_2') = P[x_1(e) \leq x_1', x_2'(e) \leq x_2'],$$

then (dropping primes) $F(x_1, x_2)$ is called the *cumulative distribution function* (c.d.f.) of the pair of random variables (x_1, x_2) (dropping e). If there exists a nonnegative function $f(x_1, x_2)$ such that

(6.7.5) $$F(x_1, x_2) = \int_{-\infty}^{x_1} \int_{-\infty}^{x_2} f(t_1, t_2) \, dt_2 \, dt_1,$$

then

$$f(x_1, x_2) = \frac{\partial^2 F(x_1, x_2)}{\partial x_1 \, \partial x_2},$$

and $f(x_1, x_2)$ is called the *joint probability density function* (p.d.f.) of the pair of random variables (x_1, x_2). The probability that this pair of random variables represents a point in a region E (that is, the probability that event E occurs) is given by

(6.7.6) $$P[(x_1, x_2) \in E] = \int_E \int f(x_1, x_2) \, dx_1 \, dx_2.$$

If we let

(6.7.6a) $$f_1(x_1) = \int_{-\infty}^{\infty} f(x_1, x_2) \, dx_2$$

and

(6.7.6b) $$f_2(x_2) = \int_{-\infty}^{\infty} f(x_1, x_2) \, dx_1,$$

then $f_1(x_1)$ and $f_2(x_2)$ are called the *marginal probability density functions* of x_1 and x_2 respectively. This means that $f_1(x_1)$ is the p.d.f. of x_1 (ignoring x_2) and $f_2(x_2)$ is the p.d.f. of x_2 (ignoring x_1).

Geometrically, if we think of $f(x_1, x_2)$ as a function describing the manner in which the total probability 1 is continuously "smeared" in the x_1, x_2-plane, then the integral in (6.7.6) represents the amount of probability contained in the region E, $f_1(x_1)$ is the p.d.f. one obtains by projecting the probability density in the x_1, x_2-plane orthogonally onto the x_1-axis, and $f_2(x_2)$ is similarly obtained by orthogonal projection of the probability density onto the x_2-axis.

If $f(x_1, x_2)$ factors into the two marginal p.d.f.'s, that is, if

$$(6.7.7) \qquad f(x_1, x_2) = f_1(x_1)f_2(x_2),$$

then x_1 and x_2 are said to be *independent continuous random variables*.

(c) The Mean Value of Functions of Two Random Variables

If (x_1, x_2) is a pair of random variables and if $g(x_1, x_2)$ is a function of (x_1, x_2), the mean value or expectation of $g(x_1, x_2)$, say $E(g(x_1, x_2))$, is given by

$$(6.7.8) \qquad E(g(x_1, x_2)) = \sum_{i=1}^{k} g(x_1^{(i)}, x_2^{(i)})p(x_1^{(i)}, x_2^{(i)}),$$

in the case of discrete random variables, and

$$(6.7.9) \qquad E(g(x_1, x_2)) = \int_{-\infty}^{\infty} \int_{-\infty}^{\infty} g(x_1, x_2)f(x_1, x_2)\, dx_1\, dx_2$$

in the case of continuous random variables.

The reader should verify the following:

THEOREM 6.7.1. *If x_1 and x_2 are independent random variables and if $g_1(x_1)$ and $g_2(x_2)$ depend only on x_1 and x_2 respectively, then*

$$(6.7.10) \qquad E(g_1(x_1)g_2(x_2)) = E(g_1(x_1))E(g_2(x_2)).$$

If we choose $g(x_1, x_2)$ as $(x_1 - \mu_1)(x_2 - \mu_2)$, we obtain the *covariance* between x_1 and x_2, that is,

$$(6.7.11) \qquad \text{Cov}\,(x_1, x_2) = E[(x_1 - \mu_1)(x_2 - \mu_2)].$$

In the case where x_1 and x_2 are independent, we have

$$(6.7.12) \qquad \text{Cov}\,(x_1, x_2) = E(x_1 - \mu_1)E(x_2 - \mu_2) = 0.$$

In many problems we deal with linear functions of two (or more) independent random variables. The following theorem is of particular importance in this connection.

THEOREM 6.7.2. *Let x_1 and x_2 be independent random variables such that the mean and variance of x_1 are μ_1 and σ_1^2, and the mean and variance of x_2 are μ_2 and σ_2^2. Then, if c_1 and c_2 are constants, $c_1 x_1 + c_2 x_2$ is a random variable having mean value $c_1 \mu_1 + c_2 \mu_2$ and variance $c_1^2 \sigma_1^2 + c_2^2 \sigma_2^2$.*

To prove this theorem, it is sufficient to consider the case of continuous random variables. (The proof for discrete random variables is obtained by replacing integral signs by signs of summation.) For the mean value of $c_1 x_1 + c_2 x_2$, we have

$$E(c_1 x_1 + c_2 x_2)$$

$$= \int_{-\infty}^{\infty} \int_{-\infty}^{\infty} (c_1 x_1 + c_2 x_2) f_1(x_1) f_2(x_2) \, dx_1 \, dx_2$$

$$= c_1 \int_{-\infty}^{\infty} x_1 f_1(x_1) \, dx_1 \int_{-\infty}^{\infty} f_2(x_2) \, dx_2 + c_2 \int_{-\infty}^{\infty} x_2 f_2(x_2) \, dx_2 \int_{-\infty}^{\infty} f_1(x_1) \, dx_1$$

$$= c_1 E(x_1) + c_2 E(x_2)$$

$$= c_1 \mu_1 + c_2 \mu_2.$$

For the variance of $c_1 x_1 + c_2 x_2$, we have similarly (omitting some straightforward details),

$$\text{Var}\,(c_1 x_1 + c_2 x_2)$$

$$= E[c_1(x_1 - \mu_1) + c_2(x_2 - \mu_2)]^2$$

$$= c_1^2 E(x_1 - \mu_1)^2 + c_2^2 E(x_2 - \mu_2)^2 + 2c_1 c_2 E[(x_1 - \mu_1)(x_2 - \mu_2)]$$

$$= c_1^2 \sigma_1^2 + c_2^2 \sigma_2^2,$$

since $E(x_1 - \mu_1)(x_2 - \mu_2) = 0$, as shown in (6.7.12).

We remark that, if x_1, x_2 are not independent (and the reader should verify),

$$E(c_1 x_1 + c_2 x_2) = c_1 \mu_1 + c_2 \mu_2,$$

and

$$\text{Var}\,(c_1 x_1 + c_2 x_2) = c_1^2 \sigma_1^2 + c_2^2 \sigma_2^2 + 2c_1 c_2 \,\text{Cov}\,(x_1, x_2).$$

In a straightforward manner, it is easy to prove the following theorem, which gives the extension of the results of this section.

THEOREM 6.7.3. *Let $x_1, x_2, \ldots, x_n$ be n random variables such that the mean and variance of x_i, is μ_i and σ_i^2 respectively, and where the co-variance of x_i and x_j is σ_{ij}, that is $E[(x_i - \mu_i)(x_j - \mu_j)] = \sigma_{ij}$, $i \neq j$. If $c_1, \ldots, c_n$ are constants, then the random variable $L = c_1 x_1 + \cdots + c_n x_n$ has mean value*

$$(6.7.13) \qquad\qquad E(L) = c_1 \mu_1 + \cdots + c_n \mu_n$$

and variance

$$(6.7.14) \quad \text{Var}(L) = c_1^2\sigma_1^2 + \cdots + c_n^2\sigma_n^2 + 2c_1c_2\sigma_{12} + 2c_1c_3\sigma_{13} + \cdots$$
$$+ 2c_{n-1}c_n\sigma_{n-1,n}.$$

Further, if $x_1, \ldots, x_n$ *are mutually independent, then* $\sigma_{ij} = 0$*, so that the mean of L is as in* (6.7.13)*, but the variance of L is*

$$(6.7.15) \qquad\qquad \text{Var}(L) = c_1^2\sigma_1^2 + \cdots + c_n^2\sigma_n^2.$$

(d) Conditional Distributions

Suppose a pair of discrete random variables (x_1, x_2) has a p.f. $p(x_1, x_2)$ and marginal p.f.'s $p_1(x_1)$ and $p_2(x_2)$, as defined in Section 6.7a. Suppose we assign to one of the random variables, say x_1, a value x_1' such that $p_1(x_1') \neq 0$ and ask for the probability that the other random variable x_2 has a particular value, say x_2'. The required probability is a *conditional probability* which we may denote by $p(x_2 = x_2' \mid x_1 = x_1')$, or, more briefly, by $p(x_2' \mid x_1')$, and is defined as follows:

$$p(x_2' \mid x_1') = \frac{p(x_1', x_2')}{p_1(x_1')}.$$

Note that the sum of $p(x_2 \mid x_1')$ over all possible values of x_2 for fixed x_1', is 1. Thus $p(x_2 \mid x_1')$, $x_2 = x_{21}, \ldots, x_{2k_2}$, is a p.f. and is called the *conditional probability function* of x_2 given that $x_1 = x_1'$. Dropping the prime, we call $p(x_2 \mid x_1)$, as defined by

$$(6.7.16) \qquad\qquad p(x_2 \mid x_1) = \frac{p(x_1, x_2)}{p_1(x_1)},$$

where $p_1(x_1) \neq 0$, the *conditional* p.f. of x_2 given x_1.

Note that we can write (6.7.16) as

$$(6.7.17) \qquad\qquad p(x_1, x_2) = p_1(x_1)p(x_2 \mid x_1),$$

thus providing a two-step procedure for finding $p(x_1, x_2)$ by first determining $p_1(x_1)$, then $p(x_2 \mid x_1)$, and multiplying the two together.

Example 6.7(b). In Example 6.7(*a*), suppose we wish to find the conditional p.f. of x_2 given x_1, that is, $p(x_2 \mid x_1)$. The p.f. of x_1 is given by

$$p_1(x_1) = \frac{\binom{13}{x_1}\binom{39}{13 - x_1}}{\binom{52}{13}}.$$

Hence

$$p(x_2 \mid x_1) = \frac{\binom{13}{x_2}\binom{26}{13 - x_1 - x_2}}{\binom{39}{13 - x_1}},$$

where the sample space of x_2 is $\{0, 1, \ldots, 13 - x_1\}$. The interpretation of $p(x_2 \mid x_1)$ is that, if the hand of 13 cards contains x_1 spades, then the value of $p(x_2 \mid x_1)$ as given above is the probability that the hand also contains x_2 hearts.

In the case of a pair of continuous random variables (x_1, x_2) having p.d.f. $f(x_1, x_2)$ and marginal p.d.f.'s $f_1(x_1)$ and $f_2(x_2)$, the conditional p.d.f. $f(x_2 \mid x_1)$ of x_2 given x_1 is defined as

$$(6.7.16a) \qquad f(x_2 \mid x_1) = \frac{f(x_1, x_2)}{f_1(x_1)},$$

where $f_1(x_1) \neq 0$, which is the analogue of (6.7.16) for a pair of continuous random variables. Note that $f(x_2 \mid x_1)$ has all of the properties of an ordinary p.d.f.

Rewriting (6.7.16a) as

$$(6.7.17a) \qquad f(x_1, x_2) = f_1(x_1)f(x_2 \mid x_1),$$

we have the analogue of (6.7.17) for obtaining the p.d.f. of a pair of continuous random variables in two steps.

Example 6.7(c). Let (x_1, x_2) be a point taken "at random" in the triangle bounded by the x_1-axis, the x_2-axis and the line having equation $x_1 + x_2 = 1$, so that the probability that (x_1, x_2) falls into any region, say E, within this triangle is inversely proportional to the area of E. To find the p.d.f. of x_2 given x_1, we proceed as follows:

It is evident that (x_1, x_2) has a p.d.f. $f(x_1, x_2)$ defined as

$$\begin{aligned} f(x_1, x_2) &= 2 \qquad \text{inside the triangle having} \\ &\qquad\qquad \text{vertices. } (0, 0), (0, 1), (1, 0); \\ &= 0 \qquad \text{otherwise.} \end{aligned}$$

The marginal p.d.f. of x_1 is given by

$$f_1(x_1) = \int_{-\infty}^{\infty} f(x_1, x_2) \, dx_2 = 2\int_0^{1-x_1} dx_2 = 2(1 - x_1).$$

Hence

$$\begin{aligned} f(x_2 \mid x_1) &= \frac{2}{2(1 - x_1)} = \frac{1}{1 - x_1} \qquad \text{for } 0 < x_2 < 1 - x_1, \\ &= 0 \qquad \text{otherwise.} \end{aligned}$$

Since $p(x_2 \mid x_1)$ is a p.f., we define the mean and variance of x_2 given x_1 as

$$(6.7.18) \qquad E(x_2 \mid x_1) = \sum_{x_2} x_2 p(x_2 \mid x_1)$$

and

$$(6.7.19) \qquad \text{Var}\,(x_2 \mid x_1) = \sum_{x_2} [x_2 - E(x_2 \mid x_1)]^2 p(x_2 \mid x_1),$$

and so on, for other functions of x_2 given x_1.

Similarly, for the case of a pair of continuous random variables, we have

$$(6.7.18a) \qquad E(x_2 \mid x_1) = \int_{-\infty}^{\infty} x_2 f(x_2 \mid x_1)\, dx_2$$

and

$$(6.7.19a) \qquad \text{Var}\,(x_2 \mid x_1) = \int_{-\infty}^{\infty} [x_2 - E(x_2 \mid x_1)]^2 f(x_2 \mid x_1)\, dx_2.$$

(e) Correlation Between Two Random Variables

The reader will note from (6.7.11) that the covariance between the random variables x_1 and x_2 is a quantity measured in [(units of x_1) × (units of x_2)]. A somewhat more convenient measure of how x_1 and x_2 "co-vary," or are dependent on each other, is the *theoretical* or *population correlation coefficient* ρ. This dimensionless measure of dependence is defined by

$$(6.7.20) \qquad \rho = \text{Cov}\,(x_1, x_2)/\sigma_1 \sigma_2,$$

where σ_1 and σ_2 are the population standard deviations of x_1 and x_2 respectively. It can be shown that $-1 \leq \rho \leq 1$.

Now from (6.7.12) and using (6.7.20), we have that if x_1 and x_2 are independent random variables, then $\rho = 0$. The converse need not be true however, as the following example shows.*

Example 6.7(d). Two random variables x_1 and x_2 have joint probability function given by

$$(6.7.21) \qquad f(x_1, x_2) = \begin{cases} 1/3 & \text{if } (x_1, x_2) = (0, 0),\ (1, 1),\ (2, 0) \\ 0 & \text{otherwise.} \end{cases}$$

It is easy to see that

$$(6.7.22) \qquad f_1(x_1) = \begin{cases} 1/3 & \text{if } x_1 = 0,\ 1,\ 2 \\ 0 & \text{otherwise} \end{cases}$$

---

* When and if the reader consults Appendix III, it will be noted that if the joint distribution of x_1 and x_2 is the bivariate normal, then $\rho = 0$ implies that x_1 and x_2 are independent. Also, if x_1 and x_2 are bivariate normal, ρ is a measure of the linear dependence between x_1 and x_2.

and that

(6.7.23)
$$f_2(x_2) = \begin{cases} 2/3 & \text{if } x_2 = 0 \\ 1/3 & \text{if } x_2 = 1. \end{cases}$$

Hence $f(0, 0) \neq f_1(0)f_2(0)$, etc. so that x_1 and x_2 are not independent. Further, simple calculations show that

(6.7.24)
$$\mu_1 = E(x_1) = 1, \qquad \mu_2 = E(x_2) = 1/3$$
$$\sigma_1^2 = E(x_1 - \mu_1)^2 = 2/3, \qquad \sigma_2^2 = E(x_2 - \mu_2)^2 = 2/9.$$

Also

$$\begin{aligned}
\text{Cov } (x_1, x_2) &= \sum\sum (x_1 - 1)(x_2 - 1/3)f(x_1, x_2) \\
&= 1/3[(0 - 1)(0 - 1/3) + (1 - 1)(1 - 1/3) \\
&\qquad\qquad\qquad\qquad\qquad\qquad + (2 - 1)(0 - 1/3)] \\
&= 1/3[1/3 + 0 - 1/3] \\
&= 0.
\end{aligned}$$

Hence $\rho = 0$, yet x_1 and x_2 are *not* independent.

6.8   EXTENSION TO SEVERAL RANDOM VARIABLES

The notions of (i) the probability function of a pair of discrete random variables and (ii) the probability density function of a pair of continuous random variables extend without special difficulties to sets of three or more random variables. Thus, in the case of n discrete random variables $(x_1, \ldots, x_n)$, we would have a set of k possible points

$$(x_1^{(1)}, \ldots, x_n^{(1)}), \ldots, (x_1^{(k)}, \ldots, x_n^{(k)})$$

in an n-dimensional space with associated probabilities

$$p(x_1^{(1)}, \ldots, x_n^{(1)}), \ldots, p(x_1^{(k)}, \ldots, x_n^{(k)})$$

respectively, which are all positive and whose sum is 1. If all of the probability is projected orthogonally onto the x_1-axis, we obtain the marginal probability function of x_1, say $p_1(x_1)$. Similarly, we obtain marginal probability functions $p_2(x_2), \ldots, p_n(x_n)$ of the remaining random variables $x_2, \ldots, x_n$.

If $p(x_1, \ldots, x_n)$ factors into the product of the marginal probability functions, that is, if

(6.8.1)
$$p(x_1, \ldots, x_n) = p_1(x_1) \cdots p_n(x_n),$$

we say that $x_1, \ldots, x_n$ are *mutually independent*.

In the case of n continuous random variables $x_1, \ldots, x_n$, we would have a probability density function $f(x_1, \ldots, x_n)$ which is nonnegative throughout the entire n-dimensional space of the variables. The probability that

$(x_1, \ldots, x_n)$ falls into any region or set E (that is, the probability that event E occurs) is given by

$$P[(x_1, \ldots, x_n) \in E] = \int \cdots \int_E f(x_1, \ldots, x_n) \, dx_1 \cdots dx_n.$$

By setting

$$f_1(x_1) = \int_{-\infty}^{\infty} \cdots \int_{-\infty}^{\infty} f(x_1, \ldots, x_n) \, dx_2 \cdots dx_n$$

with similar definitions for $f_2(x_2), \ldots, f_n(x_n)$, we obtain the marginal probability density functions $f_1(x_1), \ldots, f_n(x_n)$ of $x_1, \ldots, x_n$ respectively.
If $f(x_1, \ldots, x_n)$ factors as follows,

$$(6.8.1a) \qquad f(x_1, \ldots, x_n) = f_1(x_1) \cdots f_n(x_n),$$

then $x_1, \ldots, x_n$, are said to be *mutually independent* random variables.

The extensions of Theorem 6.7.1 and 6.7.2 to the case of n-independent random variables are straightforward and are left to the reader.

The concepts of a conditional p.f. and of a conditional p.d.f. of x_n given $x_1, \ldots, x_{n-1}$ are straightforward and are also left to the reader for development.

6.9 RANDOM SAMPLING

(a) Random Sampling from a Probability Distribution

In the special case where $x_1, \ldots, x_n$ are mutually independent and all have identical probability functions $p(x)$ (or identical p.d.f.'s $f(x)$ in case $x_1, \ldots, x_n$ are continuous random variables), the set of n random variables $(x_1, \ldots, x_n)$ is said to be a *random sample* from $p(x)$ (or from $f(x)$), and $x_1, \ldots, x_n$ are called *elements of the sample*. Thus, the sample elements $x_1, \ldots, x_n$ have equal means which we will call μ and equal variances which we will call σ^2. Sometimes we say that $(x_1, \ldots, x_n)$ is a random sample from a *population* having p.f. $p(x)$ [or p.d.f. $f(x)$], and μ and σ^2 are called the *population mean* and *population variance* respectively.

Or, sometimes we say that $x_1, \ldots, x_n$ are n independent observations on a random variable x, where $E(x) = \mu$ and Var $(x) = \sigma^2$. Now suppose we consider the sample sum T, defined by

$$(6.9.1) \qquad T = x_1 + \cdots + x_n.$$

The reader may verify that (see Theorem 6.7.3) the mean and variance of T are

$$(6.9.2) \qquad E(T) = n\mu$$

and

$$(6.9.3) \qquad\qquad \text{Var } (T) = n\sigma^2,$$

respectively. The quantity T is a random variable, and (6.9.2) and (6.9.3) give some important properties of T.

Now the reader should bear in mind that T is the sum of n independent measurements x_i, $i = 1, \ldots, n$ and not equal to the variable $L = nx$. While it is true that $E(T) = E(L) = n\mu$, the variance of T is $n\sigma^2 \neq n^2\sigma^2 = $ Var (L). This is because L is found by observing a single x, and multiplying it by n, whereas, to repeat and reemphasize, T is found by observing n "$x's$" (which we are assuming to be independent) called $x_1, \ldots, x_n$ and then summing the results, i.e., $T = x_1 + \cdots + x_n$.

Continuing, suppose we still assume that $(x_1, \ldots, x_n)$ are n independent observations on x; if we let $\bar{x}$ be the sample mean defined by

$$(6.9.4) \qquad\qquad \bar{x} = \frac{1}{n}(x_1 + \cdots + x_n),$$

then it can be seen from the fact that $\bar{x} = (1/n)T$ that the mean and variance of $\bar{x}$ are given by [see (6.5.7)]

$$(6.9.5) \qquad\qquad E(\bar{x}) = \frac{1}{n} E(T) = \mu,$$

$$(6.9.6) \qquad\qquad \text{Var } (\bar{x}) = \frac{1}{n^2} \sigma_T^2 = \frac{\sigma^2}{n}.$$

Formulas (6.9.5) and (6.9.6) state particularly important properties of $\bar{x}$. The fact that $E(\bar{x}) = \mu$ simply means the distribution of $\bar{x}$ has the same mean as the population from which the sample is drawn. The fact that Var $(\bar{x})$ is inversely proportional to n shows that, as n increases, the distribution of $\bar{x}$ becomes more highly concentrated about its mean μ.

> *Example 6.9(a).* Suppose $(x_1, \ldots, x_n)$ are random variables representing weights of n plastic items successively produced by a molding machine. If the mean and variance of weights of an indefinitely large population of items molded by this particular machine using this kind of plastic are μ and σ^2, then the mean and variance of T, the total weight $x_1 + \cdots + x_n$ of the sample of n elements $(x_1, \ldots, x_n)$ are $n\mu$ and $n\sigma^2$ respectively.

> *Example 6.9(b).* Suppose $x_1, \ldots, x_n$ are random variables representing errors made in independently measuring the length of a bar n times when its "true" length is known. If it is assumed that $x_1, \ldots, x_n$ are independent random variables each with mean 0 and variance σ^2, it will be seen that the mean and variance of the average error $\bar{x}$ are 0 and σ^2/n respectively.

(b) Law of Large Numbers for a Sample Mean

By using the Chebychev inequality (6.6.1), we can make a stronger statement about the concentration of probability in the distribution of $\bar{x}$ around its mean μ than by merely saying that the variance $\bar{x}$ is σ^2/n and, thus, inversely proportional to n.

More precisely, for any given $\epsilon > 0$, suppose we consider

$$P(|\bar{x} - \mu| > \epsilon),$$

that is, the probability that $\bar{x}$ will fall *outside* the interval

$$[\mu - \epsilon, \mu + \epsilon].$$

We can write

(6.9.7) $$P(|\bar{x} - \mu| > \epsilon) = P\left[|\bar{x} - \mu| > \left(\frac{\epsilon\sqrt{n}}{\sigma}\right)\frac{\sigma}{\sqrt{n}}\right].$$

Applying the Chebychev inequality (6.6.1) and noting that $\epsilon\sqrt{n}/\sigma$ in (6.9.7) plays the role of λ in (6.6.1) and $\sigma/\sqrt{n}$ (the standard deviation of $\bar{x}$) in (6.9.7) plays the role of σ in (6.6.1), then

(6.9.8) $$P(|\bar{x} - \mu| > \epsilon) \leq \frac{\sigma^2}{\epsilon^2 n}.$$

This states that, for an arbitrarily small positive number ϵ, the sample size n can be chosen sufficiently large to make the probability as small as we please so that the sample mean $\bar{x}$ will not differ from the population mean μ by more than ϵ. The statement in (6.9.8) is sometimes called the (weak) *law of large numbers*.

(c) Random Sampling From a Finite Population

In Section 6.9a we discussed random sampling from a probability distribution [having a p.f. $p(x)$ or a p.d.f. $f(x)$]. This is sometimes referred to as *sampling from an infinite population* since the elements $x_1, \ldots, x_n$, being independent random variables all having the same p.f. $p(x)$ [or p.d.f. $f(x)$], can be thought of as measurements taken on n objects successively drawn at random from an indefinitely large population in which the x values of the objects have p.f. $p(x)$ [or p.d.f. $f(x)$].

Now suppose the population being sampled has only a finite number of objects say $0_1, 0_2, \ldots, 0_N$, such that the x-values of these objects are $X_1, X_2, \ldots, X_N$ respectively. Let the mean μ of this population be defined as the mean of $X_1, \ldots, X_N$, that is,

(6.9.9) $$\mu = \frac{1}{N}(X_1 + \cdots + X_N),$$

and let the variance σ^2 of the population be defined as

(6.9.10) $$\sigma^2 = \frac{1}{N-1} \sum_{\alpha=1}^{N} (X_\alpha - \mu)^2.$$

Note that the mean μ and variance σ^2 are defined for the population whose X values are $X_1, \ldots, X_N$ in the same way that the mean $\bar{x}$ and variance s^2 are defined in (5.3.1) and (5.3.6) for a sample whose x values are $x_1, \ldots, x_n$.

If we consider a sample $x_1, \ldots, x_n$ of size n drawn from our finite population of N objects (without replacement), there are $\binom{N}{n}$ possible samples which could be drawn, each of which would have a sum T and a mean $\bar{x}$. If each of these samples is equally likely to be drawn, then the mean and variance of all $\binom{N}{n}$ possible sample sums are given by

(6.9.11) $$E(T) = n\mu,$$

(6.9.12) $$\text{Var }(T) = \left(\frac{1}{n} - \frac{1}{N}\right) n^2 \sigma^2$$

and the mean and variance of all $\binom{N}{n}$ possible sample means are

(6.9.13) $$E(\bar{x}) = \mu,$$

(6.9.14) $$\text{Var }(\bar{x}) = \left(\frac{1}{n} - \frac{1}{N}\right) \sigma^2.$$

Note that the formulas for $E(T)$ and $E(\bar{x})$ given by (6.9.11) and (6.9.13) are identical with those for $E(T)$ and $E(\bar{x})$ for the case of sampling from a probability distribution (that is, from an indefinitely large population) given by (6.9.2) and (6.9.5). Furthermore, as $N \to \infty$, the formulas for Var (T) and Var $(\bar{x})$ given by (6.9.12) and (6.9.14) become the same as those for Var (T) and Var $(\bar{x})$ for the case of sampling from a probability distribution (that is, an indefinitely large population) as given by (6.9.3) and (6.9.6).

It is sufficient to verify (6.9.11) and (6.9.12) since (6.9.13) and (6.9.14) follow from (6.9.11) and (6.9.12) by using the fact that $\bar{x} = (1/n)T$.

To verify (6.9.11), consider the sample sum $T = x_1 + \cdots + x_n$ for all $\binom{N}{n}$ possible samples. We wish to take the mean of all of these T's, which is what we mean by $E(T)$. We have,

(6.9.15) $$E(T) = \frac{T_1 + \cdots + T_{\binom{N}{n}}}{\binom{N}{n}},$$

where $T_1, \ldots, T_{\binom{N}{n}}$ would be the $\binom{N}{n}$ possible sample sums if they were arranged in some order. Now each of the X's, that is, $X_1, \ldots, X_N$, occurs in $\binom{N-1}{n-1}$ T's. Hence we may write (6.9.15) in terms of $X_1, \ldots,$ X_N as follows:

$$(6.9.15a) \qquad E(T) = \frac{\binom{N-1}{n-1}(X_1 + \cdots + X_N)}{\binom{N}{n}}.$$

But $\binom{N-1}{n-1} \Big/ \binom{N}{n} = n/N$ and $X_1 + \cdots + X_N = N\mu$. Hence we have

$$E(T) = n\mu,$$

thus verifying (6.9.11).

Now consider (6.9.12). We can write

$$(6.9.16) \qquad \mathrm{Var}\,(T) = \frac{1}{\binom{N}{n}}\left[(T_1 - n\mu)^2 + \cdots + \left(T_{\binom{N}{n}} - n\mu\right)^2\right]$$

and using (6.9.15), we find

$$\mathrm{Var}\,(T) = \frac{1}{\binom{N}{n}}\left(T_1^2 + \cdots + T_{\binom{N}{n}}^2\right) - n^2\mu^2.$$

Now for any sample $x_1, \ldots, x_n$, the T for this sample is $(x_1 + \cdots + x_n)$ and T^2 is $(x_1^2 + \cdots + x_n^2 + 2x_1x_2 + \cdots + 2x_{n-1}x_n)$. Thus any X^2, say X_α^2, occurs in $\binom{N-1}{n-1}$ T^2's, and any product of two X's say $X_\alpha X_\beta$, occurs in $\binom{N-2}{n-2}$ of the T^2's. Hence, we can express $\mathrm{Var}\,(T)$ in terms of $X_1^2, \ldots,$ $X_N^2, X_1 X_2, \ldots, X_{N-1}X_N$ as follows,

$$(6.9.17) \qquad \mathrm{Var}\,(T) = \frac{1}{\binom{N}{n}}\left[\binom{N-1}{n-1}(X_1^2 + \cdots + X_N^2)\right.$$

$$\left. + 2\binom{N-2}{n-2}(X_1X_2 + \cdots + X_{N-1}X_N)\right] - n^2\mu^2,$$

which can be written as

$$(6.9.18) \quad \mathrm{Var}\,(T) = \frac{1}{\binom{N}{n}}\left\{\left[\binom{N-1}{n-1} - \binom{N-2}{n-2}\right](X_1^2 + \cdots + X_N^2)\right.$$

$$\left. + \binom{N-2}{n-2}(X_1 + \cdots + X_N)^2\right\} - n^2\mu^2,$$

by adding and subtracting $\binom{N-2}{n-2}(X_1^2 + \cdots + X_N^2)$ inside the bracket in (6.9.17) and noting that

$$2\binom{N-2}{n-2}(X_1 X_2 + \cdots + X_{N-1} X_N) + \binom{N-2}{n-2}(X_1^2 + \cdots + X_N^2)$$

$$= \binom{N-2}{n-2}(X_1 + \cdots + X_N)^2.$$

But since

$$\binom{N-1}{n-1} \Big/ \binom{N}{n} = \frac{n}{N} \quad \text{and} \quad \binom{N-2}{n-2} \Big/ \binom{N}{n} = \frac{n(n-1)}{N(N-1)}$$

and $X_1 + \cdots + X_N = N\mu$, we find that (6.9.18) reduces to

$$\text{Var}(T) = \frac{n(N-n)}{N(N-1)} [(X_1^2 + \cdots + X_N^2) - N\mu^2].$$

But we see from (6.9.10) that σ^2 can be written as

$$\sigma^2 = \frac{1}{N-1} (X_1^2 + \cdots + X_N^2 - N\mu^2).$$

Hence

$$\text{Var}(T) = \left(\frac{1}{n} - \frac{1}{N}\right) n^2 \sigma^2,$$

which is (6.9.12).

Example 6.9(c). If N objects are numbered $1, 2, \ldots, N$ respectively, and if a sample of n objects is drawn at random from this population of N objects, what is the mean and variance of the sum T of numbers drawn in the sample?

In this example the X values $X_1, \ldots, X_N$ of the objects may be taken as $1, 2, \ldots, N$ respectively. Thus,

$$\mu = \frac{1 + 2 + \cdots + N}{N} = \frac{N+1}{2},$$

$$\sigma^2 = \frac{1}{N-1} (1^2 + \cdots + N^2 - N\mu^2)$$

$$= \frac{1}{N-1} \left[\frac{N(N+1)(2N+1)}{6} - N\left(\frac{N+1}{2}\right)^2\right];$$

that is

$$\mu = \tfrac{1}{2}(N+1),$$
$$\sigma^2 = \tfrac{1}{12}N(N+1).$$

Hence,

$$E(T) = \frac{n}{2}(N+1) \quad \text{and} \quad \text{Var}(T) = \frac{1}{12}\left(\frac{1}{n} - \frac{1}{N}\right) n^2 N(N+1).$$

6.10 THE MOMENT GENERATING FUNCTION

Referring to (6.5.1) and (6.5.2), suppose we set $g(x) = e^{xt}$ and find its expectation. We obtain

(6.10.1) $$E(e^{xt}) = \sum_{i=1}^{k} e^{x_i t} p(x_i)$$

if x is a discrete random variable, and

(6.10.2) $$E(e^{xt}) = \int_{-\infty}^{\infty} e^{xt} f(x)\, dx$$

in case* x is a continuous random variable. In either case, the function of t so obtained is called the *moment generating function* of the random variable x, and is denoted by $M_x(t)$, that is,

(6.10.3) $$M_x(t) = E(e^{xt}).$$

Note that $M_x(t)$ can be written as

$$M_x(t) =: E\left(1 + xt + \frac{x^2 t^2}{2!} + \cdots + \frac{x^k t^k}{k!} + \cdots\right)$$

$$= 1 + tE(x) + \frac{t^2}{2!} E(x^2) + \cdots + \frac{t^k}{k!} E(x^k) + \cdots,$$

and from Section 6.5, we then have

(6.10.4) $$M_x(t) = 1 + t\mu_1' + \frac{t^2}{2!}\mu_2' + \cdots + \frac{t^k}{k!}\mu_k' + \cdots,$$

if the moments μ_1', μ_2', ... are all finite. That is, the coefficient of $t^k/k!$ in the expansion of the moment generating function (m.g.f.) is the kth moment about the origin of x.

If we differentiate $M_x(t)$ k times, we obtain (assuming differentiability under integral or summation signs),

(6.10.5) $$\frac{d^k}{dt^k} M_x(t) = E(x^k e^{xt}),$$

and if we then set $t = 0$, we have

(6.10.6) $$\frac{d^k}{dt^k} M_x(0) = E(x^k) = \mu_k'.$$

* Moment generating functions need not always exist. The sum in (6.10.1) or the integral in (6.10.2) may not converge. In such cases, one may define the *characteristic function* of x, which always exists, by replacing t by it, where $i = \sqrt{-1}$. For further details, see Cramer (1946).

For example, suppose x has the Poisson distribution. Then

$$M_x(t) = \sum_{x=0}^{\infty} e^{xt} \frac{e^{-\mu}\mu^x}{x!}$$

$$= e^{-\mu} \sum_{x=0}^{\infty} \frac{(\mu e^t)^x}{x!}$$

$$= e^{-\mu} e^{\mu e^t} = e^{\mu(e^t-1)}.$$

Thus,

$$\frac{dM_x(t)}{dt} = \mu e^{\mu(e^t-1)+t}$$

$$\frac{d^2M_x(t)}{dt^2} = \mu\{(\mu e^t + 1)e^{\mu(e^t-1)+t}\}.$$

Hence

$$\mu_1' = \frac{dM_x(0)}{dt} = \mu$$

and

$$\mu_2' = \frac{d^2M_x(0)}{dt^2} = \mu[(\mu + 1)] = \mu^2 + \mu.$$

From (6.2.2a), we have that

$$\text{Var}(x) = E(x^2) - [E(x)]^2$$

$$= \mu_2' - (\mu)^2 = \mu.$$

Thus, the mean and variance of a Poisson random variable x are both equal to μ.

We now mention some important properties of the m.g.f. of x, $M_x(t)$. First, if we are interested in the random variable cx, where c is any constant, then by definition

(6.10.7) $M_{cx}(t) = E(e^{cxt}) = E(e^{xct}) = M_x(ct).$

Second, the m.g.f. for $x + a$, where a is constant, is

(6.10.8) $M_{x+a}(t) = E(e^{(x+a)t}) = E(e^{at}e^{xt}) = e^{at}M_x(t).$

Note that (6.10.8) enables us to find moments of x about its mean. For if we set $a = -\mu$, we have

$$M_{x-\mu}(t) = E(e^{(x-\mu)t})$$

and straightforward differentiation shows that

(6.10.9) $$\frac{d^kM_{x-\mu}(0)}{dt^k} = \mu_k.$$

Third, suppose x_1 and x_2 are two independent random variables and c_1 and c_2 are constants. Let $M_{x_1}(t)$ and $M_{x_2}(t)$ be the m.g.f.'s of x_1 and x_2 respectively. Then, the m.g.f. of the random variable

(6.10.10) $$L = c_1 x_1 + c_2 x_2$$

is

$$M_L(t) = E(e^{(c_1 x_1 + c_2 x_2)t}).$$

Using Theorem 6.7.1 of Section 6.7, we have

(6.10.11) $$M_L(t) = E(e^{x_1 c_1 t})E(e^{x_2 c_2 t}) = M_{x_1}(c_1 t)M_{x_2}(c_2 t).$$

Indeed, if we are interested in the linear combination

$$L = \sum_{i=1}^{k} c_i x_i,$$

where x_i, $i = 1, \ldots, k$, are independent random variables, we find similarly that

(6.10.12) $$M_L(t) = \prod_{i=1}^{k} M_{x_i}(c_i t).$$

The moment generating function has an important property given by the following uniqueness theorem, which we state without proof:

THEOREM 6.10.1.   *If two random variables x and y have the same moment generating function M(t), then their c.d.f.'s are identical.*

6.11 SOME PROOFS AND DERIVATIONS

(a) Mean and Variance of the Hypergeometric Distribution

Given the hypergeometric distribution having p.f.

(6.11.1) $$h(x) = \frac{\dbinom{N\theta}{x}\dbinom{N(1-\theta)}{n-x}}{\dbinom{N}{n}},$$

we wish to show that its mean μ and variance σ^2 are given by

$$\mu = n\theta,$$

$$\sigma^2 = \frac{N-n}{N-1}n\theta(1-\theta).$$

To find the mean we must evaluate the sum

$$\mu = \Sigma\, xh(x),$$

where the sum extends over the entire sample space of x.

Since $h(x)$ is a probability distribution, we have

$$\Sigma\, h(x) = 1,$$

and hence

(6.11.2) $$\Sigma \binom{N\theta}{x}\binom{N(1-\theta)}{n-x} = \binom{N}{n}.$$

To evaluate $\Sigma\, xh(x)$, we must consider the sum

(6.11.3) $$\Sigma\, x\binom{N\theta}{x}\binom{N(1-\theta)}{n-x}.$$

But $x\binom{N\theta}{x} = N\theta\binom{N\theta-1}{x-1}$ for $x \neq 0$, and hence the sum in (6.11.3) may be expressed as

(6.11.4) $$N\theta \Sigma \binom{N\theta-1}{x-1}\binom{N(1-\theta)}{(n-1)-(x-1)}.$$

It will be seen from (6.11.2) that the sum in (6.11.4) has the value $N\theta\binom{N-1}{n-1}$. Hence

(6.11.5) $$\mu = \frac{N\theta\binom{N-1}{n-1}}{\binom{N}{n}}$$

which reduces to $\mu = n\theta$.

Using (6.2.2a) for σ^2, it is to be noted that

(6.11.6) $$\sigma^2 = \Sigma\, x^2 h(x) - (n\theta)^2$$

$$= \Sigma\, x(x-1)h(x) + n\theta - (n\theta)^2.$$

The problem of obtaining the value of σ^2 for the hypergeometric distribution reduces to the evaluation of $\Sigma\, x(x-1)h(x)$ which, in turn, reduces to the evaluation of

(6.11.7) $$\Sigma\, x(x-1)\binom{N\theta}{x}\binom{N(1-\theta)}{n-x}.$$

But $x(x-1)\binom{N\theta}{x} = (N\theta)(N\theta-1)\binom{N\theta-2}{x-2}$ for $x \geq 2$, and, hence, (6.11.7) reduces to

$$(N\theta)(N\theta-1)\Sigma \binom{N\theta-2}{x-2}\binom{N(1-\theta)}{(n-2)-(x-2)} = (N\theta)(N\theta-1)\binom{N-2}{n-2}.$$

Hence, we find

$$\sigma^2 = (N\theta)(N\theta - 1)\frac{\binom{N-2}{n-2}}{\binom{N}{n}} + n\theta - (n\theta)^2,$$

which simplifies to

(6.11.8) $$\sigma^2 = \frac{N-n}{N-1}n\theta(1-\theta).$$

(b) Mean and Variance of the Binomial Distribution

Given the binomial distribution having p.f.

(6.11.9) $$b(x) = \binom{n}{x}\theta^x(1-\theta)^{n-x}, \qquad x = 0, 1, \ldots, n$$

we wish to show that its mean μ and variance σ^2 are given by

$$\mu = n\theta,$$
$$\sigma^2 = n\theta(1-\theta).$$

The mean is defined as

(6.11.10) $$\mu = \sum_{x=0}^{n} x\binom{n}{x}\theta^x(1-\theta)^{n-x}.$$

But

$$x\binom{n}{x} = n\binom{n-1}{x-1}.$$

Hence

$$\mu = n\theta \sum_{x=1}^{n}\binom{n-1}{x-1}\theta^{x-1}(1-\theta)^{(n-1)-(x-1)}$$

$$= n\theta[(1-\theta) + \theta]^{n-1}.$$

That is,

(6.11.11) $$\mu = n\theta.$$

The variance σ^2 is given by (using the same device as in (6.11.6), viz., $x^2 = x(x-1) + x$)

(6.11.12) $$\sigma^2 = \sum_{x=0}^{n} x(x-1)b(x) + n\theta - (n\theta)^2,$$

since $\sum_{x=0}^{n} xb(x) = \mu$. Inserting the expression for $b(x)$, we obtain for the first term of the right-hand side of (6.11.12),

$$\sum_{x=0}^{n} x(x-1)\binom{n}{x}\theta^x(1-\theta)^{n-x},$$

which reduces to

$$n(n-1)\theta^2 \sum_{x=2}^{n} \binom{n-2}{x-2} \theta^{x-2}(1-\theta)^{(n-2)-(x-2)} = n(n-1)\theta^2[(1-\theta)+\theta]^{n-2}$$
$$= n(n-1)\theta^2.$$

Hence we obtain

$$\sigma^2 = n(n-1)\theta^2 + n\theta - (n\theta)^2,$$

which reduces to

(6.11.13) $$\sigma^2 = n\theta(1-\theta).$$

(c) Mean and Variance of the Poisson Distribution

Given the Poisson distribution having p.f.

(6.11.14) $$p(x) = e^{-\mu} \frac{\mu^x}{x!}, \qquad x = 0, 1, 2, \ldots,$$

we wish to show that its mean and variance both have the value μ (the parameter appearing in the distribution itself).

To obtain the value of the mean, we must evaluate the sum

$$\sum_{x=0}^{\infty} x e^{-\mu} \frac{\mu^x}{x!} = \sum_{x=1}^{\infty} x \frac{e^{-\mu}\mu^x}{x!},$$

which reduces to

$$\mu e^{-\mu} \sum_{(x-1)=0}^{\infty} \frac{\mu^{(x-1)}}{(x-1)!} = \mu e^{-\mu}e^{\mu} = \mu.$$

That is, the mean of the Poisson distribution (6.11.14) is μ.

The variance of the Poisson distribution (6.11.14) is given by

$$\sum_{x=0}^{\infty} x(x-1)e^{-\mu}\frac{\mu^x}{x!} + \mu - \mu^2.$$

But the sum reduces to

$$\mu^2 e^{-\mu} \sum_{(x-2)=0}^{\infty} \frac{\mu^{(x-2)}}{(x-2)!} = \mu^2 e^{-\mu}e^{\mu} = \mu^2.$$

Hence, the variance of the Poisson distribution is also μ.

6.12 TRANSFORMED RANDOM VARIABLES AND THEIR DISTRIBUTIONS

It is often the case that a statistical investigation begins with the consideration of a random variable, say x, with related distribution function $f(x)$, but that interest soon focuses on a specific function of x, say $y = g(x)$. Since x is a random variable, then, of course, $y = g(x)$ is also a random variable, and the natural question that now arises is: what is the distribution of the random variable y? Further, what is the sample space of y? We sometimes refer to y as a *transformed* variable, and the transformation g is, of course, a transformation on the random variable x.

Or it may be that we are working with a pair of random variables, say x_1 and x_2, but are actually interested in the random variable $z = g(x_1, x_2)$, and would like to know the distribution of z. Here, z is a transformed variable, and the transformation is on the pair of random variables (x_1, x_2). [A simple illustration is $z = g(x_1, x_2) = (x_1 + x_2)/2$, the sample mean of x_1 and x_2.] Of course, it may also be the case that there is interest in the pair of random variables $z = g(x_1, x_2)$ and $w = v(x_1, x_2)$; for example, $z = (x_1 + x_2)/2$ and $w = (x_1 - x_2)/2$. Here, we may be motivated not only to find the joint distribution of the transformed variables z and w, but also we may wish to find their marginal distributions, using the procedures of Section 6.7.

It turns out that questions of the above type may be answered in a variety of ways. In this section, we discuss and illustrate a method for attacking this sort of problem, and do this for both discrete and continuous cases. Our discussion will be mainly in terms of the "one and two variables" cases, but extensions to three or more variables will be indicated.

(a) Discrete Random Variables

(*i*) *One Variable.* We begin our discussion with the following simple illustration. After paying a certain entrance fee, players are to throw a die, and the number of dots, say x, that appears on the uppermost face is recorded, whereupon the player that tosses the die receives an amount $y = 5x - 3$, in dollars. What is the distribution of the payoffs y?

Now the transformation $y = 5x - 3$ *maps* the sample space $S_x = \{1, 2, 3, 4, 5, 6\}$ of the random variable x into the sample space S_y of the random variable y, where $S_y = \{2, 7, 12, 17, 22, 27\}$. Notice that, if given that the element y_0 belongs to the sample space S_y of y, (denoted by $y_0 \in S_y$), there is only one element in S_x that corresponds to it (indeed it is the element $x = (y_0 + 3)/5$), and similarly, given an element $x_0 \in S_x$, there is only one corresponding element in S_y, namely $y = 5x_0 - 3$. We say that the transformation $y = g(x) = 5x - 3$ is a one-one transformation on x, with (single) *inverse* $x = u(y) = (y + 3)/5$.

We seek the probability function of y, and we have that the probability of the event x attaining the value x_0 is

(6.12.1) $\qquad P(x = x_0) = p(x_0) = 1/6 \qquad$ for all $x_0 \in S_x$.

Hence, the probability function for y, say $h(y)$, may be obtained as follows: if $y_0 \in S_y$, then

(6.12.2) $\quad h(y_0) = P(y = y_0) = P(5x - 3 = y_0) = P(x = (y_0 + 3)/5)$.

But the expression on the extreme right hand side of (6.12.2) is the probability of the random variable x attaining the value $(y_0 + 3)/5$, and may be evaluated by using the probability function of x, that is, we have

(6.12.3) $\qquad h(y_0) = p((y_0 + 3)/5) = 1/6 \qquad$ for $y_0 \in S_y$

or, dropping subscripts,

(6.12.3a) $\quad h(y) = 1/6 \qquad$ for $\qquad y \in S_y = \{2, 7, 12, 17, 22, 27\}$.

The above is an illustration of the application of the following theorem.

THEOREM 6.12.1. *Suppose the discrete random variable* x, *with sample space* S_x *has probability function (p.f.)* $p(x)$. *Suppose that* $y = g(x)$, *is a one-one transformation on* x, *with inverse* $x = u(y)$, *into the sample space of* y, *say* S_y. *Then* $h(y)$, *the p.f. of the random variable* y, *is given by*

(6.12.4) $\qquad h(y) = p(u(y)), \qquad$ for $\qquad y \in S_y$.

Example 6.12(a). Suppose x has the binomial distribution

(6.12.5) $\quad b(x) = \binom{n}{x} \theta^x (1 - \theta)^{n-x}, \qquad x \in S_x = \{0, 1, 2, \ldots, n\}$.

Let $y = x^2$. Find the distribution of y. It is clear that $S_y = \{0, 1, 4, 9, \ldots, n^2\}$. Further, notice that in this example, $y = x^2$ defines a one-one transformation on x, with inverse $x = \sqrt{y}$, for there are no negative values in S_x. Using (6.12.4), then, we have that the p.f. of y is

(6.12.6) $\qquad h(y) = b(\sqrt{y}) = \binom{n}{\sqrt{y}} \theta^{\sqrt{y}} (1 - \theta)^{n-\sqrt{y}}$

for all $y \in S_y = \{0, 1, 4, 9, \ldots, n^2\}$.

(ii) *Two or More Variables.* Suppose now that $p(x_1, x_2)$ is the joint probability function of the random variables x_1 and x_2, i.e.,

$$P(x_1 = x_{10}, x_2 = x_{20}) = p(x_{10}, x_{20}).$$

The sample space S_{x_1, x_2} of (x_1, x_2) is a two-dimensional set of points and is such that if $(x_1, x_2) \in S_{x_1, x_2}$, then, $p(x_1, x_2) > 0$ and $\Sigma\Sigma\, p(x_1, x_2) = 1$, where $\Sigma\Sigma$ denotes the operation of summation over all points $(x_1, x_2) \in S_{x_1, x_2}$. Further, suppose $y_1 = g_1(x_1, x_2)$ and $y_2 = g_2(x_1, x_2)$ defines a one-one transformation which maps points in S_{x_1, x_2} into a two-dimensional set S_{y_1, y_2}, the sample space of the pair of random variables (y_1, y_2). Let the (single) inverse of the transformation be denoted by $x_1 = u_1(y_1, y_2)$, $x_2 = u_2(y_1, y_2)$. That is to say, for each element or point $(x_{10}, x_{20}) \in S_{x_1, x_2}$, there is one and only one element $(y_{10}, y_{20}) \in S_{y_1, y_2}$ such that $y_{10} = g_1(x_{10}, x_{20})$ and $y_{20} = g_2(x_{10}, x_{20})$, and vice versa, that is, for each element $(y_{10}, y_{20}) \in S_{y_1, y_2}$ there is one and only one element $(x_{10}, x_{20}) \in S_{x_1, x_2}$ such

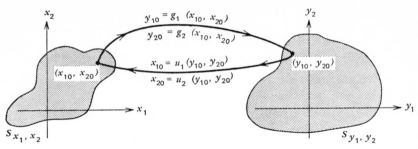

Figure 6.12.1. The one-one transformation $y_1 = g_1(x_1, x_2)$, $y_2 = g_2(x_1, x_2)$ with inverse $x_1 = u_1(y_1, y_2)$, $x_2 = u_2(y_1, y_2)$, maps (x_{10}, x_{20}) into (y_{10}, y_{20}), and (y_{10}, y_{20}) into (x_{10}, x_{20}).

that $x_{10} = u_1(y_{10}, y_{20})$ and $x_{20} = u_2(y_{10}, y_{20})$; see Figure 6.12.1. Indeed, if (x_{10}, x_{20}) corresponds to (y_{10}, y_{20}) under the one-one transformation $y_i = g_i(x_1, x_2)$, $i = 1, 2$, as in Figure 6.12.1 then we have that

$$(6.12.7) \quad P(y_1 = y_{10}, y_2 = y_{20}) = P(x_1 = x_{10}, x_2 = x_{20})$$
$$= P(x_1 = u_1(y_{10}, y_{20}), x_2 = u_2(y_{10}, y_{20})),$$

that is

$$(6.12.7a) \quad P(y_1 = y_{10}, y_2 = y_{20})$$
$$= p(u_1(y_{10}, y_{20}), u_2(y_{10}, y_{20})), (y_{10}, y_{20}) \in S_{y_1, y_2}.$$

Dropping the "0" part of the subscripts, we have from (6.12.7a) that the probability function h of (y_1, y_2) is

$$(6.12.7b) \qquad h(y_1, y_2) = p(u_1(y_1, y_2), u_2(y_1, y_2)), (y_1, y_2) \in S_{y_1, y_2}.$$

We summarize the above in the following theorem.

THEOREM 6.12.2. *Suppose the random variables* (x_1, x_2) *have joint probability function* $p(x_1, x_2)$, *and that* $y_1 = g_1(x_1, x_2)$, $y_2 = g_2(x_1, x_2)$ *is a one-one transformation on* (x_1, x_2), *with inverse* $x_1 = u_1(y_1, y_2)$, $x_2 = u_2(y_1, y_2)$. *Suppose further that this transformation maps the sample space of* (x_1, x_2), *say* S_{x_1, x_2}, *into the sample space of* (y_1, y_2), *say* S_{y_1, y_2}. *Then, the probability function of* (y_1, y_2) *is given by*

$$(6.12.8) \qquad h(y_1, y_2) = p(u_1(y_1, y_2), u_2(y_1, y_2)), (y_1, y_2) \in S_{y_1, y_2}.$$

We remark that we may obtain the marginal distribution of y_1 by summing out the variable y_2 in (6.12.8); see Section 6.7. That is, fixing y_1 at y_{10}, we may obtain

$$(6.12.9) \qquad h_1(y_{10}) = P(y_1 = y_{10}) = \sum^1 h(y_{10}, y_2),$$

where here the summation $\sum^1$ denotes the operation of summation over all points in the sample space S_{y_1, y_2} for which the first coordinate $y_1 = y_{10}$. Hence Theorem 6.12.2 suggests a possible procedure for finding the distribution of a random variable of interest, say $y_1 = g_1(x_1, x_2)$, as follows:

(a) define another variable, say $y_2 = g_2(x_1, x_2)$ so that g_1, g_2 constitutes a one-one transformation on (x_1, x_2);

(b) find the joint probability function h of (y_1, y_2) using Theorem 6.12.2;

(c) find the marginal of y_1 by summing out y_2 in $h(y_1, y_2)$.

Example 6.12(b). Suppose x_1 and x_2 are two independent observations on a Poisson random variable with parameter μ, and that we wish to find the distribution of $y_1 = x_1 + x_2$. We proceed as follows. Let $y_2 = x_2$. Then the transformation $y_1 = x_1 + x_2$, $y_2 = x_2$ is one-one, and has inverse $x_1 = y_1 - y_2$, $x_2 = y_2$, and maps the sample space $S_{x_1, x_2} = \{(x_1, x_2) \mid (x_1, x_2)$ nonnegative integers$\}$ into the sample space $S_{y_1, y_2} = \{(y_1, y_2) \mid 0 \leq y_2 \leq y_1; y_1, y_2$ nonnegative integers$\}$. This may be seen as follows. First, since x_1 and x_2 are independent and each are Poisson variables, the sample space of S_{x_1, x_2} is as defined above. Now from $y_1 = x_1 + x_2$, $y_2 = x_2$, we easily find that $x_2 = y_2$, $x_1 = y_1 - y_2$. Since x_2 is Poisson, we have $0 \leq x_2$, so that $0 \leq y_2$. Since x_1 is Poisson, we have $0 \leq x_1$, so that $0 \leq y_1 - y_2$, that is $y_2 \leq y_1$. Combining these statements we have $0 \leq y_2 \leq y_1$ (see Figure 6.12.2).

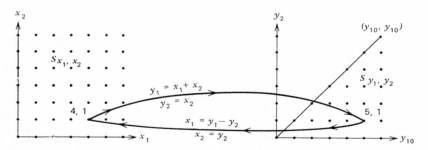

Figure 6.12.2. The transformation $y_1 = x_1 + x_2, y_2 = x_2$ maps $S_{x_1, x_2} = \{(x_1, x_2) \mid x_1, x_2$ nonnegative integers$\}$ in one-one fashion into $S_{y_1, y_2} = \{(y_1, y_2) \mid 0 \leq y_2 \leq y_1; y_1, y_2$ nonnegative integers$\}$.

Again, since x_1 and x_2 are independent Poisson, with mean μ, we have that

$$(6.12.10) \qquad p(x_1, x_2) = \frac{e^{-\mu} \mu^{x_1}}{x_1!} \cdot \frac{e^{-\mu} \mu^{x_2}}{x_2!}, \qquad (x_1, x_2) \in S_{x_1, x_2}.$$

From Theorem 6.12.2, then, we have

$$(6.12.11) \qquad h(y_1, y_2) = \frac{e^{-2\mu} \mu^{y_1}}{(y_1 - y_2)! \, y_2!}, \qquad (y_1, y_2) \in S_{y_1, y_2}.$$

Now since we are interested in y_1 only, we wish to use (6.12.11) to find the marginal of y_1. We have, fixing $y_1 = y_{10}$, that

(6.12.12) $$h_1(y_{10}) = \sum^1 \frac{e^{-2\mu} \mu^{y_{10}}}{(y_{10} - y_2)! \, y_2!} ,$$

where here the summation is over all points in sample space for which $y_1 = y_{10}$. Consulting Figure 6.12.2, we see that this summation implies that we are to sum over y_2, where $0 \le y_2 \le y_{10}$. Hence

(6.12.12a) $$h_1(y_{10}) = e^{-2\mu} \mu^{y_{10}} \sum_{y_2=0}^{y_{10}} \frac{1}{(y_{10} - y_2)! \, y_2!} .$$

Multiplying by 1 in the form $y_{10}!/y_{10}!$, we have

(6.12.12b) $$h_1(y_{10}) = \frac{e^{-2\mu} \mu^{y_{10}}}{y_{10}!} \left\{ \sum_{y_2=0}^{y_{10}} \frac{y_{10}!}{y_2! \, (y_{10} - y_2)!} \right\} ,$$

and the summation inside the braces { } of (6.12.12b) is clearly the binomial expansion of $(1 + 1)^{y_{10}} = 2^{y_{10}}$. Thus,

(6.12.12c) $$h_1(y_{10}) = \frac{e^{-2\mu} (2\mu)^{y_{10}}}{y_{10}!}$$

where y_{10} is any integer satisfying $0 \le y_{10}$. Dropping the "0" part of all subscripts, we have

(6.12.12d) $\quad h_1(y_1) = \dfrac{e^{-2\mu} (2\mu)^{y_1}}{y_1!}$, $y_1 \in S_{y_1} = \{y_1 \mid 0 \le y_1, \, y_1 \text{ integer}\}$,

that is, $y_1 = x_1 + x_2$ is a Poisson random variable, mean 2μ, when (x_1, x_2) is a random sample of 2 independent observations on a Poisson variable with mean μ.

The extension to more than two variables is now obvious. If $p(x_1, \ldots, x_n)$ is the joint probability function of the random variables $(x_1, \ldots, x_n)$, and interest lies in the functions $y_t = g_t(x_1, \ldots, x_n)$, $t = 1, \ldots, m$, where $m < n$, then define $n - m$ functions $y_{m+h} = g_{m+h}(x_1, \ldots, x_n)$, $h = 1, \ldots, n - m$ so that the transformation $y_i = g_i(x_1, \ldots, x_n)$, $i = 1, \ldots, m, \ldots, n$ is a one to one transformation from $S_{x_1, \ldots, x_n}$, the sample space of $(x_1, \ldots, x_n)$, into $S_{y_1, \ldots, y_n}$, the sample space of $y_1, \ldots, y_n$. We then may find the joint distribution of $y_1, \ldots, y_n$, say h, where h is given by

(6.12.13) $\quad h(y_1, \ldots, y_n) = p(u_1(y_1, \ldots, y_n), \ldots,$

$$u_n(y_1, \ldots, y_n)), \, (y_1, \ldots, y_n) \in S_{y_1, \ldots, y_n}$$

where $x_1 = u_1(y_1, \ldots, y_n), \ldots, x_n = u_n(y_1, \ldots, y_n)$ denotes the inverse of the transformation. If interest lies in $(y_1, \ldots, y_m)$, $m < n$, we simply "sum" out $y_{m+1}, \ldots, y_n$ from (6.12.13). For the case where the transformation $y_i = g_i(x_1, \ldots, x_n)$, $i = 1, \ldots, n$, $n \ge 1$, is not one-one, see Section 6.12(c).

(b) Continuous Random Variables

The case of transformations on continuous random variables, is not different, in principle, from the case of discrete random variables, as one would expect. (For example, the same care must be taken with the sample space of the transformed variables, etc.) However, one new ingredient emerges when dealing with continuous random variables, namely, account must be taken of the so-called *Jacobian* of the transformation. This arises in a natural way, which we now discuss.

(i) One Variable. Suppose x is a continuous random variable with probability density function (p.d.f.) $f(x)$, and denote the sample space of x by S_x, i.e., if $x \in S_x$, $f(x) > 0$, and $\int_{S_x} f(x)\, dx = 1$. Further suppose that interest is in the random variable $y = g(x)$, where $g(x)$ is a one-one transformation on x, with single inverse $x = u(y)$, which maps S_x into the sample space, S_y, of y. Denote the cumulative distribution function (c.d.f.) of x by F; and the cumulative distribution function of y by H. We have that for $y_0 \in S_y$,

$$(6.12.14) \qquad H(y_0) = P(y \leq y_0) = P(g(x) \leq y_0).$$

Two cases may now arise, namely (i) $g(x)$ is an increasing function of x, in which case the event $g(x) \leq y_0$ is equivalent to the event $x \leq u(y_0)$, or (ii) $g(x)$ is a decreasing function of x, in which case the event $g(x) \leq y_0$ is equivalent to the event $x \geq u(y_0)$. Hence, we have, from (6.12.14), that

$$(6.12.15) \quad H(y_0) = \begin{cases} P(x \leq u(y_0)) & \text{if } g \text{ is increasing in } x \\ 1 - P(x < u(y_0)) & \text{if } g \text{ is decreasing in } x, \end{cases}$$

or, dropping subscripts, we have for $y \in S_y$,

$$(6.12.15a) \qquad H(y) = \begin{cases} F(u(y)) & \text{if } g \text{ is increasing} \\ 1 - F(u(y)) & \text{if } g \text{ is decreasing}. \end{cases}$$

Further, in either case, we find on differentiating the above, that the p.d.f. of y may be written as

$$(6.12.16) \qquad h(y) = f(u(y)) \cdot |J| \qquad y \in S_y$$

where the factor $|J|$ is the absolute value of the Jacobian of the transformation $y = g(x)$, given by $J = u'(y) = du(y)/dy = dx/dy$. The absolute value is taken to ensure that we preserve the property that $h(y) > 0$ for all $y \in S_y$. We summarize the important result (6.12.16) in the following theorem.

THEOREM 6.12.3. *Suppose x has p.d.f. $f(x)$, c.d.f. $F(x)$, and that $y = g(x)$ is a one-one transformation on x, (single) inverse $x = u(y)$, which maps the sample space S_x into S_y. Then the p.d.f. of y is given by the equation (6.12.16).*

Very often in succeeding chapters we will deal with the case of $y = g(x) = ax + b$ and hence state the result as a corollary to the above theorem.

COROLLARY 6.12.3(a). *If $y = ax + b$, then the p.d.f. of y is*

$$(6.12.17) \qquad h(y) = f\left(\frac{y - b}{a}\right) \Big/ |a| \qquad \text{for } y \in S_y.$$

Example 6.12(c). Suppose x is uniform on $[0, 1]$, that is, we have that

$$f(x) = 1 \qquad \text{if } x \in S_x = \{x \mid 0 \le x \le 1\}.$$

Let $y = x^{-\frac{1}{2}}$. Then clearly y maps S_x into $S_y = \{y \mid 1 \le y < \infty\}$, in one-one fashion, with inverse $x = y^{-2}$. The Jacobian is $J = dx/dy = -2y^{-3}$, and hence $|J| = 2y^{-3}$. Using (6.12.16), we have

$$h(y) = 1 \times 2y^{-3} \qquad \text{if } y \in S_y.$$

Example 6.12(d). Suppose x has p.d.f. given by

$$f(x) = \frac{1}{\sqrt{2\pi}} e^{-x^2/2}, \qquad -\infty < x < \infty,$$

i.e., $S_x = \{x \mid -\infty < x < \infty\}$. Let $y = 3x + 4$. Then $x = -\frac{1}{3}(y - 4)$ (and $J = -\frac{1}{3}$). Using (6.12.16), or (6.12.17), we have

$$h(y) = \frac{1}{\sqrt{2\pi}} e^{-(y-4)^2/9\times 2} \Big/ |-3|, \qquad y \in S_y = \{y \mid \quad \infty < y < \infty\}$$

or

$$h(y) = \frac{1}{3\sqrt{2\pi}} e^{-(y-4)^2/18}, \qquad y \in S_y.$$

(ii) Two or More Random Variables. Suppose now that a pair of continuous random variables (x_1, x_2), has joint p.d.f. $f(x_1, x_2)$, where of course, $f(x_1, x_2) > 0$ for all points $(x_1, x_2) \in S_{x_1.x_2}$, the sample space of (x_1, x_2), with $\iint_{S_{x_1.x_2}} f(x_1, x_2)\, dx_1\, dx_2 = 1$. Suppose further that $y_1 = g_1(x_1, x_2)$, $y_2 = g_2(x_1, x_2)$ defines a one-one transformation from $S_{x_1.x_2}$ into the two-dimensional set $S_{y_1.y_2}$, the sample space of the continuous random variables y_1 and y_2, with inverse $x_1 = u_1(y_1, y_2)$, $x_2 = u_2(y_1, y_2)$. To repeat and reemphasize, the transformation $y_i = g_i(x_1, x_2)$ is assumed one-one, which means that the transformation establishes a one-one correspondence between, say, the element (point) $(x_{10}, x_{20}) \in S_{x_1.x_2}$, and $(y_{10}, y_{20}) \in S_{y_1.y_2}$

where $y_{10} = g_1(x_{10}, x_{20})$, $y_{20} = g_2(x_{10}, x_{20})$, or $x_{10} = u_1(y_{10}, y_{20})$, $x_{20} = u_2(y_{10}, y_{20})$; see Figure 6.12.1. We denote the p.d.f. of (y_1, y_2) by $h(y_1, y_2)$.

Now let A be an event (subset) of S_{x_1, x_2} and suppose the transformation $y_i = g_i(x_1, x_2)$, $i = 1, 2$, maps A into B, where B is an event in S_{y_1, y_2}. Because the transformation is one-one, that is, a point in A is mapped into a unique point in B, and vice versa, we have that "the events A and B are equivalent," that is,

$$P((y_1, y_2) \in B) = P((x_1, x_2) \in A)$$

or

$$\iint_B h(y_1, y_2) \, dy_1 \, dy_2 = \iint_A f(x_1, x_2) \, dx_1 \, dx_2.$$

If we make the one-one transformation $x_i = u_i(y_1, y_2)$, $i = 1, 2$ in the integral on the right in the above equation, we have, from a theorem in mathematical analysis, that

$$6.12.18) \quad \iint_B h(y_1, y_2) \, dy_1 \, dy_2 = \iint_B f(u_1(y_1, y_2), u_2(y_1, y_2)) \, |J| \, dy_1 \, dy_2$$

where $J = \dfrac{\partial(x_1, x_2)}{\partial(y_1, y_2)}$ is the *Jacobian of the transformation* and given by

$$(6.12.19) \quad J = \frac{\partial(x_1, x_2)}{\partial(y_1, y_2)} = \det \begin{pmatrix} \dfrac{\partial x_1}{\partial y_1} & \dfrac{\partial x_1}{\partial y_2} \\ \dfrac{\partial x_2}{\partial y_1} & \dfrac{\partial x_2}{\partial y_2} \end{pmatrix}.$$

[In general, $\det \begin{pmatrix} a_{11} & a_{12} \\ a_{21} & a_{22} \end{pmatrix}$ stands for the *determinant* of the so called *two by two matrix* $\begin{pmatrix} a_{11} & a_{12} \\ a_{21} & a_{22} \end{pmatrix}$ and has value $a_{11} a_{22} - a_{21} a_{12}$. Sometimes the matrix is written as (a_{ij}), with the understanding that i and j runs over the subscripts 1, 2 and that a_{ij} is the element in the ith row and jth column of the matrix. In (6.12.19), the matrix (a_{ij}) that is involved has typical element $a_{ij} = \dfrac{\partial x_i}{\partial y_j}$. For a $n \times n$ matrix (a_{ij}), with $n \geq 3$, we define $\det (a_{ij})$ as

$$(6.12.20) \quad \det (a_{ij}) = \sum \pm a_{1i_1} a_{2i_2} \cdots a_{ni_n},$$

where the summation is over all $n!$ permutations $(i_1, i_2, \ldots, i_n)$ of the n integers $(1, 2, \ldots, n)$, with a plus sign if the permutation is an *even* one, and a minus sign if the permutation is an *odd* one. A permutation is deemed to be even if it may be put in the natural order $(1, \ldots, n)$ by an even number of inversions. For example, the permutation 231 is even, since the inversions needed to bring about the natural order 1, 2, 3 are two in number, viz,

$$231 \to 213 \to 123,$$

that is, first inverting the "3" with the "1," and second, inverting the "2" with the "1." Of course, if a permutation is not even, it is odd; for example, the permutation 213 needs only the inversion of the "2" with the "1" to bring it into the order 123, and so we call the permutation 213 an odd permutation. We will define the permutation $123 \cdots n$ as being even.

Hence, if $n = 3$, then from (6.12.20) we have that

(6.12.21)
$$\det (a_{ij}) = \det \begin{pmatrix} a_{11} & a_{12} & a_{13} \\ a_{21} & a_{22} & a_{23} \\ a_{31} & a_{32} & a_{33} \end{pmatrix}$$
$$= a_{11}a_{22}a_{33} + a_{12}a_{23}a_{31} + a_{13}a_{21}a_{32}$$
$$- (a_{13}a_{22}a_{31} + a_{11}a_{23}a_{32} + a_{12}a_{21}a_{33}).$$

For example, the fourth term in (6.12.21) is $-a_{13}a_{22}a_{31}$, and is given a minus sign since $(i_1 \quad i_2 \quad i_3) = 321$ is an odd permutation—indeed 321 needs the 3 inversions $321 \to 312$, $312 \to 132$ and $132 \to 123$ to bring it into the natural permutation 123.]

Returning to (6.12.18) and (6.12.19), we have that since (6.12.18) holds for all sets $B \subset S_{y_1,y_2}$ that

(6.12.22) $h(y_1, y_2) = f(u_1(y_1, y_2), u_2(y_1, y_2)) |J|,$ $(y_1, y_2) \in S_{y_1,y_2}.$

We summarize by stating the following theorem.

THEOREM 6.12.4.   *If the pair of random variables (x_1, x_2) has joint p.d.f. $f(x_1, x_2)$, and if $y_i = g_i(x_1, x_2)$ is a one-one transformation on (x_1, x_2), with inverse $x_i = u_i(y_1, y_2)$, $i = 1, 2$, which maps the sample space of (x_1, x_2), S_{x_1,x_2}, in one-one fashion into S_{y_1,y_2}, the sample space of (y_1, y_2), then the p.d.f. h of the pair of random variables (y_1, y_2) is given by (6.12.22) and (6.12.19).*

We note here in passing that a helpful property of J that is often used, is

(6.12.23)
$$J = \frac{\partial(x_1, x_2)}{\partial(y_1, y_2)} = 1 \bigg/ \frac{\partial(y_1, y_2)}{\partial(x_1, x_2)}$$

where of course,

$$\frac{\partial(y_1, y_2)}{\partial(x_1, x_2)} = \det \left(\frac{\partial y_i}{\partial x_j} \right), \qquad i, j = 1, 2.$$

Also, note that the above theorem may be used as a technique for determining the distribution of a variable of interest $y_1 = g_1(x_1, x_2)$. For by choosing $y_2 = g_2(x_1, x_2)$ so that $y_i = g_i(x_1, x_2)$, $i = 1, 2$, constitutes a one-one transformation on (x_1, x_2), we may find $h(y_1, y_2)$ from (6.12.22), and once obtained, we may utilize $h(y_1, y_2)$ to find the marginal of y_1 by integrating out y_2 from $h(y_1, y_2)$ in the prescribed manner; see Section 6.7.

Example 6.12(e).   Suppose (x_1, x_2) has joint p.d.f. f given by

$$f(x_1, x_2) = 1 \qquad \text{if } (x_1, x_2) \in S_{x_1.x_2} = \{(x_1, x_2) \mid 0 \le x_i \le 1, i = 1, 2\}.$$

Find the joint p.d.f. of (y_1, y_2) where

$$y_1 = x_1 + x_2, \quad y_2 = x_1 - x_2.$$

We note that the inverse of the above transformation is $x_1 = (y_1 + y_2)/2$, $x_2 = (y_1 - y_2)/2$, and that $S_{x_1.x_2}$ is mapped into $S_{y_1.y_2}$, where $S_{y_1.y_2}$ is the two-dimensional set in y_1, y_2 space bounded by the lines $y_1 = -y_2$, $y_1 = y_2$, $y_1 = 2 - y_2$ and $y_1 = 2 + y_2$. This may be seen as follows:

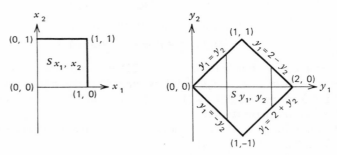

Figure 6.12.3. The transformation $y_1 = x_1 + x_2, y_2 = x_1 - x_2$ maps $S_{x_1.x_2}$ into $S_{y_1.y_2}$ in one-one fashion, and conversely, $S_{y_1.y_2}$ is mapped into $S_{x_1.x_2}$ by the inverse transformation $x_1 = (y_1 + y_2)/2, x_2 = (y_1 - y_2)/2$.

We have that

$$\begin{cases} 0 \le x_1 \le 1 \\ 0 \le x_2 \le 1 \end{cases} \quad \text{if and only if} \quad \begin{cases} 0 \le (y_1 + y_2)/2 \le 1 \\ 0 \le (y_1 - y_2)/2 \le 1 \end{cases}$$

and the latter holds if and only if

$$-y_2 \le y_1 \le 2 - y_2$$
$$y_2 \le y_1 \le 2 + y_2.$$

Further, the Jacobian of this transformation

$$J = \frac{\partial(x_1, x_2)}{\partial(y_1, y_2)} = \det \begin{pmatrix} \dfrac{\partial x_1}{\partial y_1} & \dfrac{\partial x_1}{\partial y_2} \\ \dfrac{\partial x_2}{\partial y_1} & \dfrac{\partial x_2}{\partial y_2} \end{pmatrix} = \det \begin{pmatrix} \tfrac{1}{2} & \tfrac{1}{2} \\ \tfrac{1}{2} & -\tfrac{1}{2} \end{pmatrix} = -\tfrac{1}{2}$$

or $|J| = 1/2$. Hence we have that

$$h(y_1, y_2) = f\left(\frac{y_1 + y_2}{2}, \frac{y_1 - y_2}{2}\right) \times \frac{1}{2} \qquad \text{if } (y_1, y_2) \in S_{y_1, y_2},$$

that is

$$h(y_1, y_2) = \tfrac{1}{2} \qquad \text{if } (y_1, y_2) \in S_{y_1, y_2}.$$

Note that the marginal of y_1 is (see Figure 6.12.3)

$$h(y_1) = \begin{cases} \displaystyle\int_{-y_1}^{y_1} h(y_1, y_2)\, dy_2 & \text{if } 0 < y_1 < 1 \\[2ex] \displaystyle\int_{y_1 - 2}^{2 - y_1} h(y_1, y_2)\, dy_2 & \text{if } 1 \le y_1 < 2, \end{cases}$$

whereupon

$$h_1(y_1) = \begin{cases} y_1 & \text{if } 0 < y_1 < 1 \\ 2 - y_1 & \text{if } 1 < y_1 < 2. \end{cases}$$

Similarly, we may find that the marginal of y_2 is

$$h_2(y_2) = \begin{cases} \displaystyle\int_{-y_2}^{y_2 + 2} h(y_1, y_2)\, dy_1 = y_2 + 1 & \text{if } -1 < y_2 < 0 \\[2ex] \displaystyle\int_{y_2}^{2 - y_2} h(y_1, y_2)\, dy_1 = 1 - y_2 & \text{if } 0 \le y_2 < 1. \end{cases}$$

In the case of $n \ge 3$ continuous random variables $(x_1, \ldots, x_n)$, a simple extension of the argument that led to (6.12.18) yields the following:

THEOREM 6.12.5. *If the continuous random variables $(x_1, \ldots, x_n)$ have joint p.d.f. $f(x_1, \ldots, x_n)$, and if $y_i = g_i(x_1, \ldots, x_n)$, $i = 1, \ldots, n$ is a one-one transformation on the x's with inverse $x_i = u_i(y_1, \ldots, y_n)$, $i = 1, \ldots, n$, which maps the sample space $S_{x_1, \ldots, x_n}$ into $S_{y_1, \ldots, y_n}$, then the p.d.f. h of $(y_1, \ldots, y_n)$ is given by*

$$(6.12.24) \quad h(y_1, \ldots, y_n)$$
$$= f(u_1(y_1, \ldots, y_n), \ldots, u_n(y_1, \ldots, y_n)) |J|, \ (y_1, \ldots, y_n) \in S_{y_1 \ldots v_n}$$

where $J = \dfrac{\partial(x_1, \ldots, x_n)}{\partial(y_1, \ldots, y_n)} = \det\left(\dfrac{\partial x_i}{\partial y_j}\right)$ is the Jacobian of the transformation
[*see* (6.12.20)].

Note that if we define det $(a) = a$, (6.12.24) is the general form of h for $n \ge 1$. Now returning to the case $n \ge 3$, it may be that we start off with interest in the m variables $y_j = g_j(x_1, \ldots, x_n)$, $j = 1, \ldots, m < n$. By defining $n - m$ further functions $y_{t+m} = g_{t+m}(x_1, \ldots, x_n)$, where $t = 1, \ldots, n - m$, in such a way that $y_i = g_i(x_1, \ldots, x_n)$, $i = 1, \ldots, m$, $\ldots, n$ is a one-one transformation on $(x_1, \ldots, x_n)$, we may find $h(y_1, \ldots, y_n)$ by using Theorem 6.12.5 above, and then integrate out the variables $(y_{m+1}, \ldots, y_n)$ from h.

(c) Non One-One Transformations

Very often we are in the position where variables of interest do not permit the definition of a suitable one-one transformation. We discuss and illustrate this situation in this subsection for the case of continuous random variables, but the extension to discrete variables will be seen to be immediate.

(*i*) *One Variable.* We begin with the following situation. Suppose x is a continuous random variable, p.d.f. $f(x)$ and c.d.f. $F(x)$, with sample space $S_x = \{x \mid -a < x < a\}$. Suppose further that interest lies in the random variable $y = x^2$. Clearly $y = x^2$ does not, in this case, constitute a one-one transformation. Indeed, given an outcome $y_0 \in S_y$, where $S_y = \{y \mid 0 < y < a^2\}$, we do not know whether the value $y = y_0$ has arisen because of the occurrence of the event $x = \sqrt{y_0}$, or because of the occurrence of the event $x = -\sqrt{y_0}$.

Now if H denotes the c.d.f. of y, we have for all $y_0 \in S_y$,

$$(6.12.25) \qquad H(y_0) = P(y \leq y_0) = P(x^2 \leq y_0),$$

that is,

$$(6.12.25a) \quad H(y_0) = P(-\sqrt{y_0} \leq x \leq \sqrt{y_0}) = F(\sqrt{y_0}) - F(-\sqrt{y_0}).$$

Dropping subscripts and differentiating we have

$$(6.12.26) \qquad h(y) = f(\sqrt{y}) \frac{d\sqrt{y}}{dy} - f(-\sqrt{y}) \frac{d(-\sqrt{y})}{dy}$$

or

$$(6.12.26a) \qquad h(y) = f(\sqrt{y}) \frac{1}{2\sqrt{y}} + f(-\sqrt{y}) \frac{1}{2\sqrt{y}},$$

which we shall write as

$$(6.12.26b) \qquad h(y) = f(\sqrt{y}) |J_1| + f(-\sqrt{y}) |J_2|,$$

where $J_1 = d(\sqrt{y})/dy = 1/2\sqrt{y}$ is the Jacobian of the transformation $y = x^2$ whose inverse is $x = u_1(y) = \sqrt{y}$, while $J_2 = d(-\sqrt{y})/dy = -1/2\sqrt{y}$ is the Jacobian of the transformation $y = x^2$ whose inverse is $x = u_2(y) = -\sqrt{y}$. [Here, $|J_1| = |J_2| = 1/2\sqrt{y}$.] Now we have written

(6.12.26a) in the form (6.12.26b) to help motivate the following interpretation of (6.12.26a).

The transformation $y = x^2$ is not one-one, and indeed, may be called a two-to-one transformation. This is because for each point $x \in S_x$, there corresponds one point in S_y, but to each point $y \in S_y$ there corresponds two points in S_x, namely $+\sqrt{y}$ and $-\sqrt{y}$. Accordingly, suppose we define two mutually exclusive subsets of S_x, say A_1 and A_2, where $A_1 \cup A_2 = S_x$, such that $A_1 = \{x \mid 0 < x < a\}$ and $A_2 = \{x \mid -a < x \leq 0\}$. Now suppose we define a one-one transformation from A_1 into S_y, namely, for $x \in A_1$, $y = x^2$ with (single) inverse $x = u_1(y) = +\sqrt{y}$, and further define a one-one transformation from A_2 into S_y, namely, if $x \in A_2$, $y = x^2$ with (single) inverse $x = u_2(y) = -\sqrt{y}$. Now since A_1 and A_2 are mutually exclusive, and using the rules of probability regarding the union of two mutually exclusive events, (6.12.26b) can be regarded as saying: the p.d.f. of y, $h(y)$, is found by summing two terms, each of the "one-one transformation" form (6.12.16), which takes into account the fact that, as mentioned before, the outcome y is the union of the disjoint events $x = \sqrt{y}$ and $x = -\sqrt{y}$. With this as guide, the extension to more than one variable is immediate, and we proceed now to discuss this case.

(ii) *More than One Variable.* Suppose now that $(x_1, \ldots, x_n)$ has joint p.d.f. $f(x_1, \ldots, x_n)$, and sample space $S_{x_1, \ldots, x_n}$. Let $y_i = g_i(x_1, \ldots, x_n)$, $i = 1, \ldots, n$, be a non one-to-one transformation on $(x_1, \ldots, x_n)$, with $(y_1, \ldots, y_n) \in S_{y_1, \ldots, y_n}$, the sample space of the random variables $(y_1, \ldots, y_n)$. Suppose it is possible to find subsets of $S_{x_1, \ldots, x_n}$, say $A_1, \ldots, A_j, \ldots, A_k$, such that $\bigcup\limits_{j=1}^{k} A_j = S_{x_1, \ldots, x_n}$, and $A_i \cap A_j = \phi$, the null set, for all $i \neq j = \{1, \ldots, k\}$, and such that $y_j = g_j(x_1, \ldots, x_n)$, $j = 1, \ldots, n$, constitutes a one-one transformation from A_j into $S_{y_1, \ldots, y_n}$ with single inverse $x_1 = u_{1j}(y_1, \ldots, y_n), \ldots, x_n = u_{nj}(y_1, \ldots, y_n)$, with Jacobian $J_j = \dfrac{\partial(u_{1j}, \ldots, \partial u_{nj})}{\partial(y_1, \ldots, y_n)} \neq 0$ for $j = 1, \ldots, k$. That is, a point $(y_1, \ldots, y_n) \in S_{y_1, \ldots, y_n}$ is mapped into a *single* point $(x_1, \ldots, x_n) \in A_j$ by the transformation $x_i = u_{ij}(y_1, \ldots, y_n)$, $i = 1, \ldots, n$. We then may state the following theorem.

THEOREM 6.12.6. *Under the conditions of the above paragraph, the p.d.f. h of $(y_1, \ldots, y_n)$ is given by*

$$(6.12.27) \quad h(y_1, \ldots, y_n) = \sum_{j=1}^{k} f(u_{1j}(y_1, \ldots, y_n), \ldots, u_{nj}(y_1, \ldots, y_n)) |J_j|$$

for $(y_1, \ldots, y_n) \in S_{y_1, \ldots, y_n}$.

The interpretation of (6.12.27) is as follows. Since $y_i = g_i(x_1, \ldots, x_n)$ is not one-one, we do not know whether a given outcome $(y_1, \ldots, y_n)$ has arisen because of the outcome $(x_1, \ldots, x_n)$ belonging to A_1 or A_2 or $\ldots$ or A_k. Since the A_j are disjoint, and since the transformation $y_i = g_i(x_1, \ldots, x_n)$, with inverse $x_i = u_{ij}(y_1, \ldots, y_n)$ is one-one from A_j into $S_{y_1, \ldots, y_n}$, then (6.12.27) says that the "probability," that is, the p.d.f. of $(y_1, \ldots, y_n)$ may be obtained by summing the "probabilities," i.e., p.d.f.'s of the "one-one transformation" of form (6.12.24), which take into account the probability that an outcome $(y_1, \ldots, y_n) \in S_{y_1, \ldots, y_n}$ may be due to $(x_1, \ldots, x_n)$ occuring in A_1 or A_2 or $\ldots$ or A_k.

Example 6.12(f). Suppose (x_1, x_2) is a random sample of 2 independent observations on a random variable x, where x has p.d.f. $f(x)$, and sample space $S_x = \{x \mid a < x < b\}$. (It could be that $a = -\infty$, and $b = \infty$ etc). Hence, (x_1, x_2) has sample space $S_{x_1, x_2} = \{(x_1, x_2) \mid a < x_1, x_2 < b\}$ and p.d.f.

$$f(x_1, x_2) = f(x_1)f(x_2), \qquad (x_1, x_2) \in S_{x_1, x_2}.$$

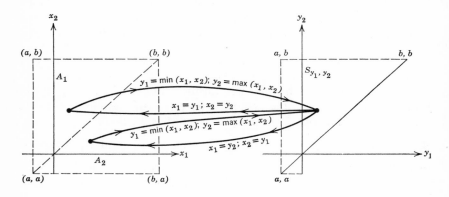

Now suppose we define $y_1 = \min(x_1, x_2)$, $y_2 = \max(x_1, x_2)$. Obviously $y_1 < y_2$, and $S_{y_1, y_2} = \{(y_1, y_2) \mid a < y_1 < y_2 < b\}$. Now given an outcome of (y_1, y_2), there are two possibilities that could give rise to this outcome— the first observation taken, x_1, could turn out to be the smallest of x_1 and x_2, *or* the first observation taken, x_1, could turn out to be the largest of x_1 and x_2. Hence, $y_1 = \min(x_1, x_2)$, $y_2 = \max(x_1, x_2)$ is not one-one. But suppose we define

$$A_1 = \{(x_1, x_2) \mid a < x_1 < x_2 < b\}, \qquad A_2 = \{(x_1, x_2) \mid a < x_2 < x_1 < b\}.$$

Then A_1 and A_2 are disjoint and $A_1 \cup A_2 = S_{x_1, x_2}$. Define the one-one transformation from A_1 into S_{y_1, y_2} as

$$y_1 = \min(x_1, x_2), \qquad y_2 = \max(x_1, x_2)$$

with single inverse

$$x_1 = u_{11}(y_1, y_2) = y_1, \quad x_2 = u_{21}(y_1, y_2) = y_2$$

and define the one-one transformation from A_2 into S_{y_1, y_2} as

$$y_1 = \min(x_1, x_2), \quad y_2 = \max(x_1, x_2).$$

with (single) inverse

$$x_1 = u_{12}(y_1, y_2) = y_2, \quad x_2 = u_{22}(y_1, y_2) = y_1.$$

Obviously, we have

$$J_1 = \det \begin{pmatrix} \dfrac{\partial x_1}{\partial y_1} & \dfrac{\partial x_1}{\partial y_2} \\ \dfrac{\partial x_2}{\partial y_1} & \dfrac{\partial x_2}{\partial y_2} \end{pmatrix} = \det \begin{pmatrix} 1 & 0 \\ 0 & 1 \end{pmatrix} = 1$$

and

$$J_2 = \det \begin{pmatrix} \dfrac{\partial x_1}{\partial y_1} & \dfrac{\partial x_1}{\partial y_2} \\ \dfrac{\partial x_2}{\partial y_1} & \dfrac{\partial x_2}{\partial y_2} \end{pmatrix} = \det \begin{pmatrix} 0 & 1 \\ 1 & 0 \end{pmatrix} = -1$$

so that $|J_1| = |J_2| = 1$. Hence, from (6.12.27) of Theorem 6.12.6,

$$h(y_1, y_2) = f(y_1, y_2) \times 1 + f(y_2, y_1) \times 1$$
$$= 2f(y_1)f(y_2)$$

for all $(y_1, y_2) \in S_{y_1, y_2}$, i.e., all $a < y_1 < y_2 < b$.

[The variables y_1 and y_2 defined above are called the order statistics of the sample (x_1, x_2). For a fuller discussion, see Chapter 13.]

We remark that (6.12.27) may be used to obtain the marginal p.d.f. of, say $y_1, \ldots, y_m$, $m < n$, by integrating out $(y_{m+1}, \ldots, y_n)$ from $h(y_1, \ldots, y_n)$. In fact, it may be that at the outset interest focuses on $y_1, \ldots, y_m$, and "filling out" the transformation by defining $y_{m+1}, \ldots, y_n$ leads to the non one-one transformation $y_1, \ldots, y_n$, with $y_i = g_i(x_1, \ldots, x_n)$. We would then proceed by determining $A_1, \ldots, A_k$ as above and apply Theorem 6.12.6 to find $h(y_1, \ldots, y_n)$. Once this is obtained, we would, as mentioned above, integrate out $y_{m+1}, \ldots, y_n$ to find the (marginal) p.d.f. of $y_1 = g_1(x_1, \ldots, x_n), \ldots, y_m = g_m(x_1, \ldots, x_n)$.

PROBLEMS

6.1. A box of ten items has three defectives. If the items are drawn one at a time and tested, let x be a random variable denoting the number of items tested in order to find the third (and last defective) in the box. If $p(x)$ is the probability function of x:

(a) Write down the expression for $p(x)$ and determine $p(x)$ numerically for all possible values of x.

(b) Graph both $p(x)$ and the c.d.f. $F(x)$.

(c) Compute the mean μ and standard deviation σ of x.

6.2. If x is a continuous random variable with p.d.f. $f(x)$, the pth *percentile* of x (sometimes called the pth percentile of the population) is defined as that value x_p for which

$$\int_{-\infty}^{x_p} f(x) \, dx = \frac{p}{100} .$$

The 50th percentile is called the *population median*. Suppose a continuous random variable has cumulative distribution

$$F(x) = \begin{cases} 0, & x \le 0, \\ x^n, & 0 < x \le 1, \\ 1, & x > 1, \end{cases}$$

where $n \ge 1$:

(a) Find the probability density function $f(x)$.

(b) Find the median of x.

(c) Find the mean and variance of x.

6.3. If a defective spot (point) occurs in a glass disk R inches in radius, assume that it is equally likely to occur anywhere on the disk. Let x be a random variable indicating the distance between the point of occurrence of a defective spot and the center of the disk.

(a) Find the expression for $F(x)$ and $f(x)$.

(b) What is the median of x?

(c) Find the mean and variance of the random variable x.

6.4. A continuous random variable x has the probability density function

$$f(x) = \begin{cases} 3x^2, & 0 < x < 1, \\ 0, & \text{otherwise.} \end{cases}$$

(a) Find the c.d.f. $F(x)$.

(b) What are the numerical values of $F(1/3)$, $F(9/10)$, and $P(1/3 < x \le 1/2)$?

(c) Find the value of a so that $P(x \le a) = 1/4$ (a is the 25th percentile of x).

(d) Find the mean μ and variance σ^2 of x.

6.5. A resistor is composed of two component parts soldered together in series; the total resistance of the resistor equals the sum of the resistances of the component parts. The first part is drawn from a production lot having a mean of 200 ohms and a standard deviation of 2 ohms, and the second part is drawn from a lot having a mean of 150 ohms and standard deviation of 3 ohms. Find the mean and standard deviation of the resistance of the assembled resistor.

6.6. In packaging corn flakes into 8-ounce packages, assume the population of net weights generated by the automatic filling machine (properly set) has a distribution with mean of 8.15 ounces and a standard deviation of .08 ounce. Assume the population of paper boxes to receive the fillings to have a distribution with mean 1.45 ounces and standard deviation of .06 ounce.

(a) The population of filled boxes will have a distribution with what mean and variance?

(b) If these boxes are packaged 24 per carton and if the population of empty cartons has mean 28.00 ounces and standard deviation of 1.20 ounces, what is the mean and variance of the population of filled cartons?

6.7. A laminated strip is built up by randomly selecting two layers of material A, three layers of material B, and four layers of material C. The thicknesses of individual layers of material A have mean .0100 inch and standard deviation .0005 inch, and the respective numbers for material B are .0050 inch and .0003 inch and those for C are .0025 inch and .0001 inch. A large lot of such laminated strips is manufactured. Find the mean and standard deviation of the thicknesses of the strips in this lot.

6.8. Mass-produced articles are fitted into cardboard containers, one article per container. Twelve of these filled containers are then packed in wooden boxes. Suppose the mean and standard deviation of the weights of the population of articles are 20.6 pounds and .8 pound respectively, those of the cardboard containers are 1.8 pounds and .1 pound respectively, and those of the wooden boxes are 3.6 pounds and .4 pound respectively.

(a) What are the values of the mean and standard deviation of the population of weights of filled boxes ready to ship?

(b) T is the total weight of 25 filled boxes taken at random. Give the mean and variance of T.

(c) $\bar{x}$ is the average weight of these 25 boxes. Give the mean and variance of $\bar{x}$.

6.9. A certain type of $\frac{1}{2}$-inch rivet is classified as acceptable by a consumer if its diameter lies between .4950 and .5050 inch. It is known that a mass-production process is such that $100\theta_1\%$ of the rivets have diameters less than .4950 inch, $100\theta_2\%$ have diameters that lie in the "acceptable" region, and $100\theta_3\%$ have diameters greater than .5050 inch, where, of course, $\theta_3 = 1 - \theta_1 - \theta_2$. If a random sample of n rivets is taken from the process what is the probability $p(x_1, x_2)$ that x_1 rivets have diameters less than .4950 inch, x_2 have diameters measuring between .4950 and .5050 inch, and $x_3 = n - x_1 - x_2$ have diameters greater than .5050 inch?

(a) Find the marginal distribution function of x_2 and explain the result in words.

(b) What are the mean and the variance of x_2? of x_1?

(c) Find the covariance of x_1 and x_2.

6.10. Suppose 420 "true" dice are rolled simultaneously.

(a) If x is a random variable denoting the total number of aces which turn up, find the values of the mean and standard deviation of x.

(b) If y is a random variable denoting the total number of dots which turn up, find the values of the mean and standard deviation of y.

6.11. A process randomly generates digits 0, 1, 2, 3, 4, 5, 6, 7, 8, 9 with equal probabilities. If T is a random variable representing the sum of n digits taken from the process, find the mean and the variance of T.

6.12. A person plays ten hands of bridge during the evening and T represents the total number of spades he obtains during the evening. Find the mean and the variance of T.

6.13. A sample of size n is drawn from a lot of 10,000 articles known to contain 100 defectives. With the use of the Chebychev inequality determine how large n should be in order for the probability to exceed .96 that the percentage of defectives in the sample will lie within the interval (.1, 1.9).

6.14. With the use of the Chebychev inequality determine how many random digits should be generated in order for the probability to exceed .9000 that the mean of the random digits will lie within the interval (3.5, 5.5).

6.15. Suppose an insurance company has among all of its insurance policies 50,000 policies of $5000 of American men aged 51. The probability of an American male aged 51 dying within one year is (approximately) .01. With the use of the Chebychev inequality decide for what value of k does the probability exceed .99 that the total death claims from the beneficiaries of this group (for the one year) will fall in the interval ($2,500,000 $- k$, $2,500,000 $+ k$).

6.16. If 2500 coins in a sack are poured out on a table, decide with the use of the Chebychev inequality for what value of k does the probability exceed .96 that the number of heads will lie in the interval $(1250 - k, 1250 + k)$?

6.17. One cigarette from each of four brands, A, B, C, D, is partially smoked by a blindfolded person. As soon as he has taken a few puffs on a cigarette, he states the letter of the brand to which he considers it to belong. (Of course, he can use each letter only once.) Let x be the random variable denoting the number of cigarettes correctly identified. If the identification is done at random (that is, he is equally likely to assign any letter to any cigarette), write down the probability distribution of x in table form. Find the mean and variance of x.

6.18. A point is taken at random from the interval $(0, 1)$, all points being equally likely. A second point is then taken in the same way. Let x be the coordinate of the point halfway between these points. x is a continuous chance quantity with a probability density function having an inverted V graph as shown in the following figure:

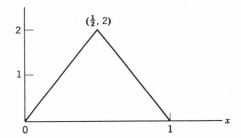

Write down the formula for $f(x)$. Find the mean and variance of x. Find the formula for $F(x)$ and graph $F(x)$.

6.19. By using the moment generating function of a random variable x having the binomial distribution (6.11.9), show that the mean and variance of x are $n\theta$ and $n\theta(1 - \theta)$ respectively.

6.20. Suppose $(x_1, \ldots, x_n)$ is a sample from a distribution whose mean is μ and variance is σ^2 and whose moment generating function exists. Show, by using the method of moment generating functions, that the mean and variance of the sample sum T are $n\mu$ and $n\sigma^2$ respectively. Also by the method of moment-generating functions, show that the mean and variance of the sample mean $\bar{x}$ are μ and σ^2/n respectively.

6.21. If (x_1, x_2) is a pair of random variables such that

$$p_2(x_2) = \frac{\mu^{x_2} e^{-\mu}}{x_2!}, \qquad x_2 = 0, 1, 2, \ldots,$$

and

$$p(x_1 \mid x_2) = \binom{x_2}{x_1} \theta^{x_1}(1 - \theta)^{x_2 - x_1}, \qquad x_1 = 0, 1, \ldots, x_2,$$

show that $p_1(x_1)$ is a Poisson distribution.

6.22. Let x_1 be a number taken at random on the interval $(0, 1)$, and x_2 a number taken at random on the interval $(x_1, 1)$. Show that the distribution of x_2 has p.d.f.

$$p_2(x_2) = -\log (1 - x_2), \qquad 0 < x_2 \leq 1$$
$$= 0, \qquad \text{otherwise.}$$

6.23. A lot contains N articles of which $N\theta$ are defective. Articles are drawn successively at random and without replacement until k defectives are drawn. Let x be a random variable denoting the number of articles which must be drawn to achieve this objective. Show that the p.f. of x is given by

$$p(x) = \binom{x - 1}{k - 1}\binom{N - x}{N\theta - k} \bigg/ \binom{N}{N\theta}$$

and that

$$E(x) = \frac{k(N + 1)}{N\theta + 1}.$$

6.24. A device fails to operate on a single trial with probability θ. Let x be a random variable denoting the number of trials required to obtain a total of k failures. If results of successive trials are independent, show that the p.f. of x is given by

$$p(x) = \binom{x - 1}{k - 1}\theta^k(1 - \theta)^{x-k}, \qquad x = k, k + 1, \ldots .$$

Show that the mean and variance of x are given by

$$E(x) = \frac{k}{\theta},$$

$$\text{Var}\,(x) = \frac{k(1 - \theta)}{\theta^2}.$$

6.25. In Problem 6.24, suppose $k = 1$. Show that if y is a random variable denoting the number of trials required to obtain one failure, then the p.f. of y is

$$p(y) = \theta(1 - \theta)^{y-1}, \qquad y = 1, 2, \ldots ,$$

and that its mean and variance is $1/\theta$ and $(1 - \theta)/\theta^2$. This distribution is called the geometric distribution. Show that $\sum_{y=1}^{\infty} p(y) = 1$.

6.26. Referring to Problem 6.25, show that

$$p(y > s + t \mid y > s) = p(y > t) = \sum_{y=t+1}^{\infty} \theta(1 - \theta)^{y-1}.$$

(The above result implies that the geometric distribution has no memory, for if the event of a failure has *not* occurred during the first s trials, then the probability that a failure will not occur in the *next* t trials is the same as the probability that it will not occur in the first t trials. In other language, the information that a failure has not occurred in the first s trials is "forgotten" in the subsequent calculations.)

6.27. Referring to Problem 6.26, let y_1 = number of trials required up to the first occurrence of a failure, y_2 = number of trials required between the first occurrence of a failure up to and including the second occurrence of a failure ,..., and y_k = number of trials required between the $(k - 1)$ occurrence of a failure up to and including the kth occurrence of a failure. (All trials are independent, as assumed in Problems 6.24–6.26.)

Let $x = y_1 + \cdots + y_k$. What is the distribution of y_i? of x? What is the mean and variance of x? Discuss the connection with Problem 6.24.

6.28. If x_1 and x_2 have joint distribution $h(x_1, x_2)$, show that

$$\text{Cov } (x_1, x_2) = E(x_1 x_2) - E(x_1)E(x_2).$$

6.29. Referring to Example 6.7(c), what is the marginal distribution of x_2? What are the mean and variances of x_1 and x_2? Compute the correlation coefficient between x_1 and x_2.

6.30. Referring to Example 6.7(c), find $E(x_1 \mid x_2)$, Var $(x_1 \mid x_2)$, $E(x_2 \mid x_1)$, Var $(x_2 \mid x_1)$.

6.31. Referring to Example 6.7(d), find $f(x_1 \mid x_2)$, $f(x_2 \mid x_1)$, and evaluate $E(x_1 \mid x_2)$, Var $(x_1 \mid x_2)$, $E(x_2 \mid x_1)$, Var $(x_2 \mid x_1)$.

6.32. The pair of random variables (x_1, x_2) has joint p.d.f. $f(x_1, x_2)$ given by $f(x_1, x_2) = 2/\pi$ for (x_1, x_2) lying inside the semicircle bounded by the x_1 axis and the curve $x_2 = \sqrt{1 - x_1^2}$, i.e., the sample space of (x_1, x_2) is $S = \{(x_1, x_2) \mid (x_1^2 + x_2^2) \leq 1, x_2 \geq 0\}$. Find the marginals of x_1 and x_2, the means and variances of x_1 and x_2, the correlation coefficient between x_1 and x_2. Also determine $f(x_1 \mid x_2)$ and $f(x_2 \mid x_1)$, and evaluate $E(x_1 \mid x_2)$, Var $(x_1 \mid x_2)$, $E(x_2 \mid x_1)$, Var $(x_2 \mid x_1)$.

6.33. Suppose (x_1, x_2) are jointly distributed as the trinomial, that is, their p.f. is given by

$$p(x_1, x_2) = \frac{n!}{x_1! \, x_2! \, (n - x_1 - x_2)!} \, \theta_1^{x_1} \theta_2^{x_2} (1 - \theta_1 - \theta_2)^{n - x_1 - x_2}$$

where $0 \leq x_1, x_2 \leq n$, and $0 \leq (x_1 + x_2) \leq n$ [c.f. with (3.5.1), with $k = 3$].

Show that the marginal of x_1 is the binomial, with parameter θ_1, sample size n, and explain how this result comes about in practice.

6.34. Suppose x has the probability function $p(x) = \theta(1 - \theta)^{x-1}$, $x = 1, 2, 3, \ldots$. What is the p.d.f. of $y = x^3$? What is the p.d.f. of $y = x^2$?

6.35. Suppose x has p.f. $p(x) = 1/6$, $x = 1, 2, \ldots, 6$. Let $y = 3x^2 + 2$. Find the p.f. of y.

6.36. Suppose z has p.f. $p(z) = 1/9$, $z = -4, -3, -2, -1, 0, 1, 2, 3, 4$. Find the p.d.f. of $w = z^2$.

6.37. Suppose x_1 and x_2 are independent random variables, with the p.f. of x_1 given by $p_i(x_i) = \binom{n_i}{x_i} \theta^{x_i}(1 - \theta)^{n-x_i}$, $x_i = 0, 1, \ldots, n_i$, for $i = 1, 2$. Find the p.f. of $y_1 = x_1 + x_2$. $\left[\textit{Hint:} \quad \text{Find the joint p.f. of } y_1 = x_1 + x_2 \text{ and } y_2 = x_2, \right.$

and "sum" out y_2. Make use of the combinatorial identity discussed in Chapter 3, namely $\displaystyle\sum_{y_2=0}^{y_1} \binom{n_1}{y_1 - y_2}\binom{n_2}{y_2} = \binom{n_1 + n_2}{y_1}.\Big]$

6.38. Suppose x_1 and x_2 are two independent random variables whose probability functions are given by $f_1(x_1) = e^{-2\mu}(2\mu)^{x_1}/x_1!$ for integer $x_1 \geq 0$, and $f_2(x_2) = e^{-\mu}\mu^{x_2}/x_2!$, respectively. Find the p.f. of $y = x_1 + x_2$. [*Hint:*

$$\sum_{y_2=0}^{y_1} \{2^{y_1-y_2}y_1!/[(y_1 - y_2)! \, y_2!]\} = (1 + 2)^{y_1} = 3^{y_1}.]$$

6.39. The random variable x has the uniform density on $(0, 1)$, i.e., its p.d.f. is given by $f(x) = 1$, $0 < x < 1$. What is the p.d.f. of the random variable $y = -3x + 4$? of $z = 3x + 4$? of $w = -2\ln x$?

6.40. The random variable x has p.d.f. $f(x) = 3x^2/64$, for $0 < x < 4$. Find the p.d.f. of $y = x^3/64$; of $w = x^2$; of $z = \sqrt{x}$.

6.41. The independent random variables x_1 and x_2 are such that

$$f_i(x_i) = \frac{1}{\Gamma(m_i)} x_i^{m_i-1} \exp\{-x_i\}, \qquad x_i > 0,$$

where $m_i > 0$, $i = 1, 2$. [Such densities are called gamma densities of order m_i, and a random variable x_i whose p.d.f. is a gamma density is called a gamma variable. The constant $\Gamma(m_i)$ is discussed in Appendix VI.] Find the joint p.d.f., say $h(y_1, y_2)$, of $y_1 = x_1 + x_2$, $y_2 = x_1/(x_1 + x_2)$. Using this show

(i) the marginal of y_1 is $h_1(y_1) = \dfrac{1}{\Gamma(m_1 + m_2)} y_1^{m_1+m_2-1} \exp\{-y_1\}$, for $y_1 > 0$,

(ii) the marginal of y_2 is $h_2(y_2) = \dfrac{\Gamma(m_1 + m_2)}{\Gamma(m_1)\Gamma(m_2)} y_2^{m_1-1}(1 - y_2)^{m_2-1}$, $0 < y_2 < 1$.

[A density of the form of $h_2(y_2)$ is called a beta distribution of first kind of order (m_1, m_2).] *Hint:* use the identities

(a) $\displaystyle\int_0^1 y_2^{m_1-1}(1 - y_2)^{m_2-1} \, dy_2 = \Gamma(m_1)\Gamma(m_2)/\Gamma(m_1 + m_2)$

and

(b) $\displaystyle\int_0^\infty y_1^{m_1+m_2-1}e^{-y_1} \, dy_1 = \Gamma(m_1 + m_2).$

Are the random variables y_1 and y_2 independent?

6.42. The random variable x has p.d.f. $f(x) = (3x + 4)/8$, for $-1 < x < 1$. Find the p.d.f. of $y = x^2$.

6.43. The random variable x has p.d.f. $f(x) = (6x + 8)/33$, for $-1 < x < 2$. Find the p.d.f. of $y = x^2$. [*Hint:* Be careful, for $-1 < x < 1$, $y = x^2$ is two-to-one, while if $1 < x < 2$, $y = x^2$ is one-to-one.]

6.44. If (x_1, x_2) is a random sample of two independent observations on a random variable x, where

$$f(x) = (2\pi)^{-\frac{1}{2}} \exp\{-x^2/2\}, \qquad -\infty < x < \infty,$$

the so-called standard normal density (see Chapter 7), then find the joint distribution of $y_1 = x_1^2 + x_2^2$ and $y_2 = x_2$. Sketch the sample spaces of (x_1, x_2) and (y_1, y_2). Integrate out y_2 to find the marginal of y_1.

6.45. A random sample of 3 independent observations (x_1, x_2, x_3) is taken on x, where x has p.d.f. $f(x)$, where $f(x) > 0$ if $x \in S_x = \{x \mid a < x < b\}$. Let y_1, y_2, y_3 define the order statistics of this sample of three, i.e., $y_1 = \min(x_1, x_2, x_3)$; $y_2 = $ middle value of (x_1, x_2, x_3); $y_3 = \max(x_1, x_2, x_3)$. Hence $S_{y_1, y_2, y_3} = \{(y_1, y_2, y_3) \mid a < y_1 < y_2 < y_3 < b\}$. Find the joint p.d.f. of y_1, y_2, y_3.

CHAPTER 7

The Normal Distribution

7.1 DEFINITION AND PROPERTIES

The normal distribution plays a fundamental role in all of mathematical statistics, and, as we shall see in later chapters, important statistical techniques are based on this distribution. The purpose of this chapter is to discuss the normal distribution, its properties, and some of its applications.

A random variable x is said to have the normal distribution $N(\mu, \sigma^2)$ if its p.d.f. is

$$(7.1.1) \quad f(x) = \frac{1}{\sigma\sqrt{2\pi}} \exp\left[-\frac{1}{2\sigma^2}(x - \mu)^2\right], \quad -\infty < x < \infty,$$

where μ and σ are constants which, as we shall now see, are the mean and standard deviation of x respectively. We note that $f(x) > 0$ for $-\infty < x < +\infty$. The proof that $f(x)$ is a probability density function, that is, $\int_{-\infty}^{\infty} f(x)\, dx = 1$, is given in Appendix I.

First, we shall show that μ is the mean of x. We have

$$(7.1.2) \quad E(x) = \int_{-\infty}^{\infty} \frac{1}{\sigma\sqrt{2\pi}}\, x \exp\left[-\frac{1}{2}\left(\frac{x - \mu}{\sigma}\right)^2\right] dx.$$

In the integral set $(x - \mu)/\sigma = z$. Then (7.1.2) takes the form

$$E(x) = \int_{-\infty}^{\infty} \frac{1}{\sqrt{2\pi}}(\mu + \sigma z)e^{-z^2/2}\, dz,$$

$$= \mu \int_{-\infty}^{\infty} \frac{1}{\sqrt{2\pi}} e^{-z^2/2}\, dz + \sigma \int_{-\infty}^{\infty} \frac{1}{\sqrt{2\pi}} z e^{-z^2/2}\, dz.$$

The integrand of the first integral in the preceding line is that of the normal density function (7.1.1), having $\mu = 0$, $\sigma = 1$, and hence the value of this integral is 1.

For the second integral, we have

$$\int_{-\infty}^{\infty} \frac{1}{\sqrt{2\pi}} z e^{-z^2/2}\, dz = -\frac{e^{-z^2/2}}{\sqrt{2\pi}}\Bigg]_{-\infty}^{+\infty} = 0.$$

Putting these results together, we find that the mean of the normal distribution having p.d.f. given by (7.1.1) is μ.

To evaluate the variance of x, we have

(7.1.3) $\text{Var}(x) = E[x - E(x)]^2$

$$= E(x - \mu)^2$$

$$= \int_{-\infty}^{\infty} (x - \mu)^2 \frac{1}{\sigma\sqrt{2\pi}} \exp\left[-\frac{1}{2}\left(\frac{x-\mu}{\sigma}\right)^2\right] dx.$$

Again, letting $(x - \mu)/\sigma = z$, we obtain

$$\text{Var}(x) = \int_{-\infty}^{\infty} \sigma^2 z^2 \frac{1}{\sqrt{2\pi}} e^{-z^2/2}\, dz$$

and integrating by parts gives

$$\text{Var}(x) = \sigma^2\left(\frac{-ze^{-z^2/2}}{\sqrt{2\pi}}\Bigg|_{-\infty}^{\infty} + \int_{-\infty}^{\infty} \frac{1}{\sqrt{2\pi}} e^{-z^2/2}\, dz\right)$$

$$= \sigma^2(0 + 1)$$

$$= \sigma^2,$$

that is, we have the result that the variance of the normal distribution, having a p.d.f. given by (7.1.1), is σ^2. [For an alternative proof of the fact that the mean and variance of the normal distribution having a p.d.f. given by (7.1.1) are μ and σ^2, see Appendix II.]

Thus we find that the p.d.f. $f(x)$ of the normal distribution $N(\mu, \sigma^2)$ given by (7.1.1) has its mean μ and standard deviation σ built in as parameters.

The reader should note that the p.d.f. $f(x)$ given by (7.1.1) is symmetric about its mean μ. With the use of the usual differentiation procedure it can be seen that $f(x)$ has its maximum for $x = \mu$ and its inflection points for $x = \mu \pm \sigma$. Examination of the second derivative of $f(x)$ shows that the graph of $f(x)$ is concave downward if $\mu - \sigma < x < \mu + \sigma$, and concave upward otherwise. The graph of the p.d.f. $f(x)$ given by (7.1.1) is shown in Figure 7.1.1.

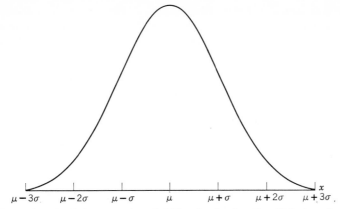

Figure 7.1.1. Graph of the normal probability density function $f(x)$ given by (7.1.1).

7.2 THE STANDARD NORMAL DISTRIBUTION

A random variable having the *standard normal distribution* $N(0, 1)$ is given a special designation throughout this book by the letter z. It is called the *standard normal variable,* and its p.d.f. is usually denoted by $\phi(z)$, where

$$(7.2.1) \qquad \phi(z) = \frac{1}{\sqrt{2\pi}} e^{-z^2/2}, \qquad -\infty < z < \infty.$$

By comparing (7.2.1) with (7.1.1) it is evident that $E(z) = 0$ and Var $(z) = 1$.

To find values of the cumulative distribution function of z, denoted by $\Phi(z)$, we use Table II, Appendix VII, which provides values of $\Phi(z)$ for $z \geq 0$ in increments of .01 for z, where

$$(7.2.2) \qquad \Phi(z) = \int_{-\infty}^{z} \phi(y)\, dy.$$

In view of the symmetry of $\phi(z)$ about $z = 0$, the reader should note that, for any $z > 0$,

$$(7.2.2a) \qquad \Phi(-z) = 1 - \Phi(z).$$

By using tables of $\Phi(z)$, we can find probabilities associated with events concerning any normal random variable. Suppose we have a random variable x which has the distribution $N(\mu, \sigma^2)$, and, for any given value x', we wish to find the probability

$$(7.2.3) \qquad P(x \leq x') = \int_{-\infty}^{x'} \frac{1}{\sigma\sqrt{2\pi}} \exp\left[-\frac{1}{2}\left(\frac{x - \mu}{\sigma}\right)^2\right] dx,$$

which is represented by the shaded area in Figure 7.2.1. We do not have tables which provide values of $P(x \leq x')$ for every possible value of μ and σ^2, nor do we need them. Suppose we let $z = (x - \mu)/\sigma$. We have

$$(7.2.4) \qquad P(x \leq x') = \int_{-\infty}^{(x'-\mu)/\sigma} \frac{1}{\sqrt{2\pi}} e^{-z^2/2} \, dz = \Phi\left(\frac{x' - \mu}{\sigma}\right).$$

Thus, in order to evaluate $P(x \leq x')$, where x has the normal distribution $N(\mu, \sigma^2)$, we can write

$$(7.2.5) \quad P(x \leq x') = P(x - \mu \leq x' - \mu) = P\left(\frac{x - \mu}{\sigma} \leq \frac{x' - \mu}{\sigma}\right)$$

$$= P\left(z \leq \frac{x' - \mu}{\sigma}\right) = \Phi\left(\frac{x' - \mu}{\sigma}\right),$$

or more briefly stated,

$$P(x \leq x') = \Phi\left(\frac{x' - \mu}{\sigma}\right).$$

That is, the probability that $x \leq x'$ is the same as the probability that $z \leq (x' - \mu)/\sigma$, where z is the standardized normal variable having a p.d.f. $\phi(z)$ given by (7.2.1). The graph of the c.d.f. $P(x \leq x')$ as given by (7.2.3) is shown in Figure 7.2.2.

The relationship between the x-scale and the z-scale is shown graphically in Figure 7.2.1 and 7.2.2.

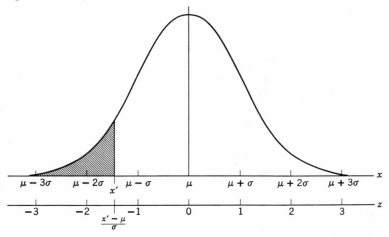

Figure 7.2.1. Graph of the normal probability density function $f(x)$ given by (7.1.1), showing the relation between the x-axis and z-axis. The shaded part of this normal density is $P(x \leq x')$ as given by (7.2.3).

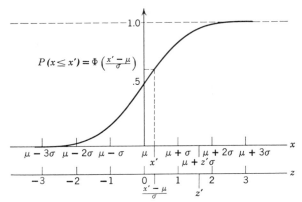

Figure 7.2.2.   Graph of the normal cumulative distribution function given by (7.2.3), showing the relation between the x-axis and the z-axis.

For convenience we often abbreviate the phrase "x is a random variable having the normal distribution $N(\mu, \sigma^2)$" by saying "x is an $N(\mu, \sigma^2)$ random variable," or "x is from an $N(\mu, \sigma^2)$ population."

   Example 7.2(a).   A manufacturer knows from experience that the diameters of 1/4-inch precision-made pins he produces have a normal distribution with mean .2500 inch and standard deviation .00025 inch. What percentage of the pins have diameters between .24951 and .25049 inch? This question is equivalent to asking for the probability that the diameter x of a pin taken at random from the production lies between .24951 and .25049 inch.
   We must find $P(.24591 \leq x \leq .25049)$, where x has the distribution $N(.25000, (.00025)^2)$. We proceed with operations indicated in (7.2.5) and obtain:

$$P(.24951 \leq x \leq .25049)$$
$$= P(.24951 - .25000 \leq x - .25000 \leq .25049 - .25000)$$
$$= P(-.00049 \leq x - .25000 \leq .00049)$$
$$= P\left(-\frac{.00049}{.00025} \leq \frac{x - .25000}{.00025} \leq \frac{.00049}{.00025}\right)$$
$$= P(-1.96 \leq z \leq 1.96) = P(z \leq 1.96) - P(z \leq -1.96)$$
$$= .975 - .025 = .95.$$

That is, 95% of the production lies between .24951 and .25049 inch.

   Example 7.2(b).   A process for producing batches of chalk is such that the "bulk density" x of a batch of chalk (in log units) is a normally distributed random variable with mean .8000 and standard deviation .0030. Find:
   1. $P(x \leq .8036)$.
   2. $P(|x - .8000| \leq .0060)$.
   3. The number c such that $P(x \leq c) = .95$.

Here x is an $N(.8000, (.003)^2)$ variable. Thus:

1. $P(x \leq .8036) = P(x - .8000 \leq .8036 - .8000)$

$$= P\left(z \leq \frac{.8036 - .8000}{.0030}\right) = P(z \leq 1.2) = .8849.$$

2. $P(|x - .8000| \leq .0060) \doteq P\left(\frac{|x - .8000|}{.0030} \leq \frac{.0060}{.0030}\right)$

$$= P(|z| \leq 2) = P(-2 \leq z \leq 2)$$
$$= P(z \leq 2) - P(z \leq -2) = .97725 - .02275$$
$$= .95450.$$

3. $$P(x \leq c) = .95,$$

$$P\left(z \leq \frac{c - .8000}{.0030}\right) = .95,$$

that is,

$$\frac{c - .8000}{.0030} = 1.645,$$

$$c = .8000 + (.0030)(1.645)$$
$$= .8000 + .004935$$
$$= .804935.$$

Throughout this book, we use the notation z_α to denote the $100\,(1 - \alpha)$ *percentage point* of the standard normal distribution, where z_α is defined by:

$$P(z \geq z_\alpha) = \int_{z_\alpha}^{\infty} \phi(z)\, dz = \alpha.$$

Or expressed another way:

$$\Phi(z_\alpha) = P(z \leq z_\alpha) = 1 - \alpha.$$

Note that $z_\alpha = -z_{1-\alpha}$. For convenience, we list a few of the most important and commonly used values of z_α in Table 7.2.1.

TABLE 7.2.1

SOME PERCENTAGE POINTS OF THE NORMAL DISTRIBUTION

α	.10	.05	.025	.01	.005
z_α	1.282	1.645	1.960	2.326	2.576

Percentage points are often called *significance points*. For example, $z_{.05} = 1.645$ is the 5% significance point of the standard normal distribution.

7.3 APPROXIMATION OF THE BINOMIAL DISTRIBUTION BY THE NORMAL DISTRIBUTION

We now turn to a remarkable application of the normal distribution, namely, that of approximating probabilities associated with the binomial distribution. We recall from Chaper 6 that, if x is a random variable having the binomial distribution $b(x)$, then its mean is $n\theta$, and its variance is $n\theta(1 - \theta)$. We now state the following:

THEOREM 7.3.1. *As* $n \to \infty$, *the distribution of the random variable*

$$z = \frac{x - n\theta}{\sqrt{n\theta(1 - \theta)}}$$

has as its limiting distribution, $N(0, 1)$.

This theorem was first proved by the French mathematician A. DeMoivre in 1733. The proof of this theorem is given in Appendix IV. Use of this theorem enables us to approximate, for large n, sums of probabilities given by the binomial distribution by appropriate integrals of the standard normal distribution. More precisely, the theorem states that, for large n, the random variable x has approximately the normal distribution $N(n\theta, n\theta(1 - \theta))$. The approximation improves as n increases and is quite good for values of θ not too close to zero or one.

Example 7.3(a). If x is a random variable having the binomial p.f.

$$b(x) = \binom{16}{x}\left(\frac{1}{2}\right)^x\left(\frac{1}{2}\right)^{16-x}, \qquad x = 0, 1, 2, \ldots, 16,$$

approximate the value of $P(6 \leq x \leq 10)$ using the normal distribution. Here, $n\theta = 8$, $n\theta(1 - \theta) = 4$. The exact value of the required probability is

$$P_B(6 \leq x \leq 10) = \sum_{x=6}^{10} \binom{16}{x}\left(\frac{1}{2}\right)^{16}.$$

By using the normal approximation, we have

$$P_B(6 \leq x \leq 10) \cong P_N(5.5 \leq x \leq 10.5)$$

$$= P_N\left(\frac{5.5 - 8}{2} \leq \frac{x - 8}{2} \leq \frac{10.5 - 8}{2}\right)$$

$$= P_N(-1.25 \leq z \leq 1.25)$$

$$= .8944 - .1056$$

$$= .7888.$$

That is, the normal approximation for $P_B(6 \leq x \leq 10)$ is .7888. The exact value of $P_B(6 \leq x \leq 10)$ is .7898.

The reason for using 5.5 and 10.5 rather than 6 and 10 in P_N (usually called *half-integer corrections for continuity*) is evident when we look at Figure 7.3.1. If we graph the probabilities given by the binomial p.f. $b(x)$ by drawing rectangles having bases equal to 1 and centered at $x = 0$, $1, \ldots, 16$ and heights given by $b(0), b(1), \ldots, b(16)$, the area under the resulting probability *histogram* is 1, since each rectangle is of area $1b(x)$, and

$$\sum_{x=0}^{16} b(x) = 1.$$

When computing $P_B(6 \leq x \leq 10) = \sum_{x=6}^{10} b(x)$, we are summing areas of rectangles, the first of which has a base with left-hand end-point 5.5, and the

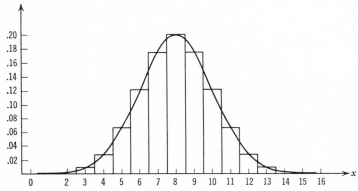

Figure 7.3.1. Graphs of the binomial probability histogram in Example 7.3(a) and of the approximating normal distribution $N(8, 4)$.

last rectangle has a base with right-hand end-point 10.5. Now if we approximate $P_B(6 \leq x \leq 10)$ by $P_N(6 \leq x \leq 10)$, that is, if we do not make the half-integer correction, we are, in effect, omitting about half of the first and half of the last rectangles from consideration, thus underestimating the required probability.

We can also approximate individual binomial probabilities. For instance,

$$P_B(x = 8) \cong P_N(7\tfrac{1}{2} \leq x \leq 8\tfrac{1}{2})$$
$$= P_N\left(-\frac{.5}{2} \leq z \leq \frac{.5}{2}\right)$$
$$= P_N(-.25 \leq z \leq .25)$$
$$= .5987 - .4013$$
$$= .1974.$$

The exact value of $P_B(x = 8)$ is .1964.

Thus, we see from Example 7.3(a) that, to approximate binomial probabilities by the normal distribution, the procedure involves setting the mean and variance of the binomial distribution equal to the mean and variance of the approximating normal distribution, and then making the half-integer corrections for continuity.

Example 7.3(b). A machine produces items in a certain plant for a mass-production process. Its record is such that 5% of the items produced are defective. (1) If a sample of 1000 items is chosen at random, what is the probability that not more than 40 defectives occur in the sample? (2) That between 40 and 60, inclusive, of the items in the sample are defective? Here, $n\theta = 1000 \times 1/20 = 50$ and $n\theta(1 - \theta) = 190/4$, that is,

$$\sqrt{n\theta(1 - \theta)} = 6.892.$$

Using the normal approximation, we have

1. $P_B(x \leq 40) \cong P_N(x \leq 40.5)$

$$= P_N\left(z \leq \frac{40.5 - 50}{6.892}\right)$$

$$= P_N\left(z \leq \frac{-9.50}{6.892}\right)$$

$$= P_N(z \leq -1.38) = .084,$$

that is,

$$P_B(x \leq 40) \cong .084.$$

2. $P_B(40 \leq x \leq 60) \cong P_N(39.5 \leq x \leq 60.5)$

$$= P_N\left(\frac{-10.5}{6.892} \leq z \leq \frac{10.5}{6.892}\right)$$

$$= P_N(-1.52 \leq z \leq 1.52) = .871.$$

that is,

$$P_B(40 \leq x \leq 60) \cong .871.$$

7.4 MEAN AND VARIANCE OF LINEAR FUNCTIONS OF INDEPENDENT NORMAL RANDOM VARIABLES

We know from Chapter 6 that, if $x_1, \ldots, x_n$ are independent random variables with means and variances $\mu_i, \sigma_i^2, i = 1, \ldots, n$, respectively, then a linear combination of the x_i, say

$$L = \sum_{i=1}^{n} c_i x_i,$$

is such that

(7.4.1) $\mu_L = E(L) = \Sigma c_i \mu_i,$

and

(7.4.2) $\sigma_L^2 = E(L - \mu_L)^2 = \sum_{i=1}^{n} c_i^2 \sigma_i^2.$

Indeed, if x_i is an $N(\mu_i, \sigma_i^2)$ random variable, then the theorems of Appendix II state that L is normally distributed, with mean μ_L and variance σ_L^2. In particular, if the x_i are $N(\mu, \sigma^2)$ random variables and are independent, that is, if $(x_1, \ldots, x_n)$ is a random sample of size n from $N(\mu, \sigma^2)$, then

$$\bar{x} = \frac{1}{n} \sum x_i = \sum \frac{1}{n} x_i$$

is a linear conbination of the x_i with $c_i = 1/n$.

Hence we have that $\bar{x}$ is itself normally distributed with mean and variance given by

(7.4.3) $$\mu_{\bar{x}} = \sum_{i=1}^{n} c_i \mu_i = \sum_{i=1}^{n} \frac{1}{n} \mu = \mu,$$

(7.4.4) $$\sigma_{\bar{x}}^2 = \sum c_i^2 \sigma_i^2 = \sum_{i=1}^{n} \frac{1}{n^2} \sigma^2 = \frac{\sigma^2}{n}.$$

This is a very important result which we restate in the following:

THEOREM 7.4.1. *If $(x_1, \ldots, x_n)$ is a random sample of size n from $N(\mu, \sigma^2)$, then $\bar{x}$ is an $N(\mu, \sigma^2/n)$ random variable.*

If T is the sample sum defined by

(7.4.5) $$T = x_1 + \cdots + x_n = \sum_{i=1}^{n} x_i,$$

the reader can similarly verify the following:

THEOREM 7.4.2. *If $(x_1, \ldots, x_n)$ is a random sample of size n from $N(\mu, \sigma^2)$, then T is an $N(n\mu, n\sigma^2)$ random variable.*

If the sample $(x_1, \ldots, x_n)$ is from any population having mean μ and variance σ^2 (both finite), it can be shown that, as $n \to \infty$, $(T - n\mu)/\sqrt{n}\sigma$ [or equivalently $(\bar{x} - \mu)\sqrt{n}/\sigma$] is a random variable having $N(0, 1)$ as its limiting distribution. That is, even though x_i, $i = 1, \ldots, n$, is a random sample of size n from a nonnormal distribution, the Central Limit Theorem states that, if n is large, T is approximately an $N(n\mu, n\sigma^2)$ random variable, with error of approximation tending to zero as $n \to \infty$. (See Appendix IV.)

In engineering statistics, there will often arise a need to compare two populations or processes involving two independent random variables x and y, in which case it is useful to examine the difference or *contrast* $L = x - y$. If we know the means and variances of x and y, we can compute the mean and variance of L from (7.4.1) and (7.4.2).

Let us then suppose that $E(x) = \mu_x$, $E(y) = \mu_y$, Var $(x) = \sigma_x^2$, Var $y = \sigma_y^2$. By using (7.4.1) and (7.4.2), we find

(7.4.6) $$\mu_L = E(L) = \mu_x - \mu_y, \checkmark$$

(7.4.7) $$\sigma_L^2 = \text{Var } (L) = \sigma_x^2 + \sigma_y^2,$$

and hence,

$$\sigma_L = \sqrt{\sigma_x^2 + \sigma_y^2}.$$

Furthermore, it follows from Appendix II that, if x and y are normally distributed, so is L.

Example 7.4(a). In manufacturing precision-made pins, assume that 1/2-inch pins have diameters which are (approximately) normally distributed with mean .4995 inch and standard deviation .0003 inch, and that matching parts which have holes to receive these pins have diameters which are (approximately) normally distributed with mean .5005 inch and standard deviation .0004 inch.

If pins and holes are matched at random, in what fraction of the matches would the pin fit?

We let d_p be the diameter in inches of a pin, and d_m be the diameter in inches of the hole of a matching part. Consider $d = d_m - d_p$. Then

$$\mu_d = \mu_{d_m} - \mu_{d_p} = .5005 - .4995 = .0010 \text{ inch}$$

and

$$\sigma_d^2 = \sigma_{d_m}^2 + \sigma_{d_p}^2 = 9 \times 10^{-8} + 16 \times 10^{-8} = 25 \times 10^{-8},$$

so that the standard deviation of d is $\sigma_d = 5 \times 10^{-4} = .0005$ inch. Furthermore, d is a linear combination of normally distributed variables and thus is itself normally distributed. Now, a pin fits a matching part if and only if $d > 0$, and the probability of this is

$$P(d > 0) = P\left(\frac{d - .0010}{.0005} > \frac{-.0010}{.0005}\right)$$

$$= P(z > -2) = .9772,$$

that is, in 97.72% of the matches the pins would fit.

Example 7.4(b). The distribution of gross weights of 8-ounce boxes of corn flakes has mean 9.60 ounces and standard deviation .80 ounce respectively. If the boxes are packaged 24 per carton, and if the population of weights of empty cartons has mean 24.00 ounces and standard deviation of 2.20 ounces, what are the mean and the variance of the population of weights of filled cartons? What percent of the filled cartons will have weights between 250 and 260 ounces?

Let T be the total weight of 24 boxes of corn flakes,

$$T = x_1 + x_2 + \cdots + x_{24}.$$

From (7.4.1) and (7.4.2) we have

$$\mu_T = \mu_x + \mu_x + \cdots + \mu_x = 24\mu_x = 24(9.60) = 230.4 \text{ ounces}$$

and

$$\sigma_T^2 = \sigma_x^2 + \sigma_x^2 + \cdots + \sigma_x^2 = 24\sigma_x^2 = 24(.64) = 15.36 \text{ ounces squared.}$$

Let $W = T + Y$, where Y is the weight of an empty carton. Then $\mu_W = \mu_T + \mu_Y = 230.4 + 24.0 = 254.4$ ounces and

$$\sigma_W^2 = \sigma_T^2 + \sigma_Y^2 = 15.36 + 4.84 = 20.2 \text{ ounces squared.}$$

To answer the second question, we must evaluate $P(250 < W < 260)$, which, when reduced in terms of the standard normal variable z, has the value $P(-.979 < z < 1.246) = \Phi(1.246) - \Phi(-.979) = .7299$. Thus, approximately 73% of the filled cartons have weights between 250 and 260 ounces. $= (0.8944 - (1 - 0.8365) = 0.8944 - 0.1635$

7.5 NORMAL PROBABILITY PAPER $= 0.7309 = 73\%$

Normal probability graph paper has the property that the graph of the c.d.f. of a normal distribution, when plotted on this paper, is a straight line. That is, by an appropriate stretch of the scale of ordinates for low and high percentages, the graph of the c.d.f. of an $N(\mu, \sigma^2)$ random variable, as shown in Figure 7.2.2, is transformed into a straight line, as shown in Figure 7.5.1.

Normal probability paper can be utilized in two ways. It provides a "rough" check on whether a sample can reasonably be regarded as having

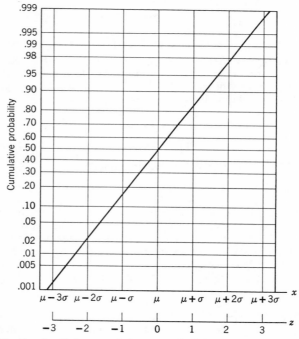

Figure 7.5.1. Graph of the cumulative normal distribution function on probability graph paper.

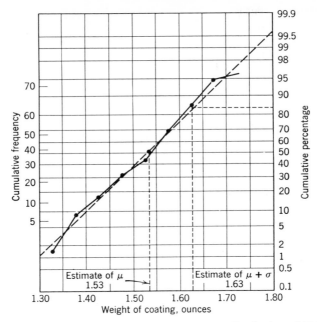

Figure 7.5.2. Graph of the cumulative relative frequency distribution of Table 5.2.1, Chapter 5, on probability graph paper.

come from a normal population and, if so, whether it can be used to obtain quick and easy estimates of μ and σ as follows. The 50th percentile of the cumulative normal distribution is μ, the mean. Consulting Table II, Appendix VII, the 84th (more precisely the 84.13th) percentile is $\mu + \sigma$. Thus, if the cumulative polygon of the sample is graphed on normal probability paper and if this polygon is approximated by a straight line, then this line is an estimate of the normal c.d.f. when graphed on normal probability paper. Furthermore, an estimate of μ is the 50th percentile of this c.d.f. (i.e., the straight line), and an estimate of σ is the difference between the 84th and 50th percentiles of the c.d.f. (i.e., the straight line).

As an example, suppose we refer to Table 5.2.1, Chapter 5. We plot the cumulative frequency polygon against the upper-class boundaries in Figure 7.5.2. We fit the polygon as best we can "by eye." Note that the points seem to be sufficiently close to a straight line so that we may consider the data to be from a normal population.

Taking the 50th and 84th percentiles (from the straight line), we see that these correspond to 1.53 and 1.63 ounces respectively; that is, the graphical estimates of μ is 1.53 and of σ is $1.63 - 1.53 = .1$ ounce. Note that these graph estimates agree closely with the computed values obtained in Section 5.4, namely $\bar{x} = 1.527$ and $s_x = .101$.

The use of normal probability paper when sample sizes are small is discussed in Section 13.7.

PROBLEMS

7.1. A random variable x has the distribution $N(1500, (200)^2)$. Find:
(a) $P(x < 1400)$;
(b) $P(x > 1700)$;
(c) A, where $P(x > A) = .05$;
(d) B, where $P(1500 - B < x < 1500 + B) = .95$.

7.2. If x has the distribution $N(15.0, 6.25)$, find:
(a) $P(x < 12.0)$;
(b) $P(x > 16.5)$;
(c) C, where $P(x < C) = .90$;
(d) D, where $P(x > D) = .025$;
(e) E, where $P(|x - 15.0| < E) = .99$.

7.3. Suppose a machine set for filling 1-pound boxes of sugar yields a population of fillings the weights of which are (approximately) normally distributed with a mean of 16.30 ounces and a standard deviation of .15 ounce. Estimate:
(a) The percentage of fillings which will be under weight (that is, less than 1 pound).
(b) The percentage of fillings within $16.3 \pm .2$ ounce.

7.4. Show that $P(a < x < a + l)$, where l is a positive constant and x has the distribution $N(\mu, \sigma^2)$, is maximized if $a = \mu - l/2$.

7.5. A process for making 1/4-inch ball bearings yields a population of ball bearings with diameters having mean .2497 inch and standard deviation of .0002 inch. If we assume approximate normality of diameters, and if specifications call for bearings with diameters within $.2500 \pm .0003$ inch:
(a) What fraction of the bearings turned out under the setup are defective (that is, do not meet diameter specifications)?
(b) If minor adjustments of the process result in changing the mean diameter but not the standard deviation, what mean should be aimed at in the process setup so as to minimize the percentage of defectives?
(c) What will be the percentage of defectives in such a setup?

7.6. In the Example 7.4(a), if "acceptable fits" are those in which the difference between hole diameter and pin diameter lies within $.0010 \pm .0005$ inch, what fraction of random matches would yield "acceptable fits"?

7.7. For each of Problems 5.6 through 5.8 do the following:
(a) On probability paper, plot the cumulative relative frequency against the upper-class boundaries, and (see Section 7.5) fit a straight line "by eye."
(b) Graphically estimate the mean and standard deviation of the distribution.

7.8. In a study of the variation of a critical dimension of electric contacts, the distribution of measurements on this dimension in a sample of 1000 contacts turned out to be as follows:

Class Midpoint (inches)	Number of Contacts (frequency)
.412	6
.417	34
.422	132
.427	179
.432	218
.437	183
.442	146
.447	69
.452	30
.457	3
Total	1000

Carry out parts (a) and (b) of Problem 7.7 for this set of data, and estimate, from the graph, what fraction of the contacts have critical dimension less than .430 inch.

7.9. If 825 random digits are taken from a random number table:
(a) What is the approximate probability that the total of the digits lies between 3700 and 3800 inclusive?
(b) For what value of K is the probability .95 that the total of the digits is $< K$?

7.10. Suppose 20% of the articles produced by a machine are defective, the defectives occurring at random during production. With the use of the normal distribution for making approximations:
(a) What is the probability that, if a sample of 400 items is taken from the production, more than 100 will be defective?
(b) For what value of K is the probability .95 that the number of defectives in the sample will fall within $80 \pm K$?

7.11. A sack of 400 nickels is emptied on a table and spread out:
(a) What is the probability (approximately) of getting between 175 and 225 heads?
(b) What is the probability (approximately) that the number of heads will be less than y? [Express the answer in terms of the function $\Phi(z)$.]

7.12. If 10% of the articles produced by a given process are defective:
(a) What is the probability (approximately) that more than 15% of a random sample of 400 items will be defective?
(b) For what value of K is the probability (approximately) .90 that the number of defectives in a sample of 400 will lie within $40 \pm K$?

7.13. It is known that the probability of dealing a bridge hand with at least one ace is approximately .7. If a person plays 100 hands of bridge, what is the approximate probability:
(a) That the number of hands he will receive containing at least one ace will be between 60 and 80 inclusive?
(b) That he will receive at most 20 hands with no aces?

7.14. A die is rolled 720 times. Using a normal distribution to approximate probabilities, estimate the probability that:

(a) More than 130 "sixes" will turn up.

(b) The number of "sixes" obtained will lie between 100 and 140 inclusive.

7.15. A mass-produced laminated item is made up of five layers. A study of the thickness of individual layers shows that the two outside layers have mean thickness of .062 inch and the three middle layers have mean thickness .042 inch. The standard deviation of thickness of outside layers is .004 inch and that for inside layers is .003 inch.

(a) If random assembly is employed, what will be the mean thicknesses of the laminated items?

(b) What will be the standard deviation of the thicknesses of the laminated items?

(c) Assuming the thicknesses of the individual sections to be approximately normally distributed, what percentage of items will have thicknesses between .240 inch and .260 inch?

(d) For what value of K will 90% of the items have thicknesses falling within .250 ± K?

7.16. An article is made up of three parts A, B, and C. The weights of the A's have an (approximately) normal distribution with mean 2.05 ounces and standard deviation .03 ounce, those of the B's have an (approximately) normal distribution with mean 3.10 ounces and standard deviation .04 ounce and those of the C's have an (approximately) normal distribution with mean 10.5 ounces and deviation .12 ounce.

(a) What fraction of the articles have weights exceeding 1 pound?

(b) The probability is .95 that four articles picked at random will have a total weight less than what value?

7.17. Suppose items of a certain kind are counted by weighing and boxed 100 to the box. The population of individual items has a mean weight of 1.45 ounces and standard deviation of .02 ounce. A batch of items weighing between 144.5 and 145.5 items is counted as 100 items. What is the probability:

(a) That a batch of 100 items will be counted correctly?

(b) That 101 items will be passed as 100?

7.18. A resistor is composed of eight component parts soldered together in series, and the total resistance of the resistor equals the sum of the resistances of the component parts. Three of the components are drawn from a production lot having a mean of 200 ohms and a standard deviation of 2 ohms; four components from a lot having a mean of 150 ohms and standard deviation of 3 ohms, and one component from a lot having a mean of 250 ohms and standard deviation of 1 ohm. If we assume (approximate) normality of the distribution of total resistances:

(a) Five percent of such resistors would have a resistance less than what value?

(b) What is the probability that a sample of four such resistors manufactured from these components will have an average resistance in excess of 1443 ohms?

CHAPTER 8

Acceptance Sampling Plans Based on Sample Means and Sample Variances

8.1 REVIEW OF ACCEPTANCE SAMPLING CONCEPTS IN ATTRIBUTE SAMPLING

The acceptance sampling plans which we have already considered in Chapter 4 are frequently called *acceptance sampling plans based on attributes*. Under these plans a decision is made to accept or reject a lot depending on the number of defective items found in a random sample from the lot. More precisely, a *single sampling plan* based on attributes for large lots is specified by two numbers (n, c), where n is the number of items drawn in the sample and c is the *allowable* number of defectives in the sample. Under such a plan a sample of size n is drawn from the lot under consideration. If the number of defectives in the sample is less than or equal to c, the lot is accepted without further inspection. If the sample contains more than c defectives, the lot is rejected. *Double sampling plans* were also considered in Chapter 4. Such plans for large lots depend on four numbers (n_1, n_2, c_1, c_2) as explained in Section 4.5.

It will be recalled that the probability of accepting a lot having a fraction of defectives equal to θ by either a single or double sampling plan was denoted by $P(A \mid \theta)$. The graph of $P(A \mid \theta)$ as a function of θ is called the *operating characteristic curve* (OC curve) of the sampling plan. For large lots where rejected lots can be *rectified* (that is, where testing is nondestructive so that defectives in a rejected lot can be identified and replaced by nondefectives) the quantity $\theta P(A \mid \theta)$ is called the *average outgoing quality* [AOQ(θ)] for incoming fraction defective θ. AOQ(θ) is a function of θ and is simply the fraction of defectives left in a large pool of inspected lots in which rejected lots are rectified, each having fraction defective equal to θ before inspection. It is to be emphasized that for the AOQ concept to apply it is necessary to have *rectifiability* of a lot. The

155

maximum ordinate of the graph of AOQ(θ) is called the *average outgoing quality limit* (AOQL), and it constitutes the upper limit of the fraction of defectives in a large pool of rectifiable lots screened of defectives by the given sampling plan.

8.2   ONE-SIDED ACCEPTANCE SAMPLING PLANS BASED ON SAMPLE MEANS

In acceptance sampling by attributes the only information about an inspected item utilized is whether the item is defective or not. For instance, if such a plan is used for inspecting a sample of steel rods for yield point a test piece might be considered as defective (or non-defective) depending on whether its yield point is below (or above) a specified value, say 50,000 pounds per square inch (psi). Thus, if the actual value of the psi is obtained for each test piece, only a fraction of this information is utilized if an attribute sampling plan is used, namely, the total number of test pieces in the sample having yield points less than 50,000 psi.

By using the actual values of the yield points in the sample rather than the number of yield point values less than 50,000 psi, we can develop a more sensitive sampling plan which, in general, requires smaller sample sizes than those for attribute sampling. To pursue the steel-rod testing example further, suppose it has been found from experience with many lots of such test pieces that the standard deviation σ of yield points in any lot is (approximately) 2500 psi, and that yield points in any given lot have been found to have approximately a normal distribution. Any test piece which has yield point less than 50,000 psi is considered to be defective. Now let us consider a sampling plan for deciding whether to accept or reject a large lot which works as follows:

1. A sample of n rods is taken from the lot and tested for yield point.
2. The lot is accepted if the sample mean $\bar{x}$ exceeds some number k.
3. It is rejected if $\bar{x}$ does not exceed k.

The numbers k and n are to be determined subject to the following risks.

(*a*) If the lot mean is 55,000 psi, the probability of accepting the lot is .10. (Consumer's risk is .10.)

(*b*) If the lot mean is 57,000 psi, the probability of accepting the lot is .95. (Producer's risk is .05.)

The solution of the problem is as follows: If we draw a sample of n rods from a lot with mean 55,000 psi (and standard deviation 2500 psi), then we know from Chapter 7, that the sample mean $\bar{x}$ has approximately the normal distribution $N(55,000, (2500/\sqrt{n})^2)$ (if the distribution of yield point values does not differ too much from a normal distribution).

Hence,

$$\frac{(\bar{x} - 55{,}000)\sqrt{n}}{2500}$$

is a z random variable [that is, one having approximately the normal distribution $N(0, 1)$], and according to (a) we must have

(8.2.1) $P(\bar{x} > k \mid \mu = 55{,}000)$

$$= P\left[\frac{(\bar{x} - 55{,}000)\sqrt{n}}{2500} > \frac{(k - 55{,}000)\sqrt{n}}{2500}\right] = .10,$$

and hence from Table II, Appendix VII, we find

(8.2.2) $$\frac{(k - 55{,}000)\sqrt{n}}{2500} = 1.282.$$

If we draw a sample of size n from a lot with mean 57,000 (and standard deviation 2500 psi), then

$$\frac{(\bar{x} - 57{,}000)\sqrt{n}}{2500}$$

is similarly a z random variable, and according to (b) we must have

(8.2.3) $P(\bar{x} > k \mid \mu = 57{,}000)$

$$= P\left[\frac{(\bar{x} - 57{,}000)\sqrt{n}}{2500} > \frac{(k - 57{,}000)\sqrt{n}}{2500}\right] = .95,$$

and hence from Table II, Appendix VII, we find

(8.2.4) $$\frac{(k - 57{,}000)\sqrt{n}}{2500} = -1.645.$$

The solution of (8.2.2) and (8.2.4) yields the required values of k and n, namely,

$$n = 14 \text{ (actual value 13.4),}$$
$$k = 55{,}876 \text{ psi.}$$

The solution we have obtained is shown graphically in Figure 8.2.1. The *acceptance set* of values of $\bar{x}$ is the interval $(55{,}876, +\infty)$, that is, the heavy portion of the $\bar{x}$-axis to the right of the value of $k \,(= 55{,}876)$. Curve I is the graph of the distribution of the random variable $\bar{x}$ for $n = 14$ if the mean pounds per square inch in the lot were 55,000 psi, and curve II is the graph of the distribution for $\bar{x}$ for $n = 14$ if the lot mean were 57,000 psi. The area under curve I to the right of k is .10, approximately, that is, for $n = 14$, $P(\bar{x} > 55{,}876 \mid \mu = 55{,}000) \cong .10$, and the area under curve II to the right of k is approximately .95, that is,

$$P(\bar{x} > 55{,}876 \mid \mu = 57{,}000) = .95.$$

To get some notion of the risks involved in a sample of size 14, if only attribute sampling were used in this example, suppose we consider the following sampling plan:

1. Draw a sample of 14 from the lot.
2. Accept the lot if the sample contains no defectives, that is, no rods with yield point $<50{,}000$ psi, otherwise reject. The OC function in this case is

$$P(A \mid \theta) = (1 - \theta)^{14},$$

where, of course, θ is the probability that a rod is defective.

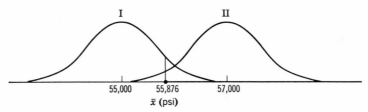

Figure 8.2.1. Acceptance set of values of $\bar{x}$ for the sampling plan of the example in Section 8.2.

If the sample is drawn from a lot in which yield points (in pounds per square inch) of rods have a normal distribution with mean 55,000 psi and standard deviation 2500 psi, θ is approximately .023, since

$$\theta = P(X < 50{,}000),$$

where

$$X = N[55{,}000, (2500)^2];$$

then

$$P(A \mid .023) = (.977)^{14} = .722.$$

The consumer's risk is thus .722.

If the sample is drawn from a lot with mean 57,000 psi and standard deviation 2500 psi, we have $\theta = .003$, since $\theta = P(X < 50{,}000)$, where $X = N[57{,}000, (2500)^2]$; then

$$P(A \mid .003) = (.997)^{14} = .959.$$

The producer's risk is .041, which is slightly less than .05. This means roughly that a sampling plan based on $\bar{x}$ for $n = 14$ is more sensitive than an attribute sampling plan based on number of defectives in a sample of 14.

8.3   GENERALIZATION

Although the foregoing discussion is presented in terms of a specific example, it can be readily extended to the more general situation where we have to determine k and n so that, for *known* values of σ and specified values of μ_1 and μ_2 ($\mu_1 < \mu_2$, say), we would have the following relation when the sample is drawn from a lot with mean μ_1 and standard deviation σ:

$$(8.3.1) \quad P(\bar{x} > k \mid \mu = \mu_1) = P\left[\frac{(\bar{x} - \mu_1)\sqrt{n}}{\sigma} > \frac{(k - \mu_1)\sqrt{n}}{\sigma}\right] = \beta,$$

and the following relation when the sample is drawn from a lot with a mean μ_2 and standard deviation σ:

$$(8.3.2) \quad P(\bar{x} > k \mid \mu = \mu_2) = P\left[\frac{(\bar{x} - \mu_2)\sqrt{n}}{\sigma} > \frac{(k - \mu_2)\sqrt{n}}{\sigma}\right] = 1 - \alpha.$$

The quantities α and β are producer's and consumer's risks. It follows from (8.3.1) and (8.3.2) that n and k would be obtained by solving the following equations:

$$(8.3.3) \quad \begin{aligned} 1 - \Phi\left[\frac{(k - \mu_1)\sqrt{n}}{\sigma}\right] &= \beta, \\ 1 - \Phi\left[\frac{(k - \mu_2)\sqrt{n}}{\sigma}\right] &= 1 - \alpha. \end{aligned}$$

In any given problem values would be given for α, β, μ_1, μ_2, σ, thus leaving only n and k to be determined from the two equations in (8.3.3).

8.4   TWO-SIDED ACCEPTANCE SAMPLING PLANS BASED ON SAMPLE MEANS

In some problems it is not only important to devise a sampling plan that will reject lots with low means with high probability but also to reject lots with high means with high probability. For example, in lots of packaged goods it is important to have a scheme which will reject lots with critically low average net weights to avoid the risk of prosecution for short-weighing, and which will reject lots with unnecessarily high average net weights in order to avoid giving away too much product. Or in the case of muzzle velocity of ammunition, it is undesirable to have velocities which are too low, and it is dangerous to the gunner to have muzzle velocities too high since this can cause the gun to explode during firing.

We now consider such a *two-sided sampling plan* based on sample means. It will be sufficient to conduct the discussion entirely in terms of a specific example.

Suppose the following sampling plan is devised to accept or reject large lots of rifle ammunition of a certain type:

1. A sample of n rounds is taken from the lot and muzzle velocities are determined for each round.

2. If the mean muzzle velocity $\bar{x}$ of the sample lies between $1750 \pm k$ the lot is accepted; otherwise the lot is rejected. Assume that it is known from experience that the standard deviation of velocities of rounds of ammunition in lots of this type, for any lot having mean velocity near 1750 feet per second, is 15 feet per second. We wish to find n and k so that:

(a) If the average muzzle velocity of the lot is 1740 feet per second, the probability of acceptance is .05.

(b) If the average muzzle velocity of the lot is 1745 feet per second, the probability of acceptance is .95.

(c) If the average muzzle velocity of the lot is 1755 feet per second, the probability of acceptance is .95.

(d) If the average muzzle velocity of the lot is 1760 feet per second, the probability of acceptance is .05.

The four conditions (a), (b), (c), (d) establish the following equations, where μ denotes average muzzle velocity of the lot.

(a') $P[1750 - k < \bar{x} < 1750 + k] = .05,$ if $\mu = 1740,$

(b') $P[1750 - k < \bar{x} < 1750 + k] = .95,$ if $\mu = 1745,$

(c') $P[1750 - k < \bar{x} < 1750 + k] = .95,$ if $\mu = 1755,$

(d') $P[1750 - k < \bar{x} < 1750 + k] = .05,$ if $\mu = 1760.$

These four equations may now be written as:

(a'') $P\left[\dfrac{(10 - k)\sqrt{n}}{15} < \dfrac{(\bar{x} - 1740)\sqrt{n}}{15} < \dfrac{(10 + k)\sqrt{n}}{15}\right] = .05,$

(b'') $P\left[\dfrac{(5 - k)\sqrt{n}}{15} < \dfrac{(\bar{x} - 1745)\sqrt{n}}{15} < \dfrac{(5 + k)\sqrt{n}}{15}\right] = .95,$

(c'') $P\left[\dfrac{(-5 - k)\sqrt{n}}{15} < \dfrac{(\bar{x} - 1755)\sqrt{n}}{15} < \dfrac{(-5 + k)\sqrt{n}}{15}\right] = .95,$

(d'') $P\left[\dfrac{(-10 - k)\sqrt{n}}{15} < \dfrac{(\bar{x} - 1760)\sqrt{n}}{15} < \dfrac{(-10 + k)\sqrt{n}}{15}\right] = .05.$

Under the assumptions we have made in (a), (b), (c), (d), the quantities

$$\frac{(\bar{x} - 1740)\sqrt{n}}{15}, \quad \frac{(\bar{x} - 1745)\sqrt{n}}{15},$$

$$\frac{(\bar{x} - 1755)\sqrt{n}}{15}, \quad \frac{(\bar{x} - 1760)\sqrt{n}}{15},$$

are all standard normal variables [they all have the normal distribution $N(0, 1)$]. Furthermore, we assume* that $[(10 + k)\sqrt{n}]/15$ will be large enough so that we may write (a''), with a high degree of approximation, as:

(a''')
$$P\left[\frac{(10 - k)\sqrt{n}}{15} < z\right] = .05.$$

Similarly, we may write with high degrees of approximation:

(b''')
$$P\left[\frac{(5 - k)\sqrt{n}}{15} < z\right] = .95,$$

(c''')
$$P\left[z < \frac{(-5 + k)\sqrt{n}}{15}\right] = .95,$$

(d''')
$$P\left[z < \frac{(-10 + k)\sqrt{n}}{15}\right] = .05.$$

Because of symmetry of the distribution of the standard normal variable z, the solution of (a''') and (b''') for n and k is equivalent to that for n and k from (c''') and (d'''). From (a''') and (b''') we, therefore, have

$$\frac{(10 - k)\sqrt{n}}{15} = +1.645,$$

$$\frac{(5 - k)\sqrt{n}}{15} = -1.645,$$

from which we find

$$n = 97,$$
$$k = 7.5.$$

Thus, the solution of our problem is to fire $n = 97$ rounds and to accept the lot if the mean muzzle velocity $\bar{x}$ lies in the interval $(1742.5, 1757.5)$, and reject otherwise.

* This assumption should be checked after n and k are determined, that is, one should verify that $P\{z > [(10 + k)\sqrt{n}]/15\}$ is approximately zero. Similar checks should be made for (b''), (c''), and (d'').

PROBLEMS

8.1. An acceptance sampling plan to be used for accepting (or rejecting) large lots of a certain type of brick is to operate as follows: (1) Draw a random sample of n bricks from the lot and find the average crushing strength $\bar{x}$ (in pounds per square inch) of the n bricks. (2) If $\bar{x} > k$, accept the lot. (3) If $\bar{x} \leq k$, reject the lot.

(a) Find n and k so that if the mean crushing strength μ in the lot is 1500 psi, the probability of acceptance is only .05, whereas the probability of acceptance is .90 if $\mu = 1550$ psi. Assume that it is known by experience that the standard deviation of crushing strengths of lots having means between 1400 psi and 1700 psi is (approximately) 80 psi.

(b) Graph the OC curve of this sampling plan (that is, graph the probability of acceptance as a function of μ).

8.2. In a large lot of mass-produced items, it is known that the standard deviation of weights of these items is .5 ounce. An acceptance sampling plan requires that a sample of n items from the lot must have $\bar{x} > 32.1$ ounces in order for the lot to be accepted; otherwise the lot is rejected.

(a) Find n so that the probability is .95 of accepting the lot if the mean of the lot is 32.2 ounces. Also, find the probability of rejecting the lot for this value of n if the mean of the lot is 32.0 ounces.

(b) Graph the OC curve of this sampling plan (as a function of the lot mean μ).

8.3. A sampling plan for accepting lots of balls of yarn is to be established as follows: (1) A test piece is taken from each of n balls randomly selected from the lot. (2) If the mean tensile strength $\bar{x}$ of the n test pieces is greater than k pounds, the lot is accepted. Otherwise the lot is rejected.

(a) Find n and k such that the probability is .95 of accepting the lot if the lot mean is 2.05 pounds, and is .05 if the lot mean is 2.00 pounds. (Assume that it is known from experience that the standard deviation of tensile strength for this type of yarn is .2 pound for any lot having average tensile strength near 2.00 pounds.)

(b) Graph the OC curve for this acceptance sampling plan (as a function of the lot mean μ).

8.4. One department of a textile manufacturer accepts lots of bobbins of cotton yarn from another department according to the following acceptance sampling plan: (1) A random sample of n bobbins of yarn is drawn and the mean breaking strength $\bar{x}$ (in pounds) of a test piece from each bobbin is found. (2) If $\bar{x} > k$, accept the lot. (3) If $\bar{x} \leq k$, reject the lot.

(a) Find n and k such that if $\mu = 6.2$ pounds, the probability of acceptance is .95, whereas if $\mu = 5.8$ pounds, the probability of acceptance is .10. Assume the standard deviation of the breaking strengths of this particular type of yarn is .5 pound for any lot having mean breaking strength near 6 pounds.

(b) Graph the OC curve for this sampling plan as a function of the mean breaking strength of the lot.

8.5. Large batches of items each having a spot weld are accepted or rejected according to the following acceptance sampling plan: (1) random samples of n items are selected and their spot welds are tested for shear strength (in pounds). The mean shear strength $\bar{x}$ is computed. (2) if $\bar{x} > k$, accept the lot. (3) If $\bar{x} \leq k$, reject the lot.

(a) Find n and k so that

$$P \text{ (acceptance} \mid \mu = 410 \text{ lb)} = .95,$$
$$P \text{ (acceptance} \mid \mu = 390 \text{ lb)} = .05.$$

It is known from experience that shear strength of this type of spot weld has a standard deviation of 40 pounds for any lot of spot welds having mean shear near 400 pounds.

(b) Graph the OC curve of the sampling plan (as a function of μ).

8.6. A company accepts or rejects batches of ore according to the percentage of nickel in a batch, using the following plan: (1) n test containers are randomly drawn from a batch of ore and the average percentage of nickel $\bar{x}$ is determined for each test container of ore. (2) If $\bar{x} > k$, the batch is accepted. (3) If $\bar{x} \leq k$, the batch is rejected.

(a) Find n and k so that, if $\mu = 4.05\%$, the probability of acceptance is .95 but if $\mu = 3.95$, the probability of acceptance is .10. Assume the standard deviation of percent nickel to be .05, for any batch of ore having a percentage of nickel near 4%.

(b) Graph the OC curve for this plan (as a function of μ).

8.7. Lots of wire strands are accepted or rejected according to the following plan: (1) n strands are randomly selected from a lot and the average tensile strength $\bar{x}$ is recorded. (2) If $\bar{x} > k$, accept the lot. (3) If $\bar{x} \leq k$, reject the lot.

(a) Find n and k so that, if $\mu = 15,600$, the probability of acceptance is .95, but if $\mu = 14,400$, the probability of acceptance is .10. Assume the standard deviation of breaking strengths is 600 psi for lots of wire strands of this type having mean tensile strength near 15,000 psi.

(b) Graph the OC curve for the plan (as a function of μ).

8.8. Lots of insecticide dispensers are accepted or rejected according to the following plan: (1) A sample of n dispensers is selected from the lot and $\bar{x}$, the average charge weight (in grams), is determined. (2) If $\bar{x} > k$, the lot is accepted. (3) If $\bar{x} \leq k$, the lot is rejected.

(a) Find n and k so that if $\mu = 465$ grams the probability of acceptance is .95, but if $\mu = 450$ grams, the probability of acceptance is .10. (Assume that the standard deviation of charge weights is 6 grams for any lot having mean charge weight near 450 grams.)

(b) Graph the OC curve for this plan (as a function of μ).

8.9. A brewery is required by law to have the contents of its cans of beer up to the level marked on the can, at the same time it does not wish to overfill the cans. An $\bar{x}$ sampling plan to control the weight of 12-ounce cans of beer in large lots is devised so as to operate as follows: (1) Draw a random sample of n cans of beer from a lot and determine $\bar{x}$, the sample mean of the filled weights of the n cans. (2) If $12.15 - d < \bar{x} < 12.15 + d$, the lot is "satisfactory." (3) If $\bar{x}$ does not satisfy (2), the lot is "unsatisfactory."

(a) Determine n so that the probability of the lot being "satisfactory" is:

.05 if lot mean is 12.05 ounces;
.95 if lot mean is 12.10 ounces;
.95 if lot mean is 12.20 ounces;
.05 if lot mean is 12.25 ounces.

Assume that it is known from previous statistical studies that the standard deviation of "filled weights" for 12-ounce cans is .05 ounce.

(b) Draw the OC curve showing the probability of accepting a lot as a function of mean filled content μ of the lot.

8.10. An electrical contractor uses the following sampling plan for checking uniformity of resistance in buying large lots of spools of copper wire: (1) He randomly samples the lot by drawing n spools and determines the resistance in ohms for a piece of wire of specified length taken from the end of each spool. He then determines the average resistance $\bar{x}$, in ohms. (2) If $.141 - d < \bar{x} < .141 + d$, the lot is accepted, otherwise the lot is rejected.

(a) Determine n and d so that the probability of accepting the lot is:

$$.05 \text{ if } \mu = .135 \text{ ohm};$$
$$.95 \text{ if } \mu = .139 \text{ ohm};$$
$$.95 \text{ if } \mu = .143 \text{ ohm};$$
$$.05 \text{ if } \mu = .147 \text{ ohm}.$$

(Assume the standard deviation of resistance measurements is .005 ohm for any lot having mean resistance near .14 ohm.)

(b) Draw the OC curve for this plan.

8.11. Hydroquinone is bought in lots according to the following sampling plan: (1) The lot is randomly sampled n times and the average melting point $\bar{x}$ is recorded in degrees centigrade. (2) If $173\ °C - d < \bar{x} < 173\ °C + d$, the lot is accepted; otherwise the lot is rejected.

(a) Determine n and d so that the probability of the lot being accepted is:

$$.05 \text{ if } \mu = 170\ °C;$$
$$.95 \text{ if } \mu = 172\ °C;$$
$$.95 \text{ if } \mu = 174\ °C;$$
$$.05 \text{ if } \mu = 176\ °C.$$

(Assume the standard deviation of melting points is 1.5 °C.)

(b) Graph the OC curve for this plan (as a function of μ).

Estimation of Population Parameters

9.1 INTRODUCTION*

We often encounter statistical problems of the following type: We have a large lot or population of objects such that, if a measurement were made on each object, we would have a distribution of these measurements. Since the measurements have not been made, this distribution is, of course, unknown. About the best we can hope to do in practice is to estimate various *characteristics* (sometimes referred to as *population parameters*) of this distribution from the information contained in measurements made in a random sample of objects from the lot or population. For instance, if we wish to estimate the mean and variance of the population distribution, it turns out that we can use the sample mean and sample variance as estimators for these quantities. If we want to estimate the median of the population, we can use the sample median. Other parameters of the population distribution can be estimated. We shall consider problems of this type in this chapter.

We shall consider two kinds of estimators for population parameters, namely *point estimators* and *interval estimators*. More specifically, suppose $(x_1, \ldots, x_n)$ is a sample from a population whose distribution has a parameter θ. If $t(x_1, \ldots, x_n)$ is a (single-valued) function of $x_1, \ldots, x_n$ which is itself a random variable, we shall refer to $t(x_1, \ldots, x_n)$ as a *statistic*. Furthermore, if

$$E(t(x_1, \ldots, x_n)) = \theta,$$

we say that $t(x_1, \ldots, x_n)$ is an *unbiased* estimator for θ. The statistic $t(x_1, \ldots, x_n)$ is sometimes referred to as a *point estimator* for the parameter θ.

---

* The reader should familiarize himself with the contents of Appendix VI before reading this chapter.

If $\underline{\theta}(x_1, \ldots, x_n)$ and $\bar{\theta}(x_1, \ldots, x_n)$ are two statistics such that

$$P[\underline{\theta}(x_1, \ldots, x_n) < \theta < \bar{\theta}(x_1, \ldots, x_n)] = 1 - \alpha,$$

we say that the random interval $(\underline{\theta}, \bar{\theta})$ is a $100(1 - \alpha)\%$ *confidence interval* for the parameter θ. The pair of statistics $(\underline{\theta}, \bar{\theta})$ is sometimes referred to as an *interval estimator* for θ. The end points $\underline{\theta}$ and $\bar{\theta}$ of the confidence interval $(\underline{\theta}, \bar{\theta})$ are sometimes called the $100(1 - \alpha)\%$ *confidence limits* of θ.

9.2 POINT ESTIMATORS FOR THE POPULATION MEAN AND VARIANCE

As important examples of point estimators, we shall consider the most commonly used point estimators for the mean and variance of a population.

Suppose we consider a population in which the variable (or measurement) x is continuous and has a p.d.f. $f(x)$. As we have seen in Chapter 6, the population mean μ and variance σ^2 are defined as follows:

$$(9.2.1) \qquad \mu = \int_{-\infty}^{\infty} xf(x)\,dx,$$

$$(9.2.2) \qquad \sigma^2 = \int_{-\infty}^{\infty} (x - \mu)^2 f(x)\,dx.$$

For the case of a population in which x is discrete and has p.f. $p(x)$, we similarly define μ and σ^2, using the operation of summation rather than integration.

If the population distribution is unknown, then μ and σ^2 are unknown. The basic question is this: how can we estimate μ and σ^2 from a random sample $(x_1, \ldots, x_n)$ drawn from the population. There are many ways of devising estimators. A simple point estimator for μ, and the most widely used, is the sample mean $\bar{x}$. The sample mean $\bar{x}$ is a random variable which has its own distribution which in turn, has its own mean and variance, namely:

$$(9.2.3) \qquad\qquad E(\bar{x}) = \mu$$

and

$$(9.2.4) \qquad\qquad \mathrm{Var}\,(\bar{x}) = \frac{\sigma^2}{n}.$$

The derivations of these results are given in Section 6.9 of Chapter 6.

It should be noted that the statistic $\bar{x}$ has a unique value for any given sample which can be represented as a point on a μ axis. If we consider an indefinitely large number of samples each of size n, then (9.2.3) essentially states that, if we were to average the $\bar{x}$'s of these samples, their average would be equal to μ; thus $\bar{x}$ is an unbiased (point) estimator for μ. Furthermore, we note from (9.2.4) that, if we were to determine the variance of all of these $\bar{x}$'s, it would be σ^2/n, which gives some indication as to how all of these $\bar{x}$'s would be distributed around the value μ. Note particularly that the larger the value of n, the more closely these $\bar{x}$'s will cluster around μ.

In a similar manner, the sample variance s^2 can be used as a point estimator for the population variance σ^2. The statistic s^2 is a random variable which has its own distribution, and the mean of this distribution is σ^2. That is,

$$(9.2.5) \qquad E(s^2) = \sigma^2,$$

and hence s^2 is an unbiased (point) estimator for σ^2. A derivation of (9.2.5) is given in Section 9.9.

Summarizing, we have:

THEOREM 9.2.1. *If $\bar{x}$ and s^2 are the mean and variance of a sample of size n from a population with unknown mean μ and unknown variance σ^2, then $\bar{x}$ is an unbiased (point) estimator for μ having variance $\sigma_{\bar{x}}^2 = \sigma^2/n$. Furthermore, s^2 is an unbiased (point) estimator for σ^2.*

9.3 INTERVAL ESTIMATORS FOR THE MEAN AND VARIANCE OF A NORMAL POPULATION

If $(x_1, \ldots, x_n)$ is a random sample from the normal distribution $N(\mu, \sigma^2)$, then we know (see Theorem 7.4.1 of Chapter 7) that $\bar{x}$ has the normal distribution $N(\mu, \sigma^2/n)$. This means that we can write

$$(9.3.1) \qquad P\left[-z_{\alpha/2} \leq \frac{\sqrt{n}(\bar{x} - \mu)}{\sigma} \leq z_{\alpha/2} \right] = 1 - \alpha.$$

Solving the set of inequalities inside the parentheses, we can rewrite (9.3.1) as

$$(9.3.2) \qquad P\left[\bar{x} - z_{\alpha/2} \frac{\sigma}{\sqrt{n}} \leq \mu \leq \bar{x} + z_{\alpha/2} \frac{\sigma}{\sqrt{n}} \right] = 1 - \alpha.$$

Equation (9.3.2) essentially states that the probability is $1 - \alpha$ that the interval $[\bar{x} - z_{\alpha/2}(\sigma/\sqrt{n}), \bar{x} + z_{\alpha/2}(\sigma/\sqrt{n})]$ or stated more briefly

$$[\bar{x} \pm z_{\alpha/2}(\sigma/\sqrt{n})],$$

contains μ. If σ is known, then the end points of the interval are known from information available from the sample and from Table II, Appendix

VII. We therefore say

(9.3.3)
$$\left[\bar{x} \pm z_{\alpha/2} \frac{\sigma}{\sqrt{n}} \right]$$

is a $100(1 - \alpha)\%$ *confidence interval* or *interval estimator* for μ. The end points of the interval (9.3.3) are sometimes referred to as *confidence limits*. Summarizing, we have:

THEOREM 9.3.1. *If $\bar{x}$ is the mean of a sample of size n from a normal distribution $N(\mu, \sigma^2)$, where σ is known, then $[\bar{x} \pm z_{\alpha/2}(\sigma/\sqrt{n})]$ is a $100(1 - \alpha)\%$ confidence interval for μ.*

Even if the population from which the sample is drawn is not quite normal, but with known σ^2, then, since $\bar{x}$ has approximately a normal distribution $N(\mu, \sigma^2/n)$ for large n (see Appendix IV), $[\bar{x} \pm z_{\alpha/2}(\sigma/\sqrt{n})]$ is approximately a $100(1 - \alpha)\%$ confidence interval for μ in large samples.

Example 9.3(a). A random sample of size 4 is taken from a population having the $N(\mu, .09)$ distribution. The observations were 12.6, 13.4, 12.8, and 13.2. To find a 95% confidence interval for μ, we first note that

$$1 - \alpha = .95, \qquad \alpha/2 = .025, \qquad z_{\alpha/2} = 1.96,$$

$$\frac{\sigma}{\sqrt{n}} = \frac{.3}{\sqrt{4}} = .15, \qquad \bar{x} = \frac{52.0}{4} = 13.00.$$

Hence the 95% confidence interval for μ is

$$[13.00 \pm 1.96(.15)] = [12.71, 13.29].$$

We turn now to the case where sampling is from a $N(\mu, \sigma^2)$ population, and both μ and σ^2 are unknown, and where we wish to estimate μ. Now if $x_1, \ldots, x_n$ are n independent observations from the $N(\mu, \sigma^2)$ population, we know (see Theorem 7.4.1) that $\bar{x} = N(\mu, \sigma^2/n)$, so that

(9.3.4)
$$E(\bar{x}) = \mu,$$

that is, $\bar{x}$ is an unbiased point estimate of μ, whether σ^2 is or is not known. Now of course we cannot use (9.3.3) as a confidence interval for μ when σ^2 is unknown. Instead we must proceed by using Theorem AVI.8 of Appendix VI. This theorem states that if $(x_1, \ldots, x_n)$ is a random sample of n independent observations from $N(\mu, \sigma^2)$, and if $\bar{x} = \frac{1}{n} \sum_{i=1}^{n} x_i$, $s^2 = (n - 1)^{-1} \sum_{i=1}^{n} (x_i - \bar{x})^2$, then the random variable

(9.3.5)
$$t_{n-1} = \frac{(\bar{x} - \mu)\sqrt{n}}{s}$$

has the Student t distribution with $n - 1$ degrees of freedom. As discussed in Appendix VI, the t distribution is symmetric around zero so that

(9.3.6) $$P(-t_{n-1;\alpha/2} \le t_{n-1} \le t_{n-1;\alpha/2}) = 1 - \alpha$$

where $P(t_{n-1} > t_{n-1;\alpha/2}) = \alpha/2$. Using (9.3.5), then, we may make the following statement:

(9.3.7) $$P\left[-t_{n-1;\alpha/2} \le \frac{\sqrt{n}(\bar{x} - \mu)}{s} \le t_{n-1;\alpha/2}\right] = 1 - \alpha,$$

which may be rewritten as follows:

(9.3.8) $$P\left(\bar{x} - t_{n-1;\alpha/2} \frac{s}{\sqrt{n}} \le \mu \le \bar{x} + t_{n-1;\alpha/2} \frac{s}{\sqrt{n}}\right) = 1 - \alpha.$$

Thus it will be seen that, if σ is unknown, we can make the following statement:

THEOREM 9.3.2. *If $\bar{x}$ and s^2 are the mean and variance of a sample of size n from a normal distribution $N(\mu, \sigma^2)$, where μ and σ^2 are unknown, then*

$$\left[\bar{x} \pm t_{n-1;\alpha/2} \frac{s}{\sqrt{n}}\right]$$

is a $100(1 - \alpha)\%$ *confidence interval for* μ.

If the population from which the sample is drawn has unknown mean and unknown variance, but is not quite normal, $[\bar{x} \pm t_{n-1;\alpha/2}(s/\sqrt{n})]$ is an approximate $100(1 - \alpha)\%$ confidence interval, often good enough for practical purposes.

Example 9.3(b). Four determinations of the percentage of methanol in a certain solution yielded $\bar{x} = 8.34\%$, $s = .03\%$. Assuming (approximate) normality of the population of determinations, to find a 95% confidence interval for μ, we note that

$$1 - \alpha = .95, \qquad \frac{\alpha}{2} = .025, \qquad n = 4$$

so that

$$n - 1 = 3, \qquad t_{3;.025} = 3.182, \qquad \frac{s}{\sqrt{n}} = \frac{.03}{\sqrt{4}} = .015.$$

Hence the 95% confidence interval for μ is

$$[8.34 \pm (3.182)(.015)] = [8.292, 8.388].$$

As mentioned in Appendix VI (c), the Student t distribution approaches the $N(0, 1)$ distribution as the degrees of freedom tend to infinity. Thus if $(n - 1)$ is large, we have that

(9.3.9) $$\frac{(\bar{x} - \mu)\sqrt{n}}{s} \simeq z.$$

Hence, if n is large, we may replace $t_{n-1;\alpha/2}$ by $z_{\alpha/2}$ in the statement of Theorem 9.3.2, and obtain the approximate $100(1-\alpha)\%$ confidence interval for μ, namely

$$(9.3.10) \qquad \left[\bar{x} \pm z_{\alpha/2} \frac{s}{\sqrt{n}}\right].$$

As a matter of fact, inspection of Table V shows that $t_{m;\alpha}$ and z_α are in "good agreement" for $m \geq 60$, and many practitioners are hence led to use (9.3.10) for $n \geq 61$.

9.4 INTERVAL ESTIMATORS FOR THE DIFFERENCE OF MEANS OF TWO NORMAL POPULATIONS

Suppose $\bar{x}_1$ and s_1^2 are the mean and variance of a sample of size n_1 from the normal distribution $N(\mu_1, \sigma_1^2)$, and $\bar{x}_2$ and s_2^2 are the mean and variance of an independent sample of size n_2 from a normal distribution $N(\mu_2, \sigma_2^2)$.

Now consider the difference of the parameters μ_1 and μ_2, say $\delta = \mu_1 - \mu_2$. The unbiased point estimator of δ is, of course, $\bar{x}_1 - \bar{x}_2$, for $E(\bar{x}_1 - \bar{x}_2) = \mu_1 - \mu_2 = \delta$. Furthermore, $\bar{x}_1 - \bar{x}_2$ has the normal distribution $N(\mu_1 - \mu_2, \sigma_1^2/n_1 + \sigma_2^2/n_2)$ (see Appendix VI). If σ_1^2 and σ_2^2 are known, then we may state that

$$(9.4.1) \qquad P\left[-z_{\alpha/2} \leq \frac{(\bar{x}_1 - \bar{x}_2) - (\mu_1 - \mu_2)}{\sqrt{\sigma_1^2/n_1 + \sigma_2^2/n_2}} \leq z_{\alpha/2}\right] = 1 - \alpha,$$

or equivalently,

$$(9.4.2) \quad P\left[(\bar{x}_1 - \bar{x}_2) - z_{\alpha/2}\sqrt{\frac{\sigma_1^2}{n_1} + \frac{\sigma_2^2}{n_2}} \leq (\mu_1 - \mu_2)\right.$$

$$\left. \leq (\bar{x}_1 - \bar{x}_2) + z_{\alpha/2}\sqrt{\frac{\sigma_1^2}{n_1} + \frac{\sigma_2^2}{n_2}}\right] = 1 - \alpha.$$

Hence we have the following:

THEOREM 9.4.1. *If $\bar{x}_1$, s_1^2 are the mean and variance of a sample of size n_1 from $N(\mu_1, \sigma_1^2)$ and $\bar{x}_2$ and s_2^2 are the mean and variance of an independent sample of size n_2 from $N(\mu_2, \sigma_2^2)$, then, if σ_1^2 and σ_2^2 are known while μ_1 and μ_2 are unknown,*

$$\left[(\bar{x}_1 - \bar{x}_2) \pm z_{\alpha/2}\sqrt{\frac{\sigma_1^2}{n_1} + \frac{\sigma_2^2}{n_2}}\right]$$

is a $100(1-\alpha)\%$ confidence interval for $\mu_1 - \mu_2$.

Example 9.4(a).   A sample of size 10 from $N(\mu_1, 25)$ yields a sample mean $\bar{x}_1 = 19.8$, while an independent sample of size 12, from $N(\mu_2, 36)$ yields a sample mean $\bar{x}_2 = 24.0$. To find a 90% confidence interval for $\mu_1 - \mu_2$, we note that

$$1 - \alpha = .90, \qquad \frac{\alpha}{2} = .05, \qquad z_{\alpha/2} = 1.645,$$

$$n_1 = 10, \qquad n_2 = 12, \qquad \sqrt{\frac{\sigma_1^2}{n_1} + \frac{\sigma_2^2}{n_2}} = \sqrt{\frac{25}{10} + \frac{36}{12}} = \sqrt{5.5} = 2.345$$

and, hence

$$[19.8 - 24.0 \pm (1.645)(2.345)] = [-9.06, -1.34]$$

is a 90% confidence interval for $\mu_1 - \mu_2$.

If n_1 and n_2 are large (in practice both strictly greater than 60), and if σ_1^2 and σ_2^2 are unknown and we *cannot assume* that $\sigma_1^2 = \sigma_2^2$, then

$$(9.4.3) \qquad \left[(\bar{x}_1 - \bar{x}_2) \pm z_{\alpha/2} \sqrt{\frac{s_1^2}{n_1} + \frac{s_2^2}{n_2}} \right]$$

is an approximate $100(1 - \alpha)$% confidence interval for $\mu_1 - \mu_2$. [We return to this point below (9.4.14).] Further, this statement holds sufficiently well for most practical purposes, even if the two populations being sampled are fairly nonnormal.

If we change the assumptions of Theorem 9.4.1 so that $\sigma_1^2 = \sigma_2^2 = \sigma^2$, where σ^2 is unknown, it is shown in Appendix VI that the quantity

$$(9.4.4) \qquad (n_1 + n_2 - 2) \frac{s_w^2}{\sigma^2} = \frac{(n_1 - 1)s_1^2 + (n_2 - 1)s_2^2}{\sigma^2}$$

has a chi-square distribution with $n_1 + n_2 - 2$ degrees of freedom. Hence, $E(s_w^2) = \sigma^2$, that is s_w^2 is an unbiased point estimator of σ^2, the common value of the population variances. It is sometimes called the pooled estimator of σ^2. Furthermore, we know that

$$(9.4.5) \qquad \frac{(\bar{x}_1 - \bar{x}_2) - (\mu_1 - \mu_2)}{\sigma \sqrt{1/n_1 + 1/n_2}}$$

has the $N(0, 1)$ distribution. Also, the quantities in (9.4.4) and (9.4.5) are independent. Again from Appendix VI, we note that

$$(9.4.6) \qquad \frac{(\bar{x}_1 - \bar{x}_2) - (\mu_1 - \mu_2)}{\sigma \sqrt{1/n_1 + 1/n_2}} \bigg/ \frac{s_w}{\sigma}$$

has the Student t distribution with $n_1 + n_2 - 2$ degrees of freedom. Therefore we can say that

$$(9.4.7) \quad P\left[-t_{n_1+n_2-2;\alpha/2} \le \frac{(\bar{x}_1 - \bar{x}_2) - (\mu_1 - \mu_2)}{s_w \sqrt{1/n_1 + 1/n_2}} \le t_{n_1+n_2-2;\alpha/2} \right] = 1 - \alpha,$$

or, alternatively,

$$(9.4.8) \quad P\left[(\bar{x}_1 - \bar{x}_2) - t_{n_1+n_2-2;\alpha/2}s_w\sqrt{\frac{1}{n_1} + \frac{1}{n_2}} \leq \mu_1 - \mu_2\right.$$

$$\left. \leq (\bar{x}_1 - \bar{x}_2) + t_{n_1+n_2-2;\alpha/2}s_w\sqrt{\frac{1}{n_1} + \frac{1}{n_2}}\right] = 1 - \alpha.$$

Hence we may summarize as follows:

THEOREM 9.4.2. *If in Theorem 9.4.1 it is assumed that $\sigma_1^2 = \sigma_2^2 = \sigma^2$, where σ^2 is unknown, then*

$$(9.4.9) \qquad \left[(\bar{x}_1 - \bar{x}_2) \pm t_{n_1+n_2-2;\alpha/2}s_w\sqrt{\frac{1}{n_1} + \frac{1}{n_2}}\right]$$

is a $100(1 - \alpha)\%$ confidence interval for $\mu_1 - \mu_2$.

Example 9.4(b). A sample of $n_1 = 5$ light bulbs of type A gave a mean length of life of $\bar{x}_1 = 1000$ hours with a standard deviation of $s_1 = 28$ hours. A sample of $n_2 = 7$ bulbs of type B yield $\bar{x}_2 = 980$ hours, and $s_2 = 32$ hours. We assume that the processes are such that $\sigma_1^2 = \sigma_2^2$. To find a 99% confidence interval for $\mu_1 - \mu_2$, we note that

$$1 - \alpha = .99, \qquad \frac{\alpha}{2} = .005,$$

$$n_1 + n_2 - 2 = 10, \qquad t_{10;.005} = 3.169,$$

and that

$$s_w^2 = \frac{(5 - 1)28^2 + (7 - 1)32^2}{10} = 928.0,$$

that is,

$$s_w = 30.46.$$

Hence the interval is

$$[(1000 - 980) \pm (3.169)(30.46)\sqrt{\tfrac{1}{5} + \tfrac{1}{7}}]$$

$$= [20 \pm 56.5]$$

$$= [-36.5, 76.5].$$

Recall from Appendix VI (c) and Section 9.3, that for large degrees of freedom m, $t_m \simeq z$. In the two-sample problem underlying Theorem 9.4.2, this means if $m = n_1 + n_2 - 2 \geq 60$, that is, if $n_1 + n_2 \geq 62$, then an approximate $100(1 - \alpha)\%$ interval for $\mu_1 - \mu_2$ is

$$(9.4.10) \qquad \left[(\bar{x}_1 - \bar{x}_2) \pm z_{\alpha/2}s_w\sqrt{\frac{1}{n_1} + \frac{1}{n_2}}\right],$$

when $\sigma_1^2 = \sigma_2^2 = \sigma^2$.

If we *cannot reasonably assume* that $\sigma_1^2 = \sigma_2^2$, we may proceed as follows. Compute the quantities c and m, where

(9.4.11)
$$c = \frac{s_1^2/n_1}{s_1^2/n_1 + s_2^2/n_2}$$

and the modified degrees of freedom for t are given by

(9.4.12)
$$\frac{1}{m} = \frac{c^2}{n_1 - 1} + \frac{(1 - c)^2}{n_2 - 1}.$$

Then

(9.4.13)
$$\left[(\bar{x}_1 - \bar{x}_2) \pm t_{m;\alpha/2} \sqrt{\frac{s_1^2}{n_1} + \frac{s_2^2}{n_2}} \right]$$

is an approximate $100(1 - \alpha)\%$ confidence interval for $\mu_1 - \mu_2$.

Now it can be shown that

(9.4.14) $\min (n_1 - 1, n_2 - 1) \le m \le n_1 + n_2 - 2.$

Hence, if n_1 and n_2 are *both strictly greater* than 60, then use of (9.4.11) and (9.4.12) leads to an m which is greater than or equal to 60. This in turn means that the probability point $t_{m;\alpha/2}$ used in the interval (9.4.13) is such that $t_{m;\alpha/2} \simeq z_{\alpha/2}$. Thus, as mentioned before, (9.4.3) is the appropriate $100(1 - \alpha)\%$ confidence interval for $(\mu_1 - \mu_2)$ when both n_1 and $n_2 > 60$ *and* $\sigma_1^2 \ne \sigma_2^2$.

> **Example 9.4(c).** Twenty-five batch yields of a plastic produced using a catalyst (method 1) are to be compared with twenty-five batch yields of the plastic produced without the catalyst (method 2). The following results were obtained (coded units):
>
> > Method 1: $n_1 = 25$; $\bar{x}_1 = 6.40$; $s_1^2 = 2.4264.$
> > Method 2: $n_2 = 25$; $\bar{x}_2 = 6.02$; $s_2^2 = 1.0176.$
>
> Assuming normality of batch yields obtained using method i, say $N(\mu_i, \sigma_i^2)$ $i = 1, 2$, find a 95% confidence interval for $(\mu_1 - \mu_2)$.
>
> We note that nothing is stated about σ_1^2 and σ_2^2. We proceed then, to first find a 95% confidence interval for σ_1^2/σ_2^2. It is shown in Section 9.6 (see Theorem 9.6.1.) that a $100(1 - \alpha)\%$ confidence interval for σ_1^2/σ_2^2 is
>
> $$\left[\frac{s_1^2}{s_2^2 F_{n_1-1, n_2-1;\alpha/2}} \; ; \quad \frac{s_1^2}{s_2^2} F_{n_2-1, n_1-1;\alpha/2} \right].$$
>
> We have that $n_1 - 1 = n_2 - 1 = 24$; $1 - \alpha = .95$; $\alpha/2 = .025$; $F_{24,24;.025} = 2.2693$; $F_{24,24;.975} = 1/F_{24,24;.025} = 1/2.2693$; $s_1^2/s_2^2 = 2.3844$. Hence, a 95% confidence interval for σ_1^2/σ_2^2 is
>
> $$\left[\frac{2.3844}{2.2693} , (2.3844)(2.2693) \right] = [1.051, 5.411].$$

This interval does not include the value "1," so that we are not justified if we assume $\sigma_1^2 = \sigma_2^2$, that is, if we assume $\sigma_1^2/\sigma_2^2 = 1$. We will proceed then, on the assumption that $\sigma_1^2 \neq \sigma_2^2$, and find a 95% confidence interval for $(\mu_1 - \mu_2)$ using the interval (9.4.13).

For this, we must first determine the value of m. (We expect m to lie between 24 and 48 inclusive—see 9.4.14.) From (9.4.11), we have that

$$c = \frac{(2.4264)/25}{(2.4264)/25 + (1.0176)/25} = .705.$$

Hence $1 - c = .295$, $c^2 = .497$, $(1 - c)^2 = .087$. Using (9.4.12) we have that

$$\frac{1}{m} = \frac{.497}{24} + \frac{.087}{24} = \frac{.584}{24},$$

or $m = 41.1$. We will use $m = 41$. Now we have $1 - \alpha = .95$; $\alpha/2 = .025$; $t_{41;.025} = 2.0195$ (by interpolation—see the instructions at the bottom of Table V). Hence, a 95% confidence interval for $(\mu_1 - \mu_2)$ is

$$\left[(6.40 - 6.02) \pm (2.0195) \sqrt{\frac{2.4264}{25} + \frac{1.0176}{25}} \right] = [.38 \pm .750].$$

That is, the 95% confidence interval for the difference $(\mu_1 - \mu_2)$ is $[-.37, 1.13]$. With confidence .95, then the sample evidence supports the statement $\mu_1 - \mu_2 = 0$, that is, $\mu_1 = \mu_2$.

Example 9.4(d). A sample of 61 strands of "type I" yarn when subjected to breaking strength tests yielded a sample mean of $\bar{x}_I = 1400$ psi and a sample standard deviation of $s_I = 120$ psi. A sample of 121 strands of "type M" yarn was also subjected to the same breaking strength tests and yielded a sample mean of $\bar{x}_M = 1250$ psi, and a sample standard deviation of $s_M = 80$ psi. Find a 95% confidence interval for $\mu_I - \mu_M$, assuming normality of breaking strengths of both types of yarn.

Since we cannot assume that $\sigma_I^2 = \sigma_M^2$, we proceed first to find a 95% confidence interval for σ_I^2/σ_M^2. We have that

$$n_I - 1 = 60; \qquad n_M - 1 = 120; \qquad 1 - \alpha = .95; \qquad \alpha/2 = .025;$$

$$F_{60,120;.025} = 1.5299; \qquad\qquad F_{60,120;.975} = 1/F_{120,60;.025} = 1/1.5810;$$

$$s_I^2/s_M^2 = (120)^2/(80)^2 = 2.25 .$$

Hence, the 95% confidence interval for σ_I^2/σ_M^2 is

$$\left[\frac{2.25}{1.5299} , (2.25)(1.5810) \right] = [1.471, 3.557].$$

Since this interval does not include the value 1, that is, since the sample evidence supports the claim that $\sigma_I^2 \neq \sigma_M^2$ at a confidence level of .95, we proceed to find a 95% confidence interval for $\mu_I - \mu_M$ using (9.4.3). We use (9.4.3) and not (9.4.13), since both n_I and n_M are strictly greater than 60. In fact, if we go through the calculations (9.4.11) and (9.4.12), the reader should verify that we would find that $m = 87$ (and $t_{87} \simeq z$).

Hence, the 95% confidence interval for $\mu_I - \mu_{II}$ is

$$\left[(1400 - 1250) \pm z_{.025}\sqrt{\frac{(120)^2}{61} + \frac{(80)^2}{121}}\,\right]$$

$$= [150 \pm 1.96\sqrt{289.0}]$$

$$= [150 \pm 33.32] = [116.7, 183.3].$$

At confidence level .95, we would support the statement that $\mu_I > \mu_{II}$.

9.5 INTERVAL ESTIMATORS FOR THE VARIANCE OF A NORMAL POPULATION

As will be seen in Appendix VI, if s^2 is the variance of a sample of n from the normal distribution $N(\mu, \sigma^2)$, then $[(n - 1)s^2]/\sigma^2$ has the chi-square distribution with $n - 1$ degrees of freedom. Thus, we can make the statement

(9.5.1) $$P\left[\chi^2_{n-1;1-\alpha/2} \leq \frac{(n - 1)s^2}{\sigma^2} \leq \chi^2_{n-1;\alpha/2}\right] = 1 - \alpha$$

or, equivalently, we have

(9.5.2) $$P\left[\frac{(n - 1)s^2}{\chi^2_{n-1;\alpha/2}} \leq \sigma^2 \leq \frac{(n - 1)s^2}{\chi^2_{n-1;1-\alpha/2}}\right] = 1 - \alpha.$$

Hence, we can summarize by the following theorem:

THEOREM 9.5.1.   *If s^2 is the sample variance in a sample of size n from the normal distribution $N(\mu, \sigma^2)$, then*

(9.5.3) $$\left[\frac{(n - 1)s^2}{\chi^2_{n-1;\alpha/2}}, \frac{(n - 1)s^2}{\chi^2_{n-1;1-\alpha/2}}\right]$$

is a $100(1 - \alpha)$% *confidence interval for* σ^2.

It should be noted that

(9.5.4) $$\left[\frac{\sqrt{n - 1}\, s}{\chi_{n-1;\alpha/2}}, \frac{\sqrt{n - 1}\, s}{\chi_{n-1;1-\alpha/2}}\right]$$

is a $100(1 - \alpha)$% confidence interval for σ, for the random interval (9.5.4) contains σ if and only if the random interval (9.5.3) contains σ^2.

If the distribution from which the sample is drawn is slightly non-normal, the interval given in (9.5.3) provides an approximate $100(1 - \alpha)$% confidence interval for σ^2, which is satisfactory for most practical purposes. However, it is true that the confidence interval for σ^2 (or σ) can be quite off the mark, as contrasted to that for μ, if the population distribution departs very much from normality.

> **Example 9.5(a).** Referring to Example 9.3(b), suppose we wish to find a 95% confidence interval for σ^2. We have $1 - \alpha = .95$, $1 - \alpha/2 = .975$, $n = 4$, so that $n - 1 = 3$, and $s^2 = .0009$. Also $\chi^2_{3;.025} = 9.348$ and $\chi^2_{3;.975} = .216$. Hence

$$\left[\frac{3(.0009)}{9.348}, \frac{3(.0009)}{.216}\right] = [.00029, .0125]$$

is a 95% confidence interval for σ^2. Also,

$$[.017, .112]$$

is a 95% confidence interval for σ.

9.6 INTERVAL ESTIMATOR FOR THE RATIO OF VARIANCES OF TWO NORMAL POPULATIONS

Suppose s_1^2 and s_2^2 are sample variances of two independent normal distributions $N(\mu_1, \sigma_1^2)$ and $N(\mu_2, \sigma_2^2)$ respectively, where σ_1^2 and σ_2^2 are unknown, and we wish to find an interval estimator for the ratio σ_1^2/σ_2^2. We may proceed as follows.

We know from Appendix VI that, if the two samples are independently drawn from the two normal distributions indicated, then the quantities

$$(9.6.1) \qquad \frac{(n_1 - 1)s_1^2}{\sigma_1^2} \quad \text{and} \quad \frac{(n_2 - 1)s_2^2}{\sigma_2^2}$$

are independent random variables having chi-square distributions with $n_1 - 1$ and $n_2 - 1$ degrees of freedom respectively. Thus, using the results of Appendix VI, we see that

$$(9.6.2) \qquad \frac{s_1^2/\sigma_1^2}{s_2^2/\sigma_2^2}$$

has the Snedecor F distribution with $(n_1 - 1, n_2 - 1)$ degrees of freedom. This means that we may write

$$(9.6.3) \qquad P\left(F_{n_1-1,n_2-1;1-\alpha/2} \leq \frac{s_1^2/s_2^2}{\sigma_1^2/\sigma_2^2} \leq F_{n_1-1,n_2-1;\alpha/2}\right) = 1 - \alpha$$

or, solving the inequalities, we may write

$$(9.6.4) \qquad P\left(\frac{s_1^2}{s_2^2}\frac{1}{F_{n_1-1,n_2-1;\alpha/2}} \leq \frac{\sigma_1^2}{\sigma_2^2} \leq \frac{s_1^2}{s_2^2}\frac{1}{F_{n_1-1,n_2-1;1-\alpha/2}}\right) = 1 - \alpha.$$

Hence we may summarize as follows:

THEOREM 9.6.1.   *If s_1^2 and s_2^2 are variances of independent samples of size n_1 and n_2 from the normal distributions $N(\mu_1, \sigma_1^2)$ and $N(\mu_2, \sigma_2^2)$ respectively, where μ_1, μ_2, σ_1^2, and σ_2^2 are unknown, then*

$$(9.6.5) \qquad \left(\frac{s_1^2}{s_2^2} \frac{1}{F_{n_1-1, n_2-1; \alpha/2}}, \frac{s_1^2}{s_2^2} \frac{1}{F_{n_1-1, n_2-1; 1-\alpha/2}} \right)$$

is a $100(1 - \alpha)\%$ *confidence interval for* σ_1^2/σ_2^2.

Note that a $100(1 - \alpha)\%$ confidence interval for σ_1/σ_2 has end points which are the square roots of the end points of (9.6.5).

The usual F tables do not provide values of $F_{m_1, m_2; \gamma}$ for $\gamma > .10$. However, we note that (see Section 9.9)

$$(9.6.6) \qquad F_{m_1, m_2; \gamma} = \frac{1}{F_{m_2, m_1; 1-\gamma}}$$

and, hence, we can use this relation to obtain some of the lower percentage points of any Snedecor F random variable.

9.7   ESTIMATION BY THE METHOD OF MAXIMUM LIKELIHOOD

We shall now briefly describe a widely used method of determining point estimators, known as the method of *maximum likelihood estimation*.

Suppose $(x_1, \ldots, x_n)$ is a random sample on a random variable x whose probability function (in the discrete case) or probability density function (in the continuous case) is $f(x; \theta)$ which depends on the population parameter θ. For example, the binomial distribution defined by (3.2.2) depends on the parameter θ. The Poisson distribution (3.3.1) depends on the parameter μ. A distribution function may depend, of course, on more than one parameter; for example, θ may stand for the pair of parameters (μ, σ), which is the case when x has the normal distribution $N(\mu, \sigma^2)$ having the probability density function defined in (7.1.1).

Since $(x_1, \ldots, x_n)$ is a sample from $f(x; \theta)$, the joint probability function (or probability density function) of the sample is $f(x_1; \theta) \cdots f(x_n; \theta)$ which we denote by $l(\theta \mid x_1, \ldots, x_n)$. The function $l(\theta \mid x_1, \ldots, x_n)$ is called the *likelihood function* of θ for the given sample $(x_1, \ldots, x_n)$.

The maximum likelihood estimator $\hat{\theta}$ for θ is that value of θ (if it exists) such that

$$(9.7.1) \qquad l(\hat{\theta} \mid x_1, \ldots, x_n) > l(\theta' \mid x_1, \ldots, x_n),$$

where θ' is any other possible value of θ. In other words, the maximum likelihood estimator is the value of θ which maximizes the likelihood function.

It is usually more convenient to work with the natural logarithm of $l(\theta \mid x_1, \ldots, x_n)$ rather than with the likelihood itself, and we denote $\log l(\theta \mid x_1, \ldots, x_n)$ by $L(\theta \mid x_1, \ldots, x_n)$. It can be seen that L has its maximum at the same value of θ that l does. Thus, the value of θ at which L is maximized is the maximum likelihood estimator $\hat{\theta}$ for θ.

> **Example 9.7(a).** It is desired to estimate the fraction θ of defectives in a lot of television tubes produced by a mass-production process. We select a sample of size n at random from the production line and observe the number of defectives. For a single tube, let x be a random variable with value 1 if the tube is defective and 0 if the tube is nondefective. The p.f. of x, say $f(x; \theta)$ is given by
>
> (9.7.2) $f(x; \theta) = \theta^x(1 - \theta)^{1-x}, \qquad x = 0, 1.$
>
> For a sample of n tubes, the observations on x would be $(x_1, \ldots, x_n)$ and, hence,
>
> $$l(\theta \mid x_1, \ldots, x_n) = \theta^T(1 - \theta)^{n-T},$$
>
> where $T = x_1 + \cdots + x_n$, which is the number of defectives in the sample. Therefore,
>
> $$L = T \log \theta + (n - T) \log (1 - \theta).$$
>
> Differentiating with respect to θ, we have
>
> (9.7.3) $$\frac{\partial L}{\partial \theta} = \frac{T}{\theta} - \frac{n - T}{1 - \theta}.$$
>
> Setting this derivative equal to zero and denoting the solution by $\hat{\theta}$, we have the maximum likelihood estimator for θ,
>
> (9.7.4) $$\hat{\theta} = \frac{T}{n}.$$
>
> In Section 9.8 we show that $\hat{\theta}$ is an unbiased estimator for θ. Maximum likelihood estimators need not always have this latter property, as the following example will show.

> **Example 9.7(b).** A random sample $(x_1, \ldots, x_n)$ of n observations is taken on x, where x is a random variable having the normal distribution $N(\mu, \sigma^2)$. The likelihood function is given by
>
> (9.7.5) $l(\theta \mid x_1, \ldots, x_n) = l(\mu, \sigma^2 \mid x_1, \ldots, x_n)$
>
> $$= \frac{1}{(\sqrt{2\pi}\sigma)^n} \exp -\frac{1}{2\sigma^2} \sum_{i=1}^{n} (x_i - \mu)^2$$
>
> where $\exp(a) = e^a$. Hence,
>
> (9.7.6) $L(\mu, \sigma^2 \mid x_1, \ldots, x_n) = -n \log \sqrt{2\pi} - \frac{n}{2} \log \sigma^2 - \frac{1}{2\sigma^2} \sum_{i=1}^{n} (x_i - \mu)^2.$

To find the maximum of this function with respect to μ and σ^2, we differentiate partially with respect to μ and with respect to σ^2, thus obtaining

(9.7.7)
$$\frac{\partial L}{\partial \mu} = \frac{1}{\sigma^2} \sum_{i=1}^{n} (x_i - \mu),$$

$$\frac{\partial L}{\partial (\sigma^2)} = -\frac{n}{2\sigma^2} + \frac{1}{2(\sigma^2)^2} \sum_{1}^{n} (x_i - \mu)^2.$$

Setting these derivatives equal to zero and denoting the solutions by $\hat{\mu}$ and $\hat{\sigma}^2$, we find

(9.7.8)
$$\hat{\mu} = \bar{x},$$

$$\hat{\sigma}^2 = \frac{1}{n} \sum_{i=1}^{n} (x_i - \bar{x})^2.$$

Note that $\bar{x}$ is unbiased for μ but that $\hat{\sigma}^2$ has expectation $[(n-1)/n]\,\sigma^2$, and hence is not an unbiased estimator for σ^2.

9.8 POINT AND INTERVAL ESTIMATORS FOR THE BINOMIAL PARAMETER

Suppose we are sampling from a population whose p.f. is given by

(9.8.1) $f(x;\ \theta) = \theta^x (1-\theta)^{1-x}, \qquad x = 0,\ 1$

[see Example 9.7(a)]. Then it may be seen that the mean μ and variance σ^2 of this population are given by

(9.8.2) $\mu = \theta, \qquad \sigma^2 = \theta(1-\theta).$

Now if $(x_1, \ldots, x_n)$ is a sample of n independent observations on x, whose p.f. is given by (9.8.1), then

$$T = x_1 + \cdots + x_n$$

has mean and variance $n\mu$ and $n\sigma^2$ respectively, that is,

(9.8.3) $E(T) = n\theta \qquad$ and $\qquad \mathrm{Var}\,(T) = n\theta(1-\theta).$

Now if we denote the statistic T/n by $\hat{\theta}$ [indeed $\hat{\theta}$ is the maximum likelihood estimator of θ, as seen in Example 9.7(a)], then from (9.8.3) we have that

(9.8.4) $E(\hat{\theta}) = \theta \qquad$ and $\qquad \mathrm{Var}\,(\hat{\theta}) = \dfrac{\theta(1-\theta)}{n},$

which is to say that $\hat{\theta}$ is an unbiased (point) estimator of θ with variance $\theta(1-\theta)/n$.

Now recall that from Theorem 7.3.1 we may write

(9.8.5)
$$z \simeq \frac{n\hat{\theta} - n\theta}{\sqrt{n\theta(1-\theta)}},$$

where z, of course, is the $N(0, 1)$ variable. Hence

(9.8.6) $$P\left[-z_{\alpha/2} \leq \frac{n\hat{\theta} - n\theta}{\sqrt{n\theta(1 - \theta)}} \leq z_{\alpha/2}\right] \cong 1 - \alpha.$$

The event mentioned in (9.8.6) is equivalent to the event

(9.8.7) $$n(\hat{\theta} - \theta)^2 \leq z_{\alpha/2}^2\theta(1 - \theta),$$

which, in turn, is equivalent to

(9.8.8) $$\theta^2(n + z_{\alpha/2}^2) - \theta(2n\hat{\theta} + z_{\alpha/2}^2) + n\hat{\theta}^2 \leq 0.$$

Now $n + z_{\alpha/2}^2$ is positive, so that the left-hand side of (9.8.8) is negative when θ lies between the roots of the equation obtained by setting the left-hand side of (9.8.8) equal to zero. The roots, of course, are given by $(-b \pm \sqrt{b^2 - 4ac})/2a$ where

(9.8.9) $$a = n + z_{\alpha/2}^2, \qquad b = -(2n\hat{\theta} + z_{\alpha/2}^2), \qquad c = n\hat{\theta}^2.$$

Hence, the (approximate) $100(1 - \alpha)\%$ confidence interval for θ is given by

(9.8.10) $$\theta_1 < \theta < \theta_2,$$

where θ_1 and θ_2 are the smaller and larger respectively of the two roots, $(-b \pm \sqrt{b^2 - 4ac})/2a$.

Example 9.8(a). A sample of 100 independent observations on x, where x has p.f. given by (9.8.1), yields $T = 60$. Thus, the unbiased point estimate of θ is $\hat{\theta} = T/n = 60/100 = .60$.

We proceed now to calculate the 95% confidence interval for θ. We have

$$1 - \alpha = .95, \qquad \frac{\alpha}{2} = .025, \qquad z_{\alpha/2} = 1.96, \qquad z_{\alpha/2}^2 = 3.84$$

and hence, from (9.8.9),

$$a = 103.84, \qquad b = -123.84, \qquad c = 36.$$

Hence the roots θ_1, θ_2 are given by

$$\frac{123.84 \pm \sqrt{(123.84)^2 - 4(103.84)(36)}}{2(103.84)}$$

that is, $\theta_1 = .50$, $\theta_2 = .69$. Thus, $(.50, .69)$ is the (approximate) 95% confidence interval for θ.

Confidence charts for θ at various confidence levels have been prepared by Clopper and Pearson and the chart for confidence level .95 is reproduced in Appendix VIII (Chart III). The upper and lower confidence curves are drawn in the $(\hat{\theta}, \theta)$-plane with the use of the equation

(9.8.11) $$a\theta^2 + b\theta + c = 0,$$

where a, b, and c are given in (9.8.9).

The quick and easy procedure, then, is to enter the chart at the observed θ on the horizontal axis and project up until the two curves for the sample size n used are intersected, and then to read off the ordinates of the points of the intersections. These points are the lower and upper confidence limits for θ.

It is interesting to note that (9.8.11), found from using the normal approximation to the binomial, may be used for n as low as 10, but for values of n that are smaller, the cumulative of the binomial itself must be used.

Example 9.8(b). Referring to Example 9.8(a), we found that there $\theta = .60$, with $n = 100$. Entering Chart III at $\theta = .60$, and reading up to the $n = 100$ curves, we find

$$\theta_1 = .50, \qquad \theta_2 = .70.$$

9.9 SOME DERIVATIONS AND PROOFS

(a) Proof That $E(s^2) = \sigma^2$

Suppose $(x_1, \ldots, x_n)$ is a random sample on a random variable x having mean μ and variance σ^2. Then the random variable

$$(9.9.1) \qquad L = c_1 x_1 + \cdots + c_n x_n,$$

where the c_i's are constants, has expectation $\mu \sum_{i=1}^{n} c_i$ and variance $\sigma^2 \sum_{i=1}^{n} c_i^2$.

In the special case

$$(9.9.2) \qquad \bar{x} = \frac{1}{n} x_1 + \cdots + \frac{1}{n} x_n,$$

we obtain

$$(9.9.3) \qquad E(\bar{x}) = \mu,$$

$$\text{Var}(\bar{x}) = \frac{\sigma^2}{n}.$$

The second member of (9.8.3) states that

$$(9.9.4) \qquad E[n(\bar{x} - \mu)^2] = \sigma^2.$$

Now consider the algebraic identity

$$(9.9.5) \qquad \sum_{i=1}^{n} (x_i - \mu)^2 = \sum_{i=1}^{n} (x_i - \bar{x})^2 + n(\bar{x} - \mu)^2.$$

Using the properties of expectations, we have

$$(9.9.6) \qquad \sum_{i=1}^{n} E(x_i - \mu)^2 = E\left[\sum_{i=1}^{n} (x_i - \bar{x})^2 \right] + E[n(\bar{x} - \mu)^2],$$

and making use of (9.9.4), we find

$$\sum_{i=1}^{n} \sigma^2 = E\left[\sum_{i=1}^{n} (x_i - \bar{x})^2 \right] + \sigma^2,$$

that is,

$$E\left[\sum_{i=1}^{n} (x_i - \bar{x})^2 \right] = (n - 1)\sigma^2.$$

Finally we have

$$E\left[\frac{1}{n-1} \sum_{i=1}^{n} (x_i - \bar{x})^2 \right] = E(s^2) = \sigma^2.$$

(b) Proof That $F_{m_1, m_2; \gamma} = 1/F_{m_2, m_1; 1-\gamma}$

From part (d) of Appendix VI, the random variable F_{m_1, m_2}, is defined as

$$(9.9.7) \qquad F_{m_1, m_2} = \frac{\chi_{m_1}^2/m_1}{\chi_{m_2}^2/m_2} = \frac{1}{F_{m_2, m_1}},$$

and $\chi_{m_1}^2$ and $\chi_{m_2}^2$ are two independent chi-square random variables with m_1 and m_2 degrees of freedom respectively.

Now let us denote the upper $100\gamma\%$ point of F_{m_1, m_2} by $F_{m_1, m_2; \gamma}$, that is,

$$(9.9.8) \qquad P(F_{m_1, m_2} \geq F_{m_1, m_2; \gamma}) = \gamma.$$

Using (9.9.7), this is equivalent to

$$(9.9.9) \qquad P\left(\frac{1}{F_{m_2, m_1}} \geq F_{m_1, m_2; \gamma} \right) = \gamma$$

or

$$(9.9.10) \qquad P\left(F_{m_2, m_1} \leq \frac{1}{F_{m_1, m_2; \gamma}} \right) = \gamma,$$

which is to say that

$$(9.9.11) \qquad P\left(F_{m_2, m_1} \geq \frac{1}{F_{m_1, m_2; \gamma}} \right) = 1 - \gamma.$$

The relation (9.9.11) states that the upper $100(1 - \gamma)\%$ point of F_{m_2, m_1} is $1/F_{m_1, m_2; \gamma}$, that is,

$$(9.9.12) \qquad F_{m_2, m_1; 1-\gamma} = \frac{1}{F_{m_1, m_2; \gamma}},$$

hence (9.6.6) follows.

PROBLEMS

9.1. A sample of 25 bulbs is taken from a large lot of 40-watt bulbs, and the sample mean of the bulb lives is 1410 hours. Assuming normality of bulb lives and that the standard deviation of bulb lives of the mass-production process involved is 200 hours, find a 95% confidence interval for the mean life of the bulbs in the lot.

9.2. A certain type of electronic condensor is manufactured by the E.C.A. company, and over a large number of years, the life times of these parts are found to be normally distributed with standard deviation $\sigma = 225$ hours. A random sample of 30 of these condensors yielded a sample mean of 1407.5 hours. Find a 99% confidence interval for μ, the mean life of E.C.A. condensors. What can you say about the statement "$\mu = 1400$"?

9.3. In estimating the mean of a normal distribution $N(\mu, \sigma^2)$ having known standard deviation σ by using a confidence interval based on a sample of size n, at least how large should n be in order for the .99 confidence interval for μ to be of length not greater than L?

9.4. A sample, size 10, from $N(\mu_1, 225)$ yields a sample mean $\bar{x}_1 = 170.2$, while an independent sample of size 12 from $N(\mu_2, 256)$ yields a sample mean $\bar{x}_2 = 176.7$. Find a 95% confidence interval for $\mu_1 - \mu_2$.

9.5. Suppose random samples of 25 are taken from two large lots of bulbs, say lot A and lot B, and the mean lives for the two samples were found to be as follows:

$$\bar{x}_A = 1580 \text{ hours}; \quad \bar{x}_B = 1425 \text{ hours}.$$

Assuming that the standard deviation of bulb life in each of the two lots is 200 hours, find 95% confidence limits for $\mu_A - \mu_B$.

9.6. Two machines A and B are packaging 8-ounce boxes of Corn Flakes From past experience with the machines, it is assumed that the standard deviations of weights of the filling from the machines A and B are .04 ounces and .05 ounces, respectively. One hundred boxes filled by each machine are selected at random and the following is found:

$$\text{Machine } A: \; n_A = 100; \quad \bar{x}_A = 8.18 \text{ ounces.}$$
$$\text{Machine } B: \; n_B = 100; \quad \bar{x}_B = 8.15 \text{ ounces.}$$

Find a 99% confidence interval for $\mu_A - \mu_B$, the difference of the means of populations of weights of fillings produced by machines A and B.

9.7. Four determinations of the pH of a certain solution were 7.90, 7.94, 7.91, and 7.93. Assuming normality of determinations with mean μ, standard deviation σ, find:

(a) 99% confidence limits for μ.

(b) 95% confidence limits for σ.

9.8. Ten determinations of percentage of water in a methanol solution yielded $\bar{x} = .552$ and $s = .037$. If μ is the "true" percentage of water in the methanol solution, find (a) a 90% confidence interval for μ and (b) a 95% confidence interval for σ.

9.9. A clock manufacturer wanted to estimate the variability of the precision of a certain type of clock being manufactured. To do this, a sample of eight clocks was run for exactly 48 hours. The number of seconds each clock was ahead or behind after the 48 hours, as measured by a master clock was recorded, the results being $+6$, -4, -7, $+5$, $+9$, -6, -3, $+2$. Assuming the time recorded by clocks of this kind to be normally distributed, find:

(a) A 95% confidence interval for the mean time recorded by this type of clock after 48 hours.

(b) A 95% confidence interval for σ, the standard deviation of the time recorded by this type of clock after 48 hours.

9.10. Table P9.10 gives the yield point (in units of 1000 psi) for a sample of 20 steel castings from a large lot.

TABLE P9.10

64.5	66.5	67.5	67.5
66.5	65.0	73.0	63.5
68.5	70.0	71.0	68.5
68.0	64.5	69.5	67.0
69.5	62.0	72.0	70.0

Assuming the yield points of the population of castings to be (approximately) normally distributed, find:

(a) A 95% confidence interval for μ, the mean yield point of the lot.

(b) A 95% confidence interval for σ, the standard deviation of the yield points of the lot.

9.11. A sample of size 12 from a population assumed to be normal with unknown variance σ^2 yields $s^2 = 86.2$. Determine 95% confidence limits for σ^2.

9.12. Tensile strengths were measured for 15 test pieces of cotton yarn randomly taken from the production of a given spindle. The value of s for this sample of 15 tensile strengths was found to be 11.2 pounds. Find 95% confidence limits for σ of the population. It is assumed that the population is approximately normal.

9.13. In firing a random sample of nine rounds from a given lot of ammunition, it was found that the standard deviation of muzzle velocities of the nine rounds was 38 feet per second. Assuming that muzzle velocities of rounds in this lot are (approximately) normally distributed, find 95% confidence limits for the standard deviation σ of muzzle velocity of the lot.

9.14. A sample of four tires was taken from a lot of brand-A tires. Another sample of four tires was taken from a lot of brand-B tires. These tires were tested for amount of wear, for 24,000 miles of driving on an 8-wheeled truck, the tires being rotated every 1000 miles. The tires were weighed before and after the test, and the loss in weight expressed as percentage of initial weight was used as a measure of wear. For the sample of brand-A tires it was found that $\bar{x}_A = 18.0$, $s_A = 1.3$, and for brand-B tires, $\bar{x}_B = 19.4$, $s_B = 1.5$.

Assuming (approximately) normal distributions having equal variances for percent wear of tires of each brand under these test conditions, find a 95% confidence interval for $\mu_A - \mu_B$.

9.15. A light bulb company tested ten light bulbs which contained filaments of type A and ten which contained filaments of type B. The following results were obtained for the length of life in hours of the 20 light bulbs (Steele and Torrie):

Filament A: 1293; 1380; 1614; 1497; 1340;
1643; 1466; 1094; 1270; 1028.
Filament B: 1061; 1627; 1065; 1383; 1092;
1711; 1021; 1138; 1017; 1143.

Assuming (approximate) normality and $\sigma_A^2 = \sigma_B^2 = \sigma^2$, find a 95% confidence interval for the difference between the mean life of bulbs with filament A and with filament B. Does the sample evidence support the assumption that these means are equal?

9.16. Table P9.16 gives the Vickers hardness numbers of ten shell casings from company A and of ten casings from company B:

TABLE P9.16

Company A (hardness)	Company B (hardness)
66.3	62.2
64.5	67.5
65.0	60.4
62.2	61.5
61.3	64.8
66.5	60.9
62.7	60.2
67.5	67.8
62.7	65.8
62.9	63.8

Find a 90% confidence interval for $\mu_A - \mu_B$, assuming the variance of Vickers hardness to be equal for shell cases made by the two companies. [Assume (approximate) normality and that $\sigma_A^2 = \sigma_B^2 = \sigma^2$.]

9.17. Resistance measurements (in ohms) were made on a sample of four test pieces of wire from lot A and five from lot B, with the results given in Table P9.17. If μ_1 and μ_2 are the "true" mean resistances of the wire in lots A and B,

TABLE P9.17

Lot A Resistance (ohms)	Lot B Resistance (ohms)
.143	.140
.142	.142
.143	.136
.137	.138
	.140

and assuming that possible measurements made from lot A and possible measurements from lot B have normal distributions $N(\mu_1, \sigma^2)$ and $N(\mu_2, \sigma^2)$, respectively, where μ_1, μ_2, and σ^2 are unknown, find a 95% confidence interval for $\mu_1 - \mu_2$. Use this interval to examine the statement that $\mu_1 = \mu_2$.

9.18. Breaking strengths (in pounds) were observed on a sample of five test pieces of type-A yarn and nine test pieces of type-B yarn, with the following results:

Type A (pounds): 93, 94, 75, 84, 91.

Type B (pounds): 99, 93, 99, 97, 90, 96, 93, 88, 89.

Assuming normality of breaking strengths for each type of yarn, and that the population variances of breaking strengths are equal in the two populations, find 99% confidence limits for $\mu_A - \mu_B$, where μ_A and μ_B are the population means.

9.19. "Before and after" tests are to be made on the breaking strengths (in ounces) of a certain type of yarn. Specifically, 7 determinations of the breaking strength are made on test pieces of the yarn before a spinning machine is reset, and 5 determinations of the breaking strength are made on test pieces obtained after the machine is reset, with the following results:

Before: 22.7, 25.7, 20.7, 26.7, 21.2, 19.2, 22.7.

After: 23.2, 23.7, 25.2, 23.7, 24.7.

Assuming that determinations made "after reset" and determinations made "before reset" are independently distributed as $N(\mu_1, \sigma_1^2)$ and $N(\mu_2, \sigma_2^2)$, respectively, find a 95% confidence interval for $\mu_1 - \mu_2$. [*Note:* We do not assume $\sigma_1^2 = \sigma_2^2$.]

9.20. Determinations of atomic weights of carbon from two preparations A and B yield the results given in Table P9.20. Assuming (approximate) normality,

TABLE P9.20

Preparation 1	Preparation 2
12.0072	11.9583
12.0064	12.0017
12.0054	11.9949
12.0016	12.0061
12.0077	

find a 95% confidence interval for $\mu_1 - \mu_2$, the difference of the true means of the determinations that are made from preparations A and B. [*Note:* We have not assumed equality of variances.]

9.21. A new catalyst is to be used in production of a plastic chemical. Twenty batches of the chemical are produced; 10 batches are produced with the new catalyst, 10 without. The following results are obtained:

Catalyst Present		Catalyst Absent	
7.2	7.3	7.0	7.2
7.4	7.5	7.5	7.3
7.8	7.2	7.1	7.1
7.5	8.4	7.3	7.0
7.2	7.5	7.0	7.3

Assuming normality, find a 99% confidence interval for the difference of the yields of batches obtained when using the catalyst and when not using the catalyst. On the basis of this interval, does the data support the claim that the use of the catalyst increases the average batch yield? [*Note:* We have not assumed equality of variances.]

9.22. An experiment to determine the viscosity of two different types of gasoline (lead versus no-lead) gave the results shown below:

$$\text{Lead type: } n_1 = 25; \quad \bar{x}_1 = 35.84; \quad s_1^2 = 130.4576.$$
$$\text{Non-lead type: } n_2 = 25; \quad \bar{x}_2 = 30.60; \quad s_2^2 = 53.0604.$$

Assuming normality, find a 95% confidence interval for the difference in viscosities. [*Note:* We have not assumed equality of variances.]

9.23. Two analysts A and B each make ten determinations of percentage chlorine in a batch of polymer. The values of the sample variances s_A^2 and s_B^2 turned out to be .5419 and .6065, respectively. If σ_A^2 and σ_B^2 are the variances of the populations of A's measurements and of B's measurements respectively, find 95% confidence limits for σ_A^2/σ_B^2.

9.24. For the data of Problem 9.14, find a 95% confidence interval for σ_A^2/σ_B^2, the ratio of the population variances, and comment on the assumption made in Problem 9.14, that the "true value of the ratio" is one.

9.25. In Problems 9.15–9.18 inclusive, the assumption was made that the ratio of the variance of the populations being sampled, is one. Carry out the instructions of Problem 9.24.

9.26. For the data of Problems 9.19–9.22, inclusive, find 95% confidence intervals for the ratio of the variances of the populations being sampled.

9.27. A large university wishes to estimate the mean expense account of the members of its staff who attend professional meetings. A random sample of 100 expense accounts yield a sample mean of $115.87 with a sample standard deviation of $14.34. Find a 95% confidence interval for the mean expense account amount, say μ.

9.28. A new spinning machine is installed and 64 test pieces of the yarn it produces are tested for breaking strength. The observed data yield a sample mean of 4.8 and sample standard deviation of 1.2. Find a 99% confidence interval for the true value of the breaking strength.

9.29. The heights of 105 male students of University X chosen randomly are measured and yield a sample mean of 68.05 inches, and a sample standard deviation of 2.79 inches. Find a 95% confidence interval for the mean of the heights of male students attending the (rather large) University X.

9.30. A sample of 70 employees selected at random from the employees of a large brewery yielded a mean disable time of 41.8 hours during the fiscal year of 1969, with a standard deviation of 6.4 hours. Construct a 99% confidence interval for the average disabled time of employees at this firm.

9.31. Two groups of judges were asked to rate the tastiness of a certain product The results were as follows:

$$Group\ 1:\ n_1 = 121;\quad \bar{x}_1 = 3.6;\quad s_1^2 = 1.96.$$
$$Group\ 2:\ n_2 = 121;\quad \bar{x}_2 = 3.2;\quad s_2^2 = 3.24.$$

(a) Find a 95% confidence interval for σ_1^2/σ_2^2.
(b) On the basis of the interval found in (a), find a 95% confidence interval for $\mu_1 - \mu_2$.

9.32. A sample of 500 rounds of ammunition supplied by manufacturer A yielded a mean muzzle velocity of 2,477 feet per second, with a standard deviation of 80 feet per second. A sample of 500 rounds made by another manufacturer, manufacturer B, yielded a mean muzzle velocity of 2,422 feet per second, with a standard deviation of 120 feet per second. Find a 99% confidence interval for the differences of muzzle velocity of the bullets supplied by A and B. (*Caution:* can you assume $\sigma_A^2 = \sigma_B^2$ on the basis of the above sample evidence?)

9.33. A sample of 100 workers in one large plant took a mean time of 23 minutes to complete a task, with a standard deviation of 4 minutes. Another sample of 100 workers took a mean time of 25 minutes to complete the same task, with a standard deviation of 6 minutes.
(a) Construct a 99% confidence interval for the ratio of the population variances.
(b) On the basis of part (a), determine a 99% confidence interval for the difference between the two population means.

9.34. A sample of 220 items turned out during a given week by a certain process had mean weight of 2.57 pounds and standard deviation of .57 pounds. During the next week a different lot of raw material was used, and the mean weight of 220 items turned out that week was 2.66 pounds, and the standard deviation was .48 pound.
(a) Construct a 99% confidence interval for the ratio of the population variances (*Hint:* You will need to interpolate in the $F_{m,m;.005}$ table.)
(b) On the basis of the interval calculated in (a), determine a 99% confidence interval for the differences of the population mean weights.

9.35. Two types of tires are tested with the following results:

$$Type\ A:\ n_1 = 121;\quad \bar{x}_1 = 27,465\ \text{miles};\quad s_1 = 2500\ \text{miles}.$$
$$Type\ B:\ n_2 = 122;\quad \bar{x}_2 = 29,572\ \text{miles};\quad s_2 = 3000\ \text{miles}.$$

(a) Find a 99% confidence region for σ_1^2/σ_2^2.
(b) On the basis of the interval found in (a), find a 99% confidence interval for $\mu_1 - \mu_2$.

9.36. A group of 121 students of a large university are retested on entrance to the school to give a mean score of 114 with a standard deviation of 19.6. Another group of 91 students who have spent one year at this school are given the same test, and they perform with a mean score of 121 and standard deviation of 16.8.

(a) Find a 99% confidence interval for the ratio of the class variances.

(b) Using the results of (a), determine an appropriate 99% confidence interval for the differences of mean scores of the classes.

9.37. The effectiveness of two drugs is tested on two groups of randomly selected patients with the following results (in coded units):

$$Group\ 1:\ n_1 = 27;\qquad \bar{x}_1 = 13.540;\qquad s_1 = .476.$$
$$Group\ 2:\ n_2 = 45;\qquad \bar{x}_2 = 11.691;\qquad s_2 = .519.$$

(a) Find a 95% confidence interval for σ_1^2/σ_2^2.

(b) On the basis of the result of (a), find a 95% confidence interval for $\mu_1 - \mu_2$, the difference of the mean effectiveness of the two drugs.

9.38. The athletic department of a large school selects two groups of 50 students each at random. The first group is chosen from students who voluntarily "engage in athletics"; the second group is chosen from students who do not engage in athletics. Their weights are measured with the following results:

$$Athletes:\ n_1 = 50;\qquad \bar{x}_1 = 158.26\ lbs;\qquad s_1 = 7.08\ lbs.$$
$$Non\text{-}athletes:\ n_2 = 50;\qquad \bar{x}_2 = 151.47\ lbs;\qquad s_2 = 7.92\ lbs.$$

(a) Find a 99% confidence interval for σ_1^2/σ_2^2.

(b) Using (a), find a 99% confidence interval for $\mu_1 - \mu_2$. Comment.

9.39. A random sample of 500 individuals contains 200 wearing glasses. Find 95% confidence limits for θ, the proportion of people in the population wearing glasses, using the method of Example 9.8(a), and also the method of Example 9.8(b).

9.40. Nine out of 15 students polled favored the holding of a demonstration on campus against "the war." Using Chart III, find 95% confidence limits for the proportion of all students favoring this proposal.

9.41. A new coin is tossed fifty times, and 20 of the tosses show heads and 30 show tails. Find a 99% confidence interval for the probability of obtaining a head when this coin is tossed once.

9.42. A random sample of 60 voters selected at random from a large city indicate that 70% will vote for candidate A in the upcoming mayorality election. Find a 99% confidence interval for the proportion of voters supporting candidate A. [See (9.8.10).]

9.43. If n is large (in practice $n \geq 100$), then $(\hat{\theta} - \theta)/\sqrt{\hat{\theta}(1 - \hat{\theta})/n} \simeq z$, where $\hat{\theta}$ is the observed sample proportion. Use this statement to develop an approximate $100(1 - \alpha)\%$ confidence interval for θ, for large n. Using this interval, compute 95% limits for θ for the data of Example 9.8(a), and compare with the results of that example.

9.44. Suppose now we are interested in the proportion of populations 1 and 2 that have a characteristic A. Suppose that n_1 independent observations are taken from population 1 and independently, n_2 independent observations are taken from population 2, with the result that $\hat{\theta}_1$ and $\hat{\theta}_2$ are the observed proportions in samples 1 and 2, respectively, having the characteristic A. If n_1 and n_2 are large, use the statement of 9.43 to show that

$$\frac{(\hat{\theta}_1 - \hat{\theta}_2) - (\theta_1 - \theta_2)}{\sqrt{\dfrac{\hat{\theta}_1(1 - \hat{\theta}_1)}{n_1} + \dfrac{\hat{\theta}_2(1 - \hat{\theta}_2)}{n_2}}} \cong z,$$

and hence that the interval $\left[(\hat{\theta}_1 - \hat{\theta}_2) \pm z_{\alpha/2} \sqrt{\dfrac{\hat{\theta}_1(1 - \hat{\theta}_1)}{n_1} + \dfrac{\hat{\theta}_2(1 - \hat{\theta}_2)}{n_2}} \right]$ is an approximate $100(1 - \alpha)\%$ confidence interval for $\theta_1 - \theta_2$, when n_1 and n_2 are both large.

9.45. A sample of 100 consumers showed 16 favoring Brand x of orange juice. An advertising campaign was then conducted. A sample of 200 consumers surveyed after the campaign showed 50 favoring Brand X. Find a 95% confidence interval for the difference in proportions of the population favoring Brand X before and after the advertising campaign. Comment.

9.46. If $(x_1, \ldots, x_n)$ is a random sample of size n from a population having a Poisson distribution with unknown parameter μ, find the maximum likelihood estimator for μ.

9.47. Events occur in time in such a way that the time interval between two successive events is a random variable t having p.d.f. $\theta e^{-\theta t}$, $\theta > 0$. Observations of the n successive time intervals for $n + 1$ events are $(t_1, \ldots, t_n)$. Assuming these time intervals to be independent, find the maximum likelihood estimator for θ.

9.48. If $(x_1, \ldots, x_n)$ is a random sample of n independent observations from a population having p.d.f. $f(x) = 1/\theta$ if $0 < x < \theta$, and zero otherwise, find the maximum likelihood estimator of θ. (*Hint:* do not use calculus, and write out the likelihood.) If $\hat{\theta}$ is the maximum likelihood estimator of θ, show that $\hat{\theta}$ is not unbiased for θ.

9.49 (Continuation). (*a*) Referring to Problem 9.48, show that $t = (n + 1)\hat{\theta}/n$ is unbiased for θ.

(*b*) Find the variance of t.

(*c*) "Another unbiased estimator of θ is $2\bar{x}$." Verify that $2\bar{x}$ is unbiased for θ. Find the variance of $2\bar{x}$.

(*d*) Determine the ratio $V(t)/V(2\bar{x})$. Which of the unbiased estimators do you prefer?

9.50. If $(x_1, \ldots, x_n)$ is a random sample of size n from $N(\mu, \sigma_0^2)$, where σ_0^2 is the known value of the population variance, find the maximum likelihood estimator of μ. Is the maximum likelihood estimator unbiased for μ? What is the distribution of the maximum likelihood estimator?

9.51. If $(x_1, \ldots, x_n)$ is a random sample of size n from $N(\mu_0, \sigma^2)$, where μ_0 is the known value of the population mean, find the maximum likelihood estimator of σ^2. Is the maximum likelihood estimator unbiased for σ^2? What is its distribution? What is its variance?

CHAPTER 9 A

Estimation of Population Parameters— the Bayesian Approach

9A.1 INTRODUCTION

In the previous chapter, we have discussed the problem of estimation of parameters using "classical" methods. The term "classical" is employed because of the interpretation that is attached to the probabilities used in the analyses, that is, the "relative frequency interpretation." Under this interpretation, an experiment of interest is conceptually thought to be capable of an infinite number of repetitions under the same (i.e., constant) conditions, and if the statement "$P(A) = p$" is made, we mean that the event A would occur in 100 $p\%$ of these repetitions. Indeed, this is the interpretation used in all of the methods in this book so far.

However, it is to be noted that there are other interpretations of probability. One is made use of everyday by laymen—we have all heard and made use of such statements as "probably the weather will be rainy tomorrow," or "the chances are good that an income tax reform bill will pass Congress next session" or "the surgeon looking after John is probably the best in North America." These are all examples of the uses of what has been termed "subjective probabilities" and these probabilities are based on the users' "degree of belief" in a certain statement or event. Clearly, this latter type of probability is not the long-run relative frequency of an event—indeed one of the important points to notice in the examples of this paragraph is the absence of the applicability of the notion of "the repetition" of an underlying experiment under "constant" conditions.

In this chapter, we examine the problem of the estimation of parameters, interpreting probabilities as "degrees of belief" in the events of interest. We will utilize Bayes' Theorem (see Theorem 2.6.1), and at this point we restate Bayes' Theorem in the form necessary for our setting.

Suppose we are sampling a random variable x which has the distribution $f(x \mid \theta)$, where θ is a parameter taking values in some continuous set, that is, has the nature of the continuum (see Section 6.3). We shall label this set Ω. (We refer the reader to Problem 9A.1 for the case where Ω is a discrete set.) Recall from Chapter 9 that θ may stand for a single parameter (as in the case of the binomial or Poisson distribution), in which case Ω is either the whole real line or a subset of it, or it might stand for more than one parameter [as in the case of the normal distribution, where $\theta = (\mu, \sigma^2)$]. If θ stands for k parameters, then Ω is a k-dimensional Euclidean space or a subset of it.

Now suppose that we wish to make inferences about θ, and that, prior to our taking any "sample evidence," our degrees of belief about θ may be summarized by a *prior distribution* $p(\theta)$. [An example of how such a $p(\theta)$ may arise is given in Example 9A.2(a) of the next section.] If $(x_1, \ldots, x_n)$ is a random sample of n independent observations on x, we note that the joint probability density function of θ and $(x_1, \ldots, x_n)$ is

$$(9A.1.1) \quad p(\theta, x_1, \ldots, x_n) = p(\theta)p(x_1, \ldots, x_n \mid \theta) = p(\theta) \prod_{i=1}^{n} f(x_i \mid \theta),$$

so that the (marginal) probability density function of $(x_1, \ldots, x_n)$ is

$$(9A.1.2) \qquad p(x_1, \ldots, x_n) = \int_{\Omega} p(\theta)p(x_1, \ldots, x_n \mid \theta) \, d\theta.$$

As noted in (9A.1.1), because $x_1, \ldots, x_n$ are n independent observations on x, we have that $p(x_1, \ldots, x_n \mid \theta) = \prod_{i=1}^{n} f(x_i \mid \theta)$, and in the language of Chapter 9, this is the *likelihood* of θ based on $x_1, \ldots, x_n$. Thus, we write

$$p(x_1, \ldots, x_n \mid \theta) = \prod_{i=1}^{n} f(x_i \mid \theta) = l(\theta \mid x_1, \ldots, x_n).$$

Referring to Section 2.6, we recall that *Bayes' Theorem* is simply a statement of conditional probability, and in the setting of this section, we have that the *posterior probability density function of* θ, given the sample $x_1, \ldots, x_n$, is

$$(9A.1.3) \quad p(\theta \mid x_1, \ldots, x_n) = p(\theta, x_1, \ldots, x_n)/p(x_1, \ldots, x_n).$$

From (9A.1.1) and (9A.1.2), we see that (9A.1.3) may be written as

$(9A.1.4)$

$$p(\theta \mid x_1, \ldots, x_n) = p(\theta)l(\theta \mid x_1, \ldots, x_n) \bigg/ \int_{\Omega} p(\theta)l(\theta \mid x_1, \ldots, x_n) \, d\theta.$$

From (9A.1.2), we see that the denominator of the posterior p.d.f. of θ is the marginal p.d.f. of $x_1, \ldots, x_n$; but $x_1, \ldots, x_n$ is given and so any function of $x_1, \ldots, x_n$ is also given. In particular, the denominator of (9A.1.4) is a constant, and we frequently write (9A.1.4) in the form

$$(9A.1.5) \qquad p(\theta \mid x_1, \ldots, x_n) \propto p(\theta)l(\theta \mid x_1, \ldots, x_n)$$

where the constant of proportionality, say K, is the normalizing constant necessary to make $p(\theta \mid x_1, \ldots, x_n)$ integrate to 1, i.e., K is such that

$$(9A.1.6) \qquad 1/K = \int_\Omega p(\theta)l(\theta \mid x_1, \ldots, x_n)\, d\theta.$$

The reader should take note of the somewhat different (and at first startling) fact that emerges due to the "degree of belief" interpretation of probability that we are using in this chapter—namely, that the above implies that θ, the parameter (or parameters) is (are) now a random variable(s), and that, once given $x_1, \ldots, x_n$, the sample and any function of it are constants.

Further, and very importantly, the *posterior distribution summarizes all the information we have about* θ, combining the prior information through the *prior distribution*, and the *sample* information through the *likelihood function*. We can thus make statements, given the sample evidence, as to our degrees of belief about θ, using the posterior distribution (9A.1.4).

A question that enters at this point is the choice of the prior distribution. If our prior belief (prior to observing the sample $x_1, \ldots, x_n$) about θ is "sharp," that is, if we feel that the parameter has a certain value very near θ_0 say, then our prior distribution should be located about θ_0 with small variance. If, however, we are uncertain about the possible values of the parameter θ, our prior will have a larger variance—see the examples of the ensuing sections.

Once we have the posterior of θ, we may use it in several ways. For example, we may compute the mean,

$$(9A.1.7) \qquad E(\theta \mid x_1, \ldots, x_n) = \int_\Omega \theta p(\theta \mid x_1, \ldots, x_n)\, d\theta,$$

and we note that (9A.1.7) is a function of $(x_1, \ldots, x_n)$ and represents a measure of location for our (posterior) beliefs about θ, posterior to observing the sample $(x_1, \ldots, x_n)$. Hence (9A.1.7) is often used as a point estimate for the unknown parameter θ, and is called the *Bayes estimate* of θ. Also, we may find a posterior region, say C, which is such that our (posterior) degree of belief in θ falling in C is $100(1 - \alpha)\%$, that is,

$$(9A.1.8) \quad P(\theta \in C \mid x_1, \ldots, x_n) = \int_C p(\theta \mid x_1, \ldots, x_n)\, d\theta = 1 - \alpha.$$

For example, if θ stands for a single real-valued parameter, then the region C may be the interval $C = [a, b]$, where

$$(9A.1.9) \quad \int_{-\infty}^{a} p(\theta \mid x_1, \ldots, x_n)\, d\theta = \alpha/2 = \int_{b}^{\infty} p(\theta \mid x_1, \ldots, x_n)\, d\theta,$$

so that the degree of belief we have in θ falling in C, i.e., $a \leq \theta \leq b$ is $1 - \alpha$, i.e.,

$$(9A.1.10) \quad P(a \leq \theta \leq b \mid x_1, \ldots, x_n) = \int_{a}^{b} p(\theta \mid x_1, \ldots, x_n) = 1 - \alpha.$$

It is to be noted that a and b are functions of $x_1, \ldots, x_n$. This point will be made clear in the examples of the ensuing sections.

9A.2 THE NORMAL OR GAUSSIAN SITUATION— VARIANCE KNOWN

(a) One Population

In this subsection we consider the case where sampling is on a normal random variable x, unknown mean μ, but known variance which we denote by σ^2. In the notation of the previous section, $\theta = \mu \in \Omega = (-\infty, \infty)$, and we wish to make inferences about μ.

Suppose that prior information about μ is such that the experimenter expects μ to be μ_0, and that (prior) confidence intervals for μ give rise to the choice of prior for μ as a normal, mean μ_0, variance σ_0^2, that is (see Figure 9.A1),

$$(9A.2.1) \qquad p(\mu) = \frac{1}{\sqrt{2\pi}\,\sigma_0} \exp\left[-\left(\frac{\mu - \mu_0}{\sigma_0}\right)^2 \Big/ 2\right]$$

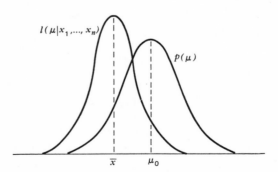

Figure 9A.1. Graphs showing an experimenter's prior for μ and the likelihood of μ based on a sample of n independent observations. [See (9A.2.1) and (9A.2.3).]

where μ_0 and σ_0 are constants chosen by the experimenter and referred to just above. (The experimenter is declaring that his prior degrees of belief about μ are such that he feels it is just as likely for $\mu < \mu_0 - \sigma_0$ as $\mu > \mu_0 + \sigma_0$, that he would give 9:1 odds that "$\mu_0 - 1.645\,\sigma_0 < \mu < \mu_0 + 1.645\,\sigma_0$" is true, etc.)

Suppose that a sample of n independent observations is taken on x and observed to be $x_1, \ldots, x_n$, so that the likelihood function, using the normality of x, is

$$(9A.2.2) \quad l(\mu \mid x_1, \ldots, x_n) = \prod_{i=1}^{n} f(x_i \mid \mu)$$

$$= \frac{1}{(2\pi\sigma^2)^{n/2}} \exp\left[-\sum_{i=1}^{n} \left(\frac{x_i - \mu}{\sigma} \right)^2 \Big/ 2 \right].$$

Since σ^2 is known, we rewrite (9A.2.2) as

$$(9A.2.3) \qquad l(\mu \mid x_1, \ldots, x_n) = c e^{-n(\bar{x}-\mu)^2/2\sigma^2},$$

where the constant c is given by

$$(9A.2.4)$$

$$c = (2\pi\sigma^2)^{-n/2} e^{-(n-1)s^2/2\sigma^2} = (2\pi\sigma^2)^{-n/2} \exp\left[-\sum_{i=1}^{n} (x_i - \bar{x})^2 \Big/ 2\sigma^2 \right].$$

Combining (9A.2.1) and (9A.2.4) via Bayes' Theorem, we find that the posterior p.d.f. for μ is such that

$$(9A.2.5) \quad p(\mu \mid x_1, \ldots, x_n) \propto \exp\left\{ -\left[\frac{(\mu - \mu_0)^2}{\sigma_0^2} + \frac{n(\mu - \bar{x})^2}{\sigma^2} \right] \Big/ 2 \right\}.$$

It can readily be shown that the exponent in (9A.2.5) may be written as

$$(9A.2.6) \qquad -\frac{1}{2}\left\{ \left[\frac{1}{\sigma_0^2} + \frac{n}{\sigma^2} \right][\mu - \bar{\bar{x}}]^2 + \frac{n}{\sigma^2 + n\sigma_0^2}(\bar{x} - \mu_0)^2 \right\}$$

where

$$(9A.2.6a) \qquad \bar{\bar{x}} = \left(\frac{1}{\sigma_0^2}\mu_0 + \frac{n}{\sigma^2}\bar{x} \right) \Big/ \left(\frac{1}{\sigma_0^2} + \frac{n}{\sigma^2} \right).$$

Now the last term in the braces { } of (9A.2.6) is constant (μ_0 is given; $\bar{x}$ is observed once $x_1, \ldots, x_n$ is observed; σ^2, σ_0^2 are known constants; n is the known sample size). Hence, absorbing constants into the constant of proportionality, we have

$$(9A.2.7) \quad p(\mu \mid x_1, \ldots, x_n) \propto \exp\left\{ -\left[\left(\frac{1}{\sigma_0^2} + \frac{n}{\sigma^2} \right)(\mu - \bar{\bar{x}})^2 \right] \Big/ 2 \right\}$$

or

$$(9A.2.7a) \quad p(\mu \mid x_1, \ldots, x_n) = \frac{(1/\sigma_0^2 + n/\sigma^2)^{1/2}}{\sqrt{2\pi}}$$

$$\times \exp\left\{-\left[\left(\frac{1}{\sigma_0^2} + \frac{n}{\sigma^2}\right)(\mu - \bar{\bar{x}})^2\right]\Big/2\right\}$$

with $\bar{\bar{x}}$ given by (9A.2.6a). We note that (9A.2.7) and (9A.2.1) are both normal probability density functions. When a prior is such that it gives rise to a posterior of same type, we will say that the prior is a *conjugate prior* for the problem. Note that the conjugate prior (9A.2.1) combines "nicely" with the likelihood to give the posterior (9A.2.7a). The prior is $N(\mu_0, \sigma_0^2)$ and the posterior is also normal, i.e., of the same type, with mean $\bar{\bar{x}}$, variance $(1/\sigma_0^2 + n/\sigma^2)^{-1}$.

Summarizing, the posterior degrees of belief we have in various events concerning μ (posterior to observing $x_1, \ldots, x_n$) may be described by using the $N(\bar{\bar{x}}, [1/\sigma_0^2 + n/\sigma^2]^{-1})$ distribution. This result is interesting for many reasons. First, the reader will note that the posterior mean of μ is

$$(9A.2.8) \quad E(\mu \mid x_1, \ldots, x_n) = \bar{\bar{x}} = \left(\frac{1}{\sigma_0^2}\mu_0 + \frac{n}{\sigma^2}\bar{x}\right)\Big/\left(\frac{1}{\sigma_0^2} + \frac{n}{\sigma^2}\right),$$

that is, the Bayes estimate of μ is $\bar{\bar{x}}$, and this estimate is a weighted combination of the prior mean of μ, i.e., μ_0, and the sample evidence about μ, $\bar{x}$. Moreover, the weights are the reciprocal of the respective variances of the prior and sample evidence, viz., σ_0^2 and σ^2/n, respectively. Second, the posterior variance of μ is

$$(9A.2.9) \quad V(\mu \mid x_1, \ldots, x_n) = (1/\sigma_0^2 + n/\sigma^2)^{-1}.$$

At this point, we define the *precision* of a random variable as the reciprocal of the variance of the random variable. We note then, that (9A.2.8) says that the Bayes estimate of μ is simply a weighted combination of the prior and sample evidence about μ, i.e., μ_0 and $\bar{x}$, the weights being the prior precision of μ and the precision of $\bar{x}$, i.e., $1/\sigma_0^2$ and n/σ^2, respectively. Further, (9A.2.9) implies that the *posterior precision* of μ is the sum of the precision of the prior evidence $(1/\sigma_0^2)$ and the precision of the sample evidence (n/σ^2).

It is interesting to note what happens as σ_0^2 tends to infinity, that is, as we have vaguer and vaguer prior information about μ. First, the Bayes estimate tends to $\bar{x}$, that is, as the prior knowledge about μ becomes more and more diffuse, the posterior mean of μ tends to a quantity that depends more and more on the *sample evidence alone*, namely $\bar{x}$, a natural and reasonable property.

Further, the posterior variance of μ, $(1/\sigma_0^2 + n/\sigma^2)^{-1}$, as σ_0^2 tends to infinity, increases to $[n/\sigma^2]^{-1} = \sigma^2/n$, the variance of $\bar{x}$ alone.

The above may be expected if one reasons as follows. If the experimenter has prior degrees of belief about μ which are very vague, he may wish to describe his feelings about μ by the so called "in-ignorance" or "uniform" prior $p(\mu)$ which gives as much probability, that is degrees of belief, in μ falling in the interval, say $(a, a + d)$, as in the interval $(b, b + d)$, etc., that is,

$$(9A.2.10) \qquad p(\mu) \propto k, \quad \text{where } k \text{ is constant.}$$

We note that (9A.2.10) is the limit, as σ_0^2 tends to infinity, of (9A.2.1). The experimenter may indeed have arrived at the choice (9A.2.10) by first choosing (9A.2.1), but then realizing that "his σ_0^2" was enormous so that (9A.2.10) actually should be employed; or it could well be that introduction to a new normal process $N(\mu, \sigma^2)$, σ^2 known, prompted only very vague feelings about μ to begin with, leading to the choice of (9A.2.10). Suppose a sample of n independent observations is now observed on this normal process, giving rise to the likelihood (9A.2.3). Combining (9A.2.10) with (9A.2.3) yields, as the reader may verify, the posterior

$$(9A.2.11) \qquad p(\mu \mid x_1, \ldots, x_n) = \frac{\sqrt{n}}{\sqrt{2\pi\sigma^2}} \exp\left[-\frac{n}{2\sigma^2}(\mu - \bar{x})^2\right],$$

and we note that this is the limit, as σ_0^2 tends to infinity, of the posterior given by (9A.2.7a). The above distribution has, of course,

$$(9A.2.12) \qquad E(\mu \mid x_1, \ldots, x_n) = \bar{x}, \qquad V(\mu \mid x_1, \ldots, x_n) = \sigma^2/n,$$

a result analogous to that of the classical approach.

We return now to the posterior (9A.2.7a). Using this posterior, we easily see that

$$(9A.2.13) \qquad (\mu - \bar{\bar{x}}) \Big/ \left(\frac{1}{\sigma_0^2} + \frac{n}{\sigma^2}\right)^{-1/2} = z,$$

where $z = N(0, 1)$, so that we have, using (9A.1.9) and (9A.1.10) that

$$(9A.2.14) \qquad P[\bar{\bar{x}} - (1/\sigma_0^2 + n/\sigma^2)^{-1/2}z_{\alpha/2} \leq \mu \leq$$
$$\bar{\bar{x}} + (1/\sigma_0^2 + n/\sigma^2)^{-1/2}z_{\alpha/2} \mid x_1, \ldots, x_n] = 1 - \alpha,$$

that is, after observing the sample $(x_1, \ldots, x_n)$, our degrees of belief in the event of μ falling in the interval

$$(9A.2.15) \qquad I_1 = [\bar{\bar{x}} \pm (1/\sigma_0^2 + n/\sigma^2)^{-1/2}z_{\alpha/2}]$$

is $100(1 - \alpha)\%$, that is, (9A.2.15) is a $100(1 - \alpha)\%$ posterior interval for μ. [See (9A.2.7a).] Now we note that as σ_0^2 grows very large, I_1 tends to the interval

(9A.2.15a) $$I_2 = [\bar{x} \pm (\sigma^2/n)^{1/2} z_{\alpha/2}]$$

[compare with (9.3.3)]. I_2 is in fact the interval that would have been obtained had we used the posterior (9A.2.11) which was found when the diffuse prior (9A.2.10) was used. The prior (9A.2.10) is often referred to as the "in-ignorance" or "uniform" or "noninformative" prior for μ.

> **Example 9A.2(a).** Management of a tire factory has decided that a certain type of tire is to be produced using a modification of an existing process. Based on previous experience with "similar" modifications of the process, the head chemist "believes that the tires to be produced using this newest modification will have true length of life until wear-out, say μ, which is normally distributed with mean $\mu_0 = 26,500$ miles and standard deviation $\sigma_0 = 1200$ miles, approximately." That is, on questioning, he expects μ to be 26,500 miles and feels 95% sure that μ lies in the interval [24,100; 28,900] and, consistent with these facts, also feels that
>
> $$P(\mu < 24,100) = P(\mu > 28,900) \cong .025,$$
>
> and so forth. That is to say, his prior distribution is such that
>
> $$[\mu_0 \pm 2\sigma_0] = [26,500 \pm 2\sigma_0] = [24,100;\ 28,900]$$
>
> or $\sigma_0 = 1200$. Hence, his prior beliefs seem to be described by $N(\mu_0, \sigma_0^2) = N(26,500, (1200)^2)$ distribution.
>
> It is also known from previous experience that over a wide range of modifications usually encountered, that y, the length of life of a tire until wear out, has the $N(\mu, (1400)^2)$ distribution, that is, $\sigma^2 = (1400)^1$.
>
> Suppose a sample of 100 tires produced by the new process yields a sample mean of $\bar{y} = 25,100$ miles. Hence, the posterior distribution of μ is [see (9A.2.7)]
>
> $$p(\mu \mid y_1, \ldots, y_{100}) = \frac{1}{\sqrt{2\pi}\sqrt{V}} \exp\left[-\frac{1}{2V}(\mu - \bar{\bar{y}})^2\right]$$
>
> where
>
> $$\begin{aligned}\bar{\bar{y}} &= \left(\frac{1}{\sigma_0^2}\mu + \frac{n}{\sigma^2}\bar{y}\right)\bigg/\left(\frac{1}{\sigma_0^2} + \frac{n}{\sigma^2}\right) \\ &= \left(\frac{1}{(1200)^2}(26,500) + \frac{100}{(1400)^2}(25,100)\right)\bigg/\left(\frac{1}{(1200)^2} + \frac{100}{(1400)^2}\right) \\ &= 25,119,\end{aligned}$$
>
> and
>
> $$\begin{aligned}V &= \left(\frac{1}{\sigma_0^2} + \frac{n}{\sigma^2}\right)^{-1} = \left(\frac{1}{(1200)^2} + \frac{100}{(1400)^2}\right)^{-1} \\ &= (139)^2,\end{aligned}$$
>
> that is, the posterior of μ is the $N(25,119;\ (139)^2)$ distribution. Hence the Bayes estimate of μ is 25,119 miles and a 95% posterior interval is
>
> $$\begin{aligned}[\bar{\bar{y}} \pm 1.96\sqrt{V}] &= [25,119 \pm 1.96 \times 139] \\ &= [24,847;\ 25,391].\end{aligned}$$

Example 9A.2(b). A new and unusual plastic is to be produced by a pilot plant using a standard extrusion method. The yield of batches of this new plastic at this plant, say x, is expected to be $N(\mu, (20)^2)$ where x is measured in tons. Because this is a new plastic, however, feelings on the part of the plant chemists are vague as to what to believe about μ, and close examination shows that over a wide range, the prior feelings about μ are best described by the noninformative prior

$$p(\mu) \propto k.$$

A sample of 50 day's production (the days selected at random) yield a sample mean of $\bar{x} = 670$ tons. From (9A.2.11), then, we have that the posterior of μ is

$$p(\mu \mid x_1, \ldots, x_{50}) = \frac{50}{\sqrt{2\pi \cdot (20^2)}} \exp\left[-\frac{50}{2 \cdot (20^2)}(\mu - 670)^2\right],$$

that is, the posterior distribution of μ is the $N(670, 8)$ distribution. Hence, the Bayes estimate of μ is 670, and the posterior probability that μ lies in the interval

$$[670 \pm (1.96)(2.82)] = [670 \pm 5.53] = [664.5, 675.5]$$

is .95, that is, this interval is a 95% posterior interval for μ.

Other Priors. The reader should note that the choice of priors is not restricted to the *conjugate* prior (9A.2.1) or the uniform prior (9A.2.10). For example, it could be that from theoretical considerations, or otherwise, it is known that μ lies in $[a, b]$, but that nothing else is known to the experimenter. For this reason we may wish to choose the prior

(9A.2.16) $p(\mu) = \dfrac{1}{b - a}$ if $a \leq \mu \leq b$,   and 0 otherwise.

Combining this with the likelihood (9A.2.3) gives the posterior for μ, $p(\mu \mid x_1, \ldots, x_n)$, which is such that

(9A.2.17) $p(\mu \mid x_1, \ldots, x_n) \propto \exp\left[-\dfrac{n}{2\sigma^2}(\mu - \bar{x})^2\right]$, $a < \mu \leq b$

or

(9A.2.17a) $p(\mu \mid x_1, \ldots, x_n) = \dfrac{\sqrt{n}}{\sqrt{2\pi}\sigma}\left[\Phi\left(\dfrac{b - \bar{x}}{\sigma/\sqrt{n}}\right) - \Phi\left(\dfrac{a - \bar{x}}{\sigma/\sqrt{n}}\right)\right]^{-1}$

$$\times \exp\left[-\dfrac{n}{2\sigma^2}(\mu - \bar{x})^2\right].$$

[$E(\mu \mid x_1, \ldots, x_n)$ and $V(\mu \mid x_1, \ldots, x_n)$ may be computed by integration by parts—see Problem 9A.2.]

(b) Two Populations

In this section we consider the case where sampling is on two random variables, x_1 and x_2, where $x_1 = N(\mu_1, \sigma_1^2)$ and $x_2 = N(\mu_2, \sigma_2^2)$, with σ_1^2 and σ_2^2 known. We suppose that both μ_1 and μ_2 are independent a priori and that their priors are of the conjugate type, that is, we will assume

$$(9A.2.18) \qquad p(\mu_1, \mu_2) = p(\mu_1)p(\mu_2)$$

where

$$p(\mu_i) = \frac{1}{\sqrt{2\pi}\,\sigma_{0i}} \exp\left[-\frac{1}{2\sigma_{0i}^2}(\mu_i - \mu_{0i})^2\right].$$

Now suppose we observe independent random samples of size n_1 and n_2 on x_1 and x_2, respectively, say $x_{11}, x_{21}, \ldots, x_{n_11}$ and $x_{12}, x_{22}, \ldots, x_{n_22}$. The likelihood of μ_1 (and μ_2) is of the form (9A.2.3) and we have

$$(9A.2.19) \quad l(\mu_1, \mu_2 \mid x_{11}, \ldots, x_{n_11}; x_{12}, \ldots, x_{n_22})$$

$$= c_1 c_2 \exp\left[-\frac{n_1}{2\sigma_1^2}(\bar{x}_1 - \mu_1)^2 - \frac{n_2}{2\sigma_2^2}(\bar{x}_2 - \mu_2)^2\right].$$

Combining (9A.2.18) with (9A.2.19) yields the posterior

$$(9A.2.20) \quad p(\mu_1, \mu_2 \mid x_{11}, \ldots, x_{n_11}; x_{12}, \ldots, x_{n_22})$$

$$= p(\mu_1 \mid x_{11}, \ldots, x_{n_11})p(\mu_2 \mid x_{12}, \ldots, x_{n_22})$$

where

$$p(\mu_j \mid x_{n_1j}, \ldots, x_{njj}) = \frac{(1/\sigma_{0j}^2 + n_j/\sigma_j^2)^{1/2}}{\sqrt{2\pi}}$$

$$\times \exp\left\{-\left[\left(\frac{1}{\sigma_{0j}^2} + \frac{n_j}{\sigma_j^2}\right)(\mu_j - \bar{\bar{x}}_b)^2\right]\Big/2\right\}$$

with

$$\bar{\bar{x}}_j = \frac{\dfrac{1}{\sigma_{0j}^2}\mu_{0j} + \dfrac{n_j}{\sigma_j^2}\bar{x}_j}{\dfrac{1}{\sigma_{0j}^2} + \dfrac{n_j}{\sigma_j^2}}$$

and

$$\bar{x}_j = n_j^{-1}\sum_{i=1}^{n_j} x_{ij},$$

which is to say that the joint posterior of μ_1 and μ_2 factors into the product of normal marginals. Specifically, the posterior of μ_1 and μ_2 is that of two independent normals, means $\bar{\bar{x}}_1$ and $\bar{\bar{x}}_2$, variances a_1 and a_2 where $a_j = (1/\sigma_{0j}^2 + n_j/\sigma_j^2)^{-1}$.

The above implies of course that the posterior of $\mu_1 - \mu_2$ is such that

(9A.2.21) $\mu_1 - \mu_2 = N(\bar{\bar{x}}_1 - \bar{\bar{x}}_2, a_1 + a_2).$

Hence the Bayes estimator of $\mu_1 - \mu_2$ is

(9A.2.22) $E(\mu_1 - \mu_2 \mid x_{11}, \ldots, x_{n_1 1}; x_{12}, \ldots, x_{n_2 2}) = \bar{\bar{x}}_1 - \bar{\bar{x}}_2$

and a $100(1 - \alpha)\%$ posterior interval for $\mu_1 - \mu_2$ is

(9A.2.23) $[(\bar{\bar{x}}_1 - \bar{\bar{x}}_2) \pm \sqrt{a_1 + a_2}\, z_{\alpha/2}].$

We again point out to the reader that if the priors of μ_1 and μ_2 are of the "uniform" type, i.e.,

(9A.2.24) $p(\mu_1, \mu_2) \propto k,$

then, as the reader may verify, combining this prior with the likelihood (9A.2.19) yields the posterior

(9A.2.25) $\displaystyle \prod_{j=1}^{2} \frac{\sqrt{n_j}}{\sqrt{2\pi\sigma_j^2}} \exp\left[-\frac{n_j}{2\sigma_j^2}(\mu_j - \bar{x}_j)^2 \right].$

That is, the posterior distribution of μ_1 and μ_2 is that of two independent normals. Hence the posterior of $\mu_1 - \mu_2$ is

$$N\left((\bar{x}_1 - \bar{x}_2), \left(\frac{\sigma_1^2}{n_1} + \frac{\sigma_2^2}{n_2} \right) \right).$$

Thus we would have that the Bayes estimate of $\mu_1 - \mu_2$ is

(9A.2.26) $E(\mu_1 - \mu_2 \mid x_{11}, \ldots, x_{n_1 1}; x_{12}, \ldots, x_{n_2 2}) = \bar{x}_1 - \bar{x}_2,$

and that the interval

(9A.2.27) $\left[(\bar{x}_1 - \bar{x}_2) \pm \sqrt{\frac{\sigma_1^2}{n_1} + \frac{\sigma_2^2}{n_2}}\, z_{\alpha/2} \right]$

is a posterior interval at level $1 - \alpha$. Note that (9A.2.27) depends only on sample information and of course this is due to the vagueness of the prior information. In fact as we let σ_{01}^2 and σ_{02}^2 tend to infinity in (9A.2.22) and (9A.2.23), we obtain the results (9A.2.26) and (9A.2.27). (This interval has been obtained before—see Theorem 9.4.1 of Chapter 9.)

Example 9A.2(c). Two plants are to produce a new type of fluorescent bulb using new equipment. Because of the "similarity" to other processes making fluorescent bulbs of different types, it is known that over wide ranges of μ_1 and μ_2 the distribution of the life of light bulbs from plants I and II are $N(\mu_1, (200)^2)$ and $N(\mu_2, (200)^2)$, respectively. Now suppose that prior information about μ_1 and μ_2 is so vague that the choice of the in-ignorance prior (9A.2.24) is indicated, so that we use $p(\mu_1, \mu_2) \propto k$. When data is taken, a random sample of $n_1 = 25$ bulbs from plant I's new production yields $\bar{y}_1 = 1410$ hours, and a random sample of $n_2 = 20$ bulbs from plant II's new production yields $\bar{y}_2 = 1260$ hours. Then the Bayes estimate for $\mu_1 - \mu_2$ for this case (see 9A.2.26) is $\bar{y}_1 - \bar{y}_2 = 1410 - 1260 = 150$ hours, and a 95% posterior interval for $\mu_1 - \mu_2$ is

$$\left[150 \pm \sqrt{\frac{200^2}{25} + \frac{200^2}{20}} \, (1.96) \right] = [150 \pm 117.60]$$

or

$$[32.4, \ 267.6].$$

Example 9A.2(d). In Example 9A.2(c), we now suppose that, before observing any data, the two plants jointly hire an expert skilled in the use of the new equipment. Suppose too that, after inspecting the plants, his prior feelings about μ_1 and μ_2 are independent and both summarized by the $N(1350, (400)^2)$ distribution. That is, we have for the priors that

$$\mu_{01} = 1350 = \mu_{02}; \qquad \sigma_{01}^2 = (400)^2 = \sigma_{02}^2,$$

and we note that, as before,

$$\sigma_1^2 = (200)^2 = \sigma_2^2.$$

Suppose now that the data is taken and is as given in the previous example, that is, $n_1 = 25$, $n_2 = 20$, $\bar{y}_1 = 1410$, $\bar{y}_2 = 1260$. Then the posterior distribution of $\mu_1 - \mu_2$ is $N(\bar{y}_1 - \bar{y}_2, a_1 + a_2)$ with the Bayes estimate $\bar{y}_1 - \bar{y}_2$ having the value

$$\frac{\dfrac{1}{(400)^2} 1350 + \dfrac{25}{(200)^2} 1410}{\dfrac{1}{(400)^2} + \dfrac{25}{(200^2)}} - \frac{\dfrac{1}{(400)^2} 1350 + \dfrac{20}{(200)^2} 1260}{\dfrac{1}{(400)^2} + \dfrac{20}{(200^2)}}$$

$$= 1409.406 - 1261.177$$

$$= 148.295 \text{ or } 148.3, \text{ say.}$$

(This is a change from "150" of the previous example, and typically, when we take into account some prior information, changes result, as in this case.) Consulting (9A.2.23) we find that a 95% posterior confidence interval for $\mu_1 - \mu_2$ is

$$[148.3 \pm \sqrt{a_1 + a_2} z_{.025}] = [148.3 \pm \sqrt{a_1 + a_2}(1.96)]$$

where

$$a_1 = (1/(400)^2 + 25/(200)^2)^{-1} = (63125 \times 10^{-8})^{-1} = 1584.1584,$$
$$a_2 = (1/(400)^2 + 20/(200)^2)^{-1} = (50625 \times 10^{-8})^{-1} = 1975.3086,$$

so that

$$\sqrt{a_1 + a_2} = \sqrt{3569.4670} = 59.75.$$

Hence the 95% posterior interval here is

$$[148.3 \pm (59.75)(1.96)] = [148.3 \pm 117.11] = [31.19, 265.41].$$

We note that this interval has somewhat smaller length than the interval of the previous example, and again, the change is due to the fact that we are incorporating prior knowledge about μ_1 and μ_2. This knowledge is not too precise ($\sigma_{0j} = 400$) so that the resulting differences between this and the previous examples do not amount to very much. The reader should go through the details to find results when $\sigma_{0j}^2 = (100)^2$. [Problem 9A.7.]

9A.3 THE NORMAL OR GAUSSIAN SITUATION—MEANS AND VARIANCES UNKNOWN

(a) One Population

In this section we assume that sampling is again on a normal variable $x = N(\mu, \sigma^2)$, where now both μ and σ^2 are unknown. In the language of section 9A.1, then $\theta = (\mu, \sigma^2)$ and the two-dimensional set Ω is the half-plane in (μ, σ^2) space, i.e.,

$$(9A.3.1) \qquad \Omega = \{(\mu, \sigma^2) \mid -\infty < \mu < \infty, \sigma^2 > 0\}.$$

We will begin with the case where the experimenter has only very vague prior information about μ and σ^2. The "uniform" or "in-ignorance" prior that is chosen for this situation is

$$(9A.3.2) \qquad p(\mu, \sigma^2) = p(\mu)p(\sigma^2), \qquad (\mu, \sigma^2) \in \Omega$$

where

$$(9A.3.2a) \qquad p(\mu) \propto k, \qquad p(\sigma^2) \propto 1/\sigma^2.$$

The reasons for the choice of the above prior for σ^2 will not be described in this text, but the interested reader is referred to pages 120–123 in Chapter 3 of the book by Jeffreys (1961). Now suppose a random sample of n independent observations on x is taken and observed to be $x_1, \ldots, x_n$. The likelihood is

$$(9A.3.3) \quad l(\mu, \sigma^2 \mid x_1, \ldots, x_n) = (2\pi)^{-n/2}(\sigma^2)^{-n/2}$$
$$\times \exp\{-[(n-1)s^2 + n(\mu - \bar{x})^2]/2\sigma^2\}$$

where $\bar{x} = n^{-1}\sum_{i=1}^{n} x_i$ and $s^2 = (n-1)^{-1}\sum_{i=1}^{n} (x_i - \bar{x})^2$ and $\exp(u) = e^u$.

Combining (9A.3.2) with (9A.3.3) yields the posterior

$$(9A.3.4) \quad p(\mu, \sigma^2 \mid x_1, \ldots, x_n) \propto (\sigma^2)^{-(n/2)-1}$$
$$\times \exp\left\{-\frac{1}{2\sigma^2}[(n-1)s^2 + n(\mu - \bar{x})^2]\right\}.$$

In the sequel we will make use of the identity

(9A.3.5) $$\int_0^\infty w^{-m-1} e^{-A/w} \, dw = \Gamma(m)/A^m$$

for constant $A > 0$ and $m > 0$. ($\Gamma(m)$ is discussed in Appendix VI before Theorem AVI.3.) We also remind the reader that, from the properties of the normal distribution,

(9A.3.6) $$\int_{-\infty}^\infty e^{-(v-a)^2/2\tau^2} \, dv = \sqrt{2\pi\tau^2}.$$

Using the identities (9A.3.5) and (9A.3.6), we find that

(9A.3.7) $$p(\mu, \sigma^2 \mid x_1, \ldots, x_n) = \frac{\sqrt{n} \, [(n-1)s^2]^{(n-1)/2}}{2^{n/2} \sqrt{\pi} \, \Gamma\left(\dfrac{n-1}{2}\right)} (\sigma^2)^{-(n/2)-1}$$

$$\times \exp\left\{ -\frac{1}{2\sigma^2} [(n-1)s^2 + n(\mu - \bar{x})^2]. \right\}$$

Suppose now that we wish to make inferences about μ alone. From (9A.3.7) we can find the marginal posterior distribution of μ, so that on using (9A.3.5), we have

$$p(\mu \mid x_1, \ldots, x_n) = \frac{\sqrt{n} \, [(n-1)s^2]^{(n-1)/2}}{2^{n/2} \sqrt{\pi} \, \Gamma\left(\dfrac{n-1}{2}\right)} \int_0^\infty (\sigma^2)^{-(n/2)-1}$$

$$\times \exp\left\{ -\frac{1}{2\sigma^2} [(n-1)s^2 + n(\mu - \bar{x})^2] \right\} d\sigma^2$$

(9A.3.8) $$= \frac{\sqrt{n} \, [(n-1)s^2]^{(n-1)/2} \Gamma\left(\dfrac{n}{2}\right)}{\sqrt{\pi} \, \Gamma\left(\dfrac{n-1}{2}\right)} [(n-1)s^2 + n(\mu - \bar{x})^2]^{-n/2}.$$

Since $x_1, \ldots, x_n$ is observed, $(n-1)s^2$ is a constant and we may simplify (9A.3.8) and obtain

(9A.3.9) $$p(\mu \mid x_1, \ldots, x_n) = \frac{\sqrt{n} \, \Gamma\left(\dfrac{n}{2}\right)}{\sqrt{\pi(n-1)} \, \Gamma\left(\dfrac{n-1}{2}\right) (s^2)^{1/2}}$$

$$\times \left[1 + \frac{n(\mu - \bar{x})^2}{(n-1)s^2} \right]^{-n/2}.$$

The reader will recall from Appendix VI that (9A.3.9) has the form of the Student t distribution, and in fact if we let $t = \sqrt{n}\,(\mu - \bar{x})/s$, (the Jacobian) then, using (6.12.17) of Corollary 6.12.3(a), we have that

$$(9A.3.10) \quad p(t \mid x_1, \ldots, x_n) = \frac{\Gamma\left(\dfrac{n}{2}\right)}{\sqrt{\pi(n-1)}\;\Gamma\left(\dfrac{n-1}{2}\right)}\left[1 + \frac{t^2}{n-1}\right]^{-n/2},$$

that is, the posterior distribution of $t = \sqrt{n}\,(\mu - \bar{x})/s$ is that of the Student t with $(n-1)$ degrees of freedom. From properties of this "t-distribution" (expected value of t_{n-1} is zero and the variance of t_{n-1} is $(n-1)/(n-3)$) we have

$$(9A.3.11) \qquad\qquad E(\mu \mid x_1, \ldots, x_n) = \bar{x},$$

with the posterior variance of μ given by

$$(9A.3.11a) \quad V(\mu \mid x_1, \ldots, x_n) = \frac{s^2}{n}\,\mathrm{Var}\,(t) = \left[\frac{\displaystyle\sum_{i=1}^{n}(x_i - \bar{x})^2}{n(n-3)}\right], \qquad n > 3.$$

Note that $\sum_{i=1}^{n}(x_i - \bar{x})^2/(n-3)$ has the form of a sample variance [except that the denominator is $(n-3)$ and not the usual $(n-1)$], and the posterior variance of μ is thus of the form of a variance of a mean of n observations, i.e., $\left[\sum_{i=1}^{n}(x_i - \bar{x})^2/(n-3)\right]\Big/n$. Further, using (9A.3.10), we have that a $100(1 - \alpha)\%$ posterior interval for μ is

$$(9A.3.12) \qquad\qquad \left[\bar{x} \pm \frac{s}{\sqrt{n}}\,t_{n-1;\alpha/2}\right],$$

that is, our degree of belief or confidence that μ will fall in the interval (9A.3.12) is $1 - \alpha$.

Example 9A.3(a). A sample of 15 test pieces of "Type A" cotton yarn produced by a newly opened textile mill is sent to a well known consumer's test organization. When tested for tensile strength, the 15 pieces yield a sample mean $\bar{x} = 903.2$ psi, with $s = 9.4$ psi.

As the test organization was very vague about how this textile mill has been performing, the priors for μ and σ^2 that are chosen by the organization are the noninformative priors $p(\mu) \propto K$ and $p(\sigma^2) \propto 1/\sigma^2$, so that the posterior of μ is [see (9A.3.10)] such that

$$\mu = \bar{x} + \frac{s}{\sqrt{n}} t_{n-1},$$

which, of course, gives rise to the posterior interval (9A.3.12). If we wish this interval to be a 95% posterior confidence interval for μ, then

$$1 - \alpha = .95; \qquad \alpha/2 = .025; \qquad n = 15; \qquad n - 1 = 14;$$
$$t_{14;.025} = 2.145; \qquad \bar{x} = 903.2; \qquad s = 9.4.$$

Hence

$$[903.2 \pm (9.4)(2.145)/\sqrt{15}] = [903.2 \pm 5.21],$$

that is,

$$[897.99, 908.41]$$

is a 95% posterior interval for μ, and, of course the Bayes' estimate is $\bar{x} = 903.2$ psi, the center of the interval.

We return now to (9A.3.7) and suppose that we are interested in making inferences about σ^2. From (9A.3.7) it is easy to see that the marginal posterior of σ^2 is

$$(9A.3.13) \quad p(\sigma^2 \mid x_1, \ldots, x_n) = \frac{[(n-1)s^2]^{(n-1)/2}}{2^{(n-1)/2}\Gamma\left(\dfrac{n-1}{2}\right)} (\sigma^2)^{-(n+1)/2}$$

$$\times \exp\left[-\frac{(n-1)s^2}{2\sigma^2}\right].$$

This resembles the form of the χ^2 distribution discussed in Appendix VI. In fact, if we put $\tau = (n-1)s^2/\sigma^2$ [the absolute value of the Jacobian of this transformation is $|d\sigma^2/d\tau| = (n-1)s^2/\tau^2$] we find that

$$(9A.3.14) \quad p(\tau \mid x_1, \ldots, x_n) = \frac{1}{2^{(n-1)/2}\Gamma\left(\dfrac{n-1}{2}\right)} \tau^{[(n-1)/2]-1}e^{-\tau/2},$$

i.e., the posterior distribution of $\tau = (n-1)s^2/\sigma^2 = \chi^2_{n-1}$. Since $\sigma^2 = (n-1)s^2/\tau$, where $\tau = \chi^2_{n-1}$, we have that

$$(9A.3.15) \qquad E(\sigma^2 \mid x_1, \ldots, x_n) = (n-1)s^2 E(1/\tau).$$

Now if $\tau = \chi^2_{n-1}$, it is easy to see that $E(1/\tau) = 1/(n-3)$, $n > 3$, when we use the fact that $\Gamma\left(\dfrac{n-3}{2} + 1\right) = \left(\dfrac{n-3}{2}\right)\Gamma\left(\dfrac{n-3}{2}\right)$. Hence

$$(9A.3.15a) \qquad E(\sigma^2 \mid x_1, \ldots, x_n) = (n-1)s^2/(n-3)$$
$$= \sum (x_i - \bar{x})^2/(n-3).$$

(Note that $V(\mu \mid x_1, \ldots, x_n) = [E(\sigma^2 \mid x_1, \ldots, x_n)]/n$.) Further, from the statement

$$P(\chi^2_{n-1;1-\alpha/2} \leq \chi^2_{n-1} \leq \chi^2_{n-1;\alpha/2}) = 1 - \alpha,$$

we have

$$P\left(\chi^2_{n-1;1-\alpha/2} \leq \tau = \frac{(n-1)s^2}{\sigma^2} \leq \chi^2_{n-1;\alpha/2} \,\Big|\, x_1, \ldots, x_n\right) = 1 - \alpha,$$

so that

(9A.3.15b)

$$P\left((n-1)s^2/\chi^2_{n-1;\alpha/2} \leq \sigma^2 \leq (n-1)s^2/\chi^2_{n-1;1-\alpha/2} \,\Big|\, x_1, \ldots, x_n\right) = 1 - \alpha.$$

In other words, we have that the posterior interval for σ^2 at confidence level $1 - \alpha$ is

(9A.3.16) $[(n-1)s^2/\chi^2_{n-1;\alpha/2}, (n-1)s^2/\chi^2_{n-1;1-\alpha/2}]$

[cf. with (9.5.3)].

Note that the interval (9A.3.16), developed here using *Bayes' Theorem with noninformative priors*, is the same as the confidence interval (9.5.3), developed (in Chapter 9) using the *frequency interpretation of probability*. Put another way, this means that in certain situations, Bayesians and non-Bayesians would come up with the same results if given the same data. The reader may well ask, what then is the difference?

The answer is one of interpretation, and should be borne in mind throughout this Chapter. To a non-Bayesian, the result (9A.3.16) implies that if samples of size n are taken from a $N(\mu, \sigma^2)$ population, over and over again, and the interval (9A.3.16) computed each time, then in $100(1 - \alpha)\%$ of the cases, the interval (9A.3.16) contains σ^2. This also means that once having taken a single sample of size n, and subsequently computing the interval (9A.3.16), then the probability is zero or one that σ^2 lies in the computed interval, zero if it does not cover σ^2, 1 if it does cover (or contain) σ^2.

However, to a Bayesian, (9A.3.16) implies something quite different. Based on a rather vague prior, the observed sample evidence is such that the Bayesian's degrees of belief in σ^2 falling in (9A.3.16), i.e., his sub-jective probability in the statement "σ^2 is contained in (9A.3.16)," is $1 - \alpha$, or put another way, he would be willing to quote odds of $1 - \alpha$ to α that σ^2 lies in (9A.3.16). To a Bayesian, once the sample evidence is at hand, the left-hand side of (9A.3.15b) is a meaningful probability, and, in fact, that probability has value $1 - \alpha$, as we have seen from the development in this chapter. To a non-Bayesian, once the sample evidence is at hand, the left hand side of (9A.3.15b) has value 0 or 1, and is a meaningful probability only in this sense.

Example 9A.3(b). Refer to Example 9A.3(a), and suppose in addition to making inferences about μ, the consumer's organization wishes to make inferences about σ^2. The use of the noninformative priors imply that the posterior of σ^2 (see (9A.3.14)) is such that

$$\sigma^2 = (n - 1)s^2/\chi^2_{n-1}.$$

Hence the Bayes estimate of σ^2 is $(n - 1)s^2/(n - 3)$, which for this set of data is $(14)(9.4)^2/12$ or 103.087. The (Bayes) interval estimate of level $1 - \alpha$ is [see (9A.3.16)]

$$[(n - 1)s^2/\chi^2_{n-1;\alpha/2}, (n - 1)s^2/\chi^2_{n-1;1-\alpha/2}].$$

Here $n = 15$; $(n - 1) = 14$; $s^2 = (9.4)^2 = 88.36$. Suppose we choose $1 - \alpha = .95$, i.e., $\alpha/2 = .025$; then $\chi^2_{14;.025} = 26.1190$ and $\chi^2_{14;.975} = 5.26872$. Hence the 95% posterior interval for σ^2 is

$$[(14)(88.36)/26.1190; (14)(88.36)/5.26872] = [47.362; 234.789].$$

We turn now to the case where the prior of μ and σ^2 is not the "inignorance" or "uniform" prior (9A.3.2), but is the conjugate prior for this problem given by

$$(9A.3.17) \qquad p(\mu, \sigma^2) = k(\sigma^2)^{-\{[(v_0+1)/2]+1\}} e^{-(1/2\sigma^2)[v_0\delta_0{}^2 + n_0(\mu - \mu_0)^2]}$$

where

$$(9A.3.17a) \qquad k = \sqrt{n_0}\, (v_0\delta_0^2)^{v_0/2} / \sqrt{\pi}\, 2^{[(v_0+1)/2]} \Gamma\left(\frac{v_0}{2}\right).$$

It is easy to see that, marginally,

$$(9A.3.18) \qquad p(\mu) = k'[1 + n_0(\mu - \mu_0)^2/v_0\delta_0^2]^{-[(v_0+1)/2]},$$

that is, the marginal prior of μ is such that $\sqrt{n_0}\,(\mu - \mu_0)/\delta_0 = t_{v_0}$, so that, the prior beliefs for μ are such that

$$(9A.3.18a) \quad E(\mu) = \mu_0, \qquad \mathrm{Var}\,(\mu) = \sigma_0^2/n_0, \quad \text{where} \quad \sigma_0^2 = \frac{v_0}{v_0 - 2}\, \delta_0^2.$$

Further, it is also easy to see that the marginal prior of σ^2 is

$$(9A.3.19) \qquad p(\sigma^2) = k''(\sigma^2)^{-(v_0/2)-1} e^{-v_0\delta_0{}^2/2\sigma^2},$$

that is, the marginal prior of σ^2 is such that $v = v_0\delta_0^2/\sigma^2 = \chi^2_{v_0}$. Thus, the prior beliefs of σ^2 are such that

$$9A.3.19a) \quad E(\sigma^2) = \sigma_0^2, \qquad \mathrm{Var}\,(\sigma^2) = 2(\sigma_0^2)^2/(v_0 - 4);$$

$$\sigma_0^2 = v_0\delta_0^2/(v_0 - 2).$$

With the results (9A.3.18a) and (9A.3.19a) to guide us, it is not hard to determine the parameters of (9A.3.17), i.e. $(\nu_0, n_0, \delta_0^2, \mu_0)$. [An illustration is given in Example 9A.3(c).] Once having determined a choice for $(\nu_0, n_0, \delta_0^2, \mu_0)$, checks may be made in various ways.*

Now suppose we take a random sample on $x = N(\mu, \sigma^2)$ and observe the sample to be $(x_1, \ldots, x_n)$. The likelihood is of course given by (9A.3.3), and combining (9A.3.3) with (9A.3.17) yields the posterior

(9A.3.20) $p(\mu, \sigma^2 \mid x_1, \ldots, x_n) \propto (\sigma^2)^{-[(n+\nu_0+1)/2]-1}$

$$\times \exp\left\{-\frac{1}{2\sigma^2}\left[\nu_0\delta_0^2 + (n-1)s^2 \right.\right.$$

$$\left.\left. + n(\mu - \bar{x})^2 + n_0(\mu - \mu_0)^2\right]\right\}$$

or

(9A.3.20a)

$$p(\mu, \sigma^2 \mid x_1, \ldots, x_n) = \frac{\sqrt{n + n_0}\, w^{(n+\nu_0)/2}}{\sqrt{\pi}\, 2^{(n+\nu_0+1)/2}\Gamma\left(\dfrac{n + \nu_0}{2}\right)}\, (\sigma^2)^{-[(n+\nu_0+1)/2]-1}$$

$$\times \exp\left\{-\frac{1}{2\sigma^2}\left[(n + n_0)(\mu - \bar{x}_u)^2 + w\right]\right\}$$

where

(9A.3.20b) $w = \nu_0\delta_0^2 + (n-1)s^2 + nn_0(\mu_0 - \bar{x})^2/(n + n_0),$

and

$$\bar{x}_u = (n_0\mu_0 + n\bar{x})/(n + n_0).$$

[The reader is cautioned not to confuse $\bar{x}_u$ with $\bar{x}$ of (9A.2.8), etc.] From the above joint posterior we find that the marginal posterior of μ is

(9A.3.21)

$p(\mu \mid x_1, \ldots, x_n)$

$$= \frac{\sqrt{n + n_0}\,\Gamma\left(\dfrac{n + \nu_0 + 1}{2}\right)}{\sqrt{\pi}\, w^{1/2}\Gamma\left(\dfrac{n + \nu_0}{2}\right)}\left[1 + \frac{(n + n_0)(\mu - \bar{x}_u)^2}{w}\right]^{-[(n+\nu_0+1)/2]},$$

---

* For example, suppose at the outset we feel that the interval $[a, b]$ is a 95% confidence interval for μ, with $P(\mu < a) = P(\mu > b) = .025$, so that, our prior beliefs about μ are such that $P[a \leq \mu \leq b] = .95$. Then μ_0 should equal $(a + b)/2$. Now our prior, marginally, for μ is such that $\sqrt{n_0}\,(\mu - \mu_0)/\delta_0 = t_{\nu_0}$, so that a $1 - \alpha = .95$ interval for μ is $[\mu_0 \pm t_{\nu_0;.025}\delta_0/\sqrt{n_0}]$. The question, "Is $(\nu_0, n_0, \delta_0^2, \mu_0)$ such that the 95% interval $[\mu_0 \pm t_{\nu_0;.025}\delta_0/\sqrt{n_0}] = [a, b]$?" may then be examined.

so that the posterior distribution of μ is such that

(9A.3.22)
$$\frac{\sqrt{n + n_0}(\mu - \bar{\bar{x}}_u)}{\sqrt{w/(n + v_0)}} = t_{n+v_0}.$$

Hence the Bayes estimate of μ is $\bar{\bar{x}}_u$, and a posterior interval at level $1 - \alpha$ for μ is

(9A.3.23)
$$[\bar{\bar{x}}_u \pm \sqrt{w/(n + v_0)(n + n_0)}\, t_{n+v_0; \alpha/2}].$$

Returning to (9A.3.20a), and integrating out μ, the marginal of σ^2 is found to be

(9A.3.24)
$$p(\sigma^2 \mid x_1, \ldots, x_n) = \frac{w^{(n+v_0)/2}}{2^{(n+v_0)/2}\Gamma\left(\dfrac{n + v_0}{2}\right)} \, (\sigma^2)^{-[(n+v_0)/2]-1} e^{-(w/2\sigma^2)},$$

which is to say that the posterior of σ^2 is such that $\sigma^2 = w/\chi^2_{n+v_0}$. Hence the Bayes estimate of σ^2 is

(9A.3.25)
$$E(\sigma^2 \mid x_1, \ldots, x_n) = w/(n + v_0 - 2),$$

and a $100(1 - \alpha)\%$ interval for σ^2 at level $1 - \alpha$ is

(9A.3.26)
$$[w/\chi^2_{n+v_0; \alpha/2}, \, w/\chi^2_{n+v_0; 1-\alpha/2}].$$

Example 9A.3(c). A manufacturer of tires has decided to bring out a new line of snow tires, equipped with studs that are impregnated into the tire at the moulding stage. It was felt that tests for amount of wear per 25,000 miles should be conducted, and to this end a standard test was planned. The weights before and after a tire is tested are recorded, and the loss in weight is then expressed as a percentage of initial weight. It is assumed (on the basis of previous experience with snow tires) that the percentage loss in weight of these tires will have an $N(\mu, \sigma^2)$ distribution, where here it is assumed that μ and σ^2 are unknown.

The chief chemist of this tire plant (with much experience in tire manufacture), when consulted, was found to have strong prior feelings about both μ and σ^2, and a conjugate prior of the form (9A.3.17) described his (prior) beliefs very well. He was of the opinion that $[18 \pm 1.2]$ was an approximate 95% confidence interval for μ, that is $E(\mu) = \mu_0 = 18$. After further questioning, it was clear that his prior distribution of μ could be described by a "t" with large degrees of freedom, since he felt that probability statements about μ could be answered to good approximation, by using a normal distribution [see Example 9A.3(g) for another method of finding v_0]. When several intervals of the form $[18 \pm a]$ were presented, it was agreed that v_0 should be chosen to have the value $v_0 = 50$. On further questioning, it was found that the chief chemist's expectation for σ^2 was 1.75, that is $E(\sigma^2) = \dfrac{v_0}{v_0-2}\,\delta_0^2 = \dfrac{50}{48}\delta_0^2 = 1.75$, so that $\delta_0^2 = 1.68$ or $\delta_0 \simeq 1.3$. Hence, the 95%

interval for μ should be of form $[18 \pm t_{50;.025}(1.3)/\sqrt{n_0}]$. Consulting t-tables, we have then that $[18 \pm 2(1.3)/\sqrt{n_0}] \simeq [18 \pm 1.2]$, which gives $1.3/\sqrt{n_0} = .6$ or $\sqrt{n_0} = 1.3/.6$ or $n_0 = 169/36 = 4.7$. Summarizing, we have $(v_0, n_0, \delta_0^2, \mu_0) = (50, 4.7, 1.68, 18)$.

Now suppose four tires of the new line variety are chosen at random and tested, and that the resulting observations have sample mean $\bar{x} = 19.4$ and sample variance of $s^2 = 2.25$. Then the Bayes estimate for μ is

$$\bar{x}_u = (n_0\mu_0 + n\bar{x})/(n + n_0) = (4.7 \times 18 + 4 \times 19.4)/8.7 = 162.2/8.7,$$

that is,

$$\bar{x}_u = 18.64.$$

To find a posterior confidence interval for μ, we need to calculate

$$w = v_0\delta_0^2 + (n - 1)s^2 + nn_0(\mu_0 - \bar{x})^2/(n + n_0)$$
$$= (50)(1.68) + 3(2.25) + [4(4.7)(18 - 19.4)^2/(4 + 4.7)]$$
$$= 94.9854.$$

From (9A.3.23) we have that

$$[\bar{x}_u \pm \sqrt{w/(n + v_0)(n + n_0)}\, t_{n+v_0;\alpha/2}]$$
$$= [18.64 \pm \sqrt{(94.9854)/(4 + 50)(4 + 4.7)}\, t_{54;.025}]$$

is a 95% (posterior) interval for μ. Interpolating in t-tables (using the $120/m$-rule quoted at the bottom of Table V of Appendix VII) gives $t_{54;.025} = 2.005$, so that the 95% posterior interval for μ is

$$[18.64 \pm .902] = [17.738, 19.542].$$

From (9A.3.25) we have that the Bayes estimate for σ^2 is

$$w/(n + v_0 - 2) = 94.9854/(4 + 50 - 2) = 1.827,$$

and a 95% posterior interval for σ^2 is [see (9A.3.26)]

(9A.3.27) $[94.9854/\chi_{54;.025}^2,\ 94.9854/\chi_{54;.975}^2]$.

Since $m = 54 < 100$, we use linear interpolation (see the bottom of Table III of Appendix VII), and obtain $\chi_{54;.025}^2 = 76.1712$ and $\chi_{54;.975}^2 = 35.6071$, so that the 95% posterior interval for σ^2 is

$$[94.9854/76.1712;\ 94.9854/35.6071] = [1.247;\ 2.668].$$

(b) Two Populations

In this subsection we again consider the case where sampling is on two normal variables, $x_1 = N(\mu_1, \sigma_1^2)$ and $x_2 = N(\mu_2, \sigma_2^2)$, where all four parameters $(\mu_1, \mu_2, \sigma_1^2, \sigma_2^2)$ are unknown. We will discuss two situations that may arise, viz,

(i) $\sigma_1^2 = \sigma_2^2 = \sigma^2$, say, where σ^2 is of course unknown, and

(ii) $\sigma_1^2 \neq \sigma_2^2$, where the σ_i^2 are unknown.

(i) $\sigma_1^2 = \sigma_2^2 = \sigma^2$, σ^2 *Unknown.* In the language of Section 9A.1 we have that $\theta = (\mu_1, \mu_2, \sigma^2)$, so that Ω is the three-dimensional set

(9A.3.28) . $\Omega = \{(\mu_1, \mu_2, \sigma^2) \mid -\infty < \mu_i < \infty, i = 1, 2, \text{ and } \sigma^2 > 0\}.$

We assume that for the situation here, a noninformative prior is appropriate, that is, μ_1, μ_2, and σ^2 are independent, and their joint prior is

(9A.3.29) $$p(\mu_1, \mu_2, \sigma^2) \propto 1/\sigma^2.$$

Now suppose two independent samples of n_1 and n_2 independent observations are taken on x_1 and x_2, respectively, and observed to be $(x_{11}, x_{21}, \ldots, x_{n_1 1})$ and $(x_{12}, x_{22}, \ldots, x_{n_2 2})$, respectively. The likelihood is

(9A.3.30) $l(\mu_1, \mu_2, \sigma^2 \mid x_{11}, \ldots, x_{n_1 1}; x_{12}, \ldots, x_{n_2 2})$

$$= [(2\pi)\sigma^2]^{-[(n_1+n_2)/2]}$$

$$\times \exp\left\{-\frac{1}{2\sigma^2}\left[n_1(\mu_1 - \bar{x}_1)^2 + n_2(\mu_2 - \bar{x}_2)^2 + \sum_{j=1}^{2}\sum_{i=1}^{n_j}(x_{ij} - \bar{x}_j)^2\right]\right\},$$

where $\bar{x}_j = \sum_{i=1}^{n_j} x_{ij}/n_j$. Combining (9A.3.29) and (9A.3.30) using Bayes' Theorem yields the posterior

(9A.3.31) $p(\mu_1, \mu_2, \sigma^2 \mid x_{11}, \ldots, x_{n_1 1}; x_{12}, \ldots, x_{n_2 2}) \propto (\sigma^2)^{-[(n_1+n_2)/2]-1}$

$$\times \exp\left\{-\frac{1}{2\sigma^2}\left[\sum_{j=1}^{2}n_j(\mu_j - \bar{x}_j)^2 + \sum_{j=1}^{2}\sum_{i=1}^{n_j}(x_{ij} - \bar{x}_j)^2\right]\right\}$$

or

(9A.3.31a) $p(\mu_1, \mu_2, \sigma^2 \mid x_{11}, \ldots, x_{n_1 1}; x_{12}, \ldots, x_{n_2 2})$

$$= \frac{\sqrt{n_1 n_2}\,[\sum\sum(x_{ij} - \bar{x}_j)^2]^{[(n_1+n_2-2)/2]}}{2^{(n_1+n_2)/2}\pi\Gamma\left(\dfrac{n_1 + n_2 - 2}{2}\right)}(\sigma^2)^{-[(n_1+n_2)/2]-1}$$

$$\times \exp\left\{-\frac{1}{2\sigma^2}\left[\sum_{j=1}^{2}n_j(\mu_j - \bar{x}_j)^2 + \sum\sum(x_{ij} - \bar{x}_j)^2\right]\right\}.$$

Integrating out μ_1 and μ_2 gives the marginal posterior of σ^2 as

(9A.3.32) $p(\sigma^2 \mid x_{11}, \ldots, x_{n_1 1}; x_{12}, \ldots\ x_{n_2 2})$

$$= \frac{[\sum\sum(x_{ij} - \bar{x}_j)^2]^{(n_1+n_2-2)/2}}{2^{(n_1+n_2-2)/2}\Gamma\left(\dfrac{n_1 + n_2 - 2}{2}\right)}(\sigma^2)^{-[(n_1+n_2-2)/2]-1}$$

$$\times \exp\left\{-\frac{\sum\sum(x_{ij} - \bar{x}_j)^2}{2\sigma^2}\right\}.$$

Now suppose we let

(9A.3.32a)
$$SS_w^2 = \sum_{j=1}^{2} \sum_{i=1}^{nj} (x_{ij} - \bar{x}_j)^2$$

$$= \sum_{i=1}^{n_1} (x_{i1} - \bar{x}_1)^2 + \sum_{i=1}^{n_2} (x_{i2} - \bar{x}_2)^2$$

$$= (n_1 - 1)s_1^2 + (n_2 - 1)s_2^2.$$

Then from (9A.3.32), it is easy to see that

(9A.3.33) $SS_w^2/\sigma^2 = \chi^2_{n_1+n_2-2}$ or $\sigma^2 = SS_w^2/\chi^2_{n_1+n_2-2}.$

Hence the Bayes estimate of σ^2 is

(9A.3.34) $E(\sigma^2 \mid x_{11}, \ldots, x_{n_11}; x_{12}, \ldots, x_{n_22})$
$$= SS_w^2/(n_1 + n_2 - 4) = [(n_1 - 1)s_1^2 + (n_2 - 1)s_2^2]/(n_1 + n_2 - 4),$$

and a $100(1 - \alpha)\%$ posterior interval for σ^2 is easily seen to be

(9A.3.35) $[SS_w^2/\chi^2_{n_1+n_2-2;\alpha/2}; SS_w^2/\chi^2_{n_1+n_2-2;1-\alpha/2}].$

Returning to (9A.3.31a), we now integrate out σ^2 and find that the joint posterior of μ_1 and μ_2 is

(9A.3.36) $p(\mu_1, \mu_2 \mid x_{11}, \ldots, x_{n_11}; x_{12}, \ldots, x_{n_22})$

$$= \frac{\sqrt{n_1 n_2}\, \Gamma\left(\dfrac{n_1 + n_2}{2}\right)}{(\Gamma(\frac{1}{2}))^2 \Gamma\left(\dfrac{n_1 + n_2 - 2}{2}\right) SS_w^2} \left[1 + \frac{\displaystyle\sum_{j=1}^{2} n_j(\mu_j - \bar{x}_j)^2}{SS_w^2} \right]^{-(n_1+n_2)/2}$$

If we let $t_j = \sqrt{n_j}\, (\mu_j - \bar{x}_j)/\sqrt{SS_w^2/(n_1 + n_2 - 2)}$, then the Jacobian of this transformation is $SS_w^2 \sqrt{n_1 n_2/(n_1 + n_2 - 2)}$ and we have

9A.3.37) $p(t_1, t_2 \mid x_{11}, \ldots, x_{n_11}; x_{12}, \ldots, x_{n_22})$

$$= \frac{\Gamma\left(\dfrac{n_1 + n_2}{2}\right)}{(\Gamma(\frac{1}{2}))^2 \Gamma\left(\dfrac{n_1 + n_2 - 2}{2}\right)(n_1 + n_2 - 2)} \left[1 + \frac{t_1^2 + t_2^2}{n_1 + n_2 - 2} \right]^{-(n_1+n_2)/2}$$

The distribution (9A.3.37) is known as the bivariate Student t distribution with $(n_1 + n_2 - 2)$ degrees of freedom.

Now suppose we are interested in $\mu_1 - \mu_2$. Rather than derive the distribution of $\mu_1 - \mu_2$ from (9A.3.36) and/or (9A.3.37), we use the following approach. The reader will recall from (9A.2.25) that, given σ_1^2 and σ_2^2, μ_1 and μ_2 have the joint posterior distribution which is that of two independent normals, $N(\bar{x}_1, \sigma_1^2/n_1)$ and $N(\bar{x}_2, \sigma_2^2/n_2)$, respectively, so that for the case discussed here, i.e., $\sigma_1^2 = \sigma_2^2 = \sigma^2$, and σ^2 known, we have

(9A.3.38) $p(\mu_1, \mu_2 \mid \sigma^2; x_{11}, \ldots, x_{n_11}; x_{12}, \ldots, x_{n_22})$

$$= \prod_{j=1}^{2} \frac{\sqrt{n_j}}{\sqrt{2\pi\sigma^2}} e^{-(\mu_j - \bar{x}_j)^2/2\sigma^2}.$$

Hence, given σ^2, the posterior distribution of $\mu_1 - \mu_2$ is

(9A.3.39) $p(\mu_1 - \mu_2 \mid \sigma^2; x_{11}, \ldots, x_{n_11}; x_{12}, \ldots, x_{n_22})$

$$= \frac{\sqrt{n_1 n_2}}{\sqrt{2\pi\sigma^2}\sqrt{n_1 + n_2}} \exp\left[-\frac{1}{2\sigma^2}\left(\frac{1}{n_1} + \frac{1}{n_2}\right)^{-1}[(\mu_1 - \mu_2) - (\bar{x}_1 - \bar{x}_2)]^2\right].$$

But from rules of probability we know that

(9A.3.40) $p(\mu_1 - \mu_2, \sigma^2 \mid x_{11}, \ldots, x_{n_11}; x_{12}, \ldots, x_{n_22})$
$$= p(\sigma^2 \mid x_{11}, \ldots, x_{n_11}; x_{12}, \ldots, x_{n_22})$$
$$\times p(\mu_1 - \mu_2 \mid \sigma^2; x_{11}, \ldots, x_{n_11}; x_{12}, \ldots, x_{n_22}).$$

Using (9A.3.32) and (9A.3.39) gives

(9A.3.40a) $p(\mu_1 - \mu_2, \sigma^2 \mid x_{11}, \ldots, x_{n_11}; x_{12}, \ldots, x_{n_22})$
$$\propto (\sigma^2)^{-[(n_1+n_2-1)/2]-1}$$

$$\times \exp\left\{-\frac{1}{2\sigma^2}\left[\left(\frac{1}{n_1} + \frac{1}{n_2}\right)^{-1}[(\mu_1 - \mu_2) - (\bar{x}_1 - \bar{x}_2)]^2 + SS_w^2\right]\right\}.$$

Hence, the (marginal) posterior of $\mu_1 - \mu_2$ is

(9A.3.41) $p(\mu_1 - \mu_2 \mid x_{11}, \ldots, x_{n_11}; x_{12}, \ldots, x_{n_22})$

$$= \int_0^\infty p(\mu_1 - \mu_2, \sigma^2 \mid x_{11}, \ldots, x_{n_11}; x_{12}, \ldots, x_{n_22}) \, d\sigma^2,$$

so that

(9A.3.42) $p(\mu_1 - \mu_2 \mid x_{11}, \ldots, x_{n_11}; x_{12}, \ldots, x_{n_22})$

$$\propto \left\{1 + [(\mu_1 - \mu_2) - (\bar{x}_1 - \bar{x}_2)]^2 \bigg/ \left[SS_w^2\left(\frac{1}{n_1} + \frac{1}{n_2}\right)\right]\right\}^{-(n_1+n_2-1)/2}$$

and, as the reader will no doubt recognize, apart from the normalizing constant, this has the form of a $t_{n_1+n_2-2}$ distribution. Specifically we have from (9A.3.42) that

(9A.3.42a)
$$\frac{(\mu_1 - \mu_2) - (\bar{x}_1 - \bar{x}_2)}{\left(\frac{1}{n_1} + \frac{1}{n_2}\right)^{1/2} [SS_w^2/(n_1 + n_2 - 2)]^{1/2}} = t_{n_1+n_2-2},$$

where $SS_w^2 = \sum\limits_{j=1}^{2} \sum\limits_{i=1}^{n_j} (x_{ij} - \bar{x}_j)^2$. Hence

(9A.3.43) $E(\mu_1 - \mu_2 \mid x_{11}, \ldots, x_{n_11}; x_{12}, \ldots, x_{n_22}) = \bar{x}_1 - \bar{x}_2,$

so that the Bayes estimate of $\mu_1 - \mu_2$ is $\bar{x}_1 - \bar{x}_2$, and

(9A.3.43a) $V(\mu_1 - \mu_2 \mid x_{11}, \ldots, x_{n_11}; x_{12}, \ldots, x_{n_22})$

$$= \left(\frac{1}{n_1} + \frac{1}{n_2}\right) \frac{SS_w^2}{n_1 + n_2 - 4}.$$

Further, from (9A.3.42a), it is easy to see that a $100(1 - \alpha)\%$ posterior interval for $\mu_1 - \mu_2$ is

(9A.3.44) $\left[\bar{x}_1 - \bar{x}_2 \pm \left(\frac{1}{n_1} + \frac{1}{n_2}\right)^{1/2} (SS_w^2/(n_1 + n_2 - 2))^{1/2} t_{n_1+n_2-2;\alpha/2} \right].$

(See also Problem 9A.24.)

Example 9A.3(d). Determinations of percentage chlorine in a batch of polymer are to be made by two analysts, say 1 and 2, to see whether they do consistent work in this situation. Based on previous experience with their performance in this particular laboratory, it is assumed that the determinations of the analysts are distributed as $N(\mu_i, \sigma^2)$, that is, $\sigma_1^2 = \sigma_2^2 = \sigma^2$. The analysts are given no information about the source of the batch of polymer, so that the priors for μ_1, μ_2, and σ^2 are of the noninformative variety, that is

$$p(\mu_1, \mu_2, \sigma^2) = p(\mu_1)p(\mu_2)p(\sigma^2),$$

where $p(\mu_i) \propto k_i$ and $p(\sigma^2) \propto \sigma^{-2}$.

Ten determinations are then made by each of the analysts with the following results:

> Analyst 1: $n_1 = 10$; $\bar{x}_1 = 12.21$; $s_1^2 = .5419.$
> Analyst 2: $n_2 = 10$; $\bar{x}_2 = 11.83$; $s_2^2 = .6065.$

Under the conditions given here, the Bayes' estimate for $\mu_1 - \mu_2$ is

$$\bar{x}_1 - \bar{x}_2 = 12.21 - 11.83 = 0.38.$$

Thus, using (9A.3.44), a 95% posterior interval for $\mu_1 - \mu_2$ is

$$[(\bar{x}_1 - \bar{x}_2) \pm (\tfrac{1}{10} + \tfrac{1}{10})^{1/2}(SS_w^2/18)^{1/2} t_{18;.025}].$$

Here [see (9A.3.32a)],

$$SS_w^2 = (n_1 - 1)s_1^2 + (n_2 - 1)s_2^2 = 9(1.1484) = 10.3356,$$

so that the 95% posterior interval for $\mu_1 - \mu_2$ is

$$[.38 \pm (1.03356/9)^{1/2}(2.101)] = [.38 \pm .712] = [-.332, 1.092].$$

The sample evidence supports "the claim that $\mu_1 - \mu_2 = 0$."
From (9A.3.35), we also have that a 95% posterior interval for σ^2 is

$$[SS_w^2/\chi_{18;.025}^2; \ SS_w^2/\chi_{18;.975}^2] = [10.3356/31.5264; \ 10.3356/8.23075]$$
$$= [.32784, 1.25573].$$

(*ii*) $\sigma_1^2 \neq \sigma_2^2$. We now discuss the case $\theta = (\mu_1, \mu_2, \sigma_1^2, \sigma_2^2)$ so that Ω is the 4-dimensional set

(9A.3.45) $\quad \Omega = \{(\mu_1, \mu_2, \sigma_1^2, \sigma_2^2) \mid -\infty < \mu_i < \infty, \sigma_i^2 > 0, i = 1, 2\}.$

We again assume that a "noninformative" prior for the situation here is appropriate, that is, μ_1, μ_2, σ_1^2, and σ_2^2 are independent, and their joint prior is

(9A.3.46) $\quad p(\mu_1, \mu_2, \sigma_1^2, \sigma_2^2) \propto 1/\sigma_1^2\sigma_2^2, \qquad (\mu_1, \mu_2, \sigma_1^2, \sigma_2^2) \in \Omega.$

Since $x_1 = N(\mu_1, \sigma_1^2)$ and $x_2 = N(\mu_2, \sigma_2^2)$, the likelihood function here, based on the usual random samples of n_1 and n_2 independent observations on x_1 and x_2, respectively, is

(9A.3.47) $\quad l(\mu_1, \mu_2, \sigma_1^2, \sigma_2^2 \mid x_{11}, \ldots, x_{n_11}; x_{12}, \ldots, x_{n_22})$

$$= \prod_{j=1}^{2} \frac{1}{(2\pi\sigma_j^2)^{n_j/2}} \exp\left\{-\frac{1}{2\sigma_j^2}[(n_j - 1)s_j^2 + n_j(\mu_j - \bar{x}_j)^2]\right\}.$$

Combining the prior (9A.3.46) with (9A.3.47) using Bayes, gives the posterior

(9A.3.48) $\quad p(\mu_1, \mu_2, \sigma_1^2, \sigma_2^2 \mid x_{11}, \ldots, x_{n_11}; x_{12}, \ldots, x_{n_22})$

$$= \prod_{j=1}^{2} \frac{\sqrt{n_j}[(n_j - 1)s_j^2]^{(n_j-1)/2}}{2^{n_j/2}\sqrt{\pi}\Gamma\left(\frac{n_j - 1}{2}\right)} (\sigma_j^2)^{-1-n_j/2} e^{-(1/2\sigma_j^2)[(n_j-1)s_j^2 + n_j(\mu_j - \bar{x}_j)^2]}.$$

Suppose we wish to make inferences about σ_1^2 and σ_2^2 and, in particular, about σ_1^2/σ_2^2. From (9A.3.48), we find

(9A.3.49) $\quad p(\sigma_1^2, \sigma_2^2 \mid x_{11}, \ldots, x_{n_11}; x_{12}, \ldots, x_{n_22})$

$$= \prod_{j=1}^{2} \frac{[(n_j - 1)s_j^2]^{(n_j-1)/2}}{2^{(n_j-1)/2}\Gamma\left(\frac{n_j - 1}{2}\right)} (\sigma_j^2)^{-[(n_j-1)/2]-1} \exp\left[-\frac{(n_j - 1)s_j^2}{2\sigma_j^2}\right],$$

that is, the posterior of σ_1^2 and σ_2^2 is that of independent $(n_1 - 1)s_1^2/\chi_{n_1-1}^2$ and $(n_2 - 1)s_2^2/\chi_{n_2-1}^2$ random variables. If we now set

$$(9A.3.50) \qquad F = \frac{s_1^2}{s_2^2} \Big/ \frac{\sigma_1^2}{\sigma_2^2}, \qquad \gamma = \sigma_2^2$$

[the absolute value of $J = \dfrac{\partial(\sigma_1^2, \sigma_2^2)}{\partial(F, \gamma)}$, the Jacobian of this transformation, is $(s_1^2 \gamma / s_2^2 F^2)$], then the joint posterior of F and γ is

$$(9A.3.51) \quad p(F, \gamma \mid x_{11}, \ldots, x_{n_11}; x_{12}, \ldots, x_{n_22})$$

$$\propto F^{[(n_1-1)/2]-1} \gamma^{-[(n_1+n_2-2)/2]-1} \exp -\left\{ \frac{(n_2 - 1)s_2^2}{2\gamma}\left[1 + \frac{n_1 - 1}{n_2 - 1} F\right]\right\}.$$

Using the identity (9A.3.5), we find then that the posterior of F is

$$(9A.3.52) \quad p(F \mid x_{11}, \ldots, x_{n_11}; x_{12}, \ldots, x_{n_22})$$

$$\propto F^{[(n_1-1)/2]-1} \Big/ \left[1 + \frac{n_1 - 1}{n_2 - 1} F\right]^{\frac{n_1+n_2-2}{2}},$$

which, apart from the normalizing constant, is the form of the F_{n_1-1, n_2-1} distribution [see (AVI.10)] that is, $F = F_{n_1-1, n_2-1}$. Now from properties of the F_{n_1-1, n_2-1} distribution, we have, since

$$(9A.3.52a) \qquad \sigma_1^2/\sigma_2^2 = \frac{s_1^2}{s_2^2 F_{n_1-1, n_2-1}} = \frac{s_1^2}{s_2^2} F_{n_2-1, n_1-1},$$

that

$$(9A.3.53) \quad E\left[\frac{\sigma_1^2}{\sigma_2^2} \Big| x_{11}, \ldots, x_{n_11}; x_{12}, \ldots, x_{n_22}\right] = \frac{(n_1 - 1) s_1^2}{(n_1 - 3) s_2^2},$$

i.e., the Bayes estimate of σ_1^2/σ_2^2 is $(n_1 - 1)s_1^2/(n_1 - 3)s_2^2$, which we may write as

$$\frac{n_2 - 1}{n_2 - 3}\left[\sum_{i=1}^{n_1}(x_{i1} - \bar{x}_1)^2/(n_1 - 3)\right] \Big/ \left[\sum_{i=1}^{n_2}(x_{i2} - \bar{x}_2)^2/(n_2 - 3)\right],$$

which, apart from the factor $(n_2 - 1)/(n_2 - 3)$, is the ratio of the Bayes estimates of σ_1^2 and σ_2^2 [see (9A.3.15a)]. Further, from the results (9A.3.50) and (9A.3.52), we may write

$$(9A.3.54)$$

$$p\left(F_{n_1-1, n_2-1;1-\alpha/2} \leq \frac{s_1^2}{s_2^2} \Big/ \frac{\sigma_1^2}{\sigma_2^2} \leq F_{n_1-1, n_2-1;\alpha/2} \Bigg| x_{11}, \ldots, x_{n_11}; x_{12} \ldots, x_{n_22}\right)$$

$$= 1 - \alpha$$

so that solving for σ_1^2/σ_2^2 yields the fact that the interval

(9A.3.55) $[s_1^2/s_2^2 F_{n_1-1, n_2-1; \alpha/2}, \ s_1^2/s_2^2 F_{n_1-1, n_2-1; 1-\alpha/2}]$

is a posterior interval for σ_1^2/σ_2^2 at level $1 - \alpha$.

Example 9A.3(e).   Resistance measurements are made on test pieces selected from lots L_1 and L_2 purchased from "supplier 1" and "supplier 2," respectively.   It is assumed that the resistances (in ohms) are distributed as $N(\mu_j, \sigma_j^2)$, $j = 1, 2$, respectively;   however, conflicting opinions as to the behaviour of material from the suppliers led ultimately to the selection of a prior for μ_1, μ_2, σ_1^2, and σ_2^2 which is such that

$$p(\mu_1, \mu_2, \sigma_1^2, \sigma_2^2) = p(\mu_1)p(\mu_2)p(\sigma_1^2)p(\sigma_2^2) \propto 1/\sigma_1^2\sigma_2^2,$$

that is, prior beliefs are such that the four parameters are treated as being independent, and their priors are "noninformative."   A sample of six test pieces from each lot is selected, and the data yields sample means and sample variances as follows:

> *Lot 1:* $n_1 = 6$, $\bar{x}_1 = .844/6$; $s_1^2 = (39 \times 10^{-6})/5$.
> *Lot 2:* $n_2 = 6$, $\bar{x}_2 = .831/6$; $s_2^2 = (174 \times 10^{-6})/5$.

Suppose we wish to make inferences about the ratio σ_1^2/σ_2^2. We have from (9A.3.53) that the posterior of σ_1^2/σ_2^2 is such that the Bayes estimate for σ_1^2/σ_2^2 is

$$(n_1 - 1)s_1^2/(n_1 - 3)s_2^2 = 5[39 \times 10^{-6}/5]/3[17 \times 10^{-6}/5] = .374$$

and a 95% posterior interval for σ_1^2/σ_2^2 [see (9A.3.55)] is

$$\left[\frac{s_1^2}{s_2^2} \frac{1}{F_{n_1-1, n_2-1; .025}} \ ; \ \frac{s_1^2}{s_2^2} F_{n_2-1, n_1-1; .025} \right]$$

$$= \left[\frac{(39 \times 10^{-6})/5}{(174 \times 10^{-6})/5} \frac{1}{F_{5.5; .025}} \ ; \ \frac{(39 \times 10^{-6})/5}{(174 \times 10^{-6})/5} F_{5.5; .025} \right]$$

$$= \left[.22414 \frac{1}{7.1464} \ ; \ (.22414)(7.1464) \right] = [.031, 1.602].$$

The sample evidence supports the claim that $\sigma_1^2/\sigma_2^2 = 1$.

We return to (9A.3.48), and suppose that we wish to contrast μ_1 and μ_2. We have that the joint posterior of μ_1 and μ_2 is, on integrating out σ_1^2 and σ_2^2,

(9A.3.56) $p(\mu_1, \mu_2 \mid x_{11}, \ldots, x_{n_1 1}; x_{12}, \ldots, x_{n_2 2})$

$$\propto \prod_{j=1}^{2} [1 + n_j(\mu_j - \bar{x}_j)^2/(n_j - 1)s_j^2]^{-(n_j/2)},$$

that is, μ_1 and μ_2 are, posterior to observing the samples $x_{11}, \ldots, x_{n_1 1}$ and $x_{12}, \ldots, x_{n_2 2}$, independent and such that

(9A.3.57) $t_j = \sqrt{n_j}(\mu_j - \bar{x}_j)/s_j, \qquad j = 1, 2$

are distributed independently as the Student t with $n_j - 1$ degrees of freedom. Hence, $\mu_j = \bar{x}_j + s_j t_{n_j-1}/\sqrt{n_j}$, so that the Bayes estimate of $\mu_1 - \mu_2$ is $\bar{x}_1 - \bar{x}_2$, i.e.,

$$(9A.3.58) \quad E[\mu_1 - \mu_2 \mid x_{11}, \ldots, x_{n_11}; x_{12}, \ldots, x_{n_22}] = \bar{x}_1 - \bar{x}_2,$$

and the variance of $\mu_1 - \mu_2$ is

$$(9A.3.59) \quad V(\mu_1 - \mu_2 \mid x_{11}, \ldots, x_{n_11}; x_{12} \cdots, x_{n_22})$$

$$= \frac{1}{n_1} \frac{\sum_{i=1}^{n_1}(x_{i1} - \bar{x}_1)^2}{n_1 - 3} + \frac{1}{n_2} \frac{\sum_{i=1}^{n_2}(x_{i2} - \bar{x}_2)^2}{n_2 - 3}.$$

The question that we now face is how to find the distribution of $\mu_1 - \mu_2$, so that we may provide ourselves with a (posterior) confidence interval for $\mu_1 - \mu_2$. Now we have that

$$(9A.3.60) \quad (\mu_1 - \mu_2) - (\bar{x}_1 - \bar{x}_2) = (s_1/\sqrt{n_1})t_{n_1-1} - (s_2/\sqrt{n_2})t_{n_2-1},$$

where the random variables t_{n_1-1} and t_{n_2-1} are independent. Dividing both sides by $\sqrt{\dfrac{s_1^2}{n_1} + \dfrac{s_2^2}{n_2}}$, we find

$$(9A.3.61) \quad t' = \frac{(\mu_1 - \mu_2) - (\bar{x}_1 - \bar{x}_2)}{\sqrt{\dfrac{s_1^2}{n_1} + \dfrac{s_2^2}{n_2}}} = (\cos \phi)t_{n_1-1} - (\sin \phi)t_{n_2-1}$$

where $\phi = \tan^{-1}[(s_2/\sqrt{n_2})/(s_1/\sqrt{n_1})]$. Notice that once given the observed values of the samples, t' is a linear combination of the independent t_{n_j-1} variables, and the weights are the constants $s_1/\sqrt{n_1}\big/\sqrt{s_1^2/n_1 + s_2^2/n_2}$ and $s_2/\sqrt{n_2}\big/\sqrt{s_1^2/n_1 + s_2^2/n_2}$.

The random variable $t' = t'(\nu_1, \nu_2; \phi)$ defined by (9A.3.61) is said to be a *Behrens-Fisher t'-variable*, with $\nu_1 = (n_1 - 1)$ and $\nu_2 = (n_2 - 1)$ degrees of freedom, and angle ϕ. The distribution of t' is symmetric about zero and for large degrees of freedom is approximately that of the $N(0, 1)$ distribution. Tables of the upper $100\gamma\%$ points of the Behrens-Fisher distribution are available [e.g., Fisher and Yates (1963)]. Table VIII of Appendix VII tabulates $t'_\gamma(\nu_1, \nu_2; \phi)$ for $\gamma = .025$ and $.005$, and various combinations of ν_1, ν_2 and ϕ which are such that

$$(9A.3.62) \quad p(t' > t'_\gamma(\nu_1, \nu_2; \phi)) = \gamma = p(t' < -t'_\gamma(\nu_1, \nu_2; \phi)).$$

A convenient approximation for finding ϕ is given in Example 9A.3(f). Using (9A.3.61) and (9A.3.62), we find that the interval

$$(9A.3.63) \qquad \left[(\bar{x}_1 - \bar{x}_2) \pm t'_{\alpha/2}(\nu_1, \nu_2; \phi) \sqrt{\frac{s_1^2}{n_1} + \frac{s_2^2}{n_2}} \right]$$

is a $100(1 - \alpha)\%$ posterior interval for $\mu_1 - \mu_2$ ($\nu_1 = n_1 - 1, \nu_2 = n_2 - 1$).

Example 9A.3(f). The variability in the amount of impurities present in a batch of a chemical used for a particular process depends upon the length of time the process is in operation. A manufacturer has just installed two new modifications on two existing production processes, and plans to take specimens, and measure the impurities and express these in percent of weight of the specimens. While assuming that the percentage impurities in each process will have a $N(\mu_j, \sigma_j^2)$ distribution, $j = 1, 2$, his prior feelings about the parameters are very vague, so that he wishes to use as prior

$$p(\mu_1, \mu_2, \sigma_1^2, \sigma_2^2) \propto 1/\sigma_1^2\sigma_2^2.$$

Now sample specimens from each process are taken with the following results:

Process 1: $n_1 = 25$; $\bar{x}_1 = 3.2$; $s_1^2 = 1.04$.
Process 2: $n_2 = 25$; $\bar{x}_2 = 3.0$; $s_2^2 = .51$.

The Bayes estimator of $\mu_1 - \mu_2$ is [see (9A.3.58)]

$$\bar{x}_1 - \bar{x}_2 = 3.2 - 3.0 = 0.2\%.$$

Suppose we now wish to find a 95% posterior interval for $\mu_1 - \mu_2$. We will need to find the angle ϕ so that we may enter Table VIII of Appendix VII to find $t'_{\alpha/2}(\nu_1, \nu_2; \phi) = t'_{.025}(24, 24; \phi)$, which is necessary to compute the interval (9A.3.63).

The rule is as follows. If $u = \sqrt{s_2^2/n_2}/\sqrt{s_1^2/n_1}$ is less than one, then ϕ, in degrees, is given by

$$(9A.3.63a) \qquad \phi = (57.296u)/(1 + .28u^2).$$

If, however, $u > 1$, then let $v = 1/u$, and compute

$$(9A.3.63b) \qquad \phi^* = (57.296v)/(1 + .28v^2),$$

and determine ϕ by letting

$$(9A.3.63c) \qquad \phi = 90° - \phi^*.$$

For this example, $n_1 = n_2 = 25$, so that $u = \sqrt{s_2^2/s_1^2} = \sqrt{104/51} = 1.428$, that is, $u > 1$. Let $v = 1/u = .7003$. Then

$$\phi^* = (57.296)(.7003)/(1 + .28(.7003)^2) = 35.30.$$

Hence $\phi = 90° - \phi^* = 90° - 35.3° = 54.70°$.

We now enter Table VIII and find, for $\nu_1 = \nu_2 = 24$, that $t'_{.025}(24, 24; 45°) = 2.056$, $t'_{.025}(24, 24; 60°) = 2.058$. Linear interpolation gives $t'_{.025}(24, 24; 54.7°) = 2.0572$. Hence, the 95% posterior interval for $(\mu_1 - \mu_2)$ is [see (9A.3.63)]

$$\left[(\bar{x}_1 - \bar{x}_2) \pm 2.0572 \sqrt{\frac{s_1^2}{n_1} + \frac{s_2^2}{n_2}} \right] = \left[0.2 \pm 2.0572 \sqrt{\frac{1.55}{25}} \right]$$

$$= [.2 \pm .512] = [-.312, .712].$$

The sample evidence supports the assertion that $\mu_1 = \mu_2$.

If both ν_1, ν_2, where $\nu_j = n_j - 1$ are large, that is, if both sample sizes n_1 and n_2 are large, then from (9A.3.61)

(9A.3.64) $\qquad\qquad t' = (\cos \phi)z_1 - (\sin \phi)z_2,$

approximately, where z_1 and z_2 are independent $N(0, 1)$ random variables. This is so because, for large degrees of freedom, the Student t variable is distributed, approximately, as a $N(0, 1)$ random variable. Now we have seen that a linear combination of normal variables is itself normal. Using (9A.3.64), we have $E(t') = 0$ and $\text{Var}(t') = \cos^2 \phi + \sin^2 \phi = 1$. Thus, if both samples are large, t' is, approximately, normal, with mean 0 and variance 1, so that

(9A.3.64a) $\quad t' = \dfrac{(\mu_1 - \mu_2) - (\bar{x}_1 - \bar{x}_2)}{\sqrt{s_1^2/n_1 + s_2^2/n_2}} \cong z, \qquad n_1 \text{ and } n_2 \text{ large.}$

The above implies that if both sample sizes are large, we may replace $t'_{\alpha/2}(\nu_1, \nu_2; \phi)$ by $z_{\alpha/2}$ in the formula (9A.3.63) for the posterior interval of $\mu_1 - \mu_2$, and thereby eliminate the need to compute ϕ.

Now suppose only one of the sample sizes is large—say it is n_1. In fact, for sake of concreteness, let us suppose that $n_1 > 61$, $n_2 \leq 61$. Because a Student t variable with large degrees of freedom is approximately a $N(0, 1)$ random variable, (9A.3.61) can be written as $t' = (\cos \phi)z - (\sin \phi)t_{n_2-1}$. In turn, z may be thought of as a Student t variable with an infinite number of degrees of freedom, so that if n_1 is large ($n_1 > 61$ in practice), and n_2 is small ($n_2 \leq 61$), and we wish to find a $1 - \alpha$ posterior interval for $\mu_1 - \mu_2$, we need to find factors of the form $t'_{\alpha/2}(\infty, \nu_2; \phi)$. These may be found from Table IX of Appendix VII, which gives values of $t_\gamma(\infty, \nu_2; \phi)$ for selected γ, ν_2, and ϕ.

We now briefly discuss the case where the prior of μ_1, μ_2, σ_1^2, σ_2^2 is such that (μ_1, σ_1^2) and (μ_2, σ_2^2) are independent and are conjugate priors of the form (9A.3.17), so that

(9A.3.65) $p(\mu_1, \mu_2, \sigma_1^2, \sigma_2^2)$

$$= \prod_{j=1}^{2} k_j (\sigma_j^2)^{-[(\nu_{0j}+1)/2]-1} \exp \left\{ -\frac{1}{2\sigma_j^2} [\nu_{0j}\delta_{0j}^2 + n_{0j}(\mu_j - \mu_{0j})^2] \right\},$$

where

$$k_j = \sqrt{n_{0j}}(\nu_{0j}\delta_{0j}^2)^{\nu_{0j}/2} \Big/ \sqrt{\pi}\, 2^{(\nu_{0j}+1)/2} \Gamma\left(\frac{\nu_{0j}}{2}\right).$$

The likelihood of $(\mu_1, \mu_2, \sigma_1^2, \sigma_2^2)$ is as given in (9A.3.47); thus we find that the joint posterior of $(\mu_1, \mu_2, \sigma_1^2, \sigma_2^2)$ is

(9A.3.66) $p(\mu_1, \mu_2, \sigma_1^2, \sigma_2^2 \mid x_{11}, \ldots, x_{n_1 1}; x_{12}, \ldots, x_{n_2 2})$

$$\propto \prod_{j=1}^{2} (\sigma_j^2)^{-[(n_j+\nu_{0j}+1)/2]-1}$$

$$\times \exp \left\{ -\frac{1}{2\sigma_j^2} [(n_j - 1)s_j^2 + \nu_{0j}\delta_{0j}^2 + n_{0j}(\mu_j - \mu_{0j})^2 + n_j(\mu_j - \bar{x}_j)^2] \right\}.$$

This leads to the following results (see Problem 9A.30).
   (i) The marginal posterior of (σ_1^2, σ_2^2) is given by

(9A.3.66a) $p(\sigma_1^2, \sigma_2^2 \mid x_{11}, \ldots, x_{n_1 1}; x_{12}, \ldots, x_{n_2 2})$

$$\propto \prod_{j=1}^{2} (\sigma_j^2)^{-[(n_j+\nu_{0j})/2]-1} e^{-w_j/2\sigma_j^2}$$

where $w_j = [\nu_{0j}\,\delta_{0j}^2 + (n_j - 1)s_j^2] + [n_j n_{0j}(\mu_{0j} - \bar{x}_j)^2]/(n_j + n_{0j})$, so that, as may be verified by the reader, the posterior distribution of σ_1^2/σ_2^2 is such that

(9A.3.66b) $\displaystyle \sigma_1^2/\sigma_2^2 = \frac{w_1}{n_1 + \nu_{01}} \Big/ \frac{w_2}{n_2 + \nu_{02}} F_{n_1+\nu_{01};n_2+\nu_{02}}$

$$= \left[\frac{w_1}{n_1 + \nu_{01}} \Big/ \frac{w_2}{n_2 + \nu_{02}} \right] \times F_{n_2+\nu_{02};n_1+\nu_{01}}.$$

This gives rise to the posterior interval for σ_1^2/σ_2^2 at level $1 - \alpha$ (see Problem 9A.30),

(9A.3.66c)

$$\left[\left[\frac{w_1/(n_1 + \nu_{01})}{w_{22}/(n + \nu_{02})} \right] \Big/ F_{n_1+\nu_{01};n_2+\nu_{02};\alpha/2} ; \left[\frac{w_1/(n_1 + \nu_{01})}{w_2/(n_2 + \nu_{02})} \right] \times F_{n_2+\nu_{02};n_1+\nu_{01};\alpha/2} \right].$$

(ii) The marginal posterior distribution of (μ_1, μ_2) is

(9A.3.67) $p(\mu_1, \mu_2 \mid x_{11}, \ldots, x_{n_11}; x_{12}, \ldots, x_{n_22})$

$$\propto \prod_{j=1}^{2} \left[1 + \frac{(n_j + n_{0j})(\mu_j - \bar{\bar{x}}_{uj})^2}{w_j} \right]^{-[(n_j + v_{0j} + 1)/2]}$$

so that $\sqrt{(n_j + n_{0j})} \, (\mu_j - \bar{\bar{x}}_{uj})/\sqrt{w_j/(n_j + v_{0j})} = t_{n_j+v_{0j}}$, and the $t_{n_j+v_{0j}}$, $j = 1, 2$ are independent, and where $\bar{\bar{x}}_{uj} = [(n_{0j}\mu_{0j} + n_j\bar{x}_j)/(n_{0j} + n_j)]$. This implies that the random variable

(9A.3.67a) $t' = \dfrac{(\mu_1 - \mu_2) - (\bar{\bar{x}}_{u1} - \bar{\bar{x}}_{u2})}{\sqrt{w_1/[(n_1 + v_{01})(n_1 + n_{01})] + w_2/[(n_2 + v_{02})(n_2 + n_{02})]}}$

$= (\cos \phi)t_{n_1+v_{01}} - (\sin \phi)t_{n_2+v_{02}}$,

where $\phi = \tan^{-1} [\sqrt{w_2/[(n_2 + v_{02})(n_2 + n_{02})]}/\sqrt{w_1/[(n_1 + v_{01})(n_1 + n_{01})]}$ is a Behrens-Fisher t' random variable, with $(n_1 + v_{01}, n_2 + v_{02})$ degrees of freedom. Hence we find that

(9A.3.67b) $\left[(\bar{\bar{x}}_{u1} - \bar{\bar{x}}_{u2}) \pm t'_{\alpha/2}(n_1 + v_{01}, n_2 + v_{02}; \phi) \right.$

$\left. \times \sqrt{w_1/[(n_1 + v_{01})(n_1 + n_{01})] + w_2/[(n_2 + v_{02})(n_2 + n_{02})]} \right]$

is a posterior interval for $\mu_1 - \mu_2$ at level $1 - \alpha$. Further, we note from the properties of a Student t or the t' distribution, that

(9A.3.68) $E[(\mu_1 - \mu_2) \mid x_{11}, \ldots, x_{n_11}; x_{12}, \ldots, x_{n_22}] = \bar{\bar{x}}_{u1} - \bar{\bar{x}}_{u2}$

so that the Bayes estimator of $\mu_1 - \mu_2$ is $\bar{\bar{x}}_{u1} - \bar{\bar{x}}_{u2}$.

Example 9A.3(g). Two fresh supplies of iron ore are brought to a refinery and an examination of the difference in % iron content is to be made by examining 4 specimens from each deposit. It is assumed that the percentages of iron in the specimens are normally distributed about the true value μ_j of percentage of iron of the jth deposit from which it is selected, $j = 1, 2$, with unknown variances, that is, if x_j is the percent of iron in a specimen from deposit j, then $x_j = N(\mu_j, \sigma_j^2)$.

The plant manager, having much experience with the two mines that have sent the deposits, has strong prior feelings about (μ_j, σ_j^2), which turn out to be well described by the conjugate prior (9A.3.65), and he arrives at a selection for $(v_{0j}, n_{0j}, \delta_{0j}^2, \mu_{0j})$, $j = 1, 2$, as follows.

In the first place, he expects σ_1^2 to be 16, and on further questioning and examination, feels that the following is true:

(9A.3.69) $P(\sigma_1^2 \leq 7(16)) = .95$ and $P(\sigma_1^2 \leq 8(16)) = .99$.

Now since the prior of μ_1, μ_2, σ_1^2, σ_2^2 is as in (9A.3.65), this implies that the prior of (μ_1, σ_1^2) is of the form (9A.3.17). This in turn implies, amongst other things, that $\nu_{01}\delta_{01}^2/\sigma_1^2 = \chi_{\nu_{01}}^2$. Using this fact in (9A.3.69), we then find, after some simple algebra, that

$$P\left(\chi_{\nu_{01}}^2 > \frac{\nu_{01}\delta_{01}^2}{7(16)}\right) = .95 \quad \text{and} \quad P\left(\chi_{\nu_{01}}^2 > \frac{\nu_{01}\delta_{01}^2}{8(16)}\right) = .99,$$

or

$$\nu_{01}\delta_{01}^2/7(16) = \chi_{\nu_{01}:.95}^2 \quad \text{and} \quad \nu_{01}\delta_{01}^2/8(16) = \chi_{\nu_{01}:.99}^2$$

or

$$7/8 = \chi_{\nu_{01}:.99}^2/\chi_{\nu_{01}:.95}^2.$$

If we now read down the ".990" and ".950" columns of the χ^2-table (Table III, Appendix VII) and take ratios of the entries, we find that the above equation is "satisfied," approximately, to two places by

$$\nu_{01} = 70 \quad [.88 \simeq (45.4418)/(51.73931) = \chi_{70;.99}^2/\chi_{70;.95}^2].$$

Now since $E(\sigma_1^2) = 16 = \nu_{01}/\delta_{01}^2(\nu_{01} - 2)$, we have that $\delta_0^2 = 68(16)/70 = 15.543$ or $\delta_0 = 3.942$. Also, it turns out that the plant manager feels 95% confident that μ_1 lies in the interval [34 ± 2.50], that is, his expectation of μ_1 is $E(\mu_1) = \mu_{01} = 34\%$.

Recalling that, in this case, the prior for μ_1 is of the form (9A.3.18), then $[\mu_{01} \pm t_{\nu_{01}:.025}\delta_{01}/\sqrt{n_{01}}]$ is the appropriate 95% prior interval; we thus have that

$$2.50 = t_{70;.025}(3.942)/\sqrt{n_{01}} \quad \text{or} \quad n_{01} = 9.986.$$

In summary, $(\nu_{01}, n_{01}, \delta_{01}^2, \mu_{01}) = (70, 9.986, 15.543, 34)$. A similar analysis for the prior of (μ_2, σ_2^2) leads to $(\nu_{02}, n_{02}, \delta_{02}^2, \mu_{02}) = (80, 12.424, 13.872, 36)$.

Suppose now that 4 sample specimens from each deposit are removed and the following results obtained:

Deposit 1: $n_1 = 4$; $\bar{x}_1 = 33.8\%$; $s_1^2 = 23.24$.

Deposit 2: $n_2 = 4$; $\bar{x}_2 = 37.2\%$; $s_2^2 = 24.32$.

We first calculate the w_j. We have [see (9A.3.66a)]

$$w_1 = 70(15.543) + 3(23.24) + 4(9.986)(34 - 33.8)^2/(4 + 9.986)$$
$$= 1{,}157.844,$$

$$w_2 = 80(13.872) + 3(24.32) + 4(12.424)(36 - 37.2)^2/(4 + 12.424)$$
$$= 1{,}187.077.$$

Now the Bayes estimator of $\mu_1 - \mu_2$ is $\bar{x}_{u1} - \bar{x}_{u2}$, which is, for the above data and priors given by

$$\bar{x}_{u1} - \bar{x}_{u2} = \frac{(9.986)(34) + 4(33.8)}{13.986} - \frac{(12.424)(36) + 4(37.2)}{16.424}$$
$$= 33.943 - 36.292 = -2.349.$$

To find a posterior interval for $\mu_1 - \mu_2$, say at level .95, we first note that the degrees of freedom involved in (9A.3.67b) are large, so that $t_{.025}'(74, 84, \phi) \cong z_{.025} = 1.96$, that is, we do not need to enter Table VIII of

Appendix VII, and hence we need not compute ϕ, where ϕ is defined in (9A.3.67a). [We remind the reader that if $n_j + \nu_{0j}$ were smaller, then we would compute ϕ and do so by the procedure discussed and illustrated in Example 9A.3(f), where ϕ is determined by (9A.3.63a), with

$$u^2 = [\{w_2/(n_2 + \nu_{02})(n_2 + n_{02})\}/\{w_1/(n_1 + \nu_{01})(n_1 + n_{01})\}].$$

Continuing, we have from (9A.3.67b) that the posterior interval for $\mu_1 - \mu_2$, at level .95, is

$$\left[-2.349 \pm 1.96 \sqrt{\frac{1157.844}{(74)(13.986)} + \frac{1187.077}{(84)(16.424)}} \right]$$

$$= [-2.349 \pm 2.758] = [-5.107, .409].$$

Further, a 95% posterior interval for σ_1^2/σ_2^2, is, from (9A.3.66c), given by

$$\left[\left[\frac{(1157.844)/74}{(1187.077)/84} \right] \Big/ F_{74,84;.025}; \left[\frac{(1157.844)/74}{(1187.077)/84} \right] F_{84,74;.025} \right].$$

By *double* interpolating using the "reciprocals of the degrees of freedom" rule mentioned at the bottom of Table VI (e.g., first find $F_{60,84;.025}$ and $F_{120,84;.025}$ by interpolation and, using these values, then find $F_{74,84;.025}$; all interpolations are performed using the above mentioned rule), we find

$$F_{74,84;.025} = 1.55364 \quad \text{and} \quad F_{84,74;.025} = 1.56351.$$

Hence, the 95% posterior interval for σ_1^2/σ_2^2 is

$$\left[\left[\frac{1157.844/74}{1187.077/84} \right] \Big/ 1.55364; \left[\frac{1157.844/74}{1187.077/84} \right] \times 1.56351 \right]$$

$$= [.713; 1.731].$$

In summary, the 95% posterior interval for $\mu_1 - \mu_2$ is $[-5.107, .409]$, and the 95% posterior interval for σ_1^2/σ_2^2 is $[.713, 1.731]$.

9A.4 THE BINOMIAL SITUATION

In this section we suppose that sampling is on a binomial random variable x, whose distribution is given by

(9A.4.1) $b(x, \theta) = \theta^x(1 - \theta)^{1-x}, \qquad x = 0 \text{ or } 1, \quad 0 < \theta < 1.$

Note that here Ω is the interval $(0, 1)$. Now suppose we are in the situation where prior to taking sample evidence, we have very vague feelings about θ and in fact our prior degrees of belief about θ are such that it is just as likely that θ falls in the interval $(\theta_1, \theta_1 + d)$ as in the interval $(\theta_2, \theta_2 + d)$, for all choices of θ_1, θ_2 and d. That is, the prior distribution of θ is the "noninformative" uniform prior

(9A.4.2) $p(\theta) = 1, \quad 0 < \theta < 1$

                                    $= 0, \quad \text{otherwise.}$

Suppose a sample of n independent observations on x is taken and observed to be $(x_1, \ldots, x_n)$. The likelihood function is then

(9A.4.3) $$l(\theta \mid x_1, \ldots, x_n) = \theta^y(1 - \theta)^{n-y},$$

where $y = \sum_{i=1}^{n} x_i$. Combining (9A.4.2) and (9A.4.3) via Bayes' Theorem yields the posterior

(9A.4.4) $$p(\theta \mid x_1, \ldots, x_n) = \frac{\Gamma(n + 2)}{\Gamma(y + 1)\Gamma(n - y + 1)} \theta^y(1 - \theta)^{n-y},$$
$$0 < \theta < 1.$$

The distribution (9A.4.4) is called a *beta distribution* of order $(y + 1, n - y + 1)$, and it can be shown (see Problem 9A.36) that

(9A.4.5) $$E(\theta \mid x_1, \ldots, x_n) = (y + 1)/(n + 2),$$

that is, the Bayes estimate of θ is $(y + 1)/(n + 2)$. Further, it can also be shown that the posterior variance of θ is

(9A.4.6) $$V(\theta \mid x_1, \ldots, x_n) = \frac{(y + 1)(n - y + 1)}{(n + 2)^2(n + 3)}.$$

We note that if n is large, $E(\theta \mid x_1, \ldots, x_n) \simeq y/n$, and

$$V(\theta \mid x_1, \ldots, x_n) = \frac{(y/n)(1 - y/n)}{n};$$

if we write y/n as p (the sample proportion), these results say that the posterior expectation of θ is p, with posterior variance $p(1 - p)/n$. The reader should compare these with the sampling theory results, which are $E(y/n) = \theta$, $\text{Var}(y/n) = \theta(1 - \theta)/n$.

To find a posterior interval of level $1 - \alpha$, we must find

$$\theta_1 = \theta_1(x_1, \ldots, x_n) \qquad \text{and} \qquad \theta_2 = \theta_2(x_1, \ldots, x_n)$$

such that

(9A.4.7) $$\int_0^{\theta_1} \frac{\Gamma(n + 2)}{\Gamma(y + 1)\Gamma(n - y + 1)} \theta^y(1 - \theta)^{n-y} \, d\theta$$
$$= \alpha/2 = \int_{\theta_2}^{1} \frac{\Gamma(n + 2)}{\Gamma(y + 1)\Gamma(n - y + 1)} \theta^y(1 - \theta)^{n-y} \, d\theta.$$

These values may be found using tables of the incomplete beta distribution. However, we may also obtain the same interval in the following way.

If, in the distribution (9A.4.4), we let

(9A.4.8) $$F = (n - y + 1)\theta/[(y + 1)(1 - \theta)]$$

then

$$\theta = [(y + 1)F/(n - y + 1)]\Big/\left[1 + \frac{y + 1}{n - y + 1}F\right],$$

so that

$$1 - \theta = \left[1 + \frac{y + 1}{n - y + 1}F\right]^{-1}.$$

The Jacobian is

$$d\theta/dF = [(y + 1)/(n - y + 1)]\Big/\left(1 + \frac{y + 1}{n - y + 1}F\right)^2.$$

Substituting in (9A.4.4), and multiplying by the Jacobian yields the posterior of F as

(9A.4.9) $$p(F \mid x_1, \ldots, x_n) \propto F^y\Big/\left[1 + \frac{y + 1}{n - y + 1}F\right]^{n+2},$$

which we rewrite as

(9A.4.9a)

$$p(F \mid x_1, \ldots, x_n) \propto F^{[2(y+1)/2]-1}\Big/\left[1 + \frac{2(y + 1)}{2(n - y + 1)}F\right]^{[2(y+1)+2(n-y+1)]/2},$$

and from (AVI.10), we have that this is the form of an $F_{2(y+1),2(n-y+1)}$ distribution. Denoting the upper and lower $\alpha/2$ points of this distribution by $F_{\alpha/2}$ and $F_{1-\alpha/2}$, respectively, we have

(9A.4.10) $$P\left(F_{1-\alpha/2} \leq \frac{n - y + 1}{y + 1}\frac{\theta}{1 - \theta} \leq F_{\alpha/2} \mid x_1, \ldots, x_n\right) = 1 - \alpha,$$

or

(9A.4.10a)

$$P\left(\frac{y + 1}{n - y + 1}F_{1-\alpha/2}\Big/\left[1 + \frac{y + 1}{n - y + 1}F_{1-\alpha/2}\right] \leq \theta\right.$$

$$\left. \leq \frac{y + 1}{n - y + 1}F_{\alpha/2}\Big/\left[1 + \frac{y + 1}{n - y + 1}F_{\alpha/2}\right] \,\Big|\, x_1, \ldots, x_n\right) = 1 - \alpha,$$

that is,

$$(9A.4.10b) \quad \left[\frac{y+1}{n-y+1} F_{1-\alpha/2} \middle/ \left[1 + \frac{y+1}{n-y+1} F_{1-\alpha/2} \right], \right.$$

$$\left. \frac{y+1}{n-y+1} F_{\alpha/2} \middle/ \left[1 + \frac{y+1}{n-y+1} F_{\alpha/2} \right] \right]$$

is a $100(1-\alpha)\%$ posterior interval for θ. It may be computed with the aid of F tables, such as those in Table VI of Appendix VII of this text.

> **Example 9A.4(a).**   A newly purchased Inca coin, is to be used to determine whether the owner of the coin or his assistant will buy coffee at the morning office break. Being a newly introduced coin, the assistant decides on the adoption of the "noninformative" prior (9A.4.2), and he proceeds to keep records of the results of the first three weeks' tosses (that is, fifteen days). This turns out to be (denoting a head by one and a tail by zero)
>
> 1 0 1 1 1 0 0 1 1 0 1 0 0 1 1.
>
> Here, $y = \sum_{i=1}^{15} x_i = 9$. Now the posterior of θ for this situation is given by (9A.4.4), that is, the posterior distribution of θ is that of a beta of order $(y+1, n-y+1) = (10, 7)$. Hence, the Bayes estimate of θ is
>
> $$E(\theta \mid x_1, \ldots, x_{15}) = 10/17 = .588.$$
>
> Using (9A.4.10b), we find that the 95% interval for θ is obtained as follows. We have $y = 9$; $n = 15$; $1 - \alpha = .95$, $\alpha/2 = .025$; $1 - \alpha/2 = .975$; $2(y+1) = 20$; $2(n-y+1) = 14$, so that we need $F_{20,14;.975} = 1/F_{14,20;.025} = 1/2.60244$, and $F_{20,14;.025} = 2.8437$. Substituting in (9A.4.10b), we have, as the reader may verify, that
>
> $$[(.588)(2.60244)^{-1}/(.412 + .588(2.60244)^{-1});$$
> $$(.588)(2.8437)/(.412 + .588(2.8437)] = [.354; .802]$$
>
> is the 95% Bayesian interval for θ. The sample evidence would support a claim that $\theta = .5$, at the .95 level.

We turn now to the case where the prior for θ is the conjugate prior for this problem, namely, the beta distribution of order (t_1, t_2); that is, the prior for θ is of the form

$$(9A.4.11) \quad p(\theta) = \frac{\Gamma(t_1 + t_2)}{\Gamma(t_1)\Gamma(t_2)} \theta^{t_1-1}(1-\theta)^{t_2-1}, \qquad 0 < \theta < 1; \quad t_1, t_2 > 0.$$

This prior distribution (see Problem 9A.37) has mean $t_1/(t_1 + t_2)$, and variance $t_1 t_2/(t_1 + t_2)^2(t_1 + t_2 + 1)$. Further, paralleling (9A.4.8), if we let $F = t_2\theta/[t_1(1-\theta)]$, then it may easily be seen that F has the $F_{2t_1, 2t_2}$

distribution (see (AVI.10)). This in turn implies (see Problem 9A.37) that the interval

$$(9A.4.11a) \quad \left[\frac{t_1}{t_2} F_{1-\alpha/2} \Big/ \left[1 + \frac{t_1}{t_2} F_{1-\alpha/2}\right], \frac{t_1}{t_2} F_{\alpha/2} \Big/ \left[1 + \frac{t_1}{t_2} F_{\alpha/2}\right]\right]$$

is a $100(1 - \alpha)\%$ prior interval for θ, where $F_{\gamma} = F_{2t_1, 2t_2; \gamma}$ is the upper $100\gamma\%$ point of the $F_{2t_1, 2t_2}$ distribution (see the arguments that led to (9A.4.10b). Statements about prior beliefs of θ as to expectation, variance, and intervals of the type (9A.4.11a) could lead to a determination of t_1 and t_2. [See Example 9A.4(b).]

Now the prior of (9A.4.11) is of the same form as the likelihood (9A.4.3) and combines "nicely" with the likelihood, on using Bayes' Theorem, to give as the posterior of θ, a beta distribution [after all, the prior (9A.4.11) is the conjugate prior]. Specifically, we have

(9A.4.12)

$$p(\theta \mid x_1, \ldots, x_n) = \frac{\Gamma(t_1 + t_2 + n)}{\Gamma(t_1 + y)\Gamma(t_2 + n - y)} \theta^{t_1 + y - 1}(1 - \theta)^{t_2 + n - y - 1}.$$

[Note that if $t_1 = t_2 = 1$, (9A.4.12) reduces to (9A.4.4), as might be expected, since, if $t_1 = t_2 = 1$, (9A.4.11) reduces to the noninformative prior (9A.4.2).] Hence the Bayes estimate of θ is

$$(9A.4.13) \qquad E(\theta \mid x_1, \ldots, x_n) = \frac{t_1 + y}{t_1 + t_2 + n},$$

and the variance of θ is

$$(9A.4.13a) \quad V(\theta \mid x_1, \ldots, x_n) = \frac{(t_1 + y)(t_2 + n - y)}{(t_1 + t_2 + n)^2(t_1 + t_2 + n + 1)}.$$

Further, we have that

$$(9A.4.14) \qquad F = \frac{t_2 + n - y}{t_1 + y} \frac{\theta}{1 - \theta} = F_{2(t_1+y), 2(t_2+n-y)},$$

and denoting $F_{2(t_1+y), 2(t_2+n-y); \gamma}$ by F_{γ}, we have that

$$(9A.4.15) \quad \left[\frac{y + t_1}{n - y + t_2} F_{1-\alpha/2} \Big/ \left(1 + \frac{y + t_1}{n - y + t_2} F_{1-\alpha/2}\right);\right.$$

$$\left.\frac{y + t_1}{n - y + t_2} F_{\alpha/2} \Big/ \left(1 + \frac{y + t_1}{n - y + t_2} F_{\alpha/2}\right)\right]$$

is a $100(1 - \alpha)\%$ posterior interval for θ.

Example 9A.4(b). A process involved in the manufacture of a certain type of incandescent bulbs is to be modified along certain prescribed lines. Opinions from the (experienced) plant manager are solicited about θ, the proportion of defective bulbs that will be produced by the modified process. It turns out that his prior feelings are well described by a prior of the type (9A.4.11), that is, a beta of order $(t_1, t_2) = (50,950)$. The values of (t_1, t_2) were arrived at as follows. The plant manager expects θ to be $E(\theta) = t_1/(t_1 + t_2) = .05$. Further, he feels that the standard deviation of θ should be, no larger than .008. From $t_1/(t_1 + t_2) = .05$ we have that $t_1/t_2 = 1/19$, or $t_2 = 19t_1$. Now as the prior variance may be written as

$$\{[t_1/(t_1 + t_2)][t_2/(t_1 + t_2)]\}/(t_1 + t_2 + 1) = (.05)(.95)/(t_1 + 19t_1 + 1)$$
$$= (.008)^2,$$

we have $t_1 = 37$, so that $t_2 = 19t_1 = 703$. Now as the reader may verify, (see Problem 9A.37), a prior 95 % interval for θ is [we use $F_{m_1, m_2; \gamma} = F_{m_1, \infty; \gamma}$ if $m_2 > 120$, etc., and (9A.4.11a)]

$$[\tfrac{1}{19}F_{74, \infty; .975}/(1 + \tfrac{1}{19}F_{74, \infty; .975}); \ \tfrac{1}{19}F_{74, \infty; .025}/(1 + \tfrac{1}{19}F_{74, \infty; .025})]$$
$$= [.0358, .0660],$$

but when the above interval is given to the plant manager, he rejects it. as being slightly too wide. (The reader is asked to calculate this interval in Problem 9A.38.) The choice of (t_1, t_2), then, continues and finally the values of (t_1, t_2) are selected to be $(50, 950)$, for this gives a standard deviation of .007, and of course, an expectation of .05 (see Problem 9A.37). But, more important, it gives as 95 % prior interval, the interval

$$[\tfrac{1}{19}F_{100, \infty; .975}/(1 + \tfrac{1}{19}F_{100, \infty; .975}); \ \tfrac{1}{19}F_{100, \infty; .025}/(1 + \tfrac{1}{19}F_{100, \infty; .025})]$$
$$= [.0377, .0637],$$

which the manager feels is a correct description of his "95 % prior interval" on θ. In fact, when other intervals of varying confidence levels are computed using the beta, order $(50, 950)$, the manager affirms that these are in accordance with his beliefs about θ and, accordingly, the prior for $p(\theta)$ that is adopted is the beta of order $(t_1, t_2) = (50, 950)$.

Now suppose a sample of 100 bulbs is taken at random from the production and it is found that seven of the 100 are defective. The posterior of θ is thus [see (9A.4.12)] the beta distribution of order $(57, 1043)$. This implies that the Bayes estimate of θ is .052, that is,

$$E(\theta \mid x_1, \ldots, x_{100}) = 57/(57 + 1043) = 57/1100 = .052,$$

and the posterior variance of θ is

$$V(\theta \mid x_1, \ldots, x_{100}) = (.052)(.948)/1101 = 45 \times 10^{-6},$$

that is, a standard deviation of $6.7 \times 10^{-3} = .0067$. Further, we have that a 95 % posterior interval for θ [see (9A.4.15)] is (we use $F_{m_1, \infty; \gamma}$ for $F_{m_1, m_2; \gamma}$ if $m_2 > 120$, etc.)

$$[\tfrac{57}{1043}F_{114, \infty; .975}/(1 + \tfrac{57}{1043}F_{114, \infty; .975}); \ \tfrac{57}{1043}F_{114, \infty; .025}/(1 + \tfrac{57}{1043}F_{114, \infty; .025})]$$
$$= [.0398, .0651].$$

PROBLEMS

9A.1. Suppose the random variable x has distribution $f(x \mid \theta)$, where θ may take on any one of the values in $\Omega = \{\theta_1, \ldots, \theta_k\}$. Suppose further that the prior probability that $\theta = \theta_i$ is $P(\theta = \theta_i) = p(\theta_i) = p_i$, $i = 1, \ldots, k$, $0 < p_i < 1$, $\sum_{i=1}^{k} p_i = 1$. If a random sample of n independent observations $x_1, \ldots, x_n$ is taken on x, what is the posterior probability that $\theta = \theta_i$, given the sample $(x_1, \ldots, x_n)$?

9A.2. Compute the Bayes estimate of μ, and its precision, if the posterior of μ is as given by (9A.2.17a).

9A.3. A hardware store has received an allotment of cans of paint with a lead base from a new supplier. Experience with this type of paint leads to the assumption that a can of the paint will cover x square feet of a primed wood surface, where $x = N(\mu, (50)^2)$. As nothing is known about the new supplier's cans of paint, the noninformative prior for μ is used, that is $p(\mu) \propto k$. A sample of 4 cans chosen at random from the lot is used on a primed wood surface, and the sample mean is found to be $\bar{x} = 404.2$ square feet. What is the posterior of μ? What is the Bayes estimate of μ, and what is its variance? Construct a 95% interval for μ.

9A.4. The modification of an existing wood-lamination process is such that it is assumed that the breaking strengths (in pounds) of strips to be produced by this newly modified process, are distributed as $N(\mu, 9)$. A sample of size 25 strips chosen at random from the new process, yields a sample mean of breaking strengths of $\bar{x} = 161.2$ lb. Assuming a noninformative prior is applicable, find the posterior of μ and hence the expectation of μ, that is, the Bayes estimate. Compute the posterior variance and find a 99% posterior interval for μ.

9A.5. Introduction of a new blast furnace leads to the planning of a plant investigation of the tensile strength (in psi) of steel beams that will be produced using the new furnace. It is known that the distribution of tensile strengths of beams will be $N(\mu, (2100)^2)$, and that prior beliefs about μ are well described by the $N(\mu_0, \sigma_0^2) = N(54,000, (1900)^2)$ distribution. A sample of 15 beams yields a sample mean of tensile strengths of 51,500 psi. Write down the posterior density of μ. Determine the Bayes estimate of μ and the posterior variance of μ. Find a 90% posterior interval for μ.

9A.6. Barbers' "stropping straps" produced by a certain manufacturer have lengths of life (in weeks) which are assumed to be normally distributed with unknown mean μ and a standard deviation of 12 weeks. A supply of 5 such straps are installed in a barber shop. The owner has strong prior feelings about μ, which turn out to be described by the $N(\mu_0, \sigma_0^2) = N(130, 15^2)$ distribution. The 5 straps turn out to have lengths of life given by 136, 126, 112, 139, and 149 weeks. Find the Bayes estimate of μ and a 99% posterior interval for μ.

9A.7. Referring to Example 9A.2(d), suppose $\sigma_{01}^2 = (100)^2 = \sigma_{02}^2$. Find the Bayes estimate of $\mu_1 - \mu_2$ and a 95% posterior interval, and compare your results with those of Example 9A.2(d).

9A.8. Light bulbs of Type A are assumed to have lengths of life which are normally distributed, mean μ_A, variance $(28)^2$, while light bulbs of Type B are assumed to have lengths of life which are normally distributed, mean μ_B, variance $(32)^2$. Prior opinions dictate the use of the priors (i) $\mu_A = N(950, 15^2)$, and (ii) $\mu_B = N(900, 10^2)$, and μ_A and μ_B independent. A sample of $n_A = 5$ bulbs selected at random from a lot of Type A bulbs yields a sample mean of $\bar{x}_A = 975$ hours, and a sample of $n_B = 8$ bulbs, selected at random from a lot of Type B bulbs, yields a sample mean of $\bar{x}_B = 930$ hours. Find the Bayes estimate of $\mu_A - \mu_B$ and a 98% posterior confidence interval.

9A.9. Two methods of producing a certain item are to be contrasted. The items produced by methods 1 and 2 are assumed to be distributed as $N(\mu_1, 25)$ and $N(\mu_2, 36)$, respectively. The prior for μ_1 and μ_2 treats μ_1 and μ_2 as independent random variables, where $\mu_1 = N(20, 5)$ and $\mu_2 = N(20, 5)$. A sample of 10 items from the production using method 1 yields $\bar{x}_1 = 19.8$, and a sample of 12 items from production using method 2 yields $\bar{x}_2 = 21.7$. Find the posterior of $\mu_1 - \mu_2$; the Bayes estimate of $\mu_1 - \mu_2$; and a 98% posterior confidence interval for $\mu_1 - \mu_2$.

9A.10. For the situation described in Problem 9A.8, suppose it is the case that the priors of μ_A and μ_B are such that $p(\mu_A, \mu_B) = p(\mu_A)p(\mu_B) \propto k$, that is, the noninformative prior is applicable. Answer the questions asked in Problem 9A.8 under this condition.

9A.11. Referring to Problem 9A.9, suppose $p(\mu_1, \mu_2) = p(\mu_1)p(\mu_2) \propto k$. Answer the questions asked in Problem 9A.9 under this condition.

9A.12. If $p(\mu, \sigma^2)$ is the conjugate prior given by (9A.3.17), show that the marginal distribution of μ is given by (9A.2.18) and that the relations (9A.3.18a) hold. Show that a $100(1 - \alpha)\%$ prior interval for μ is $[\mu_0 \pm t_{v_0; \alpha/2}\delta_0/\sqrt{n_0}]$.

9A.13. If $p(\mu, \sigma^2)$ is given by (9A.3.17), show that the marginal of σ^2 is given in (9A.3.19), and that the relations (9A.3.19a) hold. Show that a $100(1 - \alpha)\%$ prior confidence interval for σ^2 is $[v_0\delta_0^2/\chi^2_{v_0; \alpha/2}, v_0\delta_0^2/\chi^2_{v_0; 1-\alpha/2}]$.

9A.14. A new machine for producing $\frac{1}{4}$-inch rivets is to be checked to see if it is functioning properly. It is assumed that the diameters of the rivets produced by this machine are distributed as $N(\mu, \sigma^2)$. The prior of μ and σ^2 is taken to be the noninformative priors, that is, μ and σ^2 are independent and $p(\mu) \propto k$, $p(\sigma^2) \propto 1/\sigma^2$. A sample of 10 rivets yields a sample mean and a sample standard deviation of $\bar{x} = .2503$ inches and $s = .0002$ inches, respectively. Find the posterior distribution of μ. What is the Bayes estimate of μ? Calculate a 98% posterior interval for μ. Do you think this new machine should be returned by the company, or not?

9A.15. A test for breaking strength of chain links is to be performed. It is assumed that the breaking strengths of the links are normally distributed with mean μ and variance σ^2. Prior opinions are vague about μ and σ^2, and noninformative priors are adopted, that is, $p(\mu, \sigma^2) = p(\mu)p(\sigma^2) \propto 1/\sigma^2$. A sample of 6 links yields $(\bar{x}, s^2) = (8750 \text{ lb}, (175 \text{ lb})^2)$. Find the posterior for μ. Find the Bayes estimate of μ. Construct a 95% posterior interval for μ. Comment on the manufacturer's claim that the mean breaking strength of links that he produces is 9000 lb.

9A.16. For the situation described in Problem 9A.14, find the posterior distribution of σ^2. What is the Bayes estimate of σ^2? Find a 95% posterior interval for σ^2.

9A.17. For the situation described in Problem 9A.15, find the posterior distribution of σ^2, the Bayes estimate of σ^2, and the 99% posterior interval for σ^2.

9A.18. The mean lifetime of bulbs produced with a new type fusing process is to be examined. These lifetimes are assumed to be distributed as $N(\mu, \sigma^2)$. The prior of (μ, σ^2) that describes prior beliefs seems to be the conjugate prior of form (9A.3.17) with $(\nu_0, n_0, \delta_0^2, \mu_0) = (50, 7.6, 160, 1080)$. [These could have been chosen using a method similar to that of Example 9A.3(c) or 9A.3(g).] A sample of size 8 bulbs yields a sample mean $\bar{x} = 1103$, and a sample variance $s^2 = 212.3$. Find the posterior of μ, and the Bayes estimate of μ. Calculate a 95% posterior interval for μ.

9A.19. Determinations of percentage methanol of a large amount of solution stored in a certain plant is to be made, after the solution has been stored for many months. It is assumed that the determinations are normally distributed about the true percent μ of methanol, and have unknown variance σ^2. The senior chemist feels that a prior of the conjugate form (9A.3.17) best describes his prior experience and that $(\nu_0, n_0, \delta_0^2, \mu_0) = (10, 12.4, .0040, 8.00\%)$. A sample of size four yields $\bar{x} = 8.34\%$ and $s = 0.03\%$. Find the posterior of μ, the Bayes estimate of μ and a 95% posterior interval for μ.

9A.20. For the Problem 9A.18, compute the posterior of σ^2, the Bayes estimate of σ^2 and a 98% posterior interval for σ^2.

9A.21. For the Problem 9A.19, compute the posterior of σ^2, the Bayes estimate of σ^2 and a 95% posterior interval for σ^2.

9A.22. Two new methods of determining nickel content of steel are to be tried out on a certain kind of steel. Past experience is such that it is assumed that the determinations using method $j = 1, 2$ are $N(\mu_j, \sigma^2)$, that is, $\sigma_1^2 = \sigma_2^2 = \sigma^2$. Information about the μ_j and σ^2 is otherwise vague and so the noninformative prior given by (9A.3.29) is adopted. Random samples are taken independently with the following results:

$$Method\ 1: \quad n_1 = 7; \quad \bar{x}_1 = 3.285\%; \quad s_1^2 = .000033.$$

$$Method\ 2: \quad n_2 = 7; \quad \bar{x}_2 = 3.258\%; \quad s_2^2 = .000092.$$

(a) Find the posterior of $\mu_1 - \mu_2$, the Bayes estimate of $\mu_1 - \mu_2$, and a 95% posterior interval for $\mu_1 - \mu_2$. Comment on the claim of an adherent of method 1 that $\mu_1 - \mu_2 > 0$.

(b) Find the posterior of σ^2, the Bayes estimate of σ^2 and a 95% posterior interval for σ^2.

9A.23. A tract of land is divided up into 40 plots of equal area, and all plots have equal irrigation facilities, sunlight and other characteristics. Two new methods of cultivating soybeans are to be contrasted. To this end, 20 of the plots are chosen at random, and crops of soybeans using method 1 are cultivated. On the remaining plots, method 2 is used. The results are (units are in hundreds of bushels):

$$Method\ 1: \quad n_1 = 20, \quad \bar{x}_1 = 24.7, \quad s_1^2 = .81.$$

$$Method\ 2: \quad n_2 = 20, \quad \bar{x}_2 = 22.1, \quad s_2^2 = .35.$$

Assuming that yields of soybeans under methods $j = 1, 2$ are $N(\mu_j, \sigma^2)$, i.e., $\sigma_1^2 = \sigma_2^2 = \sigma^2$, and that $p(\mu_1, \mu_2, \sigma^2) \propto 1/\sigma^2$:

(a) Write down the posterior of $\mu_1 - \mu_2$, the Bayes estimate of $\mu_1 - \mu_2$, and calculate a 95% posterior interval for $\mu_1 - \mu_2$.

(b) Repeat (a) for σ^2.

9A.24. Referring to part b(i) of Section 9A.3, suppose that the prior distribution of (μ_1, μ_2, σ^2) is the conjugate prior for the situation of part b(i), that is, $p(\mu_1, \mu_2, \sigma^2)$ is such that

$$p(\mu_1, \mu_2, \sigma^2) \propto (\sigma^2)^{-(n_{01}+n_{02})/2} \exp \left\{ -\frac{1}{2\sigma^2} \left[\sum_{j=1}^{2} n_{0j}(\mu_j - \mu_{0j})^2 + w_0 \right] \right\},$$

where $0 < n_{0j}, 0 < w_0, -\infty < \mu_{0j} < \infty$.

(a) What is the marginal prior of σ^2? What is the prior expectation and variance of σ^2? Find a $100(1 - \alpha)\%$ prior interval for σ^2.

(b) Find the posterior of (μ_1, μ_2, σ^2) by combining the above conjugate prior with the likelihood (9A.3.30). Determine the marginal posterior of σ^2, and, by substituting in (9A.3.40), find the marginal posterior of $\mu_1 - \mu_2$. Also derive a $100(1 - \alpha)\%$ posterior interval for $\mu_1 - \mu_2$.

9A.25. Referring to Problem 9A.22, suppose we cannot assume that $\sigma_1^2 = \sigma_2^2 = \sigma^2$. Suppose instead that we use the prior $p(\mu_1, \mu_2, \sigma_1^2, \sigma_2^2) \propto 1/\sigma_1^2\sigma_2^2$ [see Example 9A.3(f)]. Find the posterior of $\mu_1 - \mu_2$, the Bayes estimate of $\mu_1 - \mu_2$ and a 95% posterior interval for $\mu_1 - \mu_2$.

9A.26. Repeat the work of Problem 9A.25 for the situation in Problem 9A.23.

9A.27. If the conditions of Problem 9A.25 hold for Problem 9A.22, find a 95% posterior interval for σ_1^2/σ_2^2. Comment on the "claim" that $\sigma_1^2 = \sigma_2^2$.

9A.28. Do the same as in Problem 9A.27 for the Problem 9A.23.

9A.29. Referring to Example 9A.3(f), find a 95% posterior interval for σ_1^2/σ_2^2. What is the Bayes estimate of σ_1^2/σ_2^2?

9A.30. If the joint posterior of $(\mu_1, \mu_2, \sigma_1^2, \sigma_2^2)$ is given by (9A.3.66), show that the (marginal) joint posterior of (σ_1^2, σ_2^2) is of the form given by (9A.3.66a), and that $\sigma_1^2/\sigma_2^2 = \left[\dfrac{w_1}{n_1 + v_{01}} \middle/ \dfrac{w_2}{n_2 + v_{02}} \right] \times F_{n_2+v_{02};\,n_1+v_{01}}$. Hence show that the posterior interval for σ_1^2/σ_2^2, at level $1 - \alpha$, is given by (9A.3.66c).

9A.31. If the joint posterior of $(\mu_1, \mu_2, \sigma_1^2, \sigma_2^2)$ is given by (9A.3.66), show that the (marginal) joint posterior of (μ_1, μ_2) is of the form given by (9A.3.67). Hence show that a posterior interval for $(\mu_1 - \mu_2)$ is given by (9A.3.67b), at level $1 - \alpha$.

9A.32. Two kinds of paper material are to be compared by making impact-strength measurements (in units of foot pounds). It is assumed that materials $j = 1, 2$ have impact-strengths which are distributed as $N(\mu_j, \sigma_j^2)$. Opinions of the manufacturers' staffs of both materials give rise to the use of conjugate priors with [see Example 9A.3(g)]

$$(v_{01}, n_{01}, \delta_{01}^2, \mu_{01}) = (5, 4, 1.032, 1.3)$$
$$(v_{02}, n_{02}, \delta_{02}^2, \mu_{02}) = (10, 5, .030, 1.0).$$

Samples of size 7 and 9 yields the following data:

$$\text{Material 1: } n_1 = 7; \ \bar{y}_1 = 1.225; \ s_1^2 = .025.$$
$$\text{Material 2: } n_2 = 9; \ \bar{y}_2 = 0.984; \ s_2^2 = .053.$$

(a) Determine the posterior of $\mu_1 - \mu_2$. Give the Bayes estimate of $\mu_1 - \mu_2$ and construct a 95% posterior interval for $\mu_1 - \mu_2$.

(b) Repeat the instructions of (a) for σ_1^2/σ_2^2.

9A.33. The breaking strength of two types of wood are to be compared by subjecting test pieces to stress tests. It is assumed that the breaking strengths of pieces of type j wood are normally distributed, mean μ_j, variance σ_j^2, $j = 1, 2$. Opinions based on previous experience give rise to conjugate priors [see Example 9A.3(g)] with

$$(\nu_{01}, n_{01}, \delta_{01}^2, \mu_{01}) = (6, 8, .04, 173.75)$$
$$(\nu_{02}, n_{02}, \delta_{02}^2, \mu_{02}) = (5, 7, .23, 172.50).$$

Independent samples of size 7 and 6 yield:

$$\text{Wood 1: } n_1 = 7; \ \bar{y}_1 = 171.83; \ s_1^2 = .085.$$
$$\text{Wood 2: } n_2 = 6; \ \bar{y}_2 = 172.25; \ s_2^2 = 3.24.$$

(a) Determine the posterior of $\mu_1 - \mu_2$. Give the Bayes estimate of $\mu_1 - \mu_2$.

(b) Repeat the instructions of (a) for σ_1^2/σ_2^2.

9A.34. Suppose sampling is on a random variable x, where $x = N(\mu, \sigma^2)$, and μ is known to have the value μ_0. Suppose the prior distribution of σ^2 is $p(\sigma^2)$, where $p(\sigma^2) \propto (\sigma^2)^{-(n_0+2)/2} \exp\{-w_0/2\sigma^2\}$, $\sigma^2 > 0$. What is the posterior distribution of σ^2, if a sample $(x_1, x_2, \ldots, x_n)$ of n independent observations is indeed taken on x? What is the Bayes estimate of σ^2? What form does a $100(1 - \alpha)\%$ posterior interval for σ^2 have?

9A.35 (continuation of 9A.34). Suppose also that a random variable y is $N(\tau, \beta^2)$, where τ is known to have the value τ_0. Let the prior of β^2 be $p(\beta) \propto (\beta^2)^{-(m_0+2)/2} \exp\{-v_0/2\beta^2\}$. If $y_1, \ldots, y_m$ are m independent observations on y and are independent of the $x_1, \ldots, x_n$ of Problem 9A.34, what is the posterior distribution of σ^2/β^2? The Bayes estimate of σ^2/β^2? A $100(1 - \alpha)\%$ posterior interval for σ^2/β^2?

9A.36. If the posterior of θ is given by (9A.4.4), show that $E(\theta \mid x_1, \ldots, x_n) = (y + 1)/(n + 2)$, where $y = \sum_{i=1}^{n} x_i$, and that $V(\theta \mid x_1, \ldots, x_n)$ is as given by (9A.4.6). [Hint: $\int_0^1 \theta^{m_1-1}(1 - \theta)^{m_2-1} \, d\theta = \Gamma(m_1)\Gamma(m_2)/\Gamma(m_1 + m_2)$, and $\Gamma(r + 1) = r\Gamma(r)$.]

9A.37. If the prior distribution of a parameter θ is given by (9A.4.11), verify that the mean is given by $t_1/(t_1 + t_2) = \theta_0$, say, and that the variance is

$$\theta_0(1 - \theta_0)/(t_1 + t_2 + 1).$$

Further, show that $F = t_2\theta/t_1(1 - \theta)$ has the $F_{2t_1, 2t_2}$ distribution and that $\left[\dfrac{t_1}{t_2} F_{1-\alpha/2} \middle/ \left(1 + \dfrac{t_1}{t_2} F_{1-\alpha/2}\right); \dfrac{t_1}{t_2} F_{\alpha/2} \middle/ \left(1 + \dfrac{t_1}{t_2} F_{\alpha/2}\right)\right]$ is a $100(1 - \alpha)\%$ prior interval for θ, where in general $F_\gamma = F_{2t_1, 2t_2; \gamma}$.

9A.38 (continuation of 9A.37). Referring to Example 9A.4(b), compute a 95% prior interval for θ, if the prior $p(\theta)$ is a beta of order (37,703). Further, verify that if $p(\theta)$ is a beta of order (50,950), then $E(\theta) = .05$, $V(\theta) = (.007)^2$, and that a 95% prior interval for θ is [.038, .064].

9A.39. Let the true probability of a certain coin turning up heads be θ, and suppose the prior for θ is (9A.4.2). Suppose 29 tosses of the coin are made and fifteen turn up as heads. What is the Bayes estimate of θ, and what is a 98% posterior interval for θ?

9A.40. Suppose θ is the true proportion of defective radio tubes being produced by a certain manufacturer. A new customer wishes to make inference about θ and adopts as his prior, the noninformative prior (9A.4.2). In a lot of 100 tubes, it is found that 7 are defective. What is the posterior of θ? What is the Bayes estimate of θ? Find the 95% posterior interval for θ.

9A.41. Much experience with a certain supplier of condensers has led a retail outlet to adopt a prior for θ, the proportion of defective condensers, as a beta of order (25,300). The retail outlet maintains a protective surveillance by sampling from newly received lots. If a sample of 100 condensers yields 3 defective, find the Bayes estimate of θ and a 95% posterior interval for θ.

9A.42. A student is about to enroll in a certain geology course. From speaking to his fraternity brothers and delving into records of past graduates, he believes that the adoption of a beta prior of order (2,207) summarizes his feelings about θ, the proportion of A's that will be awarded in the geology course by the instructor. Of a class of 100 enrolled in this geology course, it eventually turns out that 13 students wind up with an A grade. What is the Bayes estimate of θ? What is a 95% posterior interval for θ? What was the prior expectation of θ, and what was a 95% (prior) interval for θ? Compare the prior and posterior intervals.

9A.43. Suppose a random variable x has the Poisson distribution with parameter τ. Let $x_1, \ldots, x_n$ be a random sample on x. What form does the likelihood have? $\left(\text{Let } t = \sum_{i=1}^{n} x_i\right)$. Suppose that prior to taking the observations, only vague information about τ is known, so that the noninformative prior for this situation $p(\tau) \propto k$, a constant, is used. What is the posterior of τ? Relate it to a χ^2 distribution. Find the Bayes estimate and a $100(1 - \alpha)\%$ posterior interval.

9A.44 (continuation of 9A.43). Suppose the prior for τ of Problem 9A.43 is the conjugate prior $p(\tau) = n_0^{t_0+1}[\Gamma(t_0 + 1)]^{-1}\tau^{t_0}e^{-n_0\tau}$. Verify that $E(\tau) = (t_0 + 1)/n_0$ and Var $(\tau) = (t_0 + 1)/n_0^2$. Combine this prior with the likelihood of problem 9A.43 to find the posterior of τ. Relate this to a χ^2 distribution. Find the Bayes estimate, and the posterior variance of τ, as well as a $100(1 - \alpha)\%$ posterior interval for τ.

CHAPTER 10

Statistical Tests

10.1 INTRODUCTION

An important kind of problem in statistical inference is to determine whether a sample could reasonably have come from a population having a completely or partially specified distribution. For instance, if a sample is known to have come from some normal distribution, is it reasonable that it could have come from one having a given mean μ_0? Or if two independent samples come from normal distributions, is it reasonable that they could have come from normal distributions with equal means? Answers to questions such as these depend on the use of sample means, variances, and other statistics determined from the sample or samples. Since statistics such as these, when determined from samples, are random variables having their own probability distributions, statements based on their values must be made in terms of probabilities. In this chapter we shall consider some of the more important statistical tests based on sample means and sample variances. It will be seen that there is a close connection between statistical testing and statistical estimation.

10.2 TESTS CONCERNING THE MEAN OF A NORMAL DISTRIBUTION HAVING KNOWN VARIANCE

(a) Case of a Left-Sided Test

Suppose $\bar{x}$ is the mean of a sample of size n from a normal distribution $N(\mu, \sigma^2)$, where σ^2 is known and μ is unknown. Suppose we wish to ask whether it is reasonable that this sample could have come from the normal population $N(\mu_0, \sigma^2)$ as compared with the possibility it could have come from some normal population $N(\mu_1, \sigma^2)$, where $\mu_1 < \mu_0$. We may abbreviate this statement and say that we wish to test the statistical hypothesis

$$H_0: \quad \mu = \mu_0$$

against alternatives

$$H_1: \quad \mu = \mu_1 < \mu_0,$$

making use of the sample size n and the sample mean $\bar{x}$. H_0 is frequently called the *null hypothesis* and H_1 the *alternative hypothesis*.

It is intuitively evident that we would choose in favor of H_1 if $\bar{x}$ is sufficiently small, that is, if $\bar{x} < k$, where k is yet to be found, and in favor of H_0 if $\bar{x} \geq k$. The set of values of $\bar{x}$ for which we reject H_0 (namely those for which $\bar{x} < k$) is called the *critical region* for the test. We can make two kinds of errors:

Type-I *error*, if we reject H_0 (accept H_1) when H_0 is true.
Type-II *error*, if we accept H_0 (reject H_1) when H_1 is true.

The probability of a type-I error, often referred to as the *size* of the test, or *significance level* of the test, is

(10.2.1) $$P(\bar{x} < k \mid \mu = \mu_0)$$

and the probability of a type-II error is

(10.2.2) $$P(\bar{x} \geq k \mid \mu = \mu_1).$$

By choosing k so that

(10.2.3) $$P(\bar{x} < k \mid \mu = \mu_0) = \alpha,$$

where α is usually chosen in practice as .01, .05, or .10, we can control the type-I error so that its probability of occurrence is α.

The type-II error has probability β, where

(10.2.4) $$\beta = P(\bar{x} \geq k \mid \mu = \mu_1) = 1 - P(\bar{x} < k \mid \mu = \mu_1),$$

where k has been chosen to satisfy (10.2.3).

The quantities α and β in hypothesis testing correspond to producer's and consumer's risks in acceptance sampling (see Chapters 4 and 8).

Since σ is known and k has been chosen so that (10.2.3) is satisfied, and since $\bar{x}$ has the distribution $N(\mu_0, \sigma^2/n)$ in (10.2.3), it is seen that

(10.2.5) $$P(\bar{x} < k \mid \mu = \mu_0) = \Phi\left[\frac{(k - \mu_0)\sqrt{n}}{\sigma}\right] = \alpha.$$

Notice that we then have

(10.2.6) $$P(\bar{x} > k \mid \mu = \mu_1) = 1 - \Phi\left[\frac{(k - \mu_1)\sqrt{n}}{\sigma}\right] = \beta.$$

Thus, for specified values of α, μ_0, μ_1, σ, and n, the number k satisfies the equation

(10.2.7) $$\frac{(k - \mu_0)\sqrt{n}}{\sigma} = z_{1-\alpha}.$$

With k thus determined, the value of β satisfies the following equation:

$$(10.2.7a) \qquad \frac{(k - \mu_1)\sqrt{n}}{\sigma} = z_\beta.$$

Solving (10.2.7), we find

$$(10.2.8) \qquad k = \mu_0 + \frac{\sigma}{\sqrt{n}} z_{1-\alpha}$$

and, hence, the critical region for $\bar{x}$ is the set of values of $\bar{x}$ for which

$$(10.2.8a) \qquad \bar{x} < \mu_0 + \frac{\sigma}{\sqrt{n}} z_{1-\alpha},$$

that is, for which

$$(10.2.9) \qquad \frac{(\bar{x} - \mu_0)\sqrt{n}}{\sigma} < z_{1-\alpha}.$$

Any observed value of $\bar{x}$ in this critical region is said to be *significantly smaller than* μ_0 at the $100\alpha\%$ *level of significance*. Because of the range of the values of α in hypothesis testing (usually $\alpha \leq .10$) and the properties of the standard normal variable, it is well to point out that the critical region can also be described as the set of values of $\bar{x}$, such that

$$\bar{x} < \mu_0 - \frac{\sigma}{\sqrt{n}} z_\alpha,$$

and that (10.2.9) can be expressed as the set of values of $\bar{x}$ for which

$$\frac{(\bar{x} - \mu_0)\sqrt{n}}{\sigma} < -z_\alpha.$$

Note that β as given by (10.2.6) is a function of μ_1, say $\beta(\mu_1)$, and the graph of $\beta(\mu_1)$ as a function of μ_1 is called the *operating characteristic* (OC) *curve* of the test of H_0 against H_1. The function

$$\gamma(\mu_1) = 1 - \beta(\mu_1)$$

is called the *power function* of the test and gives the probability of rejecting $\mu = \mu_0$ if $\mu = \mu_1$. The graph of the power function plotted against μ_1 is called the *power curve* of the test. Note that the ordinate of the power curve at μ_0 is α, that is,

$$\gamma(\mu_0) = 1 - \beta(\mu_0) = \alpha.$$

Example 10.2(a).   A sample of 16 lengths of thread from a day's production of a thread on a given machine has mean tensile strength $\bar{x} = 967.8$ psi. Suppose the population of tensile strengths of lengths of thread in the day's production is $N(\mu, (128)^2)$ (it is known from experience that for this type of thread $\sigma = 128$ psi) and that we wish to test the hypothesis

$$H_0: \quad \mu = \mu_0 = 1000$$

against alternatives

$$H_1: \quad \mu = \mu_1 < 1000.$$

In this case we have from (10.2.8) that

$$k = 1000 + \frac{128}{\sqrt{16}} z_{1-\alpha}.$$

If we choose α, the probability of a type-I error, as .05, then

$$k = 1000 + 32(-1.645) = 947.36,$$

and the critical region consists of the values of $\bar{x}$ for which $\bar{x} < 947.36$. Since the obtained value of $\bar{x}$ is 967.8 which does not fall in the critical region $(-\infty, 947.36)$, we say that $\bar{x}$ is not significantly smaller than the value of μ under H_0, that is, $\mu = 1000$.

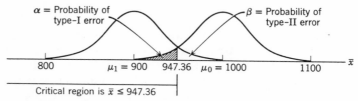

Figure 10.2.1. Graphical description of the constituents of Example 10.2(a).

The power $\gamma(\mu_1)$ of our test for H_0 against any $\mu = \mu_1 < 1000$ is the probability of rejecting H_0 (that is, rejecting the hypothesis that $\mu = 1000$) when H_1 is true (that is, when μ has some value $\mu_1 < 1000$), and is given by

$$\gamma(\mu_1) = P(\bar{x} < 947.36 \mid \mu = \mu_1) = P\left(z < \frac{947.36 - \mu_1}{128/\sqrt{16}}\right),$$

that is,

(10.2.10) $$\gamma(\mu_1) = \Phi\left[\frac{(947.36 - \mu_1)\sqrt{16}}{128}\right].$$

Figure 10.2.1 shows α and β, the value of k, μ_0 ($\mu_0 = 1000$), and μ_1 ($\mu_1 = 900$). Indeed, when $\mu_1 = 900$, the power (10.2.10) has value .931, hence $\beta(900)$ is .069, that is the probability of a type-II error for $\mu = \mu_1 = 900$ is .069, and we note again that the probability that this test will reject H_0 when indeed $\mu = 900$ is .931.

(b) Case of a Right-Sided Test

Suppose we wish to test the hypothesis

$$H_0: \quad \mu = \mu_0$$

against alternatives

$$H_1: \quad \mu = \mu_1 > \mu_0$$

on the basis of the mean $\bar{x}$ of a sample of size n from $N(\mu, \sigma^2)$, where σ^2 is known. The reader can verify that, in this case, the critical region for

$\bar{x}$ is the set of values of $\bar{x}$ for which $\bar{x} > k$, and where k satisfies

(10.2.11) $$P(\bar{x} > k \mid \mu = \mu_0) = \alpha.$$

The probability of a type-II error for this test is given by

(10.2.12) $$P(\bar{x} \leq k \mid \mu = \mu_1) = \beta.$$

Corresponding to relations (10.2.7) and (10.2.7a), we obtain

(10.2.13) $$\frac{(k - \mu_0)\sqrt{n}}{\sigma} = z_\alpha,$$

(10.2.13a) $$\frac{(k - \mu_1)\sqrt{n}}{\sigma} = z_{1-\beta}.$$

The critical region for $\bar{x}$ for the right-sided test therefore consists of values of $\bar{x}$ such that

$$\bar{x} > \mu_0 + \frac{\sigma}{\sqrt{n}} z_\alpha$$

or, equivalently,

(10.2.14) $$\frac{(\bar{x} - \mu_0)\sqrt{n}}{\sigma} > z_\alpha.$$

(c) Case of the Two-Sided Test

Now suppose we wish to test the hypothesis

$$H_0: \quad \mu = \mu_0$$

against alternatives

$$H_1: \quad \mu = \mu_1 \neq \mu_0.$$

(This problem is sometimes referred to as a *two-sided problem*, as contrasted with the *one-sided problems* of Sections 10.2 *a* and *b* above.) In this case it is evident that the critical region of the test will consist of all values of $\bar{x}$ for which $|\bar{x} - \mu_0| > k$ where, for a given probability α of a type-I error, k is chosen so that

$$P(|\bar{x} - \mu_0| > k \mid \mu = \mu_0) = \alpha,$$

that is,

(10.2.15) $$1 - P(\mu_0 - k < \bar{x} < \mu_0 + k \mid \mu = \mu_0) = \alpha.$$

Hence we must have

(10.2.16) $$1 - \left[\Phi\left(\frac{k\sqrt{n}}{\sigma}\right) - \Phi\left(-\frac{k\sqrt{n}}{\sigma}\right) \right] = \alpha.$$

But remembering that

$$\Phi(z) + \Phi(-z) = 1,$$

we have from (10.2.16), after some simplification, that

$$1 - \Phi\left(\frac{k\sqrt{n}}{\sigma}\right) = \frac{\alpha}{2};$$

hence

$$\frac{k\sqrt{n}}{\sigma} = z_{\alpha/2}.$$

Therefore,

(10.2.17) $$k = \frac{\sigma}{\sqrt{n}} z_{\alpha/2};$$

hence the critical region for $\bar{x}$ is the set of values of $\bar{x}$ for which

(10.2.18) $$|\bar{x} - \mu_0| > \frac{\sigma}{\sqrt{n}} z_{\alpha/2}.$$

If $\bar{x}$ satisfies (10.2.18), we would say that $\bar{x}$ *differs from μ_0 significantly at the $100\alpha\%$ level of significance.*

The power of our two-sided test for H_0 against H_1 is given by

(10.2.19) $$\gamma(\mu_1) = 1 - P(\mu_0 - k < \bar{x} < \mu_0 + k \mid \mu = \mu_1),$$

where $k = (\sigma/\sqrt{n}) z_{\alpha/2}$, that is,

(10.2.20) $$\gamma(\mu_1) = 1 - \left\{\Phi\left[\frac{(\mu_0 - \mu_1)\sqrt{n}}{\sigma} + z_{\alpha/2}\right] - \Phi\left[\frac{(\mu_0 - \mu_1)\sqrt{n}}{\sigma} - z_{\alpha/2}\right]\right\}.$$

The reader should note that, since

$$P\left(|\bar{x} - \mu_0| > \frac{\sigma}{\sqrt{n}} z_{\alpha/2} \mid \mu = \mu_0\right) = \alpha,$$

we have

$$P\left(|\bar{x} - \mu_0| < \frac{\sigma}{\sqrt{n}} z_{\alpha/2} \mid \mu = \mu_0\right) = 1 - \alpha,$$

which is equivalent to the statement that $[\bar{x} \pm (\sigma/\sqrt{n}) z_{\alpha/2}]$ is a $100(1 - \alpha)\%$ confidence interval for μ_0. Thus $\bar{x}$ being (*not being*) *significantly different from μ_0 at the $100\alpha\%$ level of significance* is equivalent to the statement that μ_0 *does not* (*does*) *lie in the $100(1 - \alpha)\%$ confidence interval for μ.*

Example 10.2(b). In Example 10.2(a), suppose we wish to test the hypothesis

$$H_0: \quad \mu = \mu_0 = 1000$$

against the alternatives

$$H_1: \quad \mu = \mu_1 \neq 1000$$

at the 5% level of significance.

We have, for $\alpha = .05$, that $z_{\alpha/2} = 1.96$; hence from (10.2.18) the critical region for $\bar{x}$ is the set of values of $\bar{x}$ for which

$$|\bar{x} - 1000| > \frac{128}{\sqrt{16}} (1.96),$$

that is,

$$|\bar{x} - 1000| > 62.72.$$

The observed value of $\bar{x} = 967.8$ does not fall into this critical set of values of $\bar{x}$ since $|967.8 - 1000|$ is not greater than 62.72. Hence, the observed value of $\bar{x}$, namely, 967.8 *does not* differ significantly from the value $\mu_0 = 1000$ (specified under H_0) at the 5% level of significance.

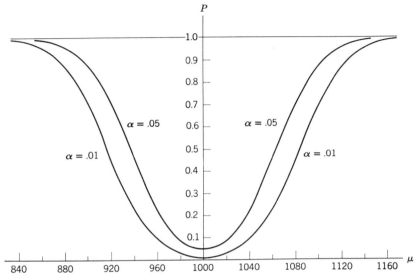

Figure 10.2.2. The power function (10.2.20) for the test of Example 10.2(b).

Looked at from the point of view of confidence intervals, this statement is equivalent to the statement that the value of $\mu_0 = 1000$ (under H_0) *is* contained in the 95% confidence interval for μ_0, namely, (967.8 ± 62.72).

A graph of the power function [see (10.2.20)] of the above test for $\alpha = .05$ and .01 is given in Figure 10.2.2.

10.3 TESTS CONCERNING THE MEAN OF A NORMAL POPULATION HAVING UNKNOWN VARIANCE

(a) The Case of a Left-Sided Test

In Section 10.2 suppose the value of σ^2 is unknown and we wish to test the hypothesis

$$H_0: \quad \mu = \mu_0$$

against alternatives

$$H_1: \quad \mu = \mu_1 < \mu_0$$

on the basis of a sample from $N(\mu, \sigma^2)$.

In this case we proceed as follows: Let n, $\bar{x}$, s^2 be the sample size, sample mean, and sample variance respectively. For probability α of a type-I error we choose the critical region in the $(\bar{x}, s)$-plane as the set of pairs of values $(\bar{x}, s)$ for which

(10.3.1)
$$\frac{(\bar{x} - \mu_0)\sqrt{n}}{s} < t_{n-1;1-\alpha},$$

where $t_{n-1;1-\alpha}$ is the value of t in the Student t distribution which satisfies

(10.3.2) $P(t_{n-1} > t_{n-1;1-\alpha}) = 1 - \alpha.$

Thus, for this critical region we have

(10.3.3) $P\left[\dfrac{(\bar{x} - \mu_0)\sqrt{n}}{s} < t_{n-1;1-\alpha} \,\middle|\, \mu = \mu_0\right] = \alpha$

as the probability of a type-I error.

If a sample of size n from $N(\mu, \sigma^2)$ has values of $(\bar{x}, s)$ which does satisfy (10.3.1), we say that $\bar{x}$ *is significantly smaller than* μ_0 *at the* $100\alpha\%$ *level of significance.*

The power of our test for H_0 against alternatives in H_1 is given by

(10.3.4) $\gamma(\mu_1) = P\left[\dfrac{(\bar{x} - \mu_0)\sqrt{n}}{s} < t_{n-1,1-\alpha} \,\middle|\, \mu = \mu_1\right],$

which can be evaluated from what are called the noncentral Student t tables. We shall not consider numerical values of $\gamma(\mu_1)$ except where $\mu_1 = \mu_0$, in which case $\gamma(\mu_0) = \alpha$.

> **Example 10.3(a).** Four determinations of copper in a certain solution yielded a sample mean $\bar{x} = 8.30\%$ with $s = .03\%$. If μ is the mean of the population of such determinations, test, at the 5% level of significance, the hypothesis
>
> $$H_0: \quad \mu = \mu_0 = 8.32$$
>
> against alternatives
>
> $$H_1: \quad \mu = \mu_1 < 8.32.$$
>
> Since $n = 4$, we have from the Student t table $t_{3;.95} = -2.353$. The value of the left-hand side of (10.3.1) is
>
> $$\frac{(8.30 - 8.32)\sqrt{4}}{0.03} = \frac{-.02}{.03}(2) = -1.333,$$
>
> which is *not* less than -2.353. Therefore, we accept the hypothesis H_0. Or, alternatively, we say that the observed sample mean 8.30 is not significantly less than the mean under H_0, namely, $\mu_0 = 8.32$ at the 5% level of significance.

(b) The Case of a Right-Sided Test

The reader will see from the left-sided test described in the foregoing, that in the right-sided test for the hypothesis

$$H_0: \quad \mu = \mu_0$$

against alternatives

$$H_1: \quad \mu = \mu_1 > \mu_0$$

on the basis of a sample from $N(\mu, \sigma^2)$, where σ^2 is unknown, the critical region for a given α consists of the set of pairs of values $(\bar{x}, s)$ for which

(10.3.5)
$$\frac{(\bar{x} - \mu_0)\sqrt{n}}{s} > t_{n-1;\alpha}.$$

(c) The Two-Sided Case

Now suppose we wish to test the hypothesis

$$H_0: \quad \mu = \mu_0$$

against alternatives

$$H_1: \quad \mu = \mu_1 \neq \mu_0$$

on the basis of a sample of size n from $N(\mu, \sigma^2)$, where σ^2 is unknown. In this case we choose the critical region as the set of pairs of values $(\bar{x}, s)$ for which

(10.3.6)
$$\left| \frac{(\bar{x} - \mu_0)\sqrt{n}}{s} \right| > t_{n-1;\alpha/2}.$$

Since

(10.3.7)
$$P\left[\left| \frac{(\bar{x} - \mu_0)\sqrt{n}}{s} \right| > t_{n-1;\alpha/2} \,\middle|\, \mu = \mu_0 \right] = \alpha,$$

the probability of a type-I error for the test is α.

If $(\bar{x}, s)$ *does (does not) satisfy the inequality in* (10.3.6), *we say that* $\bar{x}$ *does (does not) differ significantly from* μ_0 *at the* $100\alpha\%$ *level of significance,* which is equivalent to the statement that the $100(1 - \alpha)\%$ *confidence interval* $[\bar{x} \pm t_{n-1;\alpha/2} (s/\sqrt{n})]$ *for* μ *does not (does) contain the value* μ_0 *specified as the value of* μ *under* H_0.

Example 10.3(b). In Example 10.3(a) suppose we wish to test the hypothesis

$$H_0: \quad \mu = \mu_0 = 8.32$$

against alternatives

$$H_1: \quad \mu = \mu_1 \neq 8.32$$

at the 5% level of significance. Noting that $n = 4$ and that $t_{3,.025} = 3.182$, we find by using (10.3.6) that the critical region is the set of pairs of values for which

(10.3.8)
$$\left| \frac{(\bar{x} - 8.32)\sqrt{4}}{s} \right| > 3.182.$$

The observed value of $(\bar{x}, s)$ is (8.30, .03) and hence the left-hand side of (10.3.8) is

$$\left| \frac{(8.30\ -8.32)2}{.03} \right| = 1.333,$$

which is less than 3.182, and so we accept the hypothesis H_0 and say that the observed value of $\bar{x}$ does not differ significantly from 8.32 at the 5% level of significance.

(d) The Problem of Paired Comparisons

In this problem we perform n independent pairs of experiments, these experiments producing the pairs of random variables (y_{11}, y_{12}), $(y_{21}, y_{22}), \ldots, (y_{n1}, y_{n2})$. It is assumed that the differences $y_{i2} - y_{i1} = x_i$, $i = 1, \ldots, n$, are independent random variables all having identical normal distributions, namely, $N(\mu, \sigma^2)$. Thus, $(x_1, \ldots, x_n)$ is essentially a random sample of size n from $N(\mu, \sigma^2)$. Let $\bar{x}$ and s^2 be the mean and variance of this sample of x's.

The problem is to test the hypothesis

$$H_0: \quad \mu = 0$$

against alternatives

$$H_1: \quad \mu \neq 0.$$

The critical region for the test consists of the pairs $(\bar{x}, s)$ for which

$$(10.3.9) \qquad \left| \frac{\bar{x}\sqrt{n}}{s} \right| > t_{n-1;\alpha/2},$$

which is a special case of the two-sided test discussed in Section 10.3c, with $\mu_0 = 0$.

Note that we do not have to assume that y_{i1} and y_{i2} are independent. For example, y_{i1}, $i = 1, \ldots, n$ could be the observations of "gain" in weight obtained by feeding laboratory animals a certain diet, and y_{i2} could be the "gain" in weight obtained by feeding the same laboratory animals a different diet. Obviously, the results for animal i, (y_{i1}, y_{i2}) are not independent, but $x_i = y_{i1} - y_{i2}$ and $x_j = y_{j1} - y_{j2}$, $i \neq j$ are independent.

Example 10.3(c). Bennett and Franklin (1954) quote the following example of the use of the t test for a paired comparison problem. Table 10.3.1 gives the results of experiments on 21 different samples of iron ore using a standard dichromate titrimetric method and a new spectrophotometric method for the determination of the iron ore content of the 21 samples. We wish to test $\mu = 0$ (at the 5% level of significance) against $\mu \neq 0$.

TABLE 10.3.1

RESULTS OF TREATING SAMPLES OF IRON ORE
WITH STANDARD AND NEW METHODS

Sample	Standard Method y_{i1}	New Method y_{i2}	$x_i = y_{i2} - y_{i1}$
1	28.22	28.27	+.05
2	33.95	33.99	+.04
3	38.25	38.20	−.05
4	42.52	42.42	−.10
5	37.62	37.64	+.02
6	36.84	36.85	+.01
7	36.12	36.21	+.09
8	35.11	35.20	+.09
9	34.45	34.40	−.05
10	52.83	52.86	+.03
11	57.90	57.88	−.02
12	51.52	51.52	.00
13	49.59	49.52	−.07
14	52.20	52.19	−.01
15	54.04	53.99	−.05
16	56.00	56.04	+.04
17	57.62	57.65	+.03
18	34.30	34.39	+.09
19	41.73	41.78	+.05
20	44.44	44.44	.00
21	46.48	46.47	−.01

From the data of Table 10.3.1, we find that $\bar{x} = .0086$, $s^2 = .002890$, and, hence, the left-hand side of (10.3.9) has the value

$$\left| \frac{.0086\sqrt{21}}{\sqrt{.002890}} \right| = .93.$$

Using t tables, we find $t_{20;.025} = 2.086$. We therefore accept the hypothesis that $\mu = 0$, and say that the two methods do not differ significantly at the 5% level.

Before proceeding to the next section, we remind the reader that for large degrees of freedom, say m, the Student t variable with m degrees of freedom is approximately an $N(0, 1)$ variable, that is, for m large, $t_m \simeq z$. This means that if n is large (in practice $n > 60$) the critical regions of the tests discussed in this section can be defined by replacing probability points of the t_{n-1} distribution by the corresponding point of the $N(0, 1)$ distribution. For example, the critical region defined by (10.3.1) may be stated for large n, as $[(\bar{x} - \mu_0)\sqrt{n}/s] < z_{1-\alpha} = -z_\alpha$, etc.

10.4 TESTS CONCERNING THE DIFFERENCE OF MEANS OF TWO NORMAL POPULATIONS HAVING KNOWN VARIANCES

Suppose $\bar{x}_1$ is the mean of a sample of size n_1 from a normal population $N(\mu_1, \sigma_1^2)$, and $\bar{x}_2$ is the mean of an independent sample of size n_2 from a normal population $N(\mu_2, \sigma_2^2)$, where σ_1^2 and σ_2^2 are both known. We know from Appendix VI that $\bar{x}_1 - \bar{x}_2$ has the normal distribution $N(\mu_1 - \mu_2, \sigma_1^2/n_1 + \sigma_2^2/n_2)$. We wish to test the hypothesis

$$H_0: \quad \mu_1 - \mu_2 = 0$$

against the alternatives

$$H_1: \quad \mu_1 - \mu_2 = \delta < 0$$

on the basis of $\bar{x}_1 - \bar{x}_2$. This is a left-sided test based on $\bar{x}_1 - \bar{x}_2$.

The procedure for testing this hypothesis is quite similar to that discussed in Section 10.2a. If the probability of a type-I error is to be α, it is evident from the results of Section 10.2a that the critical region for $\bar{x}_1 - \bar{x}_2$ is the set of values $\bar{x}_1 - \bar{x}_2$ for which

$$(10.4.1) \qquad \frac{\bar{x}_1 - \bar{x}_2}{\sqrt{\sigma_1^2/n_1 + \sigma_2^2/n_2}} < z_{1-\alpha}.$$

Since

$$(10.4.2) \qquad P\left(\frac{\bar{x}_1 - \bar{x}_2}{\sqrt{\sigma_1^2/n_1 + \sigma_2^2/n_2}} < z_{1-\alpha} \,\Big|\, \mu_1 - \mu_2 = 0 \right) = \alpha,$$

the probability of type-I error is α. The probability β of a type-II error is the probability of accepting $\mu_1 - \mu_2 = 0$ when $\mu_1 - \mu_2 = \delta$, that is,

$$(10.4.3) \quad \beta = P\left(\frac{\bar{x}_1 - \bar{x}_2}{\sqrt{\sigma_1^2/n_1 + \sigma_2^2/n_2}} > z_{1-\alpha} \,\Big|\, \mu_1 - \mu_2 = \delta \right)$$

or

$$(10.4.4) \quad \beta = P\left[\frac{(\bar{x}_1 - \bar{x}_2) - \delta}{\sqrt{\sigma_1^2/n_1 + \sigma_2^2/n_2}} > z_{1-\alpha} - \frac{\delta}{\sqrt{\sigma_1^2/n_1 + \sigma_2^2/n_2}} \,\bigg|\, \mu_1 - \mu_2 = \delta \right]$$

which can be written as

$$(10.4.5) \qquad \beta = 1 - \Phi\left(z_{1-\alpha} - \frac{\delta}{\sqrt{\sigma_1^2/n_1 + \sigma_2^2/n_2}} \right).$$

If $\bar{x}_1 - \bar{x}_2$ satisfies (10.4.1), we say that $(\bar{x}_1 - \bar{x}_2)$ is significantly less than zero, or $\bar{x}_1$ is significantly smaller than $\bar{x}_2$ at the $100\alpha\%$ level of significance.

The case of a right-sided test can be treated in a similar manner, and the details are left to the reader.

In the two-sided case, the hypothesis to be tested is

$$H_0: \quad \mu_1 - \mu_2 = 0$$

against alternatives

$$H_1: \quad \mu_1 - \mu_2 = \delta \neq 0.$$

The critical region in this case consists of the set of values of $\bar{x}_1 - \bar{x}_2$ for which

(10.4.6)
$$\left| \frac{\bar{x}_1 - \bar{x}_2}{\sqrt{\sigma_1^2/n_1 + \sigma_2^2/n_2}} \right| > z_{\alpha/2}.$$

The probability of a type-I error for this test is α since the probability is α that (10.4.6) is satisfied if $\mu_1 - \mu_2 = 0$. The power of the test, that is, the probability of rejecting $\mu_1 - \mu_2 = 0$ when $\mu_1 - \mu_2 = \delta$ is given by

$$\gamma(\delta) = P\left(\left| \frac{\bar{x}_1 - \bar{x}_2}{\sqrt{\sigma_1^2/n_1 + \sigma_2^2/n_2}} \right| > z_{\alpha/2} \,\Big|\, \mu_1 - \mu_2 = \delta \right).$$

This may be evaluated as

(10.4.7)
$$\gamma(\delta) = 1 - \left[\Phi\left(z_{\alpha/2} - \frac{\delta}{\sqrt{\sigma_1^2/n_1 + \sigma_2^2/n_2}} \right) - \Phi\left(-z_{\alpha/2} - \frac{\delta}{\sqrt{\sigma_1^2/n_1 + \sigma_2^2/n_2}} \right) \right].$$

If $\bar{x}_1 - \bar{x}_2$ does (does not) satisfy (10.4.6), we say that $\bar{x}_1$ does (does not) differ significantly from $\bar{x}_2$ at the $100\alpha\%$ level of significance, which is equivalent to the statement that $\mu_1 - \mu_2 = 0$ is not (is) contained within the $100(1 - \alpha)\%$ confidence interval $[(\bar{x}_1 - \bar{x}_2) \pm z_{\alpha/2}\sqrt{\sigma_1^2/n_1 + \sigma_2^2/n_2}]$ for $\mu_1 - \mu_2$.

Example 10.4(a). Suppose two machines, say M_1 and M_2, are packaging 6-ounce cans of talcum powder. It is known from the past behavior of the machines that the standard deviations of weights of their respective fillings are .04 ounce and .05 ounce respectively.

Suppose 100 cans filled by each machine are emptied and the contents are carefully weighed and that the sample means are $\bar{x}_1 = 6.11$ ounces and $\bar{x}_2 = 6.14$ ounces. We wish to test at the 1% level of significance the hypothesis

$$H_0: \quad \mu_1 = \mu_2$$

against the alternatives

$$H_1: \quad \mu_1 \neq \mu_2,$$

where μ_1 and μ_2 are means of populations of weights of fillings produced by machines M_1 and M_2 respectively.

Here $\alpha/2 = .005$ and $z_{.005} = 2.576$. Furthermore, $\sigma_1^2 = .0016$ and $\sigma_2^2 = .0025$. Hence the left-hand side of (10.4.6) for this example has the observed value

$$\left| \frac{6.11 - 6.14}{\sqrt{.0016/100 + .0025/100}} \right| = 4.685,$$

which is larger than 2.576 and, hence, we reject the hypothesis and say that the machines differ significantly in their filling weights at the 1% level. Note that the 99% confidence interval for $\mu_1 - \mu_2$ is

$$\left[(6.11 - 6.14) \pm (2.576) \sqrt{\frac{.0016}{100} + \frac{.0025}{100}} \right] = [-.03 \pm .015]$$

$$= [-.045, -.015]$$

and does not contain the value 0.

10.5 TESTS CONCERNING THE DIFFERENCE OF MEANS OF TWO NORMAL DISTRIBUTIONS HAVING EQUAL BUT UNKNOWN VARIANCES

If, in Section 10.4, the variances of σ_1^2 and σ_2^2 are unknown but can be assumed equal to σ^2, say, where σ^2 is unknown, we proceed as follows.

Our problem here is that of testing the hypothesis

$$H_0: \quad \mu_1 - \mu_2 = 0$$

against the alternatives

$$H_1: \quad \mu_1 - \mu_2 = \delta < 0,$$

on the basis of the information in two samples: one from $N(\mu_1, \sigma^2)$ and one from $N(\mu_2, \sigma^2)$. We denote the sizes, means, and variances of the two samples by $n_1, \bar{x}_1, s_1^2$, and $n_2, \bar{x}_2, s_2^2$ respectively. This is a left-sided test, and it is evident from Section 9.4 on the estimation of $\mu_1 - \mu_2$ that we can define the critical region for this test as the set of values $(\bar{x}_1, \bar{x}_2, s_1, s_2)$, for which

(10.5.1) $$\frac{\bar{x}_1 - \bar{x}_2}{s_w \sqrt{1/n_1 + 1/n_2}} < t_{n_1+n_2-2;1-\alpha},$$

where

(10.5.2) $$s_w^2 = \frac{(n_1 - 1)s_1^2 + (n_2 - 1)s_2^2}{n_1 + n_2 - 2}.$$

When H_0 is true, it will be seen that the probability of a type-I error is α.

If we consider a right-sided test for the hypothesis

$$H_0: \quad \mu_1 - \mu_2 = 0$$

against alternatives

$$H_1: \quad \mu_1 - \mu_2 = \delta > 0,$$

it is evident that the critical region for this test consists of the set of values of $(\bar{x}_1, \bar{x}_2, s_1, s_2)$, for which

$$(10.5.3) \qquad \frac{\bar{x}_1 - \bar{x}_2}{s_w\sqrt{1/n_1 + 1/n_2}} > t_{n_1+n_2-2;\alpha}.$$

It will be noted that the probability of making a type-I error by this test is α.

If we wish to consider a two-sided test for the hypothesis

$$H_0: \quad \mu_1 - \mu_2 = 0$$

against alternatives

$$H_1: \quad \mu_1 - \mu_2 = \delta \neq 0,$$

the critical region for the test consists of the set of values of $(\bar{x}_1, \bar{x}_2, s_1, s_2)$, for which

$$(10.5.4) \qquad \left| \frac{\bar{x}_1 - \bar{x}_2}{s_w\sqrt{1/n_1 + 1/n_2}} \right| > t_{n_1+n_2-2;\alpha/2}.$$

Again note that the probability of a type-I error is α.

[We again remind the reader that if $\sigma_1^2 = \sigma_2^2$ and if the total sample size $n_1 + n_2$ is large (in practice, $n_1 + n_2 \geq 62$), then $t_{n_1+n_2-2} \simeq z$. This means that the critical region defined by (10.5.4) may be stated, for large $n_1 + n_2$ as $|(\bar{x}_1 - \bar{x}_2)/s_w\sqrt{1/n_1 + 1/n_2}| > z_{\alpha/2}$. A similar remark applies to (10.5.1) and 10.5.3).]

Summarizing (10.5.4), we may state: *If $\bar{x}_1$, $\bar{x}_2$, and s_w satisfy (10.5.4), we say that $\bar{x}_1$ and $\bar{x}_2$ differ significantly at the $100\alpha\%$ level of significance.*

Example 10.5(a).   Two methods of determining nickel content of steel, say M_1 and M_2, are tried out on a certain kind of steel. Samples of four determinations are made by each method, with the following results:

$$\bar{x}_1 = 3.285\%, \qquad s_1^2 = .000033,$$
$$\bar{x}_2 = 3.258\%, \qquad s_2^2 = .000092.$$

Let us denote by μ_1 and μ_2 the means and by σ_1^2 and σ_2^2 the variances of the populations of determinations made by methods M_1 and M_2 respectively.

We shall assume that $\sigma_1^2 = \sigma_2^2 = \sigma^2$, and test, at the 5% level of significance, the hypothesis

$$H_0: \quad \mu_1 = \mu_2$$

against

$$H_1: \quad \mu_1 > \mu_2.$$

This is a one-sided (right-hand) test with $\alpha = .05$, $n_1 + n_2 - 2 = 6$; hence $t_{6;.05} = 1.943$. The value of s_w^2 is

$$\frac{3(.000033) + 3(.000092)}{6} = .000063,$$

which gives $s_w = .00794$. The observed value of the left-hand side of (10.5.3) is

$$\frac{.027}{.00794\sqrt{\frac{1}{4} + \frac{1}{4}}} = 4.808,$$

which is greater than 1.943; hence we reject the hypothesis and say that $\bar{x}_1$ is significantly larger than $\bar{x}_2$ and accept the hypothesis that $\mu_1 > \mu_2$.

We remark that, if $\sigma_1^2 \neq \sigma_2^2$, we use the fact that for large values of n_1 and n_2 (in practice both should be greater than 60) the ratio

$$(10.5.5) \qquad \frac{(\bar{x}_1 - \bar{x}_2) - (\mu_1 - \mu_2)}{\sqrt{s_1^2/n_1 + s_2^2/n_2}}$$

is approximately distributed as an $N(0, 1)$ variable. If n_1 and n_2 are small, (10.5.5) is approximately distributed as a Student t variable with m degrees of freedom, where m is given by (9.4.12) of Section 9.4. (Now it could be that both n_1 and n_2 are smaller than 60, but that m, as computed by (9.4.12) is greater than 60. We may then again use the fact that (10.5.5) is approximately distributed as an $N(0, 1)$ variable.) In any case, the use of (10.5.5) would change the relevant critical regions (10.5.1), (10.5.3), and (10.5.4). The details are left to the reader as an exercise.

10.6   TESTS CONCERNING THE VARIANCE OF A NORMAL DISTRIBUTION

Suppose we have a sample of size n from a normal distribution $N(\mu, \sigma^2)$ and that we wish to make the left-sided test of the hypothesis

$$H_0: \quad \sigma^2 = \sigma_0^2$$

against alternatives

$$H_1: \quad \sigma^2 = \sigma_1^2 < \sigma_0^2,$$

on the basis of the sample variance. Recalling from Theorem A.VI.5 that, if $\sigma^2 = \sigma_0^2$, then $[(n-1)s^2]/\sigma_0^2$ is a chi-square variable with $n - 1$

degrees of freedom, we would choose as the critical region for the test the set of values of s^2 for which

(10.6.1)
$$\frac{(n-1)s^2}{\sigma_0^2} < \chi^2_{n-1;1-\alpha}.$$

The probability of a type-I error of this test is α.

In testing the hypothesis
$$H_0: \quad \sigma^2 = \sigma_0^2$$

against alternatives
$$H_1: \quad \sigma^2 = \sigma_1^2 > \sigma_0^2,$$

we have a right-sided test in which the critical region consists of the set of values of s^2 for which

(10.6.2)
$$\frac{(n-1)s^2}{\sigma_0^2} > \chi^2_{n-1;\alpha}.$$

In testing the hypothesis
$$H_0: \quad \sigma^2 = \sigma_0^2$$

against the alternatives
$$H_1: \quad \sigma^2 \neq \sigma_0^2,$$

we have a two-sided test in which the critical region consists of the set of values of s^2 for which

(10.6.3) $\dfrac{(n-1)s^2}{\sigma_0^2} < \chi^2_{n-1;1-\alpha/2}$ or $\dfrac{(n-1)s^2}{\sigma_0^2} > \chi^2_{n-1;\alpha/2}.$

The probability of a type-I error for this test is seen to be α.

Example 10.6(a). A sample of size 11 from a population which is normally distributed gives $s^2 = 154.6$. Test at the 5% level of significance the hypothesis
$$H_0: \quad \sigma^2 = 140$$

against alternatives
$$H_1: \quad \sigma^2 = \sigma_1^2 > 140.$$

This is a (one-sided) right-hand test. We note that $n = 11$ and $\alpha = .05$, so that $\chi^2_{10;.05} = 18.307$. Hence the critical region is given by the values of s^2 for which

(10.6.4)
$$\frac{10s^2}{140} > 18.307.$$

Inserting the observed value of s^2, the left-hand side of (10.6.4) is $1546/140 = 11.043$, which is not larger than 18.307. We therefore accept the hypothesis H_0 that $\sigma^2 = 140$.

The power functions of the tests described above can be found in a straightforward manner. Consider the power of the test having critical region defined by (10.6.2).

The power is the probability of rejecting the hypothesis H_0 that $\sigma^2 = \sigma_0^2$ when $\sigma^2 = \sigma_1^2 > \sigma_0^2$, and is given by

(10.6.5)
$$\gamma(\sigma_1^2) = P\left[\frac{(n-1)s^2}{\sigma_0^2} > \chi_{n-1;\alpha}^2 \mid \sigma^2 = \sigma_1^2 > \sigma_0^2\right]$$

$$= P\left[\frac{(n-1)s^2}{\sigma_1^2} > \frac{\sigma_0^2}{\sigma_1^2}\chi_{n-1;\alpha}^2 \mid \sigma^2 = \sigma_1^2\right].$$

But when sampling from $N(\mu, \sigma_1^2)$, we have seen that $[(n-1)s^2]/\sigma_1^2$ is a χ_{n-1}^2 variable. Hence the power is

(10.6.6)
$$\gamma(\sigma_1^2) = P\left(\chi_{n-1}^2 > \frac{\sigma_0^2}{\sigma_1^2}\chi_{n-1;\alpha}^2\right).$$

10.7 TESTS CONCERNING THE RATIO OF VARIANCES OF TWO NORMAL DISTRIBUTIONS

Suppose we have independent samples of sizes n_1 and n_2 from the normal distribution $N(\mu_1, \sigma_1^2)$ and $N(\mu_2, \sigma_2^2)$ respectively, and that we wish to test the hypothesis (at significance level α)

$$H_0: \quad \frac{\sigma_1^2}{\sigma_2^2} = 1$$

against the alternatives

$$H_1: \quad \frac{\sigma_1^2}{\sigma_2^2} = \lambda < 1$$

on the basis of the sample sizes n_1, n_2 and sample variances s_1^2, s_2^2. From Appendix VI, we know that $(s_1^2/\sigma_1^2)/(s_2^2/\sigma_2^2)$ is a Snedecor F ratio with $(n_1 - 1, n_2 - 1)$ degrees of freedom. Hence the critical region for this left-sided test consists of the pairs of values of (s_1^2, s_2^2) for which

(10.7.1)
$$\frac{s_1^2}{s_2^2} < F_{n_1-1,n_2-1;1-\alpha}.$$

If (s_1^2, s_2^2) satisfies (10.7.1), we say that s_1^2 is significantly smaller than s_2^2 at the $100\alpha\%$ level of significance.

The critical region of the test for the hypothesis

$$H_0: \quad \frac{\sigma_1^2}{\sigma_2^2} = 1$$

against alternatives

$$H_1: \quad \frac{\sigma_1^2}{\sigma_2^2} = \lambda > 1$$

consists of the set of values of (s_1^2, s_2^2) for which

(10.7.2)
$$\frac{s_1^2}{s_2^2} > F_{n_1-1, n_2-1; \alpha}.$$

This, of course, is a right-sided test. If (s_1^2, s_2^2) satisfies (10.7.2), we say that s_1^2 is significantly larger than s_2^2 at the $100\alpha\%$ level of significance.

The critical region of the two-sided test for the hypothesis

$$H_0: \quad \frac{\sigma_1^2}{\sigma_2^2} = 1$$

against alternatives

$$H_1: \quad \frac{\sigma_1^2}{\sigma_2^2} = \lambda \neq 1$$

consists of the set of values of (s_1^2, s_2^2) for which

(10.7.3)
$$\frac{s_1^2}{s_2^2} < F_{n_1-1, n_2-1; 1-\alpha/2} \quad \text{or} \quad \frac{s_1^2}{s_2^2} > F_{n_1-1, n_2-1; \alpha/2}.$$

If (s_1^2, s_2^2) satisfies (10.7.3), we say that s_1^2 and s_2^2 are significantly different at the $100\alpha\%$ level of significance.

Note that the probability of making a type-I error for each of the three tests discussed in the foregoing is α.

The power functions of the left-sided, the right-sided, and two-sided tests previously described are seen to be respectively,

(10.7.4)
$$\gamma(\lambda) = P\left(F_{n_1-1, n_2-1} < \frac{1}{\lambda} F_{n_1-1, n_2-1; 1-\alpha}\right), \qquad \lambda < 1,$$

(10.7.5)
$$\gamma(\lambda) = P\left(F_{n_1-1, n_2-1} > \frac{1}{\lambda} F_{n_1-1, n_2-1; \alpha}\right), \qquad \lambda > 1,$$

and

(10.7.6)
$$\gamma(\lambda) = P\left(F_{n_1-1, n_2-1} < \frac{1}{\lambda} F_{n_1-1, n_2-1; 1-\alpha/2}\right)$$
$$+ P\left(F_{n_1-1, n_2-1} > \frac{1}{\lambda} F_{n_1-1, n_2-1; \alpha/2}\right), \qquad \lambda \neq 1,$$

where $\lambda = \sigma_1^2/\sigma_2^2$.

The reader may verify that s_1^2 and s_2^2 being (not being) significantly different at the $100\alpha\%$ level of significance is equivalent to the statement that $\sigma_1^2/\sigma_2^2 = 1$ is not (is) contained in the $100(1 - \alpha)\%$ confidence interval

$$\left(\frac{s_1^2}{s_2^2 F_{n_1-1, n_2-1; \alpha/2}}, \frac{s_1^2}{s_2^2 F_{n_1-1, n_2-1; 1-\alpha/2}}\right)$$

for σ_1^2/σ_2^2.

Example 10.7(a).   Five pieces of material were subjected to treatment T_1, and six pieces of a similar material were subjected to a different treatment, say T_2. Measurements made on these two samples gave the following results for the sample variances: $s_1^2 = .00045$, and $s_2^2 = .00039$. Test at the 5% level the hypothesis

$$H_0: \quad \sigma_1^2 = \sigma_2^2 \quad \text{against alternatives} \quad H_1: \quad \sigma_1^2 > \sigma_2^2.$$

This is a right-sided test, and we note that $\alpha = .05$, $n_1 = 5$, and $n_2 = 6$. Hence we will make use of $F_{4,5;.05} = 5.199$. Consulting (10.7.2), the critical region of this test is the set of values of (s_1^2, s_2^2) for which

(10.7.7) $$\frac{s_1^2}{s_2^2} > 5.199.$$

But the observed value of s_1^2/s_2^2 in this example is $.00045/.00039 = 1.154$ which is not greater than 5.199. Hence we accept the hypothesis H_0.

10.8 SEQUENTIAL TESTS OF HYPOTHESES

All tests considered in the preceding sections are based on samples of predetermined fixed size n. In this section we shall consider a procedure for testing the hypothesis

$$H_0: \quad \text{the population sampled has p.f. (or p.d.f.) } f_0(x)$$

against the alternative

$$H_1: \quad \text{the population sampled has p.f. (or p.d.f.) } f_1(x),$$

in which the sample size is not fixed in advance.

Such a procedure, called a *sequential test*, works as follows. We take a sequence of independent observations $x_1, x_2, \ldots$, one at a time, and make one of three decisions after drawing each observation. For the mth observation, $m = 1, 2, \ldots$, the decisions are as follows:

1. If

(10.8.1) $$A < \frac{f_1(x_1) \cdots f_1(x_m)}{f_0(x_1) \cdots f_0(x_m)} < B,$$

we draw an $(m + 1)$st observation.

2. If

(10.8.2) $$\frac{f_1(x_1) \cdots f_1(x_m)}{f_0(x_1) \cdots f_0(x_m)} \leq A,$$

we stop sampling and accept H_0.

3. If

(10.8.3) $$\frac{f_1(x_1) \cdots f_1(x_m)}{f_0(x_1) \cdots f_0(x_m)} \geq B,$$

we stop sampling and accept H_1 (that is, we reject H_0).

A and B are chosen so as to make the probabilities of type-I and type-II errors equal to α and β respectively. Exact values of A and B are difficult to obtain. However, for the small values of α and β which are ordinarily used in practice, we can use as good approximations,

$$(10.8.4) \qquad A \simeq \frac{\beta}{1-\alpha} \quad \text{and} \quad B \simeq \frac{1-\beta}{\alpha}.$$

Example 10.8(a). Suppose in Example 10.2(a), we wish to test the hypothesis

$$H_0: \quad \mu = \mu_0 = 1000$$

against the alternative

$$H_1: \quad \mu = \mu_1 = 1080,$$

with $\alpha = .01$ and $\beta = .01$. The latter condition implies that the power of the test is to be .99, that is, if the alternative hypothesis H_1 is true, we wish to reject H_0 with probability .99.

From Example 10.2(a) we note that

$$f_0(x) = \frac{1}{\sigma\sqrt{2\pi}} \exp\left[-\frac{1}{2}\left(\frac{x-\mu_0}{\sigma}\right)^2\right],$$

$$f_1(x) = \frac{1}{\sigma\sqrt{2\pi}} \exp\left[-\frac{1}{2}\left(\frac{x-\mu_1}{\sigma}\right)^2\right],$$

where, of course, $\sigma = 128$, $\mu_0 = 1000$, and $\mu_1 = 1080$.

Upon drawing x_1, we find

$$\frac{f_1(x_1)}{f_0(x_1)} = \exp\left[\frac{1}{2}\left(\frac{x_1-\mu_0}{\sigma}\right)^2 - \frac{1}{2}\left(\frac{x_1-\mu_1}{\sigma}\right)^2\right]$$

$$= \exp\left[\left(\frac{\mu_1-\mu_0}{\sigma^2}\right)x_1 + \frac{\mu_0^2-\mu_1^2}{2\sigma^2}\right],$$

and if we draw $x_1, \ldots, x_m$, we find

$$\frac{f_1(x_1)\cdots f_1(x_m)}{f_0(x_1)\cdots f_0(x_m)} = \exp\left[\frac{\mu_1-\mu_0}{\sigma^2}\sum_{i=1}^{m}x_i + \frac{m(\mu_0^2-\mu_1^2)}{2\sigma^2}\right].$$

Referring to (10.8.1) and (10.8.4) and the expression above, we see that sampling continues as long as

$$\frac{\beta}{1-\alpha} < \exp\left[\frac{\mu_1-\mu_0}{\sigma^2}\sum_{i=1}^{m}x_i + \frac{m(\mu_0^2-\mu_1^2)}{2\sigma^2}\right] < \frac{1-\beta}{\alpha}.$$

By taking natural logarithms, this inequality is equivalent to the following inequality:

$$\ln\frac{\beta}{1-\alpha} < \frac{\mu_1-\mu_0}{\sigma^2}\sum_{i=1}^{m}x_i + \frac{\mu_0^2-\mu_1^2}{2\sigma^2}m < \ln\frac{1-\beta}{\alpha},$$

where m is the number of observations already taken. Rearranging terms in the inequality gives:

$$(10.8.5) \qquad \left(\frac{\sigma^2}{\mu_1-\mu_0}\right)\ln\frac{\beta}{1-\alpha} + \left(\frac{\mu_1+\mu_0}{2}\right)m$$

$$< \sum_{i=1}^{m}x_i < \left(\frac{\sigma^2}{\mu_1-\mu_0}\right)\ln\frac{1-\beta}{\alpha} + \left(\frac{\mu_1+\mu_0}{2}\right)m.$$

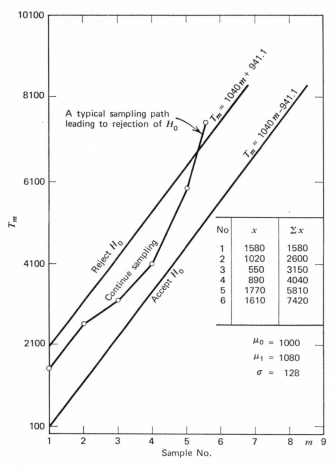

Figure 10.8.1. Decision regions for the sequential test of Example 10.8(*a*) showing a typical sampling path.

These inequalities may be displayed in a graph with $\sum_{i=1}^{m} x_i = T_m$ as ordinate and m as abscissa (see Figure 10.8.1). The lines having equations

$$(10.8.6) \qquad T_m = \frac{\sigma^2}{\mu_1 - \mu_0} \ln \frac{1 - \beta}{\alpha} + \frac{\mu_1 + \mu_0}{2} m$$

and

$$(10.8.7) \qquad T_m = \frac{\sigma^2}{\mu_1 - \mu_0} \ln \frac{\beta}{1 - \alpha} + \left(\frac{\mu_1 + \mu_0}{2}\right) m$$

are plotted, and, as the observations are taken, T_m is plotted against m.

Values for $\ln \dfrac{1 - \beta}{\alpha}$ and $\ln \dfrac{\beta}{1 - \alpha}$ for conventional values of α and β are displayed in Table 10.8.1.

TABLE 10.8.1

TABLE OF VALUES OF $\ln \dfrac{1 - \beta}{\alpha}$ (UPPER ENTRY IN CELL), AND $\ln \dfrac{\beta}{1 - \alpha}$,

FOR VARIOUS α AND β

		α				
		.005	.010	.025	.050	.100
β	.005	5.292 −5.292	4.599 −5.288	3.724 −5.272	2.990 −5.247	2.297 −5.193
	.010	5.288 −4.599	4.595 −4.595	3.679 −4.580	2.986 −4.554	2.293 −4.500
	.025	5.272 −3.724	4.580 −3.679	3.664 −3.664	2.970 −3.638	2.277 −3.583
	.050	5.247 −2.990	4.554 −2.986	3.638 −2.970	2.944 −2.944	2.251 −2.890
	.100	5.193 −2.297	4.500 −2.293	3.583 −2.277	2.890 −2.251	2.197 −2.197

Now it is true that for this type of sequential sampling scheme, the expected sample size is usually appreciably less than the sample size required for a sample with fixed n and the same α and β.

Example 10.8(a). Suppose in Example 10.2(a) we wish to test

$$H_0: \quad \mu = \mu_0 = 1000$$
$$H_1: \quad \mu = \mu_1 = 1080$$

with $\sigma = 128$, $\alpha = .01$ and $\beta = .01$. Then the two lines have equations:

$$T_m = 1040m + 941.1$$

and

$$T_m = 1040m - 941.1.$$

As long as the sequence of sample points (m, T_m), $m = 1, 2, \ldots$, falls inside the band formed by these two lines, sampling is continued. (See Figure 10.8.1.) As soon as a point falls in the upper-left region, the decision is made to stop sampling and reject H_0 (that is, to accept H_1). As soon as a point falls in the lower-right region, the decision is made to stop sampling and accept H_0 (that is, reject H_1).

If we define a new variable $y_i = x_i - \left(\dfrac{\mu_0 + \mu_1}{2}\right)$ then (10.8.5) becomes simply

$$(10.8.8) \quad \left(\frac{\sigma^2}{\mu_1 - \mu_0}\right) \ln \frac{\beta}{1 - \alpha} \leq \sum y_i \leq \left(\frac{\sigma^2}{\mu_1 - \mu_0}\right) \ln \frac{1 - \beta}{\alpha}.$$

The cumulative sum $\sum y_i$ can then be plotted in a horizontal fashion analogous to *Shewhart Charts* (these are discussed in Chapter 11) with the upper and lower control limits given by the limits of the inequality (10.8.8) as illustrated in Figure 10.8.2 for the data of Example 10.8(a).

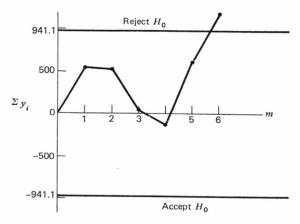

Figure 10.8.2. Decision regions for the sequential test for data given in Figure 10.8.1 where $y_i = x_i - (\mu_0 + \mu_1)/2$.

In practice it is usual not to plot $\sum y$ whenever it is negative. Such a rule has the effect of restarting the sequential plot with each new observation until a response $x > (\mu_0 + \mu_1)/2$ is recorded. Plotting is then continued so long as $\sum y$ remains positive and less than the upper critical boundary. Often, plotting is completely dispensed with and only the cumulative sum, when positive, recorded. As soon as the critical total $\dfrac{\sigma^2}{\mu_1 - \mu_0} \ln \dfrac{1 - \beta}{\alpha}$ is exceeded, the process is rejected.

It should be pointed out that we can similarly define a sequential test if the random variable on which we make our sequence of independent observations is discrete. That is, if we use probability functions $p_0(x)$ and $p_1(x)$ rather than p.d.f. $f_0(x)$ and $f_1(x)$.

Example 10.8(b).   A certain process yields mass-produced items that have $100\theta\%$ defective, θ unknown.   The manufacturer would like to make a sequential test of the hypothesis

(10.8.9) $$H_0: \quad \theta = \theta_0$$

against the alternative

(10.8.9a) $$H_1: \quad \theta = \theta_1 > \theta_0$$

at significance level α and power $1 - \beta$ if $\theta = \theta_1$, that is, the desired value of the probability of type-II error is to be β if $\theta = \theta_1$.

Let x be a random variable having the value 1 if an item drawn from the process is defective, and the value 0 if the item is nondefective.   The probability function of x is given by

$$f(x) = \theta^x (1 - \theta)^{1-x}, \quad x = 0, 1.$$

If we draw m items, let the values of x obtained be $x_1, \ldots, x_m$, in which case

$$f(x_1) \cdots f(x_m) = \theta^{T_m}(1 - \theta)^{m-T_m},$$

where $T_m = \sum\limits_{i=1}^{m} x_i$ and is the total number of defectives among the m items drawn.

Now let

$$f_1(x) = \theta_1^x (1 - \theta_1)^{1-x}$$

and

$$f_0(x) = \theta_0^x (1 - \theta_0)^{1-x}.$$

Referring to (10.8.1) and (10.8.4), we see that sampling continues as long as

$$\frac{\beta}{1 - \alpha} < \frac{f_1(x_1) \cdots f_1(x_m)}{f_0(x_1) \cdots f_0(x_m)} < \frac{1 - \beta}{\alpha},$$

that is, as long as

(10.8.10) $$\frac{\beta}{1 - \alpha} < \frac{\theta_1^{T_m}(1 - \theta_1)^{m-T_m}}{\theta_0^{T_m}(1 - \theta_0)^{m-T_m}} < \frac{1 - \beta}{\alpha}.$$

Taking logarithms to base e, we have (10.8.10) is equivalent to:

$$\log \frac{\beta}{1 - \alpha} < T_m \log \frac{\theta_1}{\theta_0} + (m - T_m) \log \frac{1 - \theta_1}{1 - \theta_0} < \log \frac{1 - \beta}{\alpha}.$$

If we plot in the (m, T_m)-plane the lines

$$T_m \log \frac{\theta_1}{\theta_0} + (m - T_m) \log \frac{1 - \theta_1}{1 - \theta_0} = \log \frac{\beta}{1 - \alpha}$$

and

$$T_m \log \frac{\theta_1}{\theta_0} + (m - T_m) \log \frac{1 - \theta_1}{1 - \theta_0} = \log \frac{1 - \beta}{\alpha},$$

then as long as the sequence of sample points $(m, T_m), m = 1, 2, \ldots$, fall inside the band formed by these two lines, we continue sampling.   As soon as a point falls above the upper line, we reject H_0 (and accept H_1) or, as soon as a point falls in the region below the lower line, we accept H_0 (and reject H_1).

10.9 A TWO-SIDED SEQUENTIAL TESTING PROCEDURE

The one-sided sequential test can be easily adapted to the problem where we wish to test the hypothesis H_0: $\mu = \mu_0$ against the two-sided alternative H_1: $\mu = \mu_0 + \delta$, where $\delta = \pm\delta_0$, say, with Type I and Type II errors of α and β, respectively. As in the case of the fixed sample size tests, it is customary to share the α risk equally across both alternative hypotheses. The testing scheme then becomes two one-sided schemes with risks $\alpha/2$ and β. Further, if we let $y = x - \mu_0$ the two-sided sequential plot can be displayed in a single graph as illustrated in Figure 10.9.1. Employing (10.8.8) the equations for the upper and lower pairs of boundary lines are given by

$$(10.9.1) \quad \left(\frac{\sigma^2}{\delta}\right) \ln \frac{\beta}{1 - \alpha/2} + \frac{\delta}{2} m < \sum y_i < \frac{\sigma^2}{\delta} \ln \frac{1 - \beta}{\alpha/2} + \frac{\delta}{2} m$$

where the δ is the quantity specified in H_1.

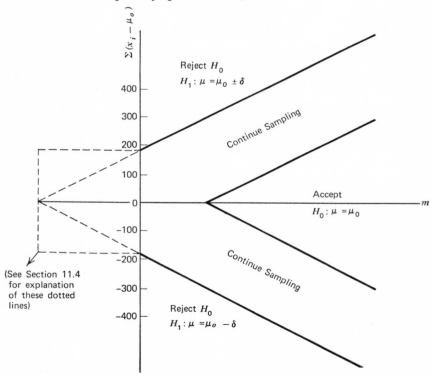

Figure 10.9.1. Two-sided Wald sequential test of hypothesis. H_0: $\mu = \mu_0 = 500$; H_1: $\mu = \mu_0 \pm \delta$, $\delta = 50$, $\sigma = 50$, $\alpha = 0.05$, $\beta = 0.10$.

Example 10.9(a). Suppose $\mu_0 = 500$, $\sigma = 50$, $\delta = \pm 50$, $\alpha = .05$ and $\beta = .10$, then we obtain for the upper pair of boundary lines in Figure 10.9.1 (here $T_m = \Sigma\, y_i$)

$$T_m = 50(3.583) + 25m = 179.2 + 25m$$
$$T_m = 50(-2.277) + 25m = -113.9 + 25m$$

and for the lower pair of boundary lines

$$T_m = -179.2 - 25m$$
$$T_m = 113.9 - 25m.$$

As illustrated in Figure 10.9.1, so long as the cumulative sum $\Sigma\, (x_i - \mu_0)$ stays within the region interior to the two pairs of parallel control lines, another observation is taken. If the cumulative sum falls outside the boundaries provided by the outer lines, the hypothesis H_0: $\mu = \mu_1$ is rejected. The hypothesis H_0 is accepted if the cumulative sum crosses into the region defined by the intersection of the boundaries which form the V-shaped region found to the right in the figure.

PROBLEMS

10.1. Suppose a certain type of 40-watt bulb has been standardized so that the mean life of the bulb is 1500 hours and the standard deviation is 200 hours. A random sample of 25 of these bulbs from lot L having mean μ was tested and found to have a mean life of 1380 hours.

(a) Test at the 1% significance level the hypothesis

$$H_0: \quad \mu = 1500$$

against the alternatives

$$H_1: \quad \mu = \mu_1 < 1500.$$

(b) What is the power of the test at $\mu = 1400$?
(c) Graph the power function.

10.2. An existing process used to manufacture paint yields daily batches which have been fairly well established to be normally distributed with mean $\mu = 800$ tons, $\sigma = 30$ tons. A modification of this process is suggested with the view of increasing production. Assuming that daily yields using the modified process are distributed as $N(\mu, 30^2)$, and a sample of 100 randomly chosen days production using the modified process has a sample mean of $\bar{x} = 812$ tons, test at the 1% level of significance

$$H_0: \quad \mu = 800$$

against the alternatives

$$H_1: \quad \mu = \mu_1 > 800.$$

What is the power of the test at $\mu = 810$? Graph the power function.

10.3(a). A machine used for producing "one-quarter" inch rivets is to be checked by taking a random sample of 10 rivets and measuring the diameters of the 10 rivets. It is feared that the "wear-off" factor of the machine will cause it to produce rivets with diameters less than $\frac{1}{4}$ of an inch. Describe the critical region in terms of $\bar{x}$, the sample mean of the 10 diameters, for a test at level of significance of 1%, of

$$H_0: \quad \mu = .25$$

against the alternatives

$$H_1: \quad \mu = \mu_1 < .25,$$

if it can be assumed that the diameters are distributed as $N(\mu, (.0015)^2)$ for a wide range of values of μ. What is the power at $\mu = .2490$? Graph the power function of the test.

10.3(b). Suppose we wish a test of $H_0:$ $\mu = .25$ against $H_1:$ $\mu = .2490$ to have size $\alpha = .01$ and a power at $\mu = .2490$ of .99, that is, $\beta = .01$. What sample size is necessary to achieve this?

10.3(c). Generalize 10.3(b), that is, suppose sampling from $N(\mu, \sigma_0^2)$ where σ_0^2 denotes the known value of the variance of the population. Suppose we wish to test $H_0:$ $\mu = \mu_0$ against $H_1:$ $\mu = \mu_1$, where $\mu_1 < \mu_0$, so that the level of the test is α and $\gamma(\mu_1) = 1 - \beta$. Show that the sample size n needed to achieve this is such that

$$\sqrt{n} = \frac{(z_\alpha + z_\beta)\sigma_0}{(\mu_0 - \mu_1)}.$$

10.4. A method for determining percent of impurity in various types of solutions is available, and is known to give determinations which have a known standard deviation of .03. Seven determinations on a certain chemical yield the values 7.18, 7.17, 7.12, 7.13, 7.14, 7.15, and 7.16. It is important that the chemical not have more than 7.13% of impurities. Assuming normality, test at the 5% level of significance, the hypothesis

$$H_0: \quad \mu = 7.13 \quad \text{against} \quad H_1: \quad \mu = \mu_1 > 7.13.$$

State the critical region explicitly and graph the power function of the test.

10.5. A certain process yields pieces of steel strand with population standard deviation of their breaking strength equal to 500 psi. A random sample of nine test pieces of strands from the process yields $\bar{x} = 12,260$ psi. If μ is the mean of the process, test at the 5% level of significance the hypothesis

$$H_0: \quad \mu = 13,500 \text{ psi}$$

against the alternatives

$$H_1: \quad \mu \neq 13,500 \text{ psi}$$

and graph the power function of the test.

10.6. Machines used in producing a particular brand of yarn are given periodic checks to help insure stable quality. A certain machine has been set in such a way that it is expected that strands of the yarn will have breaking strength of $\mu = 19.50$ ounces, with a standard deviation of $\sigma = 1.80$ ounces. (It has been found from experience that σ remains steady at 1.80 over a wide range of values of μ.) A sample of twelve pieces of yarn selected randomly yields a sample mean of breaking strengths of 18.46 ounces. Assuming normality, test at the 5% level of significance, the hypothesis

$$H_0: \quad \mu = 19.50$$

against the alternatives

$$H_1: \quad \mu \neq 19.50.$$

Graph the power function of the test.

10.7. A liquor company is concerned about the filling of its bottles with too little liquid or too much liquid. A filling machine, which has been set to fill "40-ounce" bottles is checked by selecting at random ten bottles and measuring their contents. Assuming that the machine fills the bottles with quantities that are distributed as $N(\mu, 1.44)$ for a wide range of values of μ, state the critical region for a 1% test of significance of

$$H_0: \quad \mu = 40$$

against the alternatives

$$H_1: \quad \mu \neq 40.$$

Graph the power function of the test.

10.8. A diamond-cutting machine has been set to turn out diamonds of ".5-karat" quality. Assume on the basis of past experience that the machine produces diamonds which have quality that has the $N(\mu, .0036)$ distribution. It is important to the jewelry supply house that the quality not be too low (dissatisfied customers), or two high (economic considerations). Accordingly, the machine is checked every so often. A recent sample of six diamonds yielded weights of .48, .59, .54, .50, .55, and .56 karats. Test at the 5% level of significance, the hypothesis

$$H_0: \quad \mu = .5$$

against the alternatives

$$H_1: \quad \mu \neq .5.$$

Graph the power function.

10.9. Ten determinations of percentage of water in a certain solution yielded $\bar{x} = .452\%$ and $s = .37\%$. If μ is the "true" percentage of water in the solution, test at the 5% level of significance the hypothesis

$$H_0: \quad \mu = .5\%$$

against the alternatives

$$H_1: \quad \mu = \mu_1 < .5\%.$$

10.10. A random sample of 50 water canteens of soldiers in a desert region show an average life of 4.34 years with a standard deviation of 1.93 years. Army experience is such that canteens are known to have an average life of 4.90 years. Test the hypothesis, at the 5% level of significance, that "canteen life" in the desert is 4.90 years, against the hypothesis that desert conditions decrease the life of the canteen.

10.11. Hourly wages of employees in a certain company have become a source of contention. An impartial arbitrator finds that industry-wide, hourly wages are approximately normally distributed with a mean of $11.65 per hour. The arbitrator examines the earning records of 40 workers, selected at random from the payroll list of the company. He finds that the sample mean is $11.53, with a sample standard deviation of $0.30. Test at the 1% level of significance, the assertion (hypothesis) of the company that their wages conform to industry practices, against the assertion (alternative hypothesis) that wages in this company are lower than that of the industry.

10.12. An aptitude test has been given over the past many years with an average performance of 90. A group of 30 students are preparing for this test and are taught with special emphasis on remedial reading. The thirty students obtain a sample mean of 94.2, with a sample standard deviation of 8.5. Has the remedial reading emphasis helped (at the 1% level of significance)?

10.13. A company is engaged in the "stewed-fruit" canning business. One of its brands is a medium sized tin of cooked prunes, which are advertised as containing 20 ounces of prunes. The company must be sure that it packs prunes into the tins so that average weight is not "too much" under, or over, 20 ounces (the F.D.A. would impose stiff fines in the former case). A random sample of 14 cans yields a mean of $\bar{x} = 20.82$ ounces, with a standard deviation of 2.20 ounces. Assuming that the weights of tins of prunes are distributed as $N(\mu, \sigma^2)$, test at the 5% level the hypothesis

$$H_0: \quad \mu = 20$$

against the alternatives

$$H_1: \quad \mu \neq 20.$$

10.14. Five determinations of percent of nickel in a prepared batch of ore produced the following results:

$$3.25, \ 3.27, \ 3.24, \ 3.26, \ 3.24.$$

If μ is the "true percent of nickel in the batch," test the hypothesis $H_0: \quad \mu = 3.25$ against the alternatives $H_1: \quad \mu \neq 3.25$ at the 1% level of significance.

10.15. Nine determinations were made by a technician of the melting point of manganese with the following results: 1268, 1271, 1259, 1266, 1257, 1263, 1272, 1260, 1256 (in degrees centigrade). Test at the 5% level of significance the hypothesis that the results are consistent with the published value of $1260°C$ as the true melting point of manganese.

10.16. Suppose a plot of land is surveyed by five student surveyors and they find the following areas for the plot (in acres): 7.27, 7.24, 7.21, 7.28, 7.23. On the basis of this information, test the hypothesis that the true area of the plot is 7.23 acres at the 5% level of significance.

10.17. A producer claims that the diameters of pins he manufactures have a standard deviation of .05 inch. A sample of nine pins has a sample standard deviation of .07 inch. Is this sample value of s significantly larger than the claimed value of σ, at the 5% level of significance?

10.18. (a) Referring to the data of Problem 10.9, test at the 5% level of significance, the hypothesis $\sigma = 0.4\%$ against the alternatives $\sigma < 0.4\%$.

(b) Referring to the data of Problem 10.14, test at the 5% level of significance, the hypothesis $\sigma = .01$ against the alternatives $\sigma > .01$.

(c) Referring to the data of Problem 10.15, test at the 5% level of significance the hypothesis $\sigma^2 = 40$ against the alternatives $\sigma^2 < 40$.

(d) Referring to the data of Problem 10.16, test at the 5% level of significance, the hypothesis $\sigma^2 = .9$ against the alternatives that $\sigma^2 \neq .9$.

10.19. The standard deviation s of muzzle velocities of a random sample of nine rounds of ammunition was found to be 93.2 feet per second. If the "standard" value of σ for the muzzle velocity of this type of ammunition is 70 feet per second, is the value of s significantly large at the 5% level of significance?

10.20. A method for determining percentage of iron in mixed fertilizer is available and long experience with the method shows that its determinations are normally distributed with standard deviation of 0.12 percent. A company producing a certain type of fertilizer wishes to compare the findings of its laboratory with that of a state laboratory. The results are:

Company Lab.: 8.84%, 8.86%, 9.16%.

State Lab.: 8.78%, 8.96%, 8.62%.

Test at the 5% level of significance the hypothesis that the both laboratories do equivalent analysis, against the hypothesis that the state laboratory has a downward bias relative to the company laboratory.

10.21. A machine is used to package "4-ounce" boxes of a certain brand of gelatin powder. A modification is suggested to increase the speed of the operation, but there is some concern that the modified settings will cause the machine to fill the boxes with less powder than before. Accordingly, 50 boxes are filled before and after modification with the following results:

Before: $n_1 = 50$; $\bar{x}_1 = 4.091$.

After: $n_2 = 50$; $\bar{x}_2 = 4.075$.

Assuming that the machine yields packages whose weights are $N(\mu, (.05)^2)$ for a wide range of values of μ, test at the 5% level of significance the hypothesis

$$H_0: \quad \mu(\text{after}) = \mu(\text{before})$$

against the alternatives

$$H_1: \quad \mu(\text{after}) < \mu(\text{before}).$$

Graph the power function of the test.

10.22. It is known from past experience that two machines, say A and B, used in producing a certain type of thread, have standard deviations of .04 and .03, respectively. The settings of the two machines are changed, and the concern is that they were both set alike. To check this, samples of 10 pieces of thread from machine A and 15 pieces of thread from machine B are taken at random and it is found that $\bar{x}_A = 25.34$ and $\bar{x}_B = 25.42$. Test the hypothesis

$$H_0: \quad \mu_A = \mu_B$$

against the alternatives

$$H_1: \quad \mu_A \neq \mu_B$$

at the 5% level of significance. Graph the power function of the test.

10.23. Suppose random samples of 25 are taken from two large lots of bulbs, say A and B, and that $\bar{x}_A = 1610$ hours and $\bar{x}_B = 1455$ hours. Assuming that the standard deviation of bulb life is 200 hours, test at the 5% level of significance the hypothesis

$$H_0: \quad \mu_A - \mu_B = 120$$

against the alternatives

$$H_1: \quad \mu_A - \mu_B \neq 120.$$

Graph the power function. What is the power if $\mu_A - \mu_B = \pm 100$?

10.24. Nicotine determinations were made on each of four standard units of tobacco at each of two laboratories, say A and B, with the results shown in Table P10.24.

TABLE P10.24

Laboratory *A* Nicotine Content (grams)	Laboratory *B* Nicotine Content (grams)
26, 24, 28, 27	28, 31, 23, 29

Test the hypothesis that the results of the two laboratories are not significantly different at the 5% level of significance. Assume equal variances.

10.25. An experiment to determine the viscosity of two different brands of car oil, *A* and *B*, gives the results shown in Table P10.25.

TABLE P10.25

Brand *A*, Viscosity	Brand *B*, Viscosity
10.28	10.31
10.27	10.31
10.30	10.26
10.32	10.30
10.27	10.27
10.27	10.31
10.28	10.29
10.29	10.26

(a) Test the hypothesis that $\sigma_A^2/\sigma_B^2 = 1$ at the 5% level of significance. (Assume normality of the two populations.)

(b) Using the result of (a), test the hypothesis H_0: $\mu_A - \mu_B = 0$ at the 5% level of significance, against the alternatives H_1: $\mu_A \neq \mu_B$.

10.26. Resistance measurements were made on test pieces selected from two lots L_1 and L_2, with the results shown in Table P10.26.

TABLE P10.26

Lot L_1 (ohms)	Lot L_2 (ohms)
.140	.135
.138	.140
.143	.142
.142	.136
.144	.138
.137	.140

If μ_1 and μ_2 are the means, and σ_1^2 and σ_2^2 are the variances of resistance measurements in L_1 and L_2 respectively, and assuming normality:

(a) Test the hypothesis H_0: $\sigma_1^2/\sigma_2^2 = 1$ against the alternatives H_1: $\sigma_1^2/\sigma_2^2 \neq 1$, at the 1% level of significance.

(b) Using (a), test the hypothesis H_0: $\mu_1 - \mu_2 = 0$ against the alternatives H_1: $\mu_1 - \mu_2 \neq 0$ at the 5% level of significance.

10.27. Two methods were used in a study of the latent heat of fusion of ice. Both method A (an electrical method) and method B (a method of mixtures) were conducted with the specimens cooled to $-0.72°C$. The data shown in Table P10.27 represents the change in total heat from $-0.72°C$ to water at $0°C$, in calories per gram of mass.

TABLE P10.27

Method A (calories/gram)		Method B (calories/gram)	
79.98	79.97	80.02	79.97
80.04	80.05	79.94	80.03
80.02	80.03	79.98	79.95
80.04	80.02	79.97	79.97
80.03	80.00		
80.03	80.02		
80.04			

Assuming normality of the population of determinations for each method as well as equal variances, test at the 5% level of significance the hypothesis that the mean of method A is the same as that for method B.

10.28 Two methods of preparing fish, A and B, are compared according to a specific scoring scheme. The claim has been made that $\mu_A > \mu_B$. The composite scores of two samples are shown in Table P10.28.

TABLE P10.28

Method A		Method B	
4.05	4.18	3.31	2.35
5.04	4.35	3.39	2.59
3.45	3.88	2.24	4.48
3.57	3.02	3.93	3.93
4.23	4.56	3.37	3.43
4.23	4.37	3.21	3.13

(a) Assuming normality of scores generated by A and B, test at the 1% level of significance, the hypothesis that $\sigma_A^2 = \sigma_B^2$.
(b) Using the result in (a), test the hypothesis H_0: $\mu_A = \mu_B$ against the alternatives H_1: $\mu_A > \mu_B$.

10.29. Elongation measurements are made on ten pieces of steel, five of which were treated with method A (aluminum plus calcium) and the remaining five were treated with method B (aluminum only). It is conjectured that the addition of calcium will improve elongations by at least 1%. The results of the measurements are shown in Table P10.29.

TABLE P10.29

Method A, %	Method B, %
34	28
27	29
30	25
26	23
33	30

Assuming normality:

(a) Test the hypothesis $\sigma_A^2 = \sigma_B^2$ at the 5% level of significance.

(b) Using (a), test the hypothesis that $\mu_A - \mu_B = 1\%$ against the alternative that $\mu_A - \mu_B > 1\%$.

10.30. A comparison of yields of marigolds from control plots and treated plots is carried out. Samples of 8 control plots and 8 treated plots yield the following data:

$$\textit{Treated:} \quad n_1 = 8; \quad \bar{x}_1 = 128.4; \quad s_1^2 = 117.1.$$
$$\textit{Not treated:} \quad n_2 = 8; \quad \bar{x}_2 = \;\;96.5; \quad s_2^2 = 227.7.$$

Assuming normality:

(a) Test at the 1% level of significance, $\sigma_1^2 = \sigma_2^2$.

(b) Using the result of (a), test the hypothesis H_0: $\mu_1 - \mu_2 = 0$ against the alternatives H_1: $\mu_1 - \mu_2 > 0$.

10.31. Viewing times of members of households in two different types of communities are contrasted, with the following results:

$$\textit{Community 1:} \quad n_1 = 40; \quad \bar{x}_1 = 19.2 \text{ hours/week}; \quad s_1 = 6.4.$$
$$\textit{Community 2:} \quad n_2 = 50; \quad \bar{x}_2 = 15.9 \text{ hours/week}; \quad s_2 = 3.2.$$

(a) Test at the 1% level of significance, the hypothesis that $\sigma_1^2 = \sigma_2^2$.

(b) Using the result of (a), test the hypothesis that $\mu_1 = \mu_2$ against the alternative $\mu_1 > \mu_2$. (Community 1 is a working class district in London and Community 2 is Oxford.)

10.32. It has been suspected for some time that the morning shift is more efficient than the afternoon shift. Random observations yield the following data:

$$\textit{Morning shift:} \quad n_1 = 5; \quad \bar{x}_1 = 22.9; \quad s_1^2 = .675.$$
$$\textit{Afternoon shift:} \quad n_2 = 7; \quad \bar{x}_2 = 21.5; \quad s_2^2 = 7.25.$$

Assuming normality:

(a) Test the hypothesis $\sigma_1^2 = \sigma_2^2$ at the 1% level of significance.

(b) Using (a), test the hypothesis H_0: $\mu_1 = \mu_2$ against the alternatives H_1: $\mu_1 > \mu_2$.

10.33. A sample of 220 items turned out during a given week by a certain process had mean weight of 2.46 pounds and standard deviation .57 pounds. During the next week a different lot of raw material was used, and the mean weight of a sample of 205 items turned out that week was 2.55 pounds and the standard deviation was .48 pounds. Assuming normality and equality of variances, would you conclude from these results that the mean weight of the product had increased significantly at the 5% level of significance during the second week?

10.34. Orange juice cans are to be filled using two different methods. The results are as follows:

$$Method\ 1:\ n_1 = 40;\ \bar{x}_1 = 21.78;\ s_1^2 = 3.11.$$
$$Method\ 2:\ n_2 = 40;\ \bar{x}_2 = 20.71;\ s_2^2 = 2.40.$$

(a) Test at the 5% level, H_0: $\sigma_1^2 = \sigma_2^2$ against H_1: $\sigma_1^2 \neq \sigma_2^2$.
(b) Test, using the result of (a), the hypothesis H_0: $\mu_1 = \mu_2$, against the alternatives H_1: $\mu_1 > \mu_2$.

10.35. The systolic blood pressure of a group of 70 patients showed $\bar{x}_1 = 145$ and $s_1 = 14$. A second group of 70 patients is given a certain drug, and they showed $\bar{x}_2 = 140$, with $s_2 = 9$. Assuming normality:
(a) Test at the 5% level of significance the hypothesis $\sigma_1^2 = \sigma_2^2$ against $\sigma_1^2 \neq \sigma_2^2$.
(b) Using (a), test the hypothesis H_0: $\mu_1 - \mu_2 = 0$ against the alternatives H_1: $\mu_1 - \mu_2 > 0$. Comment.

10.36. Two randomly selected groups of 70 trainees are taught a new assembly line operation by two different methods, with the following results when the groups are tested:

$$Group\ 1:\ n_1 = 70;\ \bar{x}_1 = 268.8;\ s_1 = 20.2.$$
$$Group\ 2:\ n_2 = 70;\ \bar{x}_2 = 255.4;\ s_2 = 26.8.$$

(a) Is s_2 significantly greater than s_1 at the 1% level?
(b) Using the answer of (a), test at the 5% level of significance, the hypothesis H_0: $\mu_1 = \mu_2$ against the alternative H_1: $\mu_1 > \mu_2$.

10.37. The data shown in Table P10.37 are the results of determining iron content of ore using two methods A (dichromate) and B (thioglycolate) on ten different samples of ore.

TABLE P10.37

Ore Number	Percent Iron by Method A	Percent Iron by Method B
1	28.22	28.27
2	33.95	33.99
3	38.25	38.20
4	42.52	42.42
5	37.62	37.64
6	37.84	37.85
7	36.12	36.21
8	35.11	35.20
9	34.45	34.40
10	52.83	52.86

Use the method of paired comparisons to test whether these two methods yield significantly different percentages of iron at the 5% level of significance.

10.38. Analysts I and II each make a determination of the melting point of hydroquinine on each of eight specimens of hydroquinine (in degrees centigrade) with the results shown in Table P10.38.

TABLE P10.38

Specimen Number	Analyst I (°C)	Analyst II (°C)
1	174.0	173.0
2	173.5	173.0
3	173.0	172.0
4	173.5	173.0
5	171.5	171.0
6	172.5	172.0
7	173.5	171.0
8	173.5	172.0

Using the method of paired comparisons, do the methods of determinations differ significantly at the 5% level of significance?

10.39. Over a long period of time, ten patients selected at random are given two treatments for a specific form of arthritis. The results were (coded units):

Patient	Treatment 1	Treatment 2
1	47	52
2	38	35
3	50	52
4	33	35
5	47	46
6	23	27
7	40	45
8	42	41
9	15	17
10	36	41

Is there a difference in treatments?

10.40. Two different methods of storing chicken are contrasted by applying technique 1, a freezing technique to one half of a chicken, and technique 2, a wrapping technique to the other half of the same chicken. Both halves are stored for three weeks and a certain "tenderness of the meat" test is then applied. This was done for $n = 200$ chickens, and [using the notation of section 10.3(d)] it was found that $\bar{x} = -2.430$ with $s_x = .925$. Test the hypothesis, at the 5% level of significance, that $\mu_x = -1$ against the alternative $\mu_x \neq -1$.

10.41. The dependability of analysts is sometimes measured by the variability of their work. Two analysts A and B each make 10 determinations of percent of iron content in a batch of prepared ore from a certain deposit. The sample variances obtained are $s_A^2 = .4322$ and $s_B^2 = .5006$, respectively. Are the analysts as dependable as each other? Test at the 5% level of significance.

10.42. (a) In Problem 10.24, it was assumed that $\sigma_A^2 = \sigma_B^2$. Test that assumption at the 1% level.

(b) In Problem 10.27, is the assumption that $\sigma_A^2 = \sigma_B^2$ warranted on the basis of the data? Use the significance level of .05.

(c) In Problem 10.33, is the assumption of equality of variances warranted? (Remember: interpolation in F tables should be carried out using the reciprocals of degrees of freedom.) Use the significance level of .01.

(d) Using the data of Table P10.37, test at the 1% level, the hypothesis $\sigma_A^2 = \sigma_B^2$.

(e) Using the data of Table P10.38, test at the 1% level, the hypothesis $\sigma_I^2 = \sigma_{II}^2$.

10.43. If x is a random variable which has p.f.

$$f(x \mid \mu) = \frac{e^{-\mu}\mu^x}{x!}, \qquad x = 0, 1, 2, \ldots,$$

describe how to sequentially test the hypothesis

$$H_0: \quad \mu = \mu_0 = 1.5$$

against the alternative

$$H_1: \quad \mu = \mu_1 = 2.0,$$

with $\alpha = .05$, $\beta = .01$.

10.44 If x has p.f. $f(x \mid \theta) = \binom{n}{x}\theta^x(1 - \theta)^{n-x}$, $x = 0, 1, \ldots, n$, describe how to sequentially test the hypothesis

$$H_0: \quad \theta = .10$$

against the alternative

$$H_1: \quad \theta = .20,$$

with $\alpha = .10$, $\beta = .05$.

10.45. If x has p.d.f. $f(x; \theta)$ given by

$$f(x \mid \sigma^2) = \frac{1}{\sqrt{2\pi\sigma^2}}e^{-(1/2\sigma^2)(x-\mu^\circ)^2},$$

where μ° is the known value of the population mean, describe how to sequentially test the hypothesis

$$H_0: \quad \sigma^2 = \sigma_0^2$$

against the alternative

$$H_1: \quad \sigma^2 = \sigma_1^2 > \sigma_0^2$$

with $\alpha = .10$, $\beta = .05$.

10.46. Recall from Theorem 7.3.1 of Chapter 7 that for large n we may write $z \simeq \dfrac{n\hat{\theta} - n\theta}{\sqrt{n\theta(1-\theta)}}$, or, for large n, $z \simeq \dfrac{\hat{\theta} - \theta}{\sqrt{\dfrac{\theta(1-\theta)}{n}}}$, where $\hat{\theta}$ is the observed sample proportion. What then, would be the approximate test of level α for testing

(a) $\theta = \theta_0$ against $\theta = \theta_1 > \theta_0$

(b) $\theta = \theta_0$ against $\theta \neq \theta_0$,

if n is large?

10.47. A well-known politician claims that he has 55% of the votes with him on a certain issue. A private poll of 1000 voters yields a sample proportion of $\hat{\theta} = .51$. Could the well-known politician be right?

10.48. Suppose sampling from two populations, say 1 and 2, which are such that the proportions of the populations having a certain characteristic A is θ_1 and θ_2 in populations 1 and 2, respectively. Suppose that n_1 independent observations are taken from population 1 and, independently, n_2 observations are taken from population 2, and that the sample proportions are $\hat{\theta}_1$ and $\hat{\theta}_2$, respectively. Show that if $\theta_1 = \theta_2 = \theta$, then

$$\hat{\theta}_1 - \hat{\theta}_2 \simeq \sqrt{\theta(1-\theta)} \sqrt{\frac{1}{n_1} + \frac{1}{n_2}} \, z$$

where $z = N(0, 1)$, and n_1 and n_2 are both large.

10.49. The result of Problem 10.48 may be used to test the hypothesis H_0: $\theta_1 = \theta_2 = \theta$. Under H_0, we have, from Problem 10.48, that

$$\frac{\hat{\theta}_1 - \hat{\theta}_2}{\sqrt{\theta(1-\theta)} \sqrt{\dfrac{1}{n_1} + \dfrac{1}{n_2}}} \simeq z$$

for large n_1 and n_2. However, θ is unknown, but if n_1 and n_2 are large, we may estimate it by the *pooled* estimator $\hat{\theta}_p = \dfrac{n_1\hat{\theta}_1 + n_2\hat{\theta}_2}{n_1 + n_2}$. That is, for large n_1 and n_2, we have if H_0 is true,

$$\frac{\hat{\theta}_1 - \hat{\theta}_2}{\sqrt{\hat{\theta}_p(1 - \hat{\theta}_p)} \sqrt{\dfrac{1}{n_1} + \dfrac{1}{n_2}}} \simeq z.$$

Use this to state an approximate α-level test (n_1, n_2 large) for

(a) testing H_0: $\theta_1 = \theta_2$ against H_1: $\theta_1 > \theta_2$

(b) testing H_0: $\theta_1 = \theta_2$ against H_1: $\theta_1 \neq \theta_2$.

10.50. During the month of January, an electronics firm turned out 1350 printed circuits of which 146 were found to be defective. The following March, it turned out 1300 circuits, of which 113 were found to be defective. Assuming randomness, has the production process improved from January to March? (That is, test the hypothesis $\theta_1 = \theta_2$ against the alternative $\theta_1 > \theta_2$ at the 1% level.)

CHAPTER 11

Control Charts

11.1 INTRODUCTION

In any repetitive manufacturing operation, almost any measurement which can be made on individual items varies from item to item. For instance, if measured to thousandths of an inch, the diameters of $\frac{1}{4}$-inch ball bearings successively manufactured vary from one ball bearing to the next. No matter how refined the manufacturing process, some variation remains in any practical situation. About the best the manufacturer can hope is to refine the process to the point where the measurements made on each of the successive objects behave as measurements drawn at random from a population having the desired mean and a satisfactorily small variance.

In general, if $x_1, x_2, \ldots, x_n$ are the measurements thus obtained on a sequence of n successively drawn objects, and if the manufacturing process has been sufficiently refined, this set of measurements should behave like independent random variables all having identical distributions, that is, $x_1, x_2, \ldots, x_n$ should behave like a random sample of size n from some distribution.

If measurements made on a sequence of manufactured objects behave like a sequence of independent and identically distributed random variables, the manufacturing process is said to be in a *state of statistical control* or *under statistical control* with respect to the characteristic measured. As indicated above, the best that a manufacturer can do in refining his process is to locate and eliminate causes of the wider variations of any important measurement on successively produced objects to the point where the process is in a state of statistical control with respect to the characteristic being measured. The concept of a process being in a state of statistical control is, of course, applicable not only to manufacturing processes but also to processes involving repetitive operations.

275

In most mass-production situations it is not economically feasible to measure every object produced. The manufacturer thus resorts to taking small samples of objects from the production line, making measurements on the objects in these samples, and deciding from this information whether the process is in a state of statistical control. A simple, effective, and widely used device for analyzing and interpreting fluctuations of measurements on successive random samples of objects from the production line is the *control chart*. One use for a control chart is to maintain surveillance over an important measurement on a succession of mass-produced objects which supposedly have been previously brought into a state of statistical control with known parameters. In such a situation there has usually been an evolution of the manufacturing process to the point where the distribution of the measurements is approximately known, or at least the main characteristics of the distribution, such as its mean and variance, are known. Control charts used for this purpose are called *theoretical* control charts. They are essentially devices for detecting important departures from an existing known state of statistical control.

Another major use of control charts is to assist in establishing a state of statistical control over a given measurement on a sequence of manufactured objects. Control charts used for this purpose depend on computations made from a historical sequence of samples of measurements and are called *empirical* control charts.

Procedures for constructing control charts are different for the two kinds of situations described above, and we shall now consider them in detail.

11.2 THEORETICAL CONTROL CHARTS FOR SURVEILLANCE IN THE CASE OF AN EXISTING STATE OF STATISTICAL CONTROL WITH KNOWN PARAMETERS

(a) Control Chart for any Sample Statistic

Suppose we have a manufacturing process which is already in a state of statistical control with respect to some measurement, but which could be thrown out of control at any time by some cause or set of causes. We want to construct a control chart for the given measurement so that, if the process is thrown out of control with respect to that measurement, the control chart will detect this event with reasonably high probability. More precisely, suppose a process is under statistical control and that a measurement is made on each object in a random sample of size n taken from a given part of the production line at a given time. If we compute some statistic w (for instance, sample mean $\bar{x}$ or a sample sum T) from the sample, then, in successive samples, the values of w, say $w_1, w_2, \ldots$, will behave as independent random variables having identical distributions. The common distribution will have a mean μ_w and a standard deviation σ_w, which will be known if the distribution of measurements on the successive objects manufactured under the state of statistical control is known. In many situations w will have an approximately normal distribution with mean μ_w and standard deviation σ_w. In such a case we can compute the probability that w will fall in any given interval, assuming of course, that the process remains under statistical control with respect to the measurement involved in computing w.

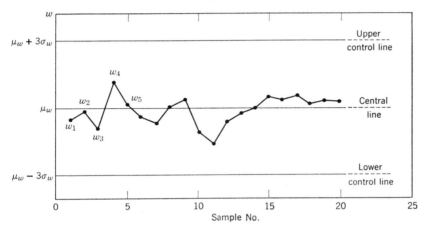

Figure 11.2.1.   A typical 3-sigma control chart for the statistic w if μ_w and σ_w are known.

For example, suppose w has a normal distribution with mean μ_w and standard deviation σ_w, then the probability is .9973 that w will fall inside the limits $(\mu_w \pm 3\sigma_w)$, or .0027 that w will fall outside these limits. If w is only approximately normally distributed, these probability statements, of course, will hold only approximately. If we set up a *3-sigma control chart* for w, as shown in Figure 11.2.1, we can plot values of w, say w_1, $w_2, \ldots, w_k$, for the successive samples against $1, 2, \ldots, k$. This chart, frequently referred to as a *Shewhart Chart*, contains three horizontal lines:

	The upper control line (UCL):	$w = \mu_w + 3\sigma_w.$
(11.2.1)	*The center control line* (CL):	$w = \mu_w.$
	The lower control line (LCL):	$w = \mu_w - 3\sigma_w.$

Now suppose some disturbance causes the measurements from which w is computed to slip an amount δ to a new state or level of statistical control, so that w now behaves like a random variable from the normal distribution

$N(\mu_w + \delta, \sigma_w^2)$. The probability that the value of w on the first sample selected under the new state of statistical control falls outside the band is given by

$$(11.2.2) \quad 1 - \frac{1}{\sigma_w \sqrt{2\pi}} \int_{\mu_w-3\sigma}^{\mu_w+3\sigma} \exp\left[-\frac{1}{2\sigma_w^2}(w - \mu_w - \delta)^2\right] dw.$$

By letting $z = (w - \mu_w)/\sigma_w$, (11.2.2) becomes

$$(11.2.3) \quad 1 - \left[\Phi\left(3 - \frac{\delta}{\sigma_w}\right) - \Phi\left(-3 - \frac{\delta}{\sigma_w}\right)\right].$$

Thus the further δ/σ_w departs from 0, the higher the probability that the quality control chart will detect the fact that some disturbance has occurred to shift the distribution of w. Once a recorded response w falls outside the 3-sigma limits it is up to the engineer to try to locate and, if possible, eliminate the cause of the disturbance.

Important statistics w frequently used in practice are the sample mean $\bar{x}$, the number of defective items m in a fixed sample of n, the sample proportion defective $p = m/n$, and for Poisson variates c, the observed number of events per unit time. Statistics such as $s = [\sum (x_i - \bar{x})^2/(n - 1)]^{1/2}$, the estimate of the standard deviation, are not often used on control charts in industry. Ease of computation, and speed in posting results on a chart, are of major importance.

Given that the parameters of the distribution of measurements are known, then the UCL and LCL, the 3-sigma control limits, are given by

$$(11.2.4)$$

Sample mean $\bar{x}$:	$\mu \pm 3\dfrac{\sigma}{\sqrt{n}}$,	normal
Sample number defective m:	$n\theta \pm 3\sqrt{n\theta(1 - \theta)}$,	binomial
Sample percent defective p:	$\theta \pm 3\sqrt{\dfrac{\theta(1 - \theta)}{n}}$,	binomial
Counts per unit time, c:	$\lambda \pm 3\sqrt{\lambda}$,	Poisson

Control charts employing these 3-sigma limits are called *theoretical control charts* since the UCL and LCL are set using known parameter values for the distributions of the measurements.

> **Example 11.2(a).** Suppose a process of manufacturing certain items is in a state of statistical control and that the fraction of defectives observed over a long period of time is θ_0. In taking successive samples of size n from the production line, the numbers of defectives are $m_1, m_2, \ldots, m_k, \ldots$. The problem is to set up a 3-sigma theoretical control chart for maintaining surveillance over the m's.

If the process is under statistical control with fraction of defectives equal to θ_0, then the quantity m, for a sample taken at random, behaves like a random variable from the binomial distribution

$$b(m) = \binom{n}{m} \theta_0^m (1 - \theta_0)^{n-m},$$

$m = 0, 1, \ldots , n$. The mean and variance of m are $n\theta_0$ and $n\theta_0(1 - \theta_0)$, hence the UCL, CL, and LCL for a 3-sigma control chart have equations, respectively, as follows:

$$m = n\theta_0 + 3\sqrt{n\theta_0(1 - \theta_0)},$$
$$m = n\theta_0,$$
$$m = n\theta_0 - 3\sqrt{n\theta_0(1 - \theta_0)}.$$

Example 11.2(b).   Suppose a machine set to fill cans with evaporated milk, when set for a given specification, will generate fillings which form a population such that the weights of the fillings have a mean of 15.0 ounces and standard deviation of .4 ounce. A 3-sigma theoretical control chart is to be constructed for maintaining surveillance over net weights of fillings in accordance with the specifications, taking the statistic w as the mean $\bar{x}$ of four fillings.

Thus $\bar{x}$ is the mean of a sample of size 4 from a population having $\mu = 15.0$ ounces and $\sigma = .4$ ounce; hence

$$\mu_{\bar{x}} = 15 \text{ ounces},$$
$$\sigma_{\bar{x}} = \frac{(.4)}{\sqrt{4}} = .2 \text{ ounce.}$$

The UCL, CL, and LCL for the required 3-sigma theoretical control chart for $\bar{x}$ have the following equations respectively:

$$\bar{x} = \mu_{\bar{x}} + 3\sigma_{\bar{x}} = 15.0 + .6 \text{ ounces},$$
$$\bar{x} = \mu_{\bar{x}} = 15.0 \text{ ounces},$$
$$\bar{x} = \mu_{\bar{x}} - 3\sigma_{\bar{x}} = 15.0 - .6 \text{ ounces.}$$

It should be pointed out that we could equally well have taken the statistic w to be the sample sum T, in which case,

$$\mu_T = 4(15.0) = 60.0 \text{ ounces},$$
$$\sigma_T = \sqrt{4}(.4) = .8 \text{ ounce},$$

and the UCL, CL, and LCL would have equations respectively,

$$T = \mu_T + 3\sigma_T = 60.0 + 2.4 \text{ ounces},$$
$$T = \mu_T = 60.0 \text{ ounces},$$
$$T = \mu_T - 3\sigma_T = 60.0 - 2.4 \text{ ounces.}$$

If $\bar{x}$ (or T) for a sample of four cans fails to fall between its upper and lower control lines, this constitutes strong evidence that some cause is operating to throw the process out of control, which should be identified and removed.

(b) Theoretical Control Chart for the Sample Variance

Suppose the variance σ^2 of the measurements on the objects manufactured by a process under statistical control is known, and that it is desired to establish a theoretical control chart on the sample variances $s_1^2, s_2^2, \ldots$, in a succession of samples of size n. For controlling s^2, it is customary to use only one control line, namely, an upper control line. If we want the probability to be only α (.001, .01, .05, or .10, say) that the value of s^2 in a sample would fall above the upper control line, we would take as the control line the horizontal line on the control chart of s^2 having equation

(11.2.5)
$$s^2 = \frac{\sigma^2 \chi_{n-1;\alpha}^2}{n-1}.$$

An upper control line for s would, of course, have equation

$$s = \frac{\sigma \chi_{n-1;\alpha}}{\sqrt{n-1}}.$$

If we wanted a theoretical control chart for s^2 with both a UCL and an LCL, the equations of these control lines would be, respectively,

$$s^2 = \frac{\sigma^2 \chi_{n-1;\alpha/2}^2}{n-1},$$

$$s^2 = \frac{\sigma^2 \chi_{n-1;1-\alpha/2}^2}{n-1}.$$

Similarly, the UCL and LCL of a theoretical control chart for s would have equations

$$s = \frac{\sigma \chi_{n-1;\alpha/2}}{\sqrt{n-1}},$$

$$s = \frac{\sigma \chi_{n-1;1-\alpha/2}}{\sqrt{n-1}}.$$

(c) Theoretical Control Chart for the Sample Range R

A procedure for controlling variability within samples which is arithmetically simpler than control charts for sample variances is to set up a control chart for the sample range R. If the process is under statistical control and if σ is known, the LCL and UCL for a 3-sigma theoretical control chart for R are horizontal lines on the chart having the following equations:

(11.2.6)
$$R = D_1\sigma,$$
$$R = D_2\sigma,$$

where D_1 and D_2 are tabulated in Table 11.3.1 for $n = 2, 3, \ldots, 10$. The lower control line for R is rarely used in practice.

(d) Some Further Remarks

Although in many practical situations the statistic w (for instance, sample mean, sample sum) is approximately normally distributed for a process under statistical control, it should be remarked that even if w is not approximately normally distributed, it is customary in practice to go ahead anyway as a matter of procedure and establish a control chart for w as shown in Figure 11.2.1. In such a case we cannot think of the UCL and LCL as .9973 probability limits, but we can still refer to them as 3-sigma control limits, the probability being "quite small" in practice that w will fall outside the limits when the process is under statistical control.

Finally, it should be noted that the concept of a theoretical control chart has a wider applicability than only to measurements on successive samples of objects from a manufacturing process. The successive samples can be generated by any repetitive process, such as throwing a pair of dice repeatedly, generating sequences of random numbers, and so on, in which case the control chart is a graphical device for checking whether the process is in a state of statistical control with known parameters.

11.3 EMPIRICAL CONTROL CHARTS FOR THE CASE IN WHICH A STATE OF STATISTICAL CONTROL DOES NOT YET EXIST

(a) Empirical Control Chart for the Sample Mean $\bar{x}$

In this case a manufacturing process is assumed to be under way but a satisfactory state of statistical control has not yet been established. Thus the measurements which would be obtained on a succession of manufactured objects would not behave like a sequence of independent random variables from a "desired" population. However, if the main causes preventing the successive measurements from behaving in this manner could be identified and removed, the remaining variations of measurements should be those which would prevail if the process were brought under control. Often the magnitude of these variations is approximately the same as those to be found within small samples of objects, where the time-spacings of objects within samples are much shorter than those of objects from one sample to the next.

The preceding discussion suggests that, if we think of μ and σ^2 as the mean and variance of the measurements on successive objects being manu-factured after the casues of the more important variations of the measure-ments are removed, than a reasonable procedure for estimating μ and σ^2 would be as follows: Take a sample of n objects closely spaced in time from the production line; repeat this periodically until we have k samples. Let $\bar{x}_1, \bar{x}_2, \ldots, \bar{x}_k$ be the means of these samples and $s_1^2, s_2^2, \ldots, s_k^2$ be the variances of these samples. We would than take as our estimate of μ, the mean of all sample means, that is,

$$(11.3.1) \qquad \bar{\bar{x}} = \frac{1}{k}(\bar{x}_1 + \cdots + \bar{x}_k),$$

and we would take as our estimate of σ^2 the pooled value of all sample variances, namely,

$$(11.3.2) \qquad s_w^2 = \frac{(n-1)s_1^2 + \cdots + (n-1)s_k^2}{n + \cdots + n - k} = \frac{1}{k}\sum_{i=1}^{k} s_i^2.$$

We would then estimate the UCL, CL, and LCL for a 3-sigma control chart for the sample mean $\bar{x}$, if a satisfactory state of statistical control could be established, to be horizontal lines having the following equations respectively:

$$(11.3.3) \qquad \begin{aligned} \bar{x} &= \bar{\bar{x}} + A_1 s_w, \\ \bar{x} &= \bar{\bar{x}}, \\ \bar{x} &= \bar{\bar{x}} - A_1 s_w, \end{aligned}$$

where A_1 depends on n and is tabulated in Table 11.3.1 for $n = 2, 3, \ldots, 10$. $A_1 s_w$ is simply an unbiased estimator for $(3/\sqrt{n})\sigma$.

To obtain the value of s_w in practical applications involves a good deal of computation. A widely used alternative procedure for estimating $(3/\sqrt{n})\sigma$ which involves simpler computations, at the expense of making somewhat less efficient use of the information contained in the samples concerning σ, is as follows: Let $R_1, R_2, \ldots, R_k$ be the ranges of the k samples, assumed to be from normal populations all having equal variances, say σ^2, and let

$$(11.3.4) \qquad \bar{R} = \frac{1}{k}(R_1 + \cdots + R_k)$$

Then an unbiased estimator for $(3/\sqrt{n})\sigma$ is $A_2\bar{R}$, where A_2 depends on n and is tabulated in Table 11.3.1 for $n = 2, 3, \ldots, 10$. Therefore, the UCL, CL, and LCL based on the sample means $\bar{x}_1, \bar{x}_2, \ldots, \bar{x}_k$ and sample ranges, $R_1, R_2, \ldots, R_k$ are horizontal lines on the $\bar{x}$ chart having equations estimated as follows:

$$\bar{x} = \bar{\bar{x}} + A_2 \bar{R},$$

(11.3.5) $$\bar{x} = \bar{\bar{x}},$$

$$\bar{x} = \bar{\bar{x}} - A_2 \bar{R}.$$

We shall call control charts with estimated control lines *empirical control charts*.

TABLE 11.3.1
VALUES OF CERTAIN CONSTANTS REQUIRED FOR CONTROL CHARTS

Sample Size		Limits for $\bar{x}$		Limits for R			
	3	$\bar{x} \pm A_1 s_w$	$\bar{x} \pm A_2 \bar{R}$	$R = D_1 \sigma$	$R = D_2 \sigma$	$R = D_3 \bar{R}$	$R = D_4 \bar{R}$
n	$\overline{\sqrt{n}}$	A_1	A_2	D_1	D_2	D_3	D_4
2	2.121	2.659	1.880	0	3.686	0	3.267
3	1.732	1.954	1.023	0	4.358	0	2.575
4	1.500	1.628	.729	0	4.698	0	2.282
5	1.342	1.427	.577	0	4.918	0	2.115
6	1.225	1.287	.483	0	5.078	0	2.004
7	1.134	1.179	.419	.205	5.203	.076	1.924
8	1.061	1.100	.373	.387	5.307	.136	1.864
9	1.000	1.032	.337	.546	5.394	.184	1.816
10	.949	.977	.308	.687	5.469	.223	1.777

It should be noted that, even before a state of statistical control is established, $A_1 s_w$ and $A_2 \bar{R}$ can be expected to provide reasonably good estimates for $3\sigma/\sqrt{n}$, where σ would be the standard deviation of measurements when satisfactory statistical control is established. The principal reason for this is that s_w and $\bar{R}$ are determined from variations of measurements *within* samples, where each sample is taken over a brief span of time, thus avoiding variations due to drifts or swings of time. On the other hand it may be more hazardous to risk $\bar{\bar{x}}$ as an estimator for the mean μ of the process if and when it comes into a state of statistical control. Hence $\bar{\bar{x}}$ should be regarded as a rough and preliminary estimator for μ, to be revised periodically on the basis of further samples as progress is made toward a state of statistical control.

(b) Empirical Control Chart for Sample Range R

In Section 11.2c we described the procedure for establishing a control chart for sample range R when σ is known. If σ is known, we may establish an empirical control chart for R, the upper and lower control lines being given by (11.2.6). If σ is unknown, the quantities $D_1 \sigma$ and $D_2 \sigma$ in (11.2.6) have $D_3 \bar{R}$ and $D_4 \bar{R}$ as unbiased estimates, so that the

3-sigma lower and upper control lines for the empirical control chart for R are horizontal lines on the chart having the equations

(11.3.6)
$$R = D_3\bar{R},$$
$$R = D_4\bar{R},$$

with D_3 and D_4 depending on n, and tabulated in Table 11.3.1 for $n = 2, 3, \ldots, 10$. Again, we point out that the upper control line is more important than the lower one for R.

11.4 THE CUMULATIVE SUM CHART

The theoretical and empirical control charts thus far discussed control the Type I error only. Thus, so long as the process is in control, that is so long as the hypothesis H_0: [the distribution of the observations is $f_0(x)$] is true, then only rarely will a plotted point based on a sample from $f_0(x)$ fall outside the 3-sigma limits. In industrial practice, however, the Type II error can have serious economic consequences since, obviously, if the alternative hypothesis H_1: [the distribution of the observations is $f_1(x) \neq f_0(x)$] is true, off quality product can result. The cumulative sum chart (a simple adaptation of the Wald Sequential Decision procedures discussed in Section 10.8) is often used as a charting technique which provides protection against both Type I and Type II errors.

Instead of plotting the series of observed measurements $w_1, w_2, \ldots, w_m$ the cumulative sum $\sum_{i=1}^{m} (w_i - E(w))$ is plotted. Thus, for normally distributed observations the $CuSum$, $\sum_{i=1}^{m} (x_i - \mu_0)$ would be plotted. So long as the mean of the process equalled μ_0 we would expect, in a long run of observations, that the $CuSum$ would vary randomly about zero. If however, the mean of the process shifted from μ_0 to, say, $\mu_0 + \delta$, the $CuSum$ would increase by δ for each new observation if δ positive, and decrease if δ negative. A persistent upward (or downward) movement of the CuSum plot would thus provide a signal that the mean had shifted. Furthermore, the chart could be used to determine when the shift occurred and to estimate δ. However it is easy to be confused when observing CuSum charts, as Figures 11.4.1a, b, c, and d illustrate. In each figure the CuSum of random normal deviated is plotted and it is clear that trends and even cyclic effects seem to occur.

It is obvious from these plots that some objective criterion must be used to determine when a trend in the CuSum chart can reasonably be declared the result of a shift in the mean of the process. For normally distributed measurements, this is accomplished by placing a "V mask" of angle 2θ

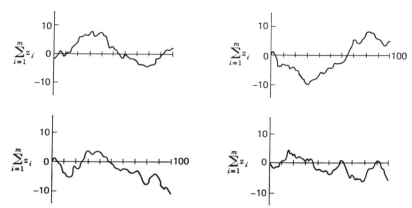

Figure 11.4.1. Examples of the cumulative sum of 100 random normal (0, 1) observations.

a lead distance d in front of the last plotted point on the CuSum chart, as illustrated in Figure 11.4.2. If a previously plotted point falls beyond the arms of the V mask, the hypothesis that the process mean $\mu = \mu_0$ is rejected. The chart can then be used to estimate the true mean, and to determine when the shift in mean occurred.

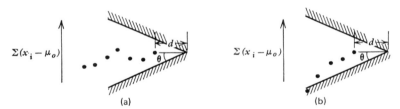

Figure 11.4.2. (a) V mask, continue sampling. (b) V mask, significant shift detected.

The V mask is identical to the "V" produced by the Two-Sided Wald Sequential Decision procedures, (see Section 10.9) except that the mask is *reversed* and *placed adjacent to the most recently plotted point*. To obtain the mask, we may start with a graph such as in Figure 10.9.1. Indeed, consulting Figure 10.9.1, we may think of cutting out a V mask by starting at the vertical axis and cutting towards the horizontal axis along the outermost lines obtained from

$$(11.4.1) \qquad T_m = \frac{\sigma^2}{\delta} \ln \frac{1 - \beta}{\alpha/2} + \frac{\delta}{2} m,$$

where $T_m = \sum_{i=1}^{m} y_i = \sum_{i=1}^{m} (x_i - \mu_0)$, and δ put successively equal to $+\delta_0$ and

$-\delta_0$, say. [In fact, the two outermost lines in Figure 10.9.1 are obtained from the data of Example 10.9(a), with $\delta_0 = 50$, etc., and are given by

(11.4.2)
$$T_m = 179.2 + 25m$$
$$T_m = -179.2 - 25m.$$

The reader should consult Example 10.9(a) again, for details.] To finish the mask, we also cut along the horizontal lines $T_m = \pm \dfrac{\sigma^2}{\delta} \ln \dfrac{1-\beta}{\alpha/2}$, and the vertical line $m = -d$, where d is given by (11.4.3) below.

The above procedure is indicated in Figure 10.9.1, by the dotted lines. Notice that the point of intersection of both lines (on the m-axis) is $(0, -d)$, where

(11.4.3)
$$d = \frac{2}{\delta}\left[\frac{\sigma^2}{\delta} \ln \frac{1-\beta}{\alpha/2}\right],$$

as may easily be seen from (11.4.1). Hence, if the cut-out V mask is reversed, and used as in Figure 11.4.2, the lead distance d indicated on Figure 11.4.2 is given by (11.4.3). Further, the V mask is of angle 2θ, where θ is such that

(11.4.4)
$$\tan \theta = \frac{\delta}{2},$$

that is, $\tan \theta$ is the slope of the lines (11.4.1). Of course, with (11.4.3) and (11.4.4), we may construct the V mask directly without the use of a graph such as in Figure 10.9.1.

Referring again to Figure 10.9.1, we recall that if points (m, T_m) fall above or below the outermost lines, we reject H_0: $\mu = \mu_0$. It is for this reason that if previously plotted points fall beyond the arms of the V-mask, we reject H_0—see the right hand panel of Figure 11.4.2, for example.

Example 11.4(a)   As in Example 10.9(a), let $\mu_0 = 500$, $\sigma = 50$, $\delta = \pm 50$, $\alpha = 0.05$, and $\beta = 0.10$. The V mask is then based on the two lines

(11.4.4)
$$T_m = 179.2 + 25m$$
$$T_m = -179.2 - 25m$$

where $T_m = \displaystyle\sum_{i=1}^{m}(x_i - \mu_0)$.

Now, very often for reasons of compactness, we "shrink" the vertical axis by plotting (m, T_m') on graph paper, where

(11.4.5)
$$T_m' = T_m/k,$$

with k an arbitrarily chosen constant. The appropriate V mask is then of

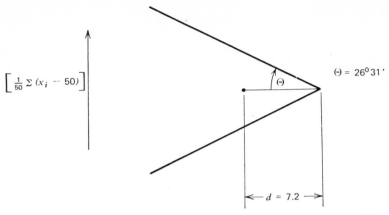

$$\left[\frac{1}{50} \, \Sigma \, (x_i - 50)\right]$$

$(-) = 26^{0}31'$

$d = 7.2$

Figure 11.4.3. V mask for CuSum chart. $\mu_0 = 500$, $\delta = 50$, $\sigma = 50$, $\alpha = 0.05$, $\beta = 0.10$.

lead distance d, angle $2\theta'$, where d is given, as before, by (11.4.3), and

(11.4.6) $$\tan \theta' = \frac{\delta}{2k} .$$

For the conditions of Example 10.9(a), then, suppose we let $k = 50$. Then

$$\tan \theta' = \frac{50}{2(50)} = 0.5, \; \theta' = 26°31'$$

and

$$d = \frac{2}{50}\left[\frac{2500}{50} (3.583)\right] = 7.2 \text{ units.}$$

The mask with angle $2\theta'$ and d as given above, is illustrated in Figure 11.4.3.

Once the V mask has been established the process is considered "under control" so long as no previous points fall beyond the arms of the V. If a point falls beyond the mask, δ can be estimated by passing a straight line through the recent data, and the turning point of the CuSum plot determined by eye. If T_m is the cumulative sum of the most recent point, and T_q, the cumulative sum at the turning point of the CuSum plot then $\hat{\delta}$, the estimate of the shift δ, is given by

$$\hat{\delta} = \frac{T_m - T_q}{(m - q)} .$$

The turning point is used as the best guess as to when the shift in mean occurred. Further details on the uses of cumulative sum charts can be found in the text *Cumulative Sum Techniques*, Woodward, R. H. and P. L. Goldsmith, Oliver and Boyd, London, 1964.

PROBLEMS

11.1. Statistical control is to be maintained on the gross weights of filled 1-pound boxes of sugar by control charts. It is decided that properly controlled gross weights of filled boxes should have a mean of 17.6 ounces and standard deviation of .3 ounce. Using these standard values of μ and σ, determine:

(a) The UCL and LCL for a 3-sigma theoretical control chart for total gross weights of successive samples of six filled boxes.

(b) The UCL and LCL for a 3-sigma theoretical control chart for the ranges of gross weights for samples of six filled boxes.

11.2. A device is used to generate random digits. It is decided that a rough check on the statistical behavior of the digits will be made by a 3-sigma theoretical control chart on sums of 100 digits taken periodically. Determine the UCL, CL, and LCL for such a control chart.

11.3. Suppose you wish to test a new kind of molded plastic die for statistical control by taking seven of these dice from the machine every hour, throwing them, and recording the sums of the dots T obtained. If the dice are to be controlled as "true", what should the center line and upper and lower control lines be for a 3-sigma chart that plots the sum of the dots (against sample number)?

11.4. Telephone calls are received at a switch board at the mean rate of 6.2 a minute. Establish a 3-sigma control chart to monitor future incoming calls on a per minute basis. What are the UCL and LCL for the number of calls every 15 seconds?

11.5. If a mass-produced plastic washer is being manufactured under acceptable statistical control, the mean of the thicknesses of the washers in the population should be .125 inch, and the standard deviation should be .005 inch. A theoretical control chart which is set up to check on whether the washer thickness is actually under statistical control works as follows: A sample of five washers is taken at random every hour and the total thickness T of the five is measured and recorded on the control chart. Determine the LCL, CL, and UCL for a 3-sigma theoretical control chart for the total thicknesses of five washers. If the process shifts so as to produce washers having an average thickness of .127 inch, what is the probability that the first sample taken after the shift occurred will produce a value of T above the UCL of the control chart? What is the probability that at least one of the first two samples taken after the shift occurred will produce a T above the UCL?

11.6. In manufacturing screws for a certain purpose, assume that specifications permit 5% of the screws to have a certain type of minor defect. If a 3-sigma theoretical control chart is to be established for maintaining surveillance over the total number of defective screws in successive samples each having 500 screws, what would the UCL, CL, and LCL be for this control chart? If the manufacturing process shifts so as to produce such defectives at the rate of 10%, what is the probability that the shift will be detected by the control chart on the first sample taken after the shift? On the first, second, or third sample after the shift?

11.7. In manufacturing enameled copper wire of a certain size, suppose the specifications will permit one insulation break per 100 feet. A 3-sigma theoretical control chart is to be set up for the number of insulation breaks in successive samples of 2500 feet. Determine the UCL, CL, and LCL for such a control chart. If the enameling process suddenly shifted so as to produce 1.5 insulation breaks per 100 feet, what is the probability the shift will be detected by the control chart on the first sample taken after the shift? On the first, second, or third sample after the shift?

11.8. Twenty-four samples of four insecticide dispensers were taken periodically, during production. The means and ranges of the charge weights (in grams) in the 24 samples are shown in Table P11.8:

TABLE P11.8

Sample Number	Sample Mean $\bar{x}$	Sample Range R	Sample Number	Sample Mean $\bar{x}$	Sample Range R
1	471.5	19	13	457.5	4
2	462.2	31	14	431.0	14
3	458.5	22	15	454.2	19
4	476.5	27	16	474.5	18
5	461.8	17	17	475.8	18
6	462.8	38	18	455.8	24
7	464.0	38	19	497.8	34
8	461.0	37	20	448.5	21
9	450.2	19	21	453.0	58
10	479.0	41	22	469.8	32
11	452.8	15	23	474.0	10
12	467.2	28	24	461.0	36

Specifications on the settings of the charging machine demand that it should produce charges having a mean of 454 grams and standard deviation of 9 grams (or less).

(a) Set up a 3-sigma theoretical control chart for means of samples of 4 under the specifications; then plot the 24 successive sample means on it and determine whether the means are behaving in accordance with the specifications.

(b) Set up a 3-sigma theoretical control chart for sample ranges under the specifications. Plot the 24 successive sample ranges and determine whether the ranges are behaving in accordance with the specifications.

11.9. It is desired to set up an empirical control chart for a certain dimension of rheostat knobs. Preliminary samples yield $\bar{x} = .1400$ inch and $\bar{R} = .008$ inch, where each of the preliminary samples has five observations. During a 3-day period, a sample of 5 knobs was taken every hour during production (a production day is 8 hours), and the mean and range of each of the 24 samples thus obtained are shown in Table P11.9 (in inches):

TABLE P11.9

Sample No.	Sample Mean $\bar{x}$	Sample Range R	Sample No.	Sample Mean $\bar{x}$	Sample Range R
1	.1396 (2)	.012	13	.1394 (1)	.010
2	.1414 (13)	.009	14	.1422 (19)	.010
3	.1415 (15)	.003	15	.1412 (12)	.009
4	.1406 (7)	.008	16	.1417 (17)	.005
5	.1416 (16)	.009	17	.1402 (5)	.006
6	.1404 (6)	.013	18	.1425 (20)	.006
7	.1400 (4)	.006	19	.1407 (8)	.011
8	.1418 (18)	.008	20	.1408 (9)	.010
9	.1410 (11)	.008	21	.1430 (22)	.006
10	.1432 (23)	.005	22	.1398 (3)	.007
11	.1448 (24)	.006	23	.1415 (14)	.008
12	.1428 (21)	.008	24	.1409 (10)	.004

[The parentheses refer to the ordering in magnitude of the $\bar{x}$'s, and are used in Chapter 14, Section 14.3, Example 14.3(a).]

(a) Using the information from these samples, construct a 3-sigma empirical control chart for sample means, and graph the sample means on this chart.

(b) Construct an empirical control chart for sample ranges and graph the samples ranges. Is the dimension under "statistical" control?

11.10. Measurement (in inches) of a certain dimension on fragmentation bomb heads manufactured during the war by the American Stove Company was

TABLE P11.10

Sample No.	Sample Mean	Sample Range	Sample No.	Sample Mean	Sample Range
1	.4402	.015	16	.4362	.015
2	.4390	.018	17	.4380	.019
3	.4448	.018	18	.4350	.008
4	.4432	.006	19	.4378	.011
5	.4428	.008	20	.4384	.009
6	.4382	.010	21	.4392	.006
7	.4358	.011	22	.4378	.008
8	.4440	.019	23	.4362	.016
.9	.4366	.010	24	.4348	.009
10	.4368	.011	25	.4338	.005
11	.4360	.011	26	.4366	.014
12	.4402	.007	27	.4346	.009
13	.4332	.008	28	.4374	.015
14	.4356	.017	29	.4339	.024
15	.4314	.010	30	.4368	.014

made on successive samples of five bomb heads giving the results shown in Table P11.10 [data from Duncan (1958)].

From the first 20 samples, set up an empirical 3-sigma control chart for $\bar{x}$ and an empirical control chart for R. Plot the values of $\bar{x}$ and R on these charts to see if the process continues in a state of control both as to average and range.

11.11. Twenty samples of five machined items were taken periodically from production and the inside diameter of a hole in each item was measured to the nearest .0001 inch. The recorded measurement x was defined as the diameter minus .7500 inch. The values of x_1, x_2, x_3, x_4, x_5, obtained for each of the 20 samples are shown in Table P11.11 [from Duncan (1958)]:

TABLE P11.11

Sample No.	x_1	x_2	x_3	x_4	x_5
1	15	11	8	15	6
2	14	16	11	14	7
3	13	6	9	5	10
4	15	15	9	15	7
5	9	12	9	8	8
6	11	14	11	12	5
7	13	12	9	6	10
8	10	15	12	4	6
9	8	12	14	9	10
10	10	10	9	14	14
11	13	16	12	15	18
12	7	10	9	11	16
13	11	7	16	10	14
14	11	7	10	10	7
15	13	9	12	13	17
16	17	10	11	9	8
17	4	14	5	11	11
18	8	9	6	13	9
19	9	10	7	10	13
20	15	10	12	12	16

Set up control charts for $\bar{x}$, R, and s for controlling the inside diameter in further production.

11.12. From previous investigations of a production of rivets it has been found that the process is in a satisfactory state of statistical control with regard to diameter of rivet heads and that the diameters are normally distributed with mean 13.42 millimeters and standard deviation .12 millimeter. Accepting the values as standards for future production, *establish* theoretical control charts for the mean and range of samples of 5 and determine whether the following values of the means and ranges in 40 samples taken from production conform to these standards. Table P11.12a gives mean diameters minus 13.00 millimeters for the 40 samples of 5, and Table P11.12*b* gives ranges in millimeters for the samples of 5 [from Hald (1952)]:

TABLE P11.12*a*

.446	.362	.414	.478	.376	.476	.424	.328	.378	.384
.478	.422	.442	.426	.358	.526	.478	.462	.430	.398
.426	.474	.380	.422	.346	.414	.480	.410	.366	.402
.400	.414	.426	.408	.432	.378	.410	.456	.344	.384

TABLE P11.12*b*

.15	.15	.31	.32	.31	.10	.25	.19	.29	.23
.30	.33	.26	.14	.22	.33	.20	.40	.14	.43
.29	.33	.17	.36	.14	.22	.18	.19	.34	.38
.27	.42	.18	.09	.23	.50	.36	.12	.28	.11

11.13. Construct a control chart for $\bar{x}$ for the following data on the blowing time of fuses, samples of size 5 being taken every hour for 12 hours. If these are the first data taken on this product, would you consider the process to be under statistical control and, hence, that the means and ranges from these samples could be used for constructing control lines for future control? The measurements in each sample of 5 have been arranged in order of magnitude [from Dixon and Massey (1957)].

42	42	19	36	42	51	60	18	15	69	64	61
65	45	24	54	51	74	60	20	30	109	91	78
75	68	80	69	57	75	72	27	39	113	93	94
78	72	81	77	59	78	95	42	62	118	109	109
87	90	81	84	78	132	138	60	84	153	112	136

11.14. For the standards established in Problem 11.12, set up a cumulative sum chart for the situation where rivet heads with mean diameters outside the interval (13.42 ± 0.10) millimeters are considered unsatisfactory. Let the Type I and Type II errors be 0.10 and 0.05, respectively.

11.15. For the standards established in Problem 11.8, set up a cumulative sum chart for the situation where charges having values outside the interval (454 ± 7) grams are unsatisfactory. Let the Type I and Type II errors be 0.05 and 0.10, respectively.

11.16. Referring to Problem 11.9, suppose dimensions of the knobs having values outside the interval (.1400 ± .0040) inches are unsatisfactory, and suppose that the standard deviation of the process is .005 inch. Let the Type I and Type II errors both be .05, and set up a CuSum chart for this situation.

CHAPTER 12

Chi-Square Goodness-of-Fit Tests

12.1 GOODNESS OF FIT

Goodness-of-fit tests arise when we wish to compare a set of observed frequencies with their expected (or theoretical) frequencies. In general, let $A_1, A_2, \ldots, A_k$ be k mutually exclusive and exhaustive outcomes of a trial and let the probabilities of these outcomes be $\theta_1, \theta_2, \ldots, \theta_k$ respectively, where $\theta_1, \ldots, \theta_k$ are all positive and $\theta_1 + \cdots + \theta_k = 1$.

Suppose n independent trials of the experiment are made, and let $f_1, f_2, \ldots, f_k$ be the number of trials which result in outcomes $A_1, A_2, \ldots, A_k$ respectively. Then $f_1, f_2, \ldots, f_k$ are random variables which have the multinomial distribution (see Section 3.4)

$$(12.1.1) \qquad \frac{n!}{f_1! f_2! \cdots f_k!} \theta_1^{f_1} \theta_2^{f_2} \cdots \theta_k^{f_k},$$

where $\sum_{i=1}^{k} f_i = n$. The expectation of f_i is $E(f_i) = n\theta_i$.

If we wish to test the null hypothesis H_0 that our sample comes from a multinomial population with probabilities $P(A_i) = \theta_i$, $i = 1, 2, \ldots, k$, where $\theta_1, \ldots, \theta_k$ are known, we use

$$(12.1.2) \qquad \chi^2 = \sum_{i=1}^{k} \frac{(f_i - n\theta_i)^2}{n\theta_i}$$

as our test statistic. If the observed values of f_i are all exactly equal to $n\theta_i$, we have a "perfect fit," and $\chi^2 = 0$. Thus, large values of χ^2 will tend to discredit the null hypothesis and smaller values of χ^2 tend to confirm the hypothesis.

For moderately large values of n, the distribution of the test statistic given in (12.1.2) is approximately the chi-square distribution having $k - 1$ degrees of freedom. In practice we usually require that $n\theta_i \geq 5$ for each i. If there is difficulty in meeting this criterion, we would combine into a single outcome several of the outcomes, say $A_{i_1}, \ldots, A_{i_r}$, having the smallest probabilities $\theta_{i_1}, \ldots, \theta_{i_r}$, so that $n(\theta_{i_1} + \cdots + \theta_{i_r}) \geq 5$. We illustrate this procedure in Examples 12.1(a) and 12.1(b).

If we test at the $100\alpha\%$ level of significance, the observed value of χ^2 is considered to be significant at this level if $\chi^2 > \chi^2_{k-1;\alpha}$; that is, we reject the hypothesis that the observations come from a distribution for which $P(A_i) = \theta_i$, $i = 1, 2, \ldots, k$, if $\chi^2 > \chi^2_{k-1;\alpha}$.

If the $\theta_1, \theta_2, \ldots, \theta_k$ are unknown but expressible in terms of a small number c of parameters which have to be estimated from $f_1, \ldots, f_k$, then χ^2 has approximately the chi-square distribution with $k - c - 1$ degrees of freedom. Of course, in this case the observed value of χ^2 is considered to be significant if it exceeds $\chi^2_{k-c-1;\alpha}$.

Example 12.1(a). An icosahedral die has two sides marked 1, two sides marked 2, . . . , and two sides marked 0, which will be designated as 10. It is desired to test whether the die is behaving like a "true" die or not on the basis of 200 throws. The results of 200 throws are shown in Table 12.1.1.

TABLE 12.1.1

RESULTS AND ANALYSIS OF 200 THROWS OF
AN ICOSAHEDRAL DIE

x_i	f_i	$n\theta_i$	$(f_i - n\theta_i)^2$	$\dfrac{(f_i - n\theta_i)^2}{n\theta_i}$
1	17	20	9	.45
2	19	20	1	.05
3	26	20	36	1.80
4	18	20	4	.20
5	16	20	16	.80
6	23	20	9	.45
7	21	20	1	.05
8	24	20	16	.80
9	20	20	0	.00
10	16	20	16	.80
	200	200		5.40

If the die is true, then the probability of each number from 1 to 10 is $2/20 = 1/10$, and we would expect that in 200 throws each number turns up $200 \times 1/10 = 20$ times on the average. The question we ask, then, is whether the set of observed f_i is compatible with the null hypothesis that the die is true, that is, that $\theta_i = 1/10$ for each i. We note that $n\theta_i \geq 5$ for each i and apply the χ^2 test. In this example, $k = 10$ and, from Table 12.1.1,

$$\sum_{i=1}^{10} \frac{(f_i - n\theta_i)^2}{n\theta_i} = 5.40.$$

Consulting Table III, Appendix VII, and choosing $\alpha = .05$, we find that $\chi^2_{9;.05} = 16.92$. Since the observed value of χ^2 is 5.40, which is less than 16.92, we accept the hypothesis that the die is "true," that is, that $\theta_i = 1/10$ for $i = 1, \ldots, 10$ or, put another way, we can say that the results of the 200 throws do not contradict the hypothesis of a "true" die at the 5% level of significance.

Example 12.1(b). It is believed that, when a certain type of uranium is placed in a radioactive counter for a given interval of time, the number x of α particles emitted during the interval behaves like a random variable having the Poisson distribution

(12.1.3) $$p(x) = \frac{e^{-\mu}\mu^x}{x!},$$

where μ is an unknown parameter.

TABLE 12.1.2

RESULTS OF AN α-EMISSION COUNTING EXPERIMENT ON URANIUM

No. of α Particles Emitted, x_i	Observed No. of Time Intervals, f_i	Expected No. of Time Intervals	$x_i f_i$
0	1	1.5	0
1	5	6.3	5
2	16	13.2	32
3	17	18.5	51
4	26	19.4	104
5	11	16.3	55
6	9	11.4	54
7	9	6.9	63
8	2	3.6	16
9	1	1.7	9
10	2	.7	20
11	1	.4	11
	100	99.9	420

In an experiment the number of emissions from a piece of uranium was determined for each of 100 time intervals of equal length, with the results shown in Table 12.1.2.

Since μ is unknown it must be estimated. With the use of the method of maximum likelihood, it is found that the "best" estimator for μ is the sample mean $\bar{x}$. From Table 12.1.2 we find $\bar{x} = 4.2$ (the average number of emissions per time interval), hence, if x has a Poisson distribution, we estimate $p(x)$ as follows:

$$p(x) \simeq \frac{e^{-4.2}(4.2)^x}{x!}.$$

Hence we expect $np(x) = 100[e^{-4.2}(4.2)^x/x!]$ time intervals to emit x α particles, $x = 0, 1, 2, \ldots$, as shown in column 3 of Table 12.1.2. Note that several of the expected frequencies are less than 5. We proceed by grouping the classes until the "expected frequencies" are greater or equal to 5. For example, we group the first and second classes, and the last four, and apply the χ^2 test to the resulting Table 12.1.3.

TABLE 12.1.3

RESULTS OF TABLE 12.1.2 AFTER SOME GROUPING

Observed No. of Samples, f_i	Expected No., $n\theta_i$	$(f_i - n\theta_i)^2$	$\dfrac{(f_i - n\theta_i)^2}{n\theta_i}$
6	7.8	3.24	.415
16	13.2	7.84	.594
17	18.5	2.25	.122
26	19.4	43.56	2.245
11	16.3	28.09	1.723
9	11.4	5.76	.505
9	6.9	4.41	.639
6	6.3	.09	.014
			6.257

As can be seen from Table 12.1.3, the chi-square test is applied to $k = 8$ classes. Furthermore, the expected frequencies are calculated by using estimates of $n\theta_1, n\theta_2, \ldots, n\theta_8$, where each θ_i is computed from the Poisson distribution (12.1.3) after estimating the single parameter μ. Hence the number of degrees of freedom involved in applying the chi-square test is $8 - 1 - 1 = 6$, using the fact that $k = 8$, $c = 1$.

The upper 5% significance point of χ_6^2 is 12.59, hence we accept the hypothesis that the data have behaved like a sample coming from a Poisson distribution, and we estimate the mean of the distribution as 4.2.

Example 12.1(c). Let us apply the chi-square test (12.1.2) to examine the hypothesis that the frequencies in Table 5.2.1 behave as if the sample comes from a normal distribution.

It will be noted from Chapter 5 that the estimates of μ and σ were $\bar{x} = 1.527$ and $s = .101$. We proceed by first extracting from Table 5.2.1 the data shown in the first three columns of Table 12.1.4, where the cells are denoted by outcomes $A_0, A_1, \ldots, A_{11}$, with A_0, A_{11} denoting the events "an observation falls in $(-\infty, 1.275)$" and "an observation falls in $(1.775, +\infty)$" respectively.

The entries in column 4 of Table 12.1.4 are found in the following way. We are testing the hypothesis that the data of Table 5.2.1 are samples from a normal distribution, and if the hypothesis is true, a "reasonable" estimate of $P(x \leq x')$ is given by

$$(12.1.4) \quad \int_{-\infty}^{x'} \frac{1}{.101\sqrt{2\pi}} \exp\left[-\frac{1}{2}\left(\frac{x - 1.527}{.101}\right)^2\right] dx = \int_{-\infty}^{\frac{x' - 1.527}{.101}} \phi(z)\, dz$$

TABLE 12.1.4

OBSERVED FREQUENCIES FROM TABLE 5.2.1 AND ESTIMATED
EXPECTED FREQUENCIES IF SAMPLE COMES FROM A NORMAL
DISTRIBUTION

(1)	(2)	(3)	(4)	(5)
	Upper		Expected	Expected
	Cell	Observed	Cumulative	Class
Outcomes	Boundaries	Frequency	Frequency	Frequency
A_0	1.275	0	.5	.5
A_1	1.325	1	1.7	1.2
A_2	1.375	5	5.0	3.3
A_3	1.425	6	11.7	6.7
A_4	1.475	13	22.8	11.1
A_5	1.525	8	36.9	14.1
A_6	1.575	17	51.2	14.3
A_7	1.625	14	62.6	11.4
A_8	1.675	7	69.6	7.0
A_9	1.725	1	73.1	3.5
A_{10}	1.775	3	74.5	1.4
A_{11}	∞	0	75.0	.5

where we have used for the values of μ and σ the estimates $\bar{x} = 1.527$ and
$s = .101$ respectively. Multiplying this estimate by $n = 75$, we obtain the
expected cumulative frequency of x'. We will let x' successively take as
values the entries in column 2 of Table 12.1.4.

Thus we obtain estimates of the cumulative frequencies (column 4).
By taking the difference of successive entries of column 4, we obtain the
expected class frequencies under the hypothesis of normality (column 5).
By grouping classes at each end of the distribution in Table 12.1.4, we obtain
the abridged Table 12.1.5.

TABLE 12.1.5

RESULTS OF TABLE 12.1.4 AFTER SOME GROUPING

Outcomes	Observed Frequency, f_i	Expected Frequency, $n\theta_i$	$\dfrac{(f_i - n\theta_i)^2}{n\theta_i}$
A_0, A_1, A_2	6	5.0	.20
A_3	6	6.7	.07
A_4	13	11.1	.33
A_5	8	14.1	2.64
A_6	17	14.3	.51
A_7	14	11.4	.59
A_8	7	7.0	.00
A_9–A_{11}	4	5.4	.36
	75	75	4.70

For this example $k = 8$, $c = 2$, hence the number of degrees of freedom is $8 - 2 - 1 = 5$. The observed value of χ^2 is 4.70, and consulting Table III, Appendix VII, we note that $\chi^2_{5;.05} = 11.07$; hence at the 5% level of significance we accept the hypothesis that the sample which produced the frequencies of column 3 of Table 12.1.4 behaves like a sample from a normal distribution.

12.2   CONTINGENCY TABLES

(a) The 2 × 2 Case; Parameters Known

Suppose the outcome of each trial in an experiment can be classified into one and only one of the four mutually exclusive and exhaustive classes $A \cap B$, $A \cap \bar{B}$, $\bar{A} \cap B$, $\bar{A} \cap \bar{B}$ with probabilities $\theta_1\tau_1$, $\theta_1\tau_2$, $\theta_2\tau_1$, $\theta_2\tau_2$ respectively, where θ_1, θ_2, τ_1, τ_2 are all positive and $\theta_1 + \theta_2 = 1$ and $\tau_1 + \tau_2 = 1$. This means that factors A and B are independent, and $P(A) = \theta_1$, $P(\bar{A}) = \theta_2$, $P(B) = \tau_1$, $P(\bar{B}) = \tau_2$.

These four classes $A \cap B$, $A \cap \bar{B}$, $\bar{A} \cap B$, $\bar{A} \cap \bar{B}$ and their probabilities can be arranged as shown in Table 12.2.1.

TABLE 12.2.1

CLASSES AND THEIR PROBABILITIES IN A 2 × 2 CONTINGENCY
TABLE IN WHICH ROWS AND COLUMNS ARE INDEPENDENT

	B	$\bar{B}$	
A	$\theta_1\tau_1$	$\theta_1\tau_2$	θ_1
$\bar{A}$	$\theta_2\tau_1$	$\theta_2\tau_2$	θ_2
	τ_1	τ_2	1

Now suppose n independent trials are performed and that f_{11} trials result in class $A \cap B$, f_{12} in $A \cap \bar{B}$, f_{21} in $\bar{A} \cap B$, and f_{22} in $\bar{A} \cap \bar{B}$. The experimental results can be arranged as shown in Table 12.2.2, where the marginal totals $f_{1.}, f_{2.}, f_{.1}, f_{.2}$ are defined as follows:

$$f_{1.} = f_{11} + f_{12}, \qquad f_{2.} = f_{21} + f_{22}, \qquad f_{.1} = f_{11} + f_{21}, \qquad f_{.2} = f_{12} + f_{22}.$$

Note that

$$f_{1.} + f_{2.} = n, \qquad f_{.1} + f_{.2} = n.$$

Under the assumption of independence of the "A" and "B" factors, the frequencies $f_{11}, f_{12}, f_{21}, f_{22}$ are random variables having the multinomial distribution

(12.2.1) $$\frac{n!}{f_{11}!\,f_{12}!\,f_{21}!\,f_{22}!}\,(\theta_1\tau_1)^{f_{11}}(\theta_1\tau_2)^{f_{12}}(\theta_2\tau_1)^{f_{21}}(\theta_2\tau_2)^{f_{22}},$$

where $0 \le f_{ij} \le n$ and $\sum_{i=1}^{2}\sum_{j=1}^{2} f_{ij} = n$.

TABLE 12.2.2

CLASSES AND THEIR FREQUENCIES IN n INDEPENDENT
TRIALS OF A 2×2 EXPERIMENT

	B	$\bar{B}$	
A	f_{11}	f_{12}	$f_{1\cdot}$
$\bar{A}$	f_{21}	f_{22}	$f_{2\cdot}$
	$f_{\cdot 1}$	$f_{\cdot 2}$	n

In fact it is easy to see that any one of the f_{ij} has the binomial distribution, size n, parameter $\theta_i\tau_j$, so that under the conditions of independence of the A and B factors, and independence of the n trials, the expectation of f_{ij} is $n\theta_i\tau_j$, that is, n times the entries of Table 12.2.1.

If we knew the values of $\theta_1, \theta_2, \tau_1, \tau_2$, we would be able to test the assumption of independence of the A and B factors through use of the following theorem.

THEOREM 12.2.1. *If $f_{11}, f_{12}, f_{21}, f_{22}$ has probability function given by (12.2.1), then for large n, the quantity*

(12.2.2) $$\chi^2 = \sum_{i=1}^{2}\sum_{j=1}^{2}\frac{(f_{ij} - n\theta_i\tau_j)^2}{n\theta_i\tau_j}$$

has, approximately, the chi-square distribution with 3 degrees of freedom.

Hence, if the observed value of χ^2 determined by (12.2.2) exceeds $\chi^2_{3;\alpha}$, we would reject the hypothesis that the A and B classifications are independent at the $100\alpha\%$ level of significance, and the sample evidence then, would support the assertion that there is a significant degree of dependence between the A and B classifications, at the $100\alpha\%$ level of significance.

Example 12.2(a). A cross breeding experiment with two strains of Drosophila is conducted to determine whether eye color and type of wing are independent characteristics or not. It is known that the probability of progeny of the strains of Drosophila used having dull color eyes (A) is 3/4, while the probability of progeny of these two strains having type B wing is 2/3. Six hundred progeny are selected at random. If the two classifications (type of eye color and type of wing) are independent, the expected numbers are given in brackets in Table 12.2(a), and the observed numbers f_{ij} are entered below the expected numbers in the same table.

TABLE 12.2(a)

	$B(2/3)$	$\bar{B}(1/3)$
$A(3/4)$	(300) 313	(150) 135
$\bar{A}(1/4)$	(100) 93	(50) 59

Does the sample evidence support the assumption of independence (at the 5% level)?

The observed value of $\chi_3^2 = \sum_{i=1}^{2}\sum_{j=1}^{2}(f_{ij} - n\theta_i\tau_j)^2/n\theta_i\tau_j$ is ($n = 600$; $\theta_1 = 3/4$, $\theta_2 = 1/4$, $\tau_1 = 2/3$, $\tau_2 = 1/3$)

$$\chi_3^2 = \frac{(313-300)^2}{300} + \frac{(135-150)^2}{150} + \frac{(93-100)^2}{100} + \frac{(59-50)^2}{50}$$

$$= 0.56333 + 1.50000 + 0.49000 + 1.62000$$
$$= 4.17333.$$

Now the $\chi_{3;.05}^2$ is 7.81473, that is, the critical region of level .05 is defined by: Reject if the observed value of $\sum_{i=1}^{2}\sum_{j=1}^{2}(f_{ij} - n\theta_i\tau_j)^2/n\theta_i\tau_j$ is greater than 7.81473. The observed value in this example is 4.17333 < 7.81473, that is, the observed value does not lie in the critical region, and so we accept the hypothesis of independence of the factors of eye color and type of wing.

(b) The 2 × 2 Case; Parameters Unknown

In general, however, we do *not* know $\theta_1, \theta_2, \tau_1, \tau_2$ and have to estimate them from the experimental results in Table 12.2.2. Again, it is easy to see that any *one* of the so-called marginal totals $f_1.$ or $f_2.$ or $f._1$ or $f._2$ have the binomial distribution. For example, $f_1.$ is binomial, parameter θ_1, sample size n. This means that $f_1.$ has expectation $n\theta_1$ so that $\hat{\theta}_1 = f_1./n$ is an unbiased point estimator of θ_1. Summarizing, we have that

$$\hat{\theta}_1 = \text{estimator for } \theta_1 = f_1./n$$
(12.2.3) $$\hat{\theta}_2 = \text{estimator for } \theta_2 = f_2./n$$
$$\hat{\tau}_1 = \text{estimator for } \tau_1 = f._1/n$$
$$\hat{\tau}_2 = \text{estimator for } \tau_2 = f._2/n,$$

where $0 \le \hat{\theta}_1, \hat{\tau}_1 \le 1$ and $\hat{\theta}_2 = 1 - \hat{\theta}_1$, $\hat{\tau}_2 = 1 - \hat{\tau}_1$.

If we have to replace θ_1, θ_2, τ_1, τ_2 in (12.2.2) by their estimators given in (12.2.3), the resulting χ^2, namely,

$$(12.2.4) \qquad \chi^2 = \sum_{i=1}^{2} \sum_{j=1}^{2} \left[\frac{(f_{ij} - f_{i.}f_{.j}/n)^2}{f_{i.}f_{.j}/n} \right]$$

for large n has approximately a chi-square distribution with $4 - 2 - 1 = 1$ degree of freedom. Here the number of unknown parameters c which have to be estimated is 2, since essentially we have to estimate only one θ and only one τ.

The quantity χ^2 given by (12.2.4) can then be used for testing the hypothesis that the A and B classifications are independent, that is, that the sample displayed in Table 12.2.2 can be regarded as having the probability distribution (12.2.1), in which the probabilities of the four classes $A \cap B$, $A \cap \bar{B}$, $\bar{A} \cap B$, $\bar{A} \cap \bar{B}$ are shown in Table 12.2.1, where θ_1, θ_2, τ_1, τ_2 are unknown.

Thus, if the observed value of χ^2 determined from an experiment by (12.2.4) exceeds $\chi^2_{1;\alpha}$, we reject the hypothesis that the A and B classifications are independent at the $100\alpha\%$ level of significance, and conclude that there is a significant degree of dependence between the A and B classifications at the $100\alpha\%$ level of significance.

Example 12.2(b). In an experiment on bars of nylon, 800 randomly chosen bars were found to be such that 360 of them had been subjected to a 60°C heat treatment and 440 to a 90°C heat treatment. Each of the bars was then further classified as brittle or nonbrittle, with results shown in Table 12.2.3.

TABLE 12.2.3

RESULTS OF AN EXPERIMENT ON EFFECT
OF HEAT TREATMENT ON BRITTLENESS
OF NYLON BARS

	Brittle (B)	Nonbrittle ($\bar{B}$)	
60° (A)	(132) 77	(228) 283	360
90° ($\bar{A}$)	(162) 217	(278) 223	440
	294	506	800

The problem is to test the hypothesis that brittleness is independent of heat treatment.

The observed frequencies $f_{11}, f_{12}, f_{21}, f_{22}$ for this experiment are the numbers 77, 283, 217, and 223, respectively, and the expected frequencies $f_{1.}f_{.1}/n, f_{1.}f_{.2}/n, f_{2.}f_{.1}/n, f_{2.}f_{.2}/n$ are 132, 228, 162, and 278, respectively. The observed value of χ^2 from these observations is

$$\chi^2 = \frac{(77 - 132)^2}{132} + \frac{(283 - 228)^2}{228} + \frac{(217 - 162)^2}{162} + \frac{(223 - 278)^2}{278} = 65.7.$$

The value of $\chi^2_{1;.05}$, that is, the critical value of chi-square with 1 degree of freedom at the 5% level of significance, is 3.84, thus indicating that we reject the hypothesis of independence between heat treatment and brittleness as suggested by the inspection of the data.

(c) The $r \times s$ Contingency Table

The results in Section 12.2a and 12.2b extend in a straightforward manner to the case of testing $r \times s$ contingency tables for independence. We have rs mutually exclusive and exhaustive classes, $A_i \cap B_j$, $i = 1, \ldots, r$, $j = 1, \ldots, s$, the A and B classifications being independent with probabilities of the events $A_i \cap B_j$ equal to $\theta_i \tau_j$, where the θ_i and τ_j are all positive with $\theta_1 + \cdots + \theta_r = 1$, and $\tau_1 + \cdots + \tau_s = 1$.

If n independent trials are performed, let f_{ij} be the number of outcomes in class $A_i \cap B_j$, where $\sum_{j=1}^{s}\sum_{i=1}^{r} f_{ij} = n$. Then the f_{ij} are random variables having the multinomial distribution

$$(12.2.5) \qquad \frac{n!}{f_{11}! f_{12}! \cdots f_{rs}!} (\theta_1\tau_1)^{f_{11}}(\theta_1\tau_2)^{f_{12}} \cdots (\theta_r\tau_s)^{f_{rs}}.$$

If the θ_i and τ_j are *known* and if n is large, the quantity

$$(12.2.6) \qquad \chi^2 = \sum_{j=1}^{s}\sum_{i=1}^{r}\left[\frac{(f_{ij} - n\theta_i\tau_j)^2}{n\theta_i\tau_j}\right]$$

has approximately a chi-square distribution with $rs - 1$ degrees of freedom.

However, if the θ_i and τ_j are unknown and are replaced by the estimators $f_{i.}/n$ and $f_{.j}/n$ respectively, where $f_{i.} = f_{i1} + \cdots + f_{is}$, and $f_{.j} = f_{1j} + \cdots + f_{rj}$, the resulting χ^2, namely,

$$(12.2.7) \qquad \chi^2 = \sum_{j=1}^{s}\sum_{i=1}^{r}\left[\frac{(f_{ij} - f_{i.}f_{.j}/n)^2}{f_{i.}f_{.j}/n}\right]$$

has, for large n, approximately a chi-square distribution with $rs - [(r - 1) + (s - 1)] - 1 = (r - 1)(s - 1)$ degrees of freedom, since the number of constants in χ^2 as given by (12.2.6) which have to be estimated are $(r - 1) + (s - 1)$, that is, $r - 1$ θ's and $s - 1$ τ's.

PROBLEMS

12.1. Suppose a coin is tossed 1000 times with the result that 462 heads and 538 tails are obtained. Are these results consistent with the hypothesis that the coin is behaving like a "true" coin at the 5% level of significance, using the chi-square test?

12.2. Table P12.2 gives the month of birth of a sample of 748 artists. Test, at the 5% level of significance, the null hypothesis that there is no seasonal variation over month of year in which artists are born.

TABLE P12.2

Month of Birth	Frequency
Jan.	66
Feb.	74
March	67
April	60
May	61
June	51
July	50
Aug.	60
Sept.	67
Oct.	59
Nov.	73
Dec.	60
Total	748

12.3. Five thumb tacks of a certain type were thrown 200 times and the number of tacks falling point up in each of the throws was counted. The experimental results are shown in Table P12.3.

TABLE P12.3

No. Falling Point Up, x	Frequency
0	5
1	27
2	41
3	67
4	43
5	17
Total	200

(a) Estimate the probability θ that a tack falls point up.
(b) Test the null hypothesis that the distribution of x, the number of tacks

falling point up, has the binomial distribution $\binom{5}{x} \theta^x (1 - \theta)^{5-x}$, $x = 0, 1, \ldots, 5$, using the estimate of θ found in Problem 12.3a.

12.4. Using the method of the example in Section 12.1 (see Table 12.1.4), fit a normal distribution to the data of Problems 5.4, 5.5, 5.6 and test the null hypothesis that the data behave like a sample from a normal population.

12.5. Houses with and without air conditioners on nine different streets in a certain city are shown in Table P12.5 [from Brownlee (1960)].

TABLE P12.5

Street	With Air Conditioners	Without Air Conditioners
1	5	18
2	8	35
3	18	25
4	3	38
5	17	24
6	11	31
7	25	17
8	19	20
9	18	18

Test (at the 5% level) the null hypothesis that having an air conditioner is independent of street locations of the houses.

12.6. A small roulette wheel was spun 380 times and yielded frequencies shown in Table P12.6 for the 38 roulette numbers (taken in pairs):

TABLE P12.6

Nos.	Frequency	Nos.	Frequency	Nos.	Frequency
0–00	24	13–14	29	25–26	21
1–2	16	15–16	21	27–28	14
3–4	19	17–18	17	29–30	25
5–6	19	19–20	25	31–32	23
7–8	25	21–22	18	33–34	16
9–10	10	23–24	20	35–36	16
11–12	22				
				Total	380

Test the hypothesis that the wheel is true, using the chi-square test at the 1% level of significance.

12.7. Pieces of vulcanite were examined according to porosity and dimensional defects, and the results are shown in Table P12.7 [data from Hald (1952)].

TABLE P12.7

	Porous	Nonporous
With defective dimensions	142	331
Without defective dimensions	1233	5099

Test the hypothesis that the two criteria of classification are independent.

12.8. In field tests of mine fuses, 216 of each of two types of fuses, A and B, chosen at random from large lots, were buried and simulated tanks ran over them. The number of "hits" and not hits" were recorded for each type of fuse, with the results shown in Table P12.8. (From Ordnance Corps Pamphlet ORD P 20 = 111):

TABLE P12.8

Fuse Type	Hit	Not Hit	Total
A	181	35	216
B	160	56	216
Total	341	91	432

Are the proportions of hits for the two types of fuses significantly different at the 5% level of significance?

12.9. The lateral deflection and range in yards obtained from the firing of each of 75 rockets are shown in Table P12.9 [from NAVORD Report 3369 (1955)].

TABLE P12.9

Lateral Deflection (yards)

Range (yards)	−250 to −51	−50 to +49	50 to 199	Total
0–1199	5	9	7	21
1200–1799	7	3	9	19
1800–2699	8	21	6	35
Total	20	33	22	75

Test at the 5% level of significance the hypothesis that lateral deflection and range are independent.

12.10. A study was made of the effect of the time of work on the quality of work in a certain plant. It was the practice in the plant for a crew to change shifts once a month. A study of 3 months of operations by one crew that remained intact for the entire period showed the numbers of defective and nondefective items turned out as given in Table P12.10 [from Duncan (1958)].

TABLE P12.10

Shift	Defective	Nondefective
1 (8:00–4:00)	52	921
2 (4:00–12:00)	61	902
3 (12:00–8:00)	73	851

Would you conclude from these data that the time of work affected significantly the quality of the results? Justify your answer.

12.11. The number of contaminated tablets were counted for 720 samples. The results are shown in Table P12.11.

TABLE P12.11

No. of Contaminated Tablets in Sample	Observed Sample Frequency
0	116
1	194
2	184
3	115
4	63
5	24
6	12
7	6
8	3
9	2
10	1

Fit a Poisson distribution to this data and test for goodness of fit.

12.12. Suppose an experiment of n independent trials results in Table 12.2.2, where it is *known* that the A and B classifications are independent. Suppose further that we wish to test $\theta_1 = \theta_{10}$, $\tau_1 = \tau_{10}$ at the $100\alpha\%$ level of significance. If n is large, how would you do this?

12.13. A certain strain of guinea pig is such that 75% of its progeny are born with white eyes, and 75% of the progeny are born with webbed feet. A sample of 160 piglets from newly born litters is classified and gives the data below. Test whether the modes of classification are independent.

	Webbed feet	Non-webbed feet
White eyes	94	33
Coloured eyes	28	5

CHAPTER 13

Order Statistics

13.1 INTRODUCTORY REMARKS

We recall from Section 6.8 that a random sample of size n from a population having a probability function $p(x)$ [or a probability density function $f(x)$] was defined as a set of n independent random variables $x_1, \ldots, x_n$ whose probability function is $p(x_1) \cdots p(x_n)$ [or whose probability density function is $f(x_1) \cdots f(x_n)$]. The random variables $x_1, \ldots, x_n$ are sometimes called *elements* of the sample. In that section we pointed out that if the mean and variance of the probability function $p(x)$ [or probability density function $f(x)$] are μ and σ^2 respectively, then the sample mean $\bar{x}$ is a random variable which has mean μ and variance σ^2/n.

In the particular case where the sample comes from the normal distribution $N(\mu, \sigma^2)$, we have seen from Theorem 7.4.1 that the sample mean $\bar{x}$ has as its distribution the normal distribution $N(\mu, \sigma^2/n)$. As a matter of fact, even if the sample is not from a normal population but a non-normal one with mean μ and variance σ^2, it can be shown that, if n is large, the sample mean $\bar{x}$ has a distribution which is approximately $N(\mu, \sigma^2/n)$. (See the central limit theorem in Appendix IV.)

We shall now consider probability distributions of statistics of a sample obtained if one orders the n elements of the sample from least to greatest, and if sampling is done on a *continuous* random variable x whose p.d.f. is $f(x)$. Suppose we let $x_1, \ldots, x_n$ be a sample of n elements from a population having continuous p.d.f. $f(x)$. We note that since x is a continuous random variable [see (6.3.3.), page 82] the probability of x assuming the specific value x' is $\lim_{\Delta x \to 0} [F(x' + \Delta x) - F(x)] = 0$. In fact by a straightforward conditional probability argument we can show that for any two of $(x_1, \ldots, x_n)$ the probability of their having the same value is zero.

Consider then the observations $(x_1, \ldots, x_n)$ from a population having a p.d.f. $f(x)$. Let

$$x_{(1)} = \text{smallest of } (x_1, \ldots, x_n),$$
$$x_{(2)} = \text{second smallest of } (x_1, \ldots, x_n),$$

.

.

.

$$x_{(k)} = k\text{-th smallest of } (x_1, \ldots, x_n),$$

.

.

.

$$x_{(n)} = \text{largest of } (x_1, \ldots, x_n).$$

The quantities $x_{(1)}, x_{(2)}, \ldots, x_{(n)}$ are random variables and are called the *order statistics* of the sample, $x_{(1)}$ is called the *smallest* element in the

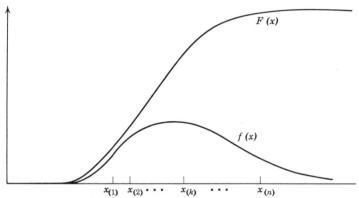

Figure 13.2.1. The order statistics of a sample of size n on a continuous random variable x having p.d.f. $f(x)$ or c.d.f. $F(x)$.

sample, $x_{(n)}$ the *largest*, $x_{(k)}$ the *kth order statistic*, and $x_{(m+1)}$ the *median* of a sample of size $n = 2m + 1$; $R = x_{(n)} - x_{(1)}$ is called the sample *range*. The order statistics $x_{(1)}, \ldots, x_{(n)}$ are represented in Figure 13.2.1.

13.2   DISTRIBUTION OF THE LARGEST ELEMENT IN A SAMPLE

As we have just stated, $x_{(n)}$ is the largest element in the sample $x_1, \ldots,$ x_n. If the sample is drawn from a population having p.d.f. $f(x)$, let $F(x)$ be the c.d.f. of the population defined by

(13.2.1)
$$F(x) = \int_{-\infty}^{x} f(x) \, dx.$$

The c.d.f. of $x_{(n)}$ is given by

$$(13.2.2) \qquad P(x_{(n)} < u) = P(x_1, \ldots, x_n \text{ are all } \leq u)$$
$$= F^n(u).$$

If we denote the c.d.f. of the largest value by $G(u)$, we have

$$(13.2.3) \qquad G(u) = F^n(u).$$

That is, if we take a random sample of n elements from a population whose p.d.f. is $f(x)$ [or whose c.d.f. is $F(x)$], then the c.d.f. $G(u)$ of the largest element in the sample, denoted briefly by u, is given by (13.2.3).

If we denote the p.d.f. of the largest element by $g(u)$ [that is, $g(u) = \frac{d}{du} G(u)$] and the probability element by $g(u)\,du$, we have

$$(13.2.4) \qquad g(u)\,du = nF^{n-1}(u)f(u)\,du.$$

This states that the probability that the largest element $x_{(n)}$ in a sample lies in the interval $(u, u + du)$ is given by the right-hand side of (13.2.4).

> **Example 13.2(a).** Suppose the mortality of a certain type of mass-produced light bulb is such that the probability that a bulb of this type taken at random from production burns out during the time interval $(t, t + dt)$ is given by
>
> $$(13.2.5) \qquad ce^{-ct}\,dt, \qquad t > 0,$$
>
> where c is some positive constant. If n bulbs of this type are taken at random, let their lives be $t_1, \ldots, t_n$. If the order statistics are $t_{(1)}, \ldots, t_{(n)}$, then $t_{(n)}$ is the life of the last bulb to burn out. We wish to determine the probability that the last one to burn out expires during the time interval $(u, u + du)$.
>
> To solve this problem we may think of a population of bulbs whose p.d.f. of length of life is
>
> $$(13.2.6) \qquad f(t) = ce^{-ct}, \qquad t > 0,$$
> $$= 0, \qquad t < 0.$$
>
> The c.d.f. is given by
>
> $$(13.2.7) \qquad F(t) = \int_{-\infty}^{t} f(t)\,dt = \int_0^t ce^{-ct}\,dt = 1 - e^{-ct}.$$
>
> Applying (13.2.4), we therefore have as the p.d.f. of u,
>
> $$g(u) = nc(1 - e^{-cu})^{n-1}e^{-cu},$$
>
> that is, the probability the last bulb to burn out expires during the time interval $(u, u + du)$ is given by $g(u)\,du$, where
>
> $$(13.2.8) \qquad g(u)\,du = nc(1 - e^{-cu})^{n-1}e^{-cu}\,du.$$

13.3 DISTRIBUTION OF THE SMALLEST ELEMENT IN A SAMPLE

We now wish to find the expression for the c.d.f. of the smallest element $x_{(1)}$ in the sample $x_1, \ldots, x_n$. That is, we want to determine $P(x_{(1)} \leq v)$ as a function of v. Denote this function by $G(v)$; we have

(13.3.1)
$$G(v) = P(x_{(1)} \leq v)$$
$$= 1 - P(x_{(1)} > v).$$

But

(13.3.2)
$$P(x_{(1)} > v) = P(x_1, \ldots, x_n \text{ are all } > v)$$
$$= [1 - F(v)]^n.$$

Therefore the c.d.f. $G(v)$ of the smallest element in the sample is given by

(13.3.3)
$$G(v) = 1 - [1 - F(v)]^n.$$

The p.d.f., say $g(v)$, of the smallest element in the sample is therefore obtained by taking the derivative of the right-hand side of (13.3.3) with respect to v. We thus find

(13.3.4)
$$g(v) = n[1 - F(v)]^{n-1}f(v),$$

and the probability element of $x_{(1)}$, that is, the probability that $x_{(1)}$ lies in $(v, v + dv)$, is given by

(13.3.5)
$$g(v)\, dv = n[1 - F(v)]^{n-1}f(v)\, dv.$$

Example 13.3(a). Suppose links of a certain type used for making chains are such that the population of individual links have breaking strengths x with p.d.f.

(13.3.6)
$$f(x) = \frac{(m + 1)(m + 2)}{c^{m+2}}\, x^m(c - x), \qquad 0 < x < c,$$
$$= 0 \text{ otherwise}$$

where c and m are certain positive constants. If a chain is made up of n links of this type taken at random from the population of links, what is the probability law of the breaking strength of the chain?

Since the breaking strength of a chain is equal to the breaking strength of its weakest link, the problem reduces to finding the p.d.f. of the smallest element $x_{(1)}$ in a sample of size n from the p.d.f. $f(x)$ given above.

First we find the c.d.f. $F(x)$ of breaking strengths of individual links by performing the following integration:

(13.3.7)
$$F(x) = \int_{-\infty}^{x} f(x)\, dx = \frac{(m + 1)(m + 2)}{c^{m+2}} \int_0^x x^m(c - x)\, dx,$$

that is

(13.3.8)
$$F(x) = (m + 2)\left(\frac{x}{c}\right)^{m+1} - (m + 1)\left(\frac{x}{c}\right)^{m+2}$$

With the use of (13.3.5) and (13.3.6) we obtain the probability element of the breaking strength of an n-link chain made from a random sample of n of these links;

$$(13.3.9) \qquad g(v)\, dv = \frac{n(m+1)(m+2)v^m}{c^{m+2}}$$

$$\times \left[1 - (m+2)\left(\frac{v}{c}\right)^{m+1} + (m+1)\left(\frac{v}{c}\right)^{m+2} \right]^{n-1} (c-v)\, dv$$

for $0 < v < c$, and $g(v) = 0$ otherwise.

13.4 DISTRIBUTION OF THE MEDIAN OF A SAMPLE AND OF THE kth-ORDER STATISTIC

Suppose we have a sample of $2m+1$ elements $x_1, \ldots, x_{2m+1}$ from a population having p.d.f. $f(x)$ [and c.d.f. $F(x)$]. If we form the order statistics $x_{(1)}, \ldots, x_{(2m+1)}$ of the sample, then $x_{(m+1)}$ is called the *sample median*. Let us find the probability element for the median, that is, the probability that $x_{(m+1)}$ lies within the interval $(w, w+dw)$, where $dw > 0$. Let us cut the x-axis into the following three disjoint intervals:

$$(13.4.1) \qquad \begin{aligned} I_1 &= (-\infty, w], \\ I_2 &= (w, w+dw], \\ I_3 &= (w+dw, +\infty). \end{aligned}$$

Then the probabilities p_1, p_2, p_3 that an element x drawn from the population with p.d.f. $f(x)$ will lie in the intervals I_1, I_2, I_3 are given by

$$(13.4.2) \qquad \begin{aligned} p_1 &= F(w), \\ p_2 &= F(w+dw) - F(w), \\ p_3 &= 1 - F(w+dw) \end{aligned}$$

respectively.

If we take a sample of size $2m+1$ from the population with p.d.f. $f(x)$, the median of the sample will lie in $(w, w+dw)$ if and only if m sample elements fall in $(-\infty, w]$, one sample element falls in $(w, w+dw]$ and m sample elements fall in $(w+dw, +\infty)$. The probability that all of this occurs is obtained by applying the multinomial probability distribution discussed in Section 3.4 of Chapter 3. This gives

$$(13.4.3) \qquad \frac{(2m+1)!}{(m!)^2}(p_1)^m(p_2)^1(p_3)^m.$$

But substituting the values of p_1, p_2, p_3 from (13.4.2) into (13.4.3), we obtain

$$(13.4.4) \qquad \frac{(2m+1)!}{(m!)^2} F^m(w)[1 - F(w+dw)]^m[F(w+dw) - F(w)].$$

Now we may write

(13.4.5) $F(w + dw) = F(w) + f(w)\,dw.$

Substituting this expression into (13.4.4), we find that (ignoring terms of order $(dw)^2$ and higher)

(13.4.6) $P(w < x_{(m+1)} < w + dw) = \dfrac{(2m + 1)!}{(m!)^2} F^m(w)[1 - F(w)]^m f(w)\,dw.$

The p.d.f. $g(w)$ of the median is the coefficient of dw on the right-hand side of (13.4.6), and the probability element of the median is $g(w)\,dw$, where

(13.4.7) $g(w)\,dw = \dfrac{(2m + 1)!}{(m!)^2} F^m(w)[(1 - F(w)]^m f(w)\,dw,$

while the sample space of the median is the same as the sample space of x in the population c.d.f. $F(x)$.

> **Example 13.4(a).** Suppose $2m + 1$ points are taken "at random" on the interval $(0, 1)$. What is the probability that the median of the $2m + 1$ points falls in $(w, w + dw)$?
>   In this example the p.d.f. of a point x taken "at random" on $(0, 1)$ is defined as
> $$f(x) = 1, \quad 0 < x < 1,$$
> $$ = 0, \text{ for all other values of } x.$$
> Then
> $$F(x) \;= 0, \qquad x \le 0,$$
> $$ = x, \qquad 0 < x < 1,$$
> $$ = 1, \qquad x \ge 1.$$
>
> The probability element $g(w)\,dw$ of the median in a sample of $2m + 1$ points is therefore given by
> (13.4.8) $g(w)\,dw = \dfrac{(2m + 1)!}{(m!)^2} w^m (1 - w)^m\,dw.$

More generally, if we have a sample of n elements, say $x_1, \ldots, x_n$, from a population having c.d.f. $f(x)$ and if $x_{(k)}$ is the kth-order statistic of the sample (the kth smallest of $x_1, \ldots, x_n$), then we can show, as in the case of the median,

(13.4.9)

$P(w < x_{(k)} < w + dw) = \dfrac{n!}{(k - 1)!\,(n - k)!} F^{k-1}(w)[1 - F(w)]^{n-k} f(w)\,dw,$

except for terms of order $(dw)^2$ and higher. Hence the probability element $g(w)\,dw$ of the kth-order statistic $x_{(k)}$ [that is, the probability that $x_{(k)}$

lies in $(w, w + dw)]$ is given by

(13.4.10) $g(w)\, dw = \dfrac{n!}{(k-1)!\,(n-k)!}\, F^{k-1}(w)[1 - F(w)]^{n-k} f(w)\, dw.$

Note that the functional form of the probability element on the right of (13.4.10) reduces to that on the right of (13.3.5) if $k = 1$, and to that on the right of (13.2.4) if $k = n$, as one would expect since in these two cases the kth-order statistic $x_{(k)}$ becomes the smallest element $x_{(1)}$ and the largest element $x_{(n)}$ respectively.

> ***Example 13.4(b).*** If n points $x_1, \ldots, x_n$ are taken "at random" on the interval $(0, 1)$ what is the probability element of the kth point $x_{(k)}$ from the left?
>
> By examining the method of solving the problem in Example 13.4(a), it is evident that the probability element of $x_{(k)}$ [that is, the probability that $x_{(k)}$ lies in the interval $(w, w + dw)$] is similarly given by:
>
> (13.4.11) $g(w)\, dw = \dfrac{n!}{(k-1)!(n-k)!}\, w^{k-1}(1 - w)^{n-k}\, dw.$

13.5   DISTRIBUTION OF THE RANGE OF A SAMPLE

If $x_{(1)}, \ldots, x_{(n)}$ are the order statistics of a sample $x_1, \ldots, x_n$, from a population with p.d.f. $f(x)$ and c.d.f. $F(x)$, the difference $x_{(n)} - x_{(1)}$ is called the sample range R. We now consider the probability distribution of R. To find the probability distribution of R, we must consider the joint probability distribution of $x_{(1)}$ and $x_{(n)}$. That is, we must first determine the probability that $x_{(n)}$ lies in the interval $(u, u + du)$ *and also* that $x_{(1)}$ lies in the interval $(v, v + dv)$, $v < u$. If we cut the x-axis into the five disjoint intervals,

$$
\begin{aligned}
I_1 &= (-\infty, v], \\
I_2 &= (v, v + dv], \\
I_3 &= (v + dv, u], \\
I_4 &= (u, u + du], \\
I_5 &= (u + du, +\infty),
\end{aligned}
$$

(13.5.1)

then the amounts of probability on these intervals are given by

$$
\begin{aligned}
p_1 &= F(v), \\
p_2 &= F(v + dv) - F(v) = f(v)\, dv, \\
p_3 &= F(u) - F(v + dv) \simeq F(u) - F(v), \\
p_4 &= F(u + du) - F(u) = f(u)\, du, \\
p_5 &= 1 - F(u + du) \simeq 1 - F(u).
\end{aligned}
$$

(13.5.2)

It will be seen that $x_{(1)}$ falls in $(v, v + dv]$ and $x_{(n)}$ falls in $(u, u + du]$ if and only if all of the following events occur:

0 elements of the sample fall in I_1,

1 element of the sample falls in I_2,

(13.5.3) $n - 2$ elements of the sample fall in I_3,

1 element of the sample falls in I_4,

0 elements of the sample fall in I_5.

Hence, it follows from the multinomial distribution in Section 3.4 of Chapter 3, that the probability that all events in (13.5.3) occur is given by

$$(13.5.4) \qquad \frac{n!}{0!\,1!\,(n-2)!\,1!\,0!}\, p_1^0 p_2^1 p_3^{n-2} p_4^1 p_5^0.$$

By substituting the values of $p_1, \ldots, p_5$ from (13.5.2), reducing this expression, and writing the result to terms of order $du\,dv$ and denoting the probability element of $x_{(1)}$ and $x_{(n)}$ by $h(u, v)\,du\,dv$ [the probability that $x_{(n)}$ falls in $(u, u + du)$ *and* that $x_{(1)}$ falls in $(v, v + dv)$] to terms of order $du\,dv$, we obtain

$$(13.5.5) \quad h(u, v)\,du\,dv = n(n - 1)[F(u) - F(v)]^{n-2} f(u) f(v)\,du\,dv,$$

where the sample space of (u, v) is given by $-\infty < v < u < +\infty$.

The range R of the sample is, of course, defined by $R = u - v$. The event $R \leq r$ is equivalent, then, to the event $u \leq v + r$. We denote the c.d.f. of R by $H(r)$, and using (13.5.5) we then have

$$H(r) = P(R \leq r) = P(u \leq v + r)$$

$$(13.5.6) \qquad = \int_{-\infty}^{\infty} \int_{v}^{v+r} n(n - 1)[F(u) - F(v)]^{n-2} f(u) f(v)\,du\,dv$$

$$= n(n - 1) \int_{-\infty}^{\infty} f(v) \left\{ \int_{v}^{v+r} [F(u) - F(v)]^{n-2} f(u)\,du \right\} dv.$$

In the inner integral, put $s = F(u)$. We note that $ds = f(u)\,du$, and when $u = v + r$, $s = F(v + r)$, and when $u = v$, $s = F(v)$. Hence we may write (13.5.6) as

$$H(r) = n(n - 1) \int_{-\infty}^{\infty} f(v) \left\{ \int_{F(v)}^{F(v+r)} [s - F(v)]^{n-2}\,ds \right\} dv$$

$$(13.5.7)$$

$$= n \int_{-\infty}^{\infty} f(v)[F(v + r) - F(v)]^{n-1}\,dv,$$

and upon differentiating (13.5.7), we then find that the probability element of R, which we denote by $h(r) \, dr$, is

$$(13.5.8) \quad h(r) \, dr = \left\{ n(n-1) \int_{-\infty}^{\infty} [F(v+r) - F(v)]^{n-2} f(v) f(v+r) \, dv \right\} dr.$$

Example 13.5(a). If a sample of n points is taken "at random" on the interval $(0, 1)$, find the distribution of the range R determined by the n points.

In this case we have for the p.d.f. of a single point x taken "at random" on $(0, 1)$,

$$f(x) = 1, \quad 0 < x < 1,$$
$$= 0, \text{ all other values of } x,$$

and

$$F(x) = 0, \quad x \leqslant 0,$$
$$= x, \quad 0 < x < 1,$$
$$= 1, \quad x \geqslant 1.$$

Then

$$(13.5.9) \qquad h(r) \, dr = \left\{ n(n-1) \int_{0}^{1-r} [(v+r) - v]^{n-2} \, dv \right\} dr,$$

that is,

$$(13.5.10) \qquad h(r) \, dr = n(n-1) r^{n-2} (1-r) \, dr.$$

Thus, if we want to find the probability that the value of R lies between two values a and b, where $0 < a < b < 1$, we have

$$(13.5.11) \qquad P(a < R < b) = n(n-1) \int_{a}^{b} (r^{n-2} - r^{n-1}) \, dr$$

$$= n(b^{n-1} - a^{n-1}) - (n-1)(b^n - a^n).$$

13.6 DISTRIBUTION-FREE TOLERANCE LIMITS

In Section 13.2 we showed that, if $x_{(n)}$ is the largest element in a sample of n elements from a population with c.d.f. $F(x)$, then

$$(13.6.1) \qquad P(u < x_{(n)} < u + du) = nF^{n-1}(u) f(u) \, du.$$

Note that $u < x_{(n)} < u + du$ if and only if $F(u) < F(x_{(n)}) < F(u + du)$. But since $f(u) \, du = dF(u)$, we have $F(u + du) = F(u) + dF(u)$. Therefore the event $u < x_{(n)} < u + du$ occurs if and only if the event $F(u) < F(x_{(n)}) < F(u) + dF(u)$ occurs, and we have

$$P[F(u) < F(x_{(n)}) < F(u) + dF(u)] = nF^{n-1}(u) \, dF(u).$$

If we denote $F(u)$ by t, say, then we have, using (13.6.1), that

$$(13.6.2) \qquad P[t < F(x_{(n)}) < t + dt] = nt^{n-1} \, dt,$$

that is, the probability that $F(x_{(n)})$ lies in the interval $(t, t + dt)$ is

$$nt^{n-1}\, dt.$$

But note that the statistical interpretation of $F(x_{(n)})$ is that it is the fraction of the objects in the population having values of x less than or equal to $x_{(n)}$, the largest element of the sample. Thus we may write

$$(13.6.3) \qquad P[F(x_{(n)}) > \beta] = n \int_{\beta}^{1} t^{n-1}\, dt = 1 - \beta^n,$$

which states that *the probability is $1 - \beta^n$ that the fraction of objects in the population having values of x less than $x_{(n)}$ is at least β*. This statement is true for *any continuous* c.d.f. $F(x)$. The interval $(-\infty, x_{(n)})$ is sometimes called a *one-sided statistical tolerance interval*.

If, for a given β and γ, we choose n as the integer most closely satisfying

$$(13.6.4) \qquad\qquad 1 - \beta^n = \gamma,$$

we say that $x_{(n)}$ is a $100\beta\%$ *upper statistical tolerance limit* with confidence coefficient γ. Values of n for values of β and γ of practical interest (.90, .95, .99) have been tabulated by Murphy (1948) and Somerville (1958).

In the usual practical situation where n is large and $1 - \beta$ is small, then

$$(13.6.5) \qquad 1 - \beta^n = 1 - [1 - (1 - \beta)]^n \cong 1 - e^{-(1-\beta)n}$$

and we may replace the expression in (13.6.4) by the approximation

$$(13.6.6) \qquad\qquad 1 - e^{-(1-\beta)n} \cong \gamma.$$

Hence for given values of β and γ, the required sample size n is approximately given by solving (13.6.6).

> **Example 13.6(a).** Suppose we choose $\beta = .95$ and ask how large n should be in order for γ to be .90. Using approximation (13.6.6), we must solve the equation
>
> $$e^{-.05n} = .10.$$
>
> From tables of e^{-x} we find that $.05n = 2.30$ or $n = 46$. This means that, if we take a sample of 46 objects from a population (or large lot), the probability is approximately .90 that at least 95% of the objects in the population have values of x less than the largest x drawn in the sample. Or stated briefly, in a sample of 46 items from a large lot, the largest element in the sample, namely $x_{(46)}$, is a 95% upper statistical tolerance limit with confidence coefficient .90.

By referring to (13.3.4), which is the p.d.f. of $x_{(1)}$, the smallest element in a sample of n elements from a population having c.d.f. $F(x)$, and following a line of reasoning similar to that used in establishing (13.6.3), we can show that

$$(13.6.7) \qquad\qquad P[1 - F(x_{(1)}) > \beta] = 1 - \beta^n,$$

which states that the probability is also $1 - \beta^n$ that at least $100\beta\%$ of the objects in the population have values of x exceeding the smallest x, namely $x_{(1)}$, in a sample of n elements from the population.

If, for a fixed β and γ, we choose n as the integer most nearly satisfying

$$(13.6.8) \qquad\qquad 1 - \beta^n = \gamma,$$

we say that $x_{(1)}$ is a $100\beta\%$ *lower statistical tolerance limit* with *confidence coefficient* γ.

If $1 - \beta$ is small, then, of course, (13.6.8) can be replaced by the approximate equation (13.6.6).

> ***Example 13.6(b).*** Referring to Example 13.6(a) if we choose $\beta = .95$, $\gamma = .90$, we find $n = 46$, hence in a sample of 46 items the smallest element drawn, namely $x_{(1)}$ is a 95% lower statistical tolerance limit having confidence coefficient .90.

Now suppose we consider the interval $(x_{(1)}, x_{(n)})$, where $x_{(1)}$ is the smallest element and $x_{(n)}$ is the largest element in the sample. The quantity $F(x_{(n)}) - F(x_{(1)})$ is the fraction of objects in the population having values of x in the *two-sided tolerance interval* $(x_{(1)}, x_{(n)})$. We wish to find the probability that $F(x_{(n)}) - F(x_{(1)}) > \beta$, where β is any number on the interval $(0, 1)$.

The joint probability element of the pair of random variables $(x_{(1)}, x_{(n)})$, that is, the probability that $x_{(n)}$ falls in the interval $(u, u + du)$ *and* $x_{(1)}$ falls in the interval $(v, v + dv)$, is given by the right-hand side of (13.5.5) which can be written as

$$(13.6.9) \qquad n(n - 1)[F(u) - F(v)]^{n-2} \, dF(u) \, dF(v).$$

From this, it is seen that the probability of the pair of random variables $F(x_{(1)})$, $F(x_{(n)})$, lying in $(s, s + ds)$ and $(t, t + dt)$ respectively, is given by

$$(13.6.10) \qquad n(n - 1)(t - s)^{n-2} \, dt \, ds,$$

where $0 < s < t < 1$.

To find the probability that $F(x_{(n)}) - F(x_{(1)}) > \beta$, where β is any number between 0 and 1, we must therefore integrate the element (13.6.10) over the triangle in the ts-plane for which $0 < s < t < 1$ and $1 > t - s > \beta$. This gives

$$(13.6.11) \quad P[F(x_{(n)}) - F(x_{(1)}) > \beta] = n(n - 1) \int_{\beta}^{1} \int_{0}^{t-\beta} (t - s)^{n-2} \, ds \, dt$$

$$= 1 - \beta^n - n(1 - \beta)\beta^{n-1}.$$

If for a given β and γ we choose n as the integer most nearly satisfying

$$(13.6.12) \qquad\qquad 1 - \beta^n - n(1 - \beta)\beta^{n-1} = \gamma,$$

we say that $(x_{(1)}, x_{(n)})$ is a *two-sided* $100\beta\%$ *statistical tolerance interval* with confidence coefficient γ. This means that the probability is γ that the interval $(x_{(1)}, x_{(n)})$, determined from a sample of size n, contains the values of x of at least $100\beta\%$ of the objects in the population (large lot). This statement is true for samples from any continuous c.d.f. $F(x)$.

If β is near 1 and n is large so that $\beta^n = [1 - (1 - \beta)]^n \cong e^{-n(1-\beta)}$, we can approximate the left side of (13.6.12) and obtain

$$(13.6.13) \qquad 1 - e^{-n(1-\beta)}[1 + n(1 - \beta)] = \gamma.$$

Note that $e^{-n(1-\beta)}[1 + n(1 - \beta)]$ is the sum of the first two terms of a Poisson distribution.

Example 13.6(c). We determine the size n of the sample for which $(x_{(1)}, x_{(n)})$ is a 95% statistical tolerance interval having confidence coefficient $.90$. We have $\beta = .95$, $\gamma = .90$, hence to find n (approximately) we must solve the equation

$$e^{-.05n}(1 + .05n) = .10.$$

Referring to tables of e^{-x}, we find $.05n \cong 3.88$ and, hence, $n = 78$.

More generally, we state without proof that, if we take any two order statistics $(x_{(c)}, x_{(c+k)})$ in a sample of n elements from a population with continuous c.d.f. $F(x)$, then $F(x_{(c+k)}) - F(x_{(c)})$ is a random variable such that

$$(13.6.14) \quad P[F(x_{(c+k)}) - F(x_{(c)}) > \beta] = \frac{n!}{(k-1)!(n-k)!} \int_\beta^1 t^{k-1}(1-t)^{n-k}\,dt.$$

Note that the integrand of the expression on the right depends only on k and n and not c or F.

Example 13.6(d). Suppose a sample of n elements is taken from a population with c.d.f. $F(x)$ and consider the interval $(x_{(2)}, x_{(n-1)})$ formed by taking the second smallest and second largest order statistics in the sample. In this case $c = 2$, and $k = n - 3$. Hence

$$(13.6.15) \quad P[F(x_{(n-1)}) - F(x_{(2)}) > \beta] = \frac{n!}{n - 4!\, 3!} \int_\beta^1 t^{n-4}(1-t)^3\,dt.$$

Thus, if for a given β and γ we choose n so that the right-hand side of (13.6.15) has the value γ, then $(x_{(2)}, x_{(n-1)})$ is a $100\beta\%$ statistical tolerance interval with confidence coefficient γ.

If we change the variable in the integrand by letting $t = 1 - u/n$, and hence $dt = -(1/n)\,du$, we obtain after some simplification

$$(13.6.16) \quad P[F(x_{(n-1)}) - F(x_{(2)}) > \beta]$$

$$= \frac{(1 - 1/n)(1 - 2/n)(1 - 3/n)}{6} \int_0^{n(1-\beta)} \left(1 - \frac{u}{n}\right)^{n-4} u^3\,du.$$

If we let $\beta \to 1$ and $n \to \infty$ so that $n(1 - \beta) \to \delta$, where $\delta > 0$, the limit of the right-hand side of (13.6.16) is

$$\frac{1}{6} \int_0^\delta u^3 e^{-u} \, du,$$

which, when successively integrated by parts, has the value

$$1 - e^{-\delta} - \frac{\delta e^{-\delta}}{1!} - \frac{\delta^2 e^{-\delta}}{2!} - \frac{\delta^3 e^{-\delta}}{3!}.$$

Hence, for large values of n and small values of $1 - \beta$, we have the approximation:

(13.6.17) $P[F(x_{(n-1)}) - F(x_{(2)}) > \beta]$

$$\cong 1 - e^{-n(1-\beta)} \left\{ 1 + n(1 - \beta) + \frac{[n(1 - \beta)]^2}{2!} + \frac{[n(1 - \beta)]^3}{3!} \right\}.$$

Thus, if for a given β and γ, we choose n so that the right-hand side of (13.6.17) has the value γ, the interval $(x_{(2)}, x_{(n-1)})$ is a $100\beta\%$ (two-sided) statistical tolerance interval with approximate confidence coefficient γ. Note that the right-hand side of (13.6.17) is 1 minus the sum of the first four terms of a Poisson distribution. As a matter of fact, if in a large sample we take the mth smallest order statistic $x_{(m)}$ and the mth largest order statistic $x_{(n-m+1)}$ where m is a small integer, say $m < n/2$, and form the interval $(x_{(m)}, x_{(n-m+1)})$, it can be shown that for a large n and small $1 - \beta$ this interval is a $100\beta\%$ statistical tolerance interval with approximate confidence coefficient γ, where

(13.6.18) $\gamma = 1 - e^{-n(1-\beta)} \left\{ 1 + [n(1 - \beta)] + \cdots + \frac{[n(1 - \beta)]^{2m-1}}{(2m - 1)!} \right\}.$

13.7   OTHER USES OF ORDER STATISTICS

(a) Probability Paper

In Section 7.5, we discovered that probability paper may be used as a "rough" check on "normality," when the samples are large and have been grouped, etc. Now for the case of small samples, a widely used procedure which also provides a rough check on "normality" is as follows.

Let $x_1, \ldots, x_n$ be n independent observations on a random variable x, whose p.d.f. is thought to be normal. Order the sample, say $x_{(1)} < x_{(2)} < \cdots < x_{(n)}$. Plot the $x_{(i)}$ against $(i - 1/2)/n$, $i = 1, \ldots, n$, on normal probability paper. If the points fall on a straight line, accept the hypothesis that the p.d.f. of x is normal, and estimate the mean μ as the 50th percentile of the sample and the standard deviation as the difference between the 84th and 50th percentiles.

Example 13.7(a).   A sample of ten observations on width of slots of certain parts to be used in an assembly, measured in one-thousands of an inch, yield, when ordered, the values

1.62, 1.91, 1.98, 2.12, 2.20, 2.32, 2.40, 2.52, 2.65, 2.86.

Figure 13.7.1 shows these observations plotted against $(i - 1/2)/10$, $i = 1, \ldots, 10$, respectively, and also, a line of "best" fit, chosen by eye. The points fall on or very near the line, and hence we accept the hypothesis of normality. The 50th percentile is 2.26 and the 84th percentile is 2.63,

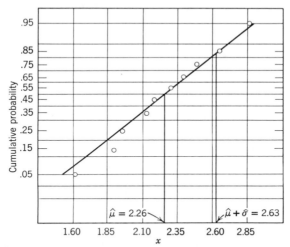

Figure 13.7.1. The ten observations of Example 13.7(b) plotted on normal probability paper.

yielding an estimate $\hat{\mu} = 2.26$ of the population mean and the estimate $\hat{\sigma} = 2.63 - 2.26 = .37$ for the population standard deviation. The actual mean and standard deviation of the sample is 2.258 and .371.

(b) The Range as an Estimate of σ in Normal Samples

Suppose a random variable x has the normal distribution with unknown standard deviation σ. If a sample of n independent observations is taken on x, then $R = x_{(n)} - x_{(1)}$ may be used as an estimate of σ. This estimate is not good for large n, but for small $n(n \leq 10)$ is called adequate. The estimate $\hat{\sigma}$ is made using the formula

(13.7.1) $\hat{\sigma} = c(n)R,$

where $c(n)$ is tabulated in Table 13.7.1.

TABLE 13.7.1

n	$c(n)$	n	$c(n)$
3	.591	7	.370
4	.486	8	.351
5	.430	9	.337
6	.395	10	.325

PROBLEMS

13.1. Suppose $F(x)$ is the fraction of bricks in a very large lot having crushing strengths of x psi or less. If 100 such bricks are drawn at random from the lot:

(a) What is the probability that the crushing strengths of all 100 bricks exceed x psi?

(b) What is the probability that the weakest brick in the sample has a crushing strength in the interval $(x, x + dx)$?

13.2. Suppose $F(x)$ is the fraction of objects in a very large lot having weights $\leq x$ pounds. If ten objects are drawn at random from the lot:

(a) What is the probability that the heaviest of these ten objects will have a weight less than or equal to u pounds?

(b) What is the probability that the lightest of the objects will have a weight less than or equal to v pounds?

13.3. A continuous random variable has the uniform distribution function on $(0, 1)$ that is,

$$f(x) = \begin{cases} 0, & x \leq 0, \\ 1, & 0 < x \leq 1, \\ 0, & 1 < x. \end{cases}$$

If $x_{(1)}, x_{(2)}, \ldots, x_{(n)}$ are the order statistics of n independent observations all having this distribution function, give the expression for the density $g(x)$ for:

(a) The largest of these n observations.

(b) The smallest of these n observations.

(c) The rth smallest of these n observations.

13.4. If ten points are picked independently and at random on the interval $(0, 1)$:

(a) What is the probability that the point nearest 1 (i.e., the largest of the ten numbers selected) will lie between .9 and 1.0?

(b) The probability is $1/2$ that the point nearest 0 will exceed what number?

13.5. Assume that the cumulative distribution function of breaking strengths (in pounds) of links used in making a certain type of chain is given by

$$F(x) = 1 - e^{-cx}, \quad x > 0,$$
$$= 0, \quad x \leq 0,$$

where c is a positive constant. What is the probability that a 100-link chain made from these links would have a breaking strength exceeding y pounds?

13.6. Suppose $n = 2m + 1$ observations are taken at random from a population with probability density function

$$f(x) = \begin{cases} 0, & x \leq a, \\ \dfrac{1}{b - a}, & a < x \leq b, \\ 0, & b < x. \end{cases}$$

Find the distribution of the median of the observations and find its mean and variance. What is the probability that the median will exceed $a + 1/4(b - a)$? Will exceed y?

13.7. In Problem 13.6, suppose $x_{(1)}$ and $x_{(n)}$ denote the smallest and largest of the n observations. What is the p.d.f. of $R = x_{(n)} - x_{(1)}$? What is the minimum sample size required to ensure that $P[R \geq (b - a)/4] = .90$?

13.8. A sample of n independent observations is taken from a population whose c.d.f. $F(x)$ is continuous but otherwise unknown. What must be the value of n to ensure that the probability is .95 of $(x_{(1)}, +\infty)$ containing at least 90% of the values of x of all objects in the population? In a sample of size 100, the probability is .95 that $(x_{(1)}, +\infty)$ contains the values of x of at least what percent of the population?

13.9. A sample of n observations is taken at random from a population with p.d.f.

$$f(x) = \begin{cases} e^{-x}, & x \geq 0, \\ 0 & \text{otherwise.} \end{cases}$$

Find the p.d.f. of the smallest observation. What is its mean and variance? What is its c.d.f.?

13.10. If x is a random variable with p.d.f. given in Problem 13.9, find the p.d.f. of the range of a sample of size n.

13.11. If $x_1, \ldots, x_n$ is a sample of size n from a population having p.d.f. $f(x)$ and c.d.f. $F(x)$, find the expression for the p.d.f. of $x_{(n-1)} - x_{(2)}$ for the $f(x)$ given in Problem 13.9.

13.12. In Example 13.6(c) solve for n if $\beta = \gamma = .95$. Also solve for n if $\beta = \gamma = .99$.

13.13. For a sample of size 100, $(x_{(1)}, x_{(100)})$ is a 95% statistical tolerance interval having what confidence coefficient?

13.14. Determine the size n of a sample for which $(x_{(2)}, x_{(n-1)})$ is a 95% statistical tolerance interval having confidence coefficient .95.

13.15. If $x_{(1)}, \ldots, x_{(n)}$ are the order statistics of a sample of size n from a population having a continuous c.d.f. $F(x)$, show that $F(x_{(n)})$ has mean $n/(n + 1)$ and variance $n/[(n + 1)^2(n + 2)]$.

13.16. In Problem 13.12 show that the mean and variance of $F(x_{(k)})$ are

$$\frac{k}{n + 1} \quad \text{and} \quad \frac{k(n - k + 1)}{(n + 1)^2(n + 2)}.$$

13.17. Suppose $x_{(1)}, \ldots, x_{(n)}$ are the order statistics of a sample from a population having the rectangular distribution with p.d.f.

$$f(x) = 0, \quad x \leq 0,$$

$$= \frac{1}{\theta}, \quad 0 < x \leq \theta,$$

$$= 0, \quad x \geq \theta,$$

where θ is an unknown parameter. Show that

$$\left(x_{(n)}, \frac{x_{(n)}}{\sqrt[n]{1-\gamma}} \right)$$

is a $100\gamma\%$ confidence interval for θ.

13.18. If sampling from a continuous $F(x)$, show that $(x_{(k)}, x_{(n-k+1)})$, $k < n/2$, is a $100\gamma\%$ confidence interval for the median of $F(x)$ (i.e., the value of x for which $F(x) = .5$), where

$$\gamma = 1 - 2\frac{\Gamma(n+1)}{\Gamma(n-k+1)\Gamma(k)}\int_0^{.5} x^{n-k}(1-x)^{k-1}\,dx.$$

CHAPTER 14

Nonparametric Tests

In Chapter 10 we discussed various statistical tests based on the assumption that the samples involved in the tests are drawn from normal populations. There are many situations in which we know very little about the shape of the population distribution from which the samples are drawn. In such cases the assumption of normality may be hazardous. There is a class of statistical tests which are valid for samples from *continuous* population distributions of any shape. These are called *nonparametric tests* and they are based on order statistics. In this chapter we shall consider several of the simplest and most widely used of these tests. Throughout this chapter we assume that population c.d.f.'s are continuous.

14.1 THE SIGN TEST

There are experimental situations in which we may think of drawing n independent pairs of sample values, say (x_1, x'_1), (x_2, x'_2), ..., (x_n, x'_n), from two c.d.f.'s $F_1(x)$ and $F_2(x)$ respectively. On the basis of these n pairs of elements we want to test the null hypothesis

$$H_0: \quad F_1(x) \equiv F_2(x)$$

against alternatives H_1: $F_1(x) > F_2(x)$ or H'_1: $F_1(x) < F_2(x)$, or H''_1: $F_1(x) \neq F_2(x)$. Let

$$v_i = \begin{cases} 1 & \text{if} \quad x_i - x'_i > 0, \\ 0 & \text{otherwise}, \end{cases}$$

$i = 1, \ldots, n$. Let $r = \sum_{i=1}^{n} v_i$.

We shall assume that $F_1(x)$ and $F_2(x)$ have derivatives $f_1(x)$ and $f_2(x)$ respectively, which are the p.d.f.'s of our two distributions.

Then if H_0 is true, that is, if $F_1(x) = F_2(x) = F(x)$, say, we have

$$P(v_i = 1) = P(x_i > x_i') = \int_{-\infty}^{\infty} \int_{x_i'}^{\infty} f(x_i)f(x_i') \, dx_i \, dx_i' = \tfrac{1}{2},$$

where $f(x)$ is the derivative of $F(x)$. Since the successive drawings are independent, r is a random variable which has the binomial distribution

(14.1.1) $$b(r) = \binom{n}{r}\left(\frac{1}{2}\right)^n,$$

$r = 0, 1, \ldots, n$.

Note that $(x_1, x_1'), (x_2, x_2'), \ldots, (x_n, x_n')$ can be pairs drawn respectively from n different pairs of p.d.f.'s, $(f_{11}(x), f_{12}(x)); (f_{21}(x), f_{22}(x)); \ldots;$ $(f_{n1}(x), f_{n2}(x))$. In this case the null hypothesis H_0 is that $f_{11}(x) \equiv f_{12}(x),$ $f_{21}(x) \equiv f_{22}(x); \ldots, f_{n1}(x) \equiv f_{n2}(x)$, and r still has the binomial distribution (14.1.1) under this null hypothesis. The extension to this general situation makes the sign test particularly useful for comparing x with x' under a variety of conditions. We say that r is significantly large at the $100\alpha\%$ level of significance if $r \geq r_{L\alpha}$, where $r_{L\alpha}$ is the smallest integer for which

(14.1.2) $$P(r \geq r_{L\alpha}) \leq \alpha$$

when H_0 is true. In a similar manner r is said to be significantly small at the $100\alpha\%$ level of significance if $r \leq r_{S\alpha}$, where $r_{S\alpha}$ is the largest integer for which

(14.1.3) $$P(r \leq r_{S\alpha}) \leq \alpha.$$

We say that r differs significantly from its expected value $1/2\ n$ if r does not fall in the interval $[r_\alpha, n - r_\alpha]$, where $r_\alpha < n/2$ is the largest integer for which

(14.1.4) $$P(r_\alpha \leq r \leq n - r_\alpha) > 1 - \alpha.$$

The Table 14.1.1 gives values of $r_{L\alpha}, r_{S\alpha}, r_\alpha$ for values of n up to 30 and $\alpha = .01$ and $.05$.

If n is larger than 30, we can approximate $r_{L\alpha}, r_{S\alpha}, r_\alpha$ by using the normal approximation (see Chapter 7) to the binomial (14.1.1), that is,

(14.1.5) $$P(r \leq r_0) = \Phi\left(\frac{r_0 + \frac{1}{2} - n/2}{\frac{1}{2}\sqrt{n}}\right).$$

Thus, if $\Phi(z_\alpha) = 1 - \alpha$, we see that

(14.1.6) $$r_{L\alpha} \cong \frac{n+1}{2} + \frac{1}{2}\sqrt{n}z_\alpha$$

TABLE 14.1.1

CRITICAL VALUES OF $r_{L\alpha}$, $r_{S\alpha}$, r_α FOR USE ON THE SIGN TEST

	$r_{L\alpha}$		$r_{S\alpha}$		r_α	
n \ α	.01	.05	.01	.05	.01	.05
5		5		0		
6		6		0		1
7	7	7	0	0		1
8	8	7	0	1	1	1
9	9	8	0	1	1	2
10	10	9	0	1	1	2
11	10	9	1	2	1	2
12	11	10	1	2	2	3
13	12	10	1	3	2	3
14	12	11	2	3	2	3
15	13	12	2	3	3	4
16	14	12	2	4	3	4
17	14	13	3	4	3	5
18	15	13	3	5	4	5
19	15	14	4	5	4	5
20	16	15	4	5	4	6
21	17	15	4	6	5	6
22	17	16	5	6	5	6
23	18	16	5	7	5	7
24	19	17	5	7	6	7
25	19	18	6	7	6	8
26	20	18	6	8	7	8
27	20	19	7	8	7	8
28	21	19	7	9	7	9
29	22	20	7	9	8	9
30	22	20	8	10	8	10

and

(14.1.7) $$r_{S\alpha} \cong \frac{n - 1}{2} - \frac{1}{2}\sqrt{n}z_\alpha$$

while

(14.1.8) $$r_\alpha \cong \frac{n + 1}{2} - \frac{1}{2}\sqrt{n}z_{\alpha/2}.$$

Example 14.1(a).   Table 14.1.2 [Kenny and Keeping (1956) Vol. 1, p. 186] shows the hemoglobin (grams/100 milliliters of blood) in anemic rats before and after 4 weeks of added iron in the diet (.5 milligram per day).

Note that $r = \sum_{i=1}^{12} v_i = 8$, and consulting Table 14.1.1, we have that $P(r \geq 10) \leq .05$, $P(r \geq 11) \leq .01$ and $P(2 \leq r \leq 10) > .99$ while $P(3 \leq r \leq 9) > .95$. Thus, no matter whether we consider a one- or a two-sided test, we do not have significance at either the 1 or 5% levels.

TABLE 14.1.2

HEMOGLOBIN IN RATS BEFORE AND AFTER CHANGE OF DIET

Rat No.	x' (Before)	x (After)	$d = x - x'$	v
1	3.4	4.9	1.5	1
2	3.0	2.3	−.7	0
3	3.0	3.1	.1	1
4	3.4	2.1	−1.3	0
5	3.7	2.6	−1.1	0
6	4.0	3.8	−.2	0
7	2.9	5.8	2.9	1
8	2.9	7.9	5.0	1
9	3.1	3.6	.5	1
10	2.8	4.1	1.3	1
11	2.8	3.8	1.0	1
12	2.4	3.3	.9	1

If we had assumed normality in the above example, and proceeded as in Example 10.3(c), we find that

$$t_{11} = 1.61,$$

which is not a significant value at the 1 or 5% levels.

Example 14.1(b). Thirty-six samples of ore were tested for their iron content by method A and method B. If we let x and x' be the iron content determined by methods A and B, and if we let $d_i = x_i - x'_i$, $i = 1, \ldots, 36$, and $v_i = 1$ or 0 according to whether $d_i > 0$ or ≤ 0, and if $r = \sum_{i=1}^{36} v_i$, then a two-sided acceptance region for a test of the hypothesis that method A is equivalent to method B at the 5% level of significance is

$$[r_{.05}, n - r_{.05}],$$

where

$$r_{.05} \cong \frac{n+1}{2} - \frac{1}{2}\sqrt{n}\, z_{.025}$$

$$\cong \tfrac{37}{2} - \tfrac{1}{2}(6)(1.96)$$

$$\cong 12.62.$$

That is, we accept the test if $12 \leq r \leq 24$ and reject otherwise.

14.2 THE MANN-WHITNEY (WILCOXON) W TEST FOR TWO SAMPLES

If the two samples are not paired as in Section 14.1 and, in fact, if the samples are not necessarily the same size, we may proceed as follows. Suppose $(x_1, \ldots, x_m)$ and $(x_1', \ldots, x_n')$ are independent samples from populations having continuous c.d.f.'s $F_1(x)$ and $F_2(x)$ respectively. Let the two samples be pooled together into a single sample of $m + n$ observations and let the order statistics of this sample be $y_{(1)}, y_{(2)}, \ldots,$ $y_{(m+n)}$. Consider the ranks (subscripts) of all y's which represent the elements of $(x_1, \ldots, x_m)$. Let the *sum of these* ranks be T, and let W be a random variable defined in terms of T as follows:

$$(14.2.1) \qquad W = mn + \frac{m(m + 1)}{2} - T.$$

Actually, W is the number of the mn possible pairs (x_i, x_j') for which $x_i < x_j'$. W is the Mann-Whitney statistic, and T is the Wilcoxon statistic.

It can be shown by rather complicated analysis (which we omit) that, if the hypothesis H_0: $F_1(x) \equiv F_2(x)$ is true, that is, if both samples come from populations having identical c.d.f.'s, then for m and n both larger than 8, W has for all practical purposes approximately a normal distribution with mean

$$(14.2.2) \qquad E(W) = \frac{mn}{2}$$

and variance

$$(14.2.3) \qquad \mathrm{Var}\,(W) = \frac{mn(m + n + 1)}{12}.$$

As a matter of fact as m and n both approach infinity, the random variable

$$(14.2.4) \qquad \frac{W - mn/2}{\sqrt{[mn(m + n + 1)]/12}}$$

has as its limiting distribution the normal distribution $N(0, 1)$.

> **Example 14.2(a).** Two chemists, A and B, make 14 and 16 determinations of plutonium, respectively, with the results shown in Table 14.2.1. (Numbers in parentheses are the rank of the observation in the combined sample.) The problem is to determine whether the two chemists are obtaining significantly different results.

TABLE 14.2.1

Chemist A	Chemist B
x	x'
263.36 (13)	286.53 (28)
254.68 (10)	254.54 (9)
248.64 (3)	284.55 (26)
272.68 (19)	253.75 (7)
261.10 (12)	283.85 (24)
287.33 (30)	252.01 (5)
268.41 (16)	245.26 (2)
287.26 (29)	275.08 (20)
276.32 (21)	286.30 (27)
243.64 (1)	272.52 (18)
256.42 (11)	282.90 (23)
282.65 (22)	266.08 (14)
250.97 (4)	267.53 (15)
284.27 (25)	252.05 (6)
	253.82 (8)
	269.81 (17)

By combining the samples into one sample and ordering the observations, we find that T, the sum of the ranks of the x observations is 216. Hence (14.2.1) becomes

$$W = mn + \frac{m(m + 1)}{2} - T$$

$$= 224 + 105 - 216 = 113.$$

Now $E(W) = 112$ and Var $(W) = [(14)(16)(31)]/12 = 578.67$, and so the standard deviation of W is 24.06. Hence the observed value of (14.2.4) becomes

$$\frac{113 - 112}{24.06} = \frac{+1}{24.06} = .042,$$

which is not significant, and so we accept the hypothesis that the results of the analyses by A and B do not differ significantly.

14.3   RUN TESTS

(a) Runs Above and Below Median

Suppose we draw a sample of $2n$ elements $(x_1, x_2, \ldots, x_{2n})$ from a continuous c.d.f. $F(x)$, and consider the problem of examining the fluctuations in sequence of $2n$ drawings for evidence of nonrandomness. One of the simplest ways of doing this is as follows: We graph the x_i against i, $i = 1, \ldots, 2n$, as shown in Figure 14.3.1.

If the order statistics of the sample are $(x_{(1)}, x_{(2)}, \ldots, x_{(2n)})$ then, since the probability is zero that two or more of the x's are equal in the sample, there will be a gap between $x_{(n)}$ and $x_{(n+1)}$. If we draw a horizontal line between $x_{(n)}$ and $x_{(n+1)}$ as shown in Figure 14.3.1, there will be n points above the line and n below. For each point above the line write a and for each point below the line write b, as shown in Figure 14.3.1.

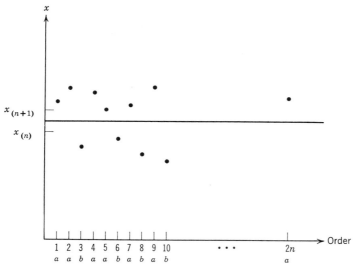

Figure 14.3.1. Chart for determining runs above and below median.

There will be n a's and n b's. The total number of distinct arrangements of a's and b's possible is $\binom{2n}{n}$. If the sample is a random sample from any continuous c.d.f. $F(x)$, then these $\binom{2n}{n}$ different possible arrangements have equal probabilities. For any given arrangement there will be clusters of one or more a's separated by clusters of one or more b's The *total number of clusters* of a's and b's, say u, is called the *numbers of runs* above or below the median line. [Note that if there had been an odd number of elements in the sample, say $2n + 1$, we could have drawn the horizontal line through the order statistic $x_{(n+1)}$, that is, the median of the sample $(x_1, \ldots, x_{2n+1})$. We would then have n points above the median and n below.] If the measurements in the sequence $x_1, \ldots, x_{2n}$, which is the order in which the measurements were drawn, have a secular drift up or down or exhibit wide swings or jumps from one general level to another, these characteristics of the sequence tend to make u significantly small. If the measurements tend to alternate too much, u tends to be significantly large. In either case u is a reasonable indicator of nonrandomness.

It can be shown that the probability function $p(u)$ of u, assuming $(x_1, \ldots, x_{2n})$ is a random sample from a continuous c.d.f., is given by

(14.3.1)

$$p(u) = 2 \frac{\left(\dfrac{\dfrac{n-1}{2}}{\dfrac{u}{2} - 1}\right)^2}{\dbinom{2n}{n}}, \qquad u = 2, 4, \ldots, 2n,$$

$$p(u) = 2 \frac{\left(\dfrac{\dfrac{n-1}{2}}{\dfrac{u}{2} - \dfrac{1}{2}}\right)\left(\dfrac{\dfrac{n-1}{2}}{\dfrac{u}{2} - \dfrac{3}{2}}\right)}{\dbinom{2n}{n}}, \qquad u = 3, 5, \ldots, 2n - 1.$$

The mean and variance of u are given by

(14.3.2) $$E(u) = n + 1$$

(14.3.3) $$\sigma_u^2 = \frac{n(n-1)}{2n-1}.$$

[See Wilks (1962) for proof of (14.3.1), (14.3.2), and (14.3.3).]

For large n, u has approximately a normal distribution with the mean and variance given in (14.3.2) and (14.3.3). As mentioned above, we will suspect nonrandomness if u is either too small or too large.

Example 14.3(a). In Problem 11.8 we are asked to construct a control chart for means of measurements on a certain dimension on rheostat knobs. Let us examine the sequence of measurements for evidence of nonrandomness with respect to runs. Ordering the observations, we find that

$$x_{(12)} = .1412 \text{ inch} \qquad \text{and} \qquad x_{(13)} = .1414 \text{ inch.}$$

We take then as central line for a test of randomness

$$x = .1413 \text{ inch.}$$

Writing a if a sample mean has value $> .1413$ and b if it is $< .1413$, we obtain the sequence

$$b\,a\,a\,b\,a\,b\,b\,a\,b\,a\,a\,a\,b\,a\,b\,a\,b\,a\,b\,b\,b\,a\,b\,a\,b.$$

Here, $u = 19$ and $n = 12$, $E(u) = 13$, $\sigma_u^2 = 12(11)/23 = 5.74$, and so $\sigma_u \cong 2.4$.

A two-tailed test for the hypothesis of randomness at the 1% level of significance is given by the following rule:

Reject the hypothesis if $|u - 13| \geq (2.4)(2.576) = 6.18$; accept the hypothesis otherwise.

In this case $|u - 13| = 6$, and so we accept the hypothesis and conclude that the sequence does not exhibit significant nonrandomness.

The general procedure for testing whether an observed value of u is significantly large or significantly small or differs significantly from its mean value of $n + 1$ one way or the other (two-tailed test) is similar to that used in the sign test, except, of course, we use the probability function of u instead of r. In most practical situations nonrandomness tends to reveal itself in significantly small values of u.

(b) The Wald-Wolfowitz Run Test

The ideas of Section 14.3a extend to the case in which we have two samples, say $x_1, \ldots, x_{n_1}$ and $x_1', \ldots, x_{n_2}'$, from two populations having continuous c.d.f.'s $F_1(x)$ and $F_2(x)$ respectively.

Suppose now we wish to test

$$H_0: \quad F_1(x) = F_2(x)$$

against alternatives

$$H_1: \quad F_1(x) \neq F_2(x).$$

The procedure is as follows. We combine the two samples into one pooled sample and order the observations in this pooled sample. For example, we might obtain a sequence such as

$$x_{(1)}, x_{(2)}, x_{(1)}', x_{(2)}', x_{(3)}, x_{(3)}', x_{(4)}', \ldots,$$

where $x_{(i)}$ is the ith-order statistic of the x-sample, etc.

Let us replace each observation by a 0 or 1, according to whether we encounter an x or an x'. The sequence above, for example, would look like

$$0\,0\,1\,1\,0\,1\,1\cdots$$

A cluster of one or more zeros or a cluster of one or more ones is called a run. Now there are $\binom{n_1 + n_2}{n_1}$ possible distinguishable arrangements (permutations) of the n_1 0's and n_2 1's, and under the hypothesis that $F_1(x) \equiv F_2(x)$ these are equiprobable. For each permutation there will be a total number u of runs, and we can use this to test the hypothesis H_0 that $F_1(x) = F_2(x)$.

Now the probability function of u is given by

$$p(u) = 2 \frac{\binom{n_1 - 1}{\frac{u}{2} - 1}\binom{n_2 - 1}{\frac{u}{2} - 1}}{\binom{n_1 + n_2}{n_1}} \quad \text{if } u \text{ is even,}$$

(14.3.4)

$$p(u) = \frac{\binom{n_1 - 1}{\frac{1}{2}(u - 1)}\binom{n_2 - 1}{\frac{1}{2}(u - 3)} + \binom{n_1 - 1}{\frac{1}{2}(u - 3)}\binom{n_2 - 1}{\frac{1}{2}(u - 1)}}{\binom{n_1 + n_2}{n_1}}, \quad \text{if } u \text{ is odd.}$$

The mean and variance of u are as follows:

$$\text{(14.3.5)} \qquad\qquad E(u) = \frac{2n_1 n_2}{n_1 + n_2} + 1$$

$$\text{(14.3.6)} \qquad\qquad \text{Var}(u) = \frac{2n_1 n_2 (2n_1 n_2 - n_1 - n_2)}{(n_1 + n_2)^2 (n_1 + n_2 - 1)}.$$

Furthermore, if n_1 and n_2 are large, u has approximately a normal distribution with (14.3.5) and (14.3.6) as mean and variance. The approximation is usually good enough for practical purposes when both n_1 and n_2 exceed 10. The reader should note that the results (14.3.4), (14.3.5), and (14.3.6) reduce to (14.3.1), (14.3.2), and (14.3.3) respectively if $n_1 = n_2 = n$.

> **Example 14.3(b).** The following two samples of measurements were obtained from sampling two populations I and II. The problem is to test, by the theory of runs, the hypothesis that the two samples are from populations having identical c.d.f.'s.

I (x)	$(n_1 = 12)$	25	30	28	34	24	25	13	32	24	30	31	35			
II (x')	$(n_2 = 15)$	44	34	22	8	47	31	40	30	32	35	18	21	35	29	22

Here $E(u) = 14.3$ and $\text{Var}(u) = 6.32$, and the standard deviation of u is 2.51. The combined sample gives, on ordering, the sequence

$$8_{x'}, \ 13_x, \ 18_{x'}, \ 21_{x'}, \ 22_{x'}, \ 22_{x'}, \ 24_x, \ 24_x, \ 25_x, \ 25_x, \ 28_x, \ 29_{x'}, \ 30_x, \ 30_x,$$

$$30_{x'}, \ 31_x, \ 31_{x'}, \ 32_x, \ 32_{x'}, \ 34_x, \ 34_{x'}, \ 35_x, \ 35_{x'}, \ 35_{x'}, \ 40_{x'}, \ 44_{x'}, \ 47_{x'},$$

where the x and x' subscripts denote measurements from the samples of x and x' respectively. We shall denote an x observation by a 1. Note that here we have encountered the following: there are two or more sample elements having the values 30, 31, 32, 34, 35. In fact 30_x and $30_{x'}$ occur at the 14th and 15th positions in the above combined ordered sample, and so on. One widely used procedure to "break" the ties is as follows.

Toss a coin—if the toss results in a head, replace 30_x by a zero, $30_{x'}$ by a one; if tails occurs, we replace 30_x by a one and $30_{x'}$ by a zero. Performing this *randomization* procedure in this example might lead to the following sequence:

$$1\ 0\ 1\ 1\ 1\ 1\ 0\ 0\ 0\ 0\ 0\ 1\ 0\ 1\ 0\ 1\ 0\ 1\ 0\ 1\ 0\ 1\ 0\ 1\ 1\ 1\ 1,$$

which gives $u = 17$ runs. Now we reject at the 1% level of significance if

$$|u - 14.3| \geq 2.51\ (2.576) = 6.5$$

and accept otherwise. But we have observed that

$$|u - 14.3| = |17 - 14.3| = 2.7,$$

and so we do not reject the hypothesis that the two samples come from populations having identical c.d.f.'s.

PROBLEMS

14.1. Table P14.1 shows measurements of the corrosion effects in various soils for coated and uncoated steel pipe [from Hoel (1954)]:

TABLE P14.1

Uncoated, x	Coated, x'
42	39
37	43
61	43
74	52
55	52
57	59
44	40
55	45
37	47
70	62
52	40
55	27
60	50
48	33
52	56
44	36
56	54
44	32
38	39
47	40

Use the sign test to test the hypothesis that the particular coating used has no effect on corrosion.

14.2. Use the sign test for Problems 15 and 16 of Chapter 10.

14.3. Two samples of 30 observations each from populations A and B are such that, when the procedure of Section 14.2 is used, the sum of the ranks of the x' measurements (from population B) is found to be 1085, hence $W = nm + [n(n + 1)]/2 - T = 280$. Is the value of W significantly small at the 1 % level?

14.4. Use the Wald-Wolfowitz test on the data of Problem 9.11.

14.5. Use the median test on the sample means of Problem 11.7 to test for randomness.

14.6. The total thickness x of the 4 pads of 36 half-ring mounts for aircraft engines taken periodically from the production line were found to be as shown in Table P14.6:

TABLE P14.6

Half-Ring No.	x (inches)	Half-Ring No.	x (inches)
1	1.5936	19	1.5910
2	1.5906	20	1.5926
3	1.5982	21	1.5890
4	1.5901	22	1.5895
5	1.5869	23	1.5936
6	1.5898	24	1.5933
7	1.5930	25	1.5920
8	1.5894	26	1.5918
9	1.5885	27	1.5908
10	1.5905	28	1.5913
11	1.5904	29	1.5922
12	1.5902	30	1.5915
13	1.5900	31	1.5927
14	1.5925	32	1.5913
15	1.5924	33	1.5911
16	1.5901	34	1.5903
17	1.5902	35	1.5916
18	1.5904	36	1.5939

Determine whether the total number of runs above and below the median in this sequence of values of x is significantly different from its expected value at the 5 % level of significance under the hypothesis that x is under statistical control.

14.7. A sequence of 100 measurements on a process supposedly under statistical control was analyzed for runs above and below the median. The total number of runs above and below the median was observed to be 39. Is this a significantly low value at the 5 % level of significance? Show calculations.

14.8. During a 45-day production period in a cement plant, test cubes were taken each day and the mean compressive strengths of the test cubes

were determined each day, with the following results in kilograms per centimeter squared (from Hald): 440, 433, 475, 418, 462, 433, 438, 469, 465, 500, 524, 491, 479, 442, 463, 423, 447, 452, 498, 497, 473, 485, 488, 440, 442, 451, 423, 475, 452, 526, 490, 511, 492, 509, 472, 428, 456, 417, 452, 468, 475, 502, 507, 477, 445.
Test the hypothesis of randomness at the 5% level of significance in this sequence by using the run test.

14.9. The observations shown in Table P14.9 were obtained by Millikan (1930) for the charge on an electron in $10^{10}e$ units:

TABLE P14.9

4.781	4.764	4.777	4.809	4.761	4.769
4.795	4.776	4.765	4.790	4.792	4.806
4.769	4.771	4.785	4.779	4.758	4.779
4.792	4.789	4.805	4.788	4.764	4.785
4.779	4.772	4.768	4.772	4.810	4.790
4.775	4.789	4.801	4.791	4.799	4.777
4.772	4.764	4.785	4.788	4.779	4.749
4.791	4.774	4.783	4.783	4.797	4.781
4.782	4.778	4.808	4.740	4.790	
4.769	4.791	4.771	4.775	4.747	

By using runs above and below the median, test the hypothesis that the variation in this sequence is behaving randomly.

14.10. Two analysts took repeated readings on the hardness of city water (Table P14.10). Determine whether one analyst has a tendency to read differently than the other, using the Mann-Whitney W-test [from Bowker and Lieberman (1959)]:

TABLE P14.10

Analyst A	Analyst B
.46	.82
.62	.61
.37	.89
.40	.51
.44	.33
.58	.48
.48	.23
.53	.25
	.67
	.88

14.11. In a trial of two types of rain gauges, 65 of type A and 12 of type B were distributed at random over a region. In a certain period 14 storms occurred, and the average amounts of rain found in the two types of gauges were as shown in Table P14.11 [from Brownlee (1960)]:

TABLE P14.11

Storm	Type A	Type B
1	1.38	1.42
2	9.69	10.37
3	.39	.39
4	1.42	1.46
5	.54	.55
6	5.94	6.15
7	.59	.61
8	2.63	2.69
9	2.44	2.68
10	.56	.53
11	.69	.72
12	.71	.72
13	.95	.93
14	.50	.53

Test the hypothesis that the two types of gauges are giving similar results with the use of the sign test.

CHAPTER 15

Regression Analysis

15.1 INTRODUCTION

In previous chapters we have been concerned with the statistical analysis of problems involving data taken on a single random variable, that is, the observation on each object in the sample consists of only a single measurement. Very often, however, we are confronted with statistical problems where the observation on each object consists of a pair, a triple, or even a higher number of measurements.

For example, in certain production processes the yield of a product η may depend upon aging time ξ_1, the amount of heat applied ξ_2, the amount of catalyst used ξ_3, etc. The response $\eta = f(\xi_1, \xi_2, \ldots, \xi_k)$ is a function of the variables $\xi_1, \xi_2, \ldots, \xi_k$, changes in the ξ's *causing* changes in η. On other occasions two or more response variables may covary together, there being no causative model relating the two responses. For example, mathematics grades and music grades on a general intelligence examination often covary, and generally high mathematics grades appear with high music grades, while low mathematics grades tend to appear with low music grades. Here the two responses are both random variables and are *correlated*. Whether variables are connected by a causative relationship, or are merely correlated, it is possible to construct mathematical models that will permit one to predict one variable given information on the other variables. In this chapter we discuss statistical procedures which enable us to deal with such problems.

In some situations a relationship may be clear from a plot of the data. For example, a gas can be kept at a consant pressure by placing it in a cylinder with a free moving piston. As the gas is heated, the volume of gas in the cylinder increases with temperature. An experiment can be performed in which the volume of the gas in the cylinder is measured at various *preselected* settings of temperature. A record and plot of such data appears in Figure 15.1.1.

Temp. °C, x	0	20	40	60	80	100
Volume cm³ y	50.0	53.7	57.3	61.0	64.8	68.3

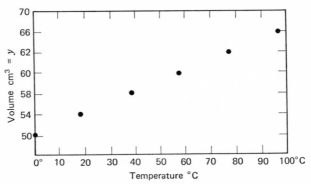

Figure 15.1.1. Experimental verification of Charles' Law.

It is obvious that the plotted points "fall" along a straight line. In fact, if a straight line were fitted by eye to these data it would be quite accurate. Students of engineering will recognize this relationship as Charles' Law. However, in problems where the plotted points do not fall nearly along a straight line or a simple curve (see Figure 15.1.2), the

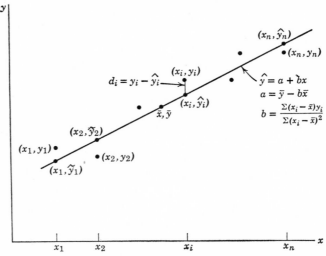

Figure 15.1.2. A line fitted to data (x_1, y_1), $i = 1, \ldots, n$ by the method of least squares.

problem of analysis becomes more complicated. The difficulty of drawing a line through a scattering of points (x_i, y_i), $i = 1, \ldots, n$, as shown in Figure 15.1.2, is that if we should try to repeat this operation on the same set of points, or if several people were to try to do it, quite different results would be obtained because of the scattering of points. What is clearly needed is some objective way of *fitting* lines or curves, that is, *fitting* various mathematical models, to observed data. The most widely used method for fitting models to data is the method of least squares described in the next section.

15.2   FITTING A STRAIGHT LINE BY LEAST SQUARES

Suppose the "true" relation between a response η (e.g., hardness of a certain plastic) and a controlled variable x (e.g., the temperature applied at the last stage of production) is a straight line. The mathematical model is then $\eta = \alpha + \beta x$ where α is the intercept and β the slope of the line. Suppose it is agreed that we will select several *different* values of x before the experiment, say $x_1, x_2, \ldots, x_n$, and record an observation of y at each of the values of x. (The values of x may be repeated.) We assume that $E(y) = \eta$ and that the variance of y is independent of the values of x used, and denote this constant value of the variance of y by σ^2.

Now the n pairs of measurements may be graphed as points in the (x, y)-plane, thus producing a scatter diagram (see Figure 15.1.2). We wish to estimate α, β, σ^2 and shall use the information contained in the n sample points to estimate these parameters.

Now, the expected value of y for each value of x is assumed to be

(15.2.1) $$E(y \mid x) = \eta = \alpha + \beta x.$$

If we further assume that y, for any given x, is a random variable which is normally distributed with expectation $\eta = \alpha + \beta x$ and variance σ^2, we have that the probability density of y, given x, is

(15.2.2) $$f(y \mid x) = \frac{1}{\sigma\sqrt{2\pi}} \exp\left[-\frac{1}{2\sigma^2}(y - \alpha - \beta x)^2\right].$$

Further, if these observed random variables $y_1, y_2, \ldots, y_n$ are statistically independent, then the likelihood, that is the joint probability density function for the sample of n pairs $(x_1, y_1), (x_2, y_2), \ldots, (x_n, y_n)$ is

(15.2.3) $$\prod_{i=1}^{n} f(y_i \mid x_i) = \prod_{i=1}^{n} \frac{1}{\sqrt{2\pi\sigma^2}} \exp\left[-\frac{1}{2\sigma^2}(y_i - \alpha - \beta x_i)^2\right]$$

$$= \left\{\left(\frac{1}{\sqrt{2\pi\sigma^2}}\right)^n \exp\left[-\frac{1}{2\sigma^2}\sum_{i=1}^{n}(y_i - \alpha - \beta x_i)^2\right]\right\}$$

where $\exp(u) = e^u$.

As discussed in Chapter 9, a widely used method for estimating values of parameters is to use as estimates those values of the parameters which maximize the likelihood function. If we apply this method to our problem to find estimators of α, β, and σ^2, we must maximize the quantity in { } in (15.2.3). But note that maximizing the quantity in { } with respect to α and β reduces to the problem of minimizing with respect to α and β, the sum of squares function

(15.2.4)
$$Q(\alpha, \beta) = \sum_{i=1}^{n} (y_i - \alpha - \beta x_i)^2$$

located in the exponent in (15.2.3)

The process of estimating α and β by minimizing the sum of squares given by (15.2.4) is called the method of least squares. Note that in the development here the use of least squares as an estimation technique is the consequence of the method of maximum likelihood *coupled* with normally distributed random variables.

The reader should also note that we may use the principle of least squares to find estimators without invoking the condition of normality. In fact, it can be proved that minimum variance unbiased estimators (of parameters) which are linear combinations of independent random variables are provided by the method of least squares. (This is a theorem due to the famous mathematician Karl Gauss.)

Further it should be noted that there are *two* mathematical models implicit in (15.2.3), one for the response $\eta = \alpha + \beta x$, and the second for the random variable y, taken to be $N(\eta, \sigma^2)$. The variable x is assumed controlled, or measured, without error, all the random variability resting in the observation y at each fixed value of x. In practice, controlling or measuring x *exactly* is impossible. However, it is assumed that the values of x are known much more precisely than those of the response y, and further that x is considered the causative variable. The analysis proceeds assuming the x variable, often called the independent variable, is fixed without error.

To find those values of α and β which minimize $Q(\alpha, \beta)$ in (15.2.4), say $\hat{\alpha}$ and $\hat{\beta}$, we proceed by taking the derivatives with respect to α and β and setting these derivatives equal to zero. We then have

(15.2.4a)
$$\left. \frac{\partial Q(\alpha, \beta)}{\partial \alpha} \right|_{\hat{\alpha}, \hat{\beta}} = 0,$$
$$\left. \frac{\partial Q(\alpha, \beta)}{\partial \beta} \right|_{\hat{\alpha}, \hat{\beta}} = 0.$$

This, in turn, produces the two equations

(15.2.5)
$$-2 \sum_{i=1}^{n} (y_i - \hat{\alpha} - \hat{\beta} x_i) = 0,$$
$$-2 \sum_{i=1}^{n} x_i(y_i - \hat{\alpha} - \hat{\beta} x_i) = 0.$$

Dividing each of these equations by -2 and performing the summations, remembering that $\sum_{i=1}^{n} y_i = n\bar{y}$ and $\sum_{i=1}^{n} x_i = n\bar{x}$, we obtain

(15.2.5a)
$$n(\hat{\alpha}) + n\bar{x}(\hat{\beta}) = n\bar{y}$$
$$n\bar{x}(\hat{\alpha}) + \sum x_i^2(\hat{\beta}) = \sum x_i y_i \,.$$

Equations (15.2.5a) are usually called the *normal equations** for estimating α and β. (As noted above, the use of the caret ˆ over the unknowns α and β denotes the estimates of these parameters.)

Thus, by solving the two equations given in (15.2.5a) with respect to $(\hat{\alpha}, \hat{\beta})$, we obtain a solution,† which we sometimes call (a, b), where a and b are given by

(15.2.6)
$$\hat{\alpha} = a = \bar{y} - b\bar{x}$$
$$\hat{\beta} = b = \frac{\sum_{i=1}^{n} x_i y_i - n\bar{x}\bar{y}}{\sum_{i=1}^{n} x_i^2 - n\bar{x}^2} = \frac{\sum_{i=1}^{n} (x_i - \bar{x})y_i}{\sum_{i=1}^{n} (x_i - \bar{x})^2} = \frac{S(XY)}{S(X^2)},$$

where $S(XY)$ and $S(X^2)$ are called the *corrected* sums of cross products and *corrected* sum of squares, respectively. It is easy to show that

$$S(X^2) = \sum x_i^2 - n\bar{x}^2; \qquad S(XY) = \sum x_i y_i - n\bar{x}\bar{y}.$$

* So long as the determinant of the coefficients,

$$\begin{vmatrix} n & n\bar{x} \\ n\bar{x} & \sum_{i=1}^{n} x_i^2 \end{vmatrix}$$

is not zero, (15.2.5a) will have a unique solution with respect to $(\hat{\alpha}, \hat{\beta})$. Now the determinant has the value $n\left(\sum_{i=1}^{n} x_i^2 - n\bar{x}^2\right)$, that is, $n(n-1)s_x^2$, which will not be zero unless all the x_i are equal. As stated at the outset, equality of all the x_i is ruled out.

† A purely algebraic solution to the problem of minimizing the sum of squares given by (15.2.4) with respect to α and β is given in Appendix V.

Given the estimates a and b, the *fitted model* is

(15.2.7)
$$\hat{y} = a + bx$$
$$= (\bar{y} - b\bar{x}) + bx$$
$$= \bar{y} + b(x - \bar{x}).$$

The above estimation procedure is often called regressing y upon x, and equation (15.2.7) is often called the fitted regression equation or line.

If the model $E(y \mid x) = \alpha + \beta x$ had been written in the equivalent alternative form $E(y \mid x) = \delta + \beta \dot{x}$ where $\dot{x} = (x - \bar{x})$ and $\delta = \alpha + \beta\bar{x}$, the normal equations would have become

(15.2.8)
$$(n)\hat{\delta} + (0)\hat{\beta} = \sum_{i=1}^{n} y_i$$
$$(0)\hat{\delta} + (\sum \dot{x}^2)\hat{\beta} = \sum_{i=1}^{n} \dot{x}_i y_i.$$

The normal equations solve on sight to give

(15.2.9) $\hat{\delta} = \sum y_i/n = \bar{y}, \qquad \hat{\beta} = \dfrac{\sum \dot{x}_i y_i}{\sum \dot{x}_i^2} = \dfrac{\sum (x_i - \bar{x})y_i}{\sum (x_i - \bar{x})^2} = \dfrac{S(XY)}{S(X^2)} = b,$

where the exchange of $\dot{x}$ for x effectively shifts the origin to the average of the x_i's. The intercept of the straight line in the new coordinate system is then simply the average $\bar{y}$. The estimate of α in the original model can be retrieved. For, on finding $(\hat{\delta}, \hat{\beta}) = (\bar{y}, b)$ from (15.2.9), we may find an estimate of α by simply evaluating $\bar{y} - b\bar{x}$, which, from (15.2.6), is the estimate a. Indeed, we shall see later that $E(a) = E(\bar{y} - b\bar{x}) = \alpha$ and $E(b) = \beta$. Rewriting the original model in this modified form simplifies computations, and such tricks are frequently performed by the experienced statistician. The quantity $\dot{x} = (x - \bar{x})$ is sometimes called the *orthogonal polynomial of degree one*.

Graphically, we can represent all possible values of x and the corresponding values of $\hat{y}$, that is, all possible pairs $(x, \hat{y})$, as a straight line in the (x, y)-plane whose equation is given by (15.2.7). As mentioned before, this line is called the regression line of y on x; its intercept is a and its slope is b. From (15.2.7) we see that the regression line passes through the point $(\bar{x}, \bar{y})$, the "center of gravity" of the scatter diagram of the n pairs of measurements $(x_1, y_1), \ldots, (x_n, y_n)$.

Since $y_1, \ldots, y_n$ are independent random variables having normal distributions $N(\alpha + \beta x_1, \sigma^2), \ldots, N(\alpha + \beta x_n, \sigma^2)$, where $x_1, \ldots, x_n$ are constants (and not random variables) we see from (15.2.6) that the statistic $\hat{\beta}$ is a linear function of the observations $y_1, \ldots, y_n$ with coefficients

(15.2.10)
$$\frac{x_1 - \bar{x}}{\sum (x_i - \bar{x})^2}, \ldots, \frac{x_n - \bar{x}}{\sum (x_i - \bar{x})^2}.$$

That is, $\hat{\beta} = b$ is a statistic of the form $c_1 y_1 + c_2 y_2 + \cdots + c_n y_n$ where the coefficients $c_i = \dfrac{x_i - \bar{x}}{\sum (x_i - \bar{x})^2}.$ Thus, we have the important result that b is a random variable having a normal distribution, that is,

(15.2.11)
$$f(b \mid x_1, \ldots, x_n) = N\left(\beta; \frac{\sigma^2}{\sum\limits_{i=1}^{n}(x_i - \bar{x})^2}\right).$$

To see that the expected value of b is β, we note that

(15.2.12)
$$\begin{aligned}
E(b) &= E(c_1 y_1 + \cdots + c_n y_n) \\
&= c_1 E(y_1) + \cdots + c_n E(y_n) \\
&= c_1(\alpha + \beta x_1) + \cdots + c_n(\alpha + \beta x_n) \\
&= \alpha(c_1 + \cdots + c_n) + \beta(c_1 x_1 + \cdots + c_n x_n) \\
&= \beta,
\end{aligned}$$

since $c_1 + \cdots + c_n = 0$, $c_1 x_1 + \cdots + c_n x_n = 1$.

To see that the variance of b is $\sigma^2/\sum (x_i - \bar{x})^2$, we have

(15.2.13)
$$\begin{aligned}
V(b) = \sigma_b^2 &= c_1^2 \sigma_{y_1}^2 + \cdots + c_n^2 \sigma_{y_n}^2 \\
&= (c_1^2 + \cdots + c_n^2)\sigma^2;
\end{aligned}$$

or

$$V(b) = \frac{\sigma^2}{\sum (x_i - \bar{x})^2} = \frac{1}{S(X^2)}\sigma^2,$$

since $c_1^2 + \cdots + c_n^2 = 1/\sum (x_i - \bar{x})^2 = 1/S(X^2)$. Also, the following result can be similarly shown: For any value of x, say x_0, the corresponding value of $\hat{y}$, say $\hat{y}_0$, given by

(15.2.14)
$$\hat{y}_0 = \bar{y} + b(x_0 - \bar{x})$$

is a random variable having the normal distribution

(15.2.15)
$$N\left\{\alpha + \beta x_0, \left[\frac{1}{n} + \frac{(x_0 - \bar{x})^2}{S(X^2)}\right]\sigma^2\right\}.$$

In particular, if we choose $x_0 = 0$, we find that the estimated intercept, $\hat{y}_0 = a$ is a random variable having the normal distribution

$$(15.2.16) \qquad N\left\{\alpha, \left[\frac{1}{n} + \frac{\bar{x}^2}{S(X^2)}\right]\sigma^2\right\}.$$

Note that the variance of the "predicted value" of y at $x = x_0$ is

$$(15.2.17) \qquad V(\hat{y}_0) = \left[\frac{1}{n} + \frac{(x_0 - \bar{x})^2}{S(X^2)}\right]\sigma^2,$$

and increases as the setting of x_0 departs farther and farther from $\bar{x}$, and that the variance of the intercept a is

$$(15.2.18) \qquad V(a) = \left[\frac{1}{n} + \frac{\bar{x}^2}{S(X^2)}\right]\sigma^2$$

and can be quite large whenever $\bar{x}$ is large.

15.3 UNBIASED ESTIMATION OF σ^2

We now turn to the problem of finding an unbiased estimator for σ^2. We will denote the point on the regression line corresponding to x_i as $\hat{y}_i$, that is $\hat{y}_i = a + bx_i = \bar{y} + b(x_i - \bar{x})$. Suppose we now take the differences, often called residuals, between the observed value of y at x_i (that is y_i) and the predicted value of y at x_i (that is $\hat{y}_i$) for $i = 1$, $2, \ldots, n$. Denoting these differences by $d_i = y_i - \hat{y}_i$, we may square and then sum the d_i, calling the result SSD (the sum of squares of deviations), that is, we have

$$
\begin{aligned}
(15.3.1) \qquad \text{SSD} &= \sum (y_i - \hat{y}_i)^2 \\
&= \sum (y_i - a - bx_i)^2.
\end{aligned}
$$

The sum of squares in (15.3.1) is often called the "residual sum of squares." Since a and b are the least squares estimators of α and β, SSD is thus the minimum of $Q(\alpha, \beta)$, where $Q(\alpha, \beta)$ is given in (15.2.4). Now suppose we write SSD as

$$
\begin{aligned}
\text{SSD} &= \sum (y_i - a - bx_i)(y_i - a - bx_i) \\
&= \sum (y_i - a - bx_i)y_i - a \sum (y_i - a - bx_i) \\
&\quad - b \sum (y_i - a - bx_i)x_i.
\end{aligned}
$$

From the equations (15.2.5), we note that the last two terms are zero, which leaves us with the result that

$$(15.3.2) \quad \begin{aligned} \text{SSD} &= \sum y_i^2 - a \sum y_i - b \sum x_i y_i \\ &= \sum y_i^2 - an\bar{y} - b \sum x_i y_i \\ &= \sum y_i^2 - (\bar{y} - b\bar{x})n\bar{y} - b \sum x_i y_i \\ &= \sum y_i^2 - n\bar{y}^2 - b(\sum x_i y_i - n\bar{x}\bar{y}) \\ &= \sum y_i^2 - n\bar{y}^2 - bS(XY) \\ &= S(Y^2) - bS(XY), \end{aligned}$$

where we have let $S(Y^2) = \sum y_i^2 - n\bar{y}^2 = \sum (y_i - \bar{y})^2$.

Thus the sum of squares of deviations can be obtained by subtracting the *correction factor* $n\bar{y}^2$ and a sum of squares due to the estimate b, that is $bS(XY)$, from the crude sum of squares.

Now from (15.2.7), we have that $\hat{y}_i = \bar{y} + b(x_i - \bar{x})$, so that squaring and adding yields, as the reader may verify,

$$(15.3.3) \quad \sum_{i=1}^{n} \hat{y}_i^2 = n\bar{y}^2 + b^2 \sum_{i=1}^{n} (x_i - \bar{x})^2$$

or

$$(15.3.4) \quad \sum_{i=1}^{n} \hat{y}_i^2 = n\bar{y}^2 + bS(XY).$$

From (15.3.2) and (15.3.4), then, we have that

$$(15.3.5) \quad \sum_{i=1}^{n} y_i^2 = n\bar{y}^2 + bS(XY) + \sum_{i=1}^{n} (y_i - \hat{y}_i)^2 = \sum_{i=1}^{n} \hat{y}_i^2 + \sum_{i=1}^{n} (y_i - \hat{y}_i)^2,$$

where the sum of squares of predicted values $\sum \hat{y}_i^2 = n\bar{y}^2 + bS(XY)$. The above partitioning of $\sum y_i^2$ is often displayed in an "analysis of variance" table, as shown in Table 15.3.1.

TABLE 15.3.1

ANALYSIS OF VARIANCE FOR A FITTED STRAIGHT LINE

Source of Variation	Sum of Squares	Degrees of Freedom	Mean Squares	Expected Mean Squares
Crude sum of squares	$\sum y_i^2$	n		
Sum of squares due to $\hat{\delta} = a + b\bar{x}$	$n\bar{y}^2$	1		$\sigma^2 + n\delta^2$
Sum of squares due to $\hat{\beta} = b$	$bS(XY)$	1	$bS(XY)/1$	$\sigma^2 + S(X^2)\beta^2$
Residual sum of squares SSD (by subtraction)	$\sum (y_i - \hat{y})^2$	$n-2$	$\text{SSD}/(n-2) = s_{e \cdot y}^2$	σ^2

$$\delta = \alpha + \beta\bar{x}; \qquad \hat{\delta} = \hat{\alpha} + \hat{\beta}\bar{x} = a + b\bar{x} = \bar{y}$$

Note that column 4 of the table gives the relevant *mean squares*, that is, the ratio of the sum of squares to the respective degrees of freedom. The last entries in the "sum of squares" column and the "degrees of freedom" column may, of course, be obtained by subtraction [see (15.3.5)].

It is instructive now to determine the expected value of SSD. We have

(15.3.6)
$$
\begin{aligned}
E(\text{SSD}) &= E\{\sum (y_i - \hat{y}_i)^2\} \\
&= E\{\sum (y_i - a - bx_i)^2\} \\
&= E\{\sum [(y_i - \bar{y}) - b(x_i - \bar{x})]^2\} \\
&= E\{\sum (y_i - \bar{y})^2 - b^2 \sum (x_i - \bar{x})^2\} \\
&= \sum E(y_i^2) - nE(\bar{y})^2 - \sum (x_i - \bar{x})^2 E(b^2) \\
&= \sum [(\alpha + \beta x_i)^2 + \sigma^2] - n\left[(\alpha + \beta \bar{x})^2 + \frac{\sigma^2}{n}\right] \\
&\qquad - \sum (x_i - \bar{x})^2 \left[\beta^2 + \frac{\sigma^2}{\sum (x_i - \bar{x})^2}\right] \\
&= (n - 2)\sigma^2 + \sum (\alpha + \beta x_i)^2 - n(\alpha + \beta \bar{x})^2 - \beta^2 \sum (x_i - \bar{x})^2 \\
&= (n - 2)\sigma^2.
\end{aligned}
$$

Thus *the quantity* SSD$/(n - 2)$, *which will be denoted by* s_{ey}^2, *is an unbiased estimator for* σ^2.

We state without proof that under the assumption used in (15.2.3), that is, the y_i's are independent $N(\alpha + \beta x_i, \sigma^2)$ random variables, then SSD$/\sigma^2$ is a random variable having the chi-square distribution with $n - 2$ degrees of freedom, and further SSD is independent of the random variables a, b, and $\hat{y}_0$. Similarly, it can be shown that $E[bS(XY)] = \sigma^2 + \beta^2 S(X^2)$.

15.4 TESTS AND CONFIDENCE INTERVALS
FOR α, β, $E(y \mid x)$

Now we have seen (see Theorem AVI.7) that the ratio of a normally distributed random variable divided by the square root of $1/\nu$ times an independent χ^2 variable with ν degrees of freedom is distributed as the Student t variable with ν degrees of freedom. In the context of our problem, the statistics $\hat{y}$, a, b are normally distributed variates, and independent of $(n - 2) s_{ey}^2 = \chi_{n-2}^2 \sigma^2$. Thus the ratios

(15.4.1)
$$
\frac{\hat{y} - E(y \mid x)}{\sqrt{\hat{V}(\hat{y})}}, \quad \frac{a - \alpha}{\sqrt{\hat{V}(a)}}, \quad \frac{b - \beta}{\sqrt{\hat{V}(b)}}
$$

are all t variables with $n - 2$ degrees of freedom, where the $\hat{V}$'s are defined by replacing σ^2 by s_{ey}^2 in equations (15.2.17), (15.2.18), and (15.2.13). For example, to test the hypothesis

$$H_0: \quad \beta = \beta_0$$

against the two-sided alternative

$$H_1: \quad \beta \neq \beta_0$$

at the $100\gamma \%$ level of significance, we may use the ratio

$$(15.4.2) \qquad t_v = \frac{b - \beta_0}{\sqrt{\dfrac{1}{S(X^2)} s_{ey}^2}},$$

which, if H_0 is true, is a Student t variable with $v = n - 2$ degrees of freedom. Hence the rejection region of significance level γ is given by:

$$\text{reject if observed } |t_v| \geq t_{v;\gamma/2}.$$

Now, since the statistics a, b, and $\hat{y}_0 = a + bx_0$ are linear statistics each independent of the estimate $s_{ey}^2 = \text{SSD}/(n - 2)$ we have, using (15.4.1), that

$$(15.4.3) \qquad \left[a \pm t_{n-2;\gamma/2} \sqrt{\left(\frac{1}{n} + \frac{\bar{x}^2}{S(X^2)} \right) s_{ey}^2} \right]$$

is a $100(1 - \gamma)\%$ confidence interval for α; also

$$(15.4.4) \qquad \left[b \pm t_{n-2;\gamma/2} \sqrt{\frac{1}{S(X^2)} s_{ey}^2} \right]$$

is a $100(1 - \gamma)\%$ confidence interval for β, and, finally

$$(15.4.5) \qquad \left[\hat{y}_0 \pm t_{n-2;\gamma/2} \sqrt{\left(\frac{1}{n} + \frac{(x_0 - \bar{x})^2}{S(X^2)} \right) s_{ey}^2} \right]$$

is a $100(1 - \gamma)\%$ confidence interval for $\alpha + \beta x_0$, the expected value of y for $x = x_0$. It should be noted that (15.4.5) reduces to (15.4.3) if we choose $x_0 = 0$ in (15.4.5).

The analysis of variance table also provides a means for testing hypotheses, and ultimately for making interval statements. Since $t_v^2 = F_{1,v}$ [see Appendix VI(d)] an alternative convenient test of the hypothesis $H_0: \beta = 0$ against the two-sided alternative $H_1: \beta \neq 0$ can be performed using the analysis of variance table by computing the ratio of the mean square for b divided by s_{ey}^2, that is, $bS(XY)/s_{ey}^2 = t_{n-2}^2 = F_{1,n-2}$, if

H_0 is true. Hence we reject H_0 if the observed $F_{1,n-2} > F_{1,n-2;\gamma}$ at the $100\gamma\%$ level of significance.

We mention without proof that a test of the *composite* hypothesis that $\alpha = \alpha_0$ *and* $\beta = \beta_0$ is provided by the ratio

(15.4.6)
$$F_{2,n-2} = \frac{[n(a - \alpha_0)^2 + 2\sum x_i(a - \alpha_0)(b - \beta_0) + \sum x_i^2(b - \beta_0)^2]}{2s_{ey}^2}.$$

The rejection region of significance level γ is given by

$$\text{reject if } F_{2,n-2} \geq F_{2,n-2;\gamma}.$$

Also, the bounds of the $100(1 - \gamma)\%$ *simultaneous confidence region* for α and β are provided by all values of α and β which satisfy the equation

$$n(a - \alpha)^2 + 2\sum x_i(a - \alpha)(b - \beta) + \sum x_i^2(b - \beta)^2 = 2s_{ey}^2 F_{2,n-2;\gamma}.$$

15.5   AN EXAMPLE

A study was instituted to determine the percent of waste solids removed in a filtration system as a function of the flow rate of the effluent being fed into the system. It was decided to use flow rates of $2, 4, \ldots, 14$ gals/min and to observe y, the percent of waste solid removed, when each of these flow rates was used. The study yielded the data displayed in Figure 15.5.1.

The mathematical model $E(y \mid x) = \alpha + \beta x$ was proposed, and the following quantities computed

$$n = 7, \quad \sum x = 56, \quad \sum y = 100.8, \quad \sum x^2 = 560, \quad \sum y^2 = 1724.80,$$
$$\sum xy = 632.6, \quad \bar{x} = 8.0, \quad \bar{y} = 14.4, \quad S(X^2) = 112,$$
$$S(Y^2) = 273.28, \quad S(XY) = -173.8.$$

The normal equations (15.2.5a) take the form

$$7a + 56b = 100.8$$
$$56a + 560b = 632.6.$$

These may be solved directly, or, using equations (15.2.6) we obtain

$$b = \frac{-173.8}{112} = -1.55; \quad a = 14.4 - (-1.55)(8.0) = 26.81.$$

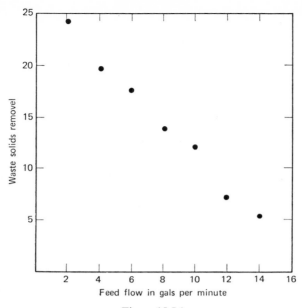

Figure 15.5.1.

Thus the best fitted straight line in the sense of least squares is

$$\hat{y} = 26.81 - 1.55x.$$

The computations for SSD, the sum of squares of deviations, may be obtained directly from equation (15.3.1), or by using (15.3.2).

TABLE 15.5.1

x	y	$\hat{y}_i$	$y_i - \hat{y}_i$	Source		Degrees of Freedom
2	24.3	23.7	0.6	Crude sum of squares Σy^2	1724.80	7
4	19.7	20.6	−0.9	Sum of squares due to $\delta = a + b\bar{x} : n\bar{y}^2 =$	1451.52	1
6	17.8	17.5	0.3	Sum of squares due to $\hat{\beta} = b : bS(XY) =$	269.70	1
8	14.0	14.4	−0.4	Residual sum of squares $= \Sigma (y_i - \hat{y}_i)^2 =$	3.58	5
10	12.3	11.3	1.0			
12	7.2	8.2	−1.0			
14	5.5	5.1	0.4	Estimate of $\sigma^2 : s_{ey}^2 = \dfrac{3.58}{5} = 0.72$		

$$\Sigma (y_i - \hat{y}_i) = 0 \qquad \Sigma (y_i - \hat{y}_i)^2 = 3.58$$

The estimate of σ^2 is therefore [see (15.3.6)] $s_{ey}^2 = 3.58/5 = 0.716$ with five degrees of freedom. Testing the hypothesis that $E(b) = \beta = 0$

against the two-sided alternative $\beta \neq 0$, we use

$$|t_5| = \left| \frac{-1.55 - 0}{\sqrt{\frac{1}{112}(0.72)}} \right| = 19.33,$$

clearly a rare event, since the $P\{|t_5| \geq 2.571\} = 0.05$. The hypothesis is thus rejected.

Applying (15.4.4), we find the 95% confidence interval for β to be

$$[-1.55 \pm 2.571\sqrt{\tfrac{1}{112}(0.72)}\,] = [-1.55 \pm 0.21],$$

that is, $[-1.76, -1.34]$. We observe that the value $\beta = 0$ specified by the null hypothesis H_0 is not contained in this confidence region.

Testing the hypothesis that the intercept $\alpha = 25$ against the one sided alternative that $\alpha > 25$ gives, using (15.4.1),

$$t_5 = \frac{26.81 - 25}{\sqrt{\left[\frac{1}{7} + \frac{(8.0)^2}{112}\right]0.72}} = \frac{1.68}{\sqrt{0.5114}} = 2.35.$$

The hypothesis is rejected since $P\{t_5 \geq 2.015\} = 0.05$. Applying (15.4.3) the 95% confidence interval for the intercept α is

$$\left[26.81 \pm 2.571\sqrt{\left[\frac{1}{7} + \frac{(8.0)^2}{112}\right]0.72}\,\right] = [26.81 \pm 1.84],$$

that is, the 95% confidence interval* for α is $[24.97, 28.65]$.

The 95% confidence limits for the response at, say, $x_0 = 4$ are given by

$$\left[20.61 \pm 2.571\sqrt{\left[\frac{1}{7} + \frac{(4 - 8.0)^2}{112}\right](0.72)}\,\right] = [20.61 \pm 1.16].$$

The width of the confidence interval for the responses will be smallest when $x = \bar{x}$, and widens as the deviation $|x - \bar{x}|$ increases. The predicted values and 95% confidence limits for various values of x are listed in Table 15.5.2. The fitted line and the 95% confidence band about the fitted line is displayed in Figure 15.5.2.

---

* This example is instructive, since the reader will note that although the hypothesis that $\alpha = 25$ was rejected by the one sided test of significance, the value $\alpha = 25$ falls within the 95% confidence interval *symmetrically* constructed about the estimate $a = 26.81$.

TABLE 15.5.2

x	2	4	6	8	10	12	14	16
Predicted response $\hat{y}$	23.71	20.61	17.51	14.41	11.31	8.21	5.11	2.01
95% confidence band	±1.49	±1.16	±0.92	±0.82	±0.92	±1.16	±1.49	±1.84

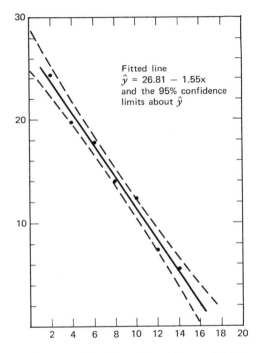

Fitted line
$\hat{y} = 26.81 - 1.55x$
and the 95% confidence
limits about $\hat{y}$

Figure 15.5.2. Plot of fitted line and 95% confidence limits for predicted values.

As noted earlier, the analysis of variance table could be used to test the hypothesis H_0: $\beta = 0$ against the two-sided alternative $\beta \neq 0$ by computing the ratio of the mean square for b to the residual mean square. This is done in Table 15.5.3.

TABLE 15.5.3

		Sum of Squares	Degrees of Freedom	Mean Squares	
Crude sum of squares:	Σy^2	1724.80	7		
Sum of squares due to					
$(a + b\bar{x})$:	$n\bar{y}^2$	1451.52	1		
Sum of squares due to b:	$bS(XY)$	269.70	1	269.70	$F_{1,5} = 269.7/.72$
Residual sum of squares:		3.58	5	0.72	$= 374.6$

Of course, we have seen that, on the basis of a two-sided t test, the data gives evidence to support the hypothesis that $\beta \neq 0$. In fact, the observed $t_5 = 19.354$. As we have noted before, $t_5^2 = F_{1,5}$, that is, the observed $F_{1,5} = (19.354)^2 = 374.57$, which is in agreement with the entry for $F_{1,5}$ in the analysis of variance Table 15.5.3.

The boundary of the 95% confidence region for α and β considered simultaneously, as provided by (15.4.6), is

$$(15.5.1) \quad 7(26.81 - \alpha)^2 + 2(56)(26.81 - \alpha)(-1.55 - \beta)$$
$$+ 560(-1.55 - \beta)^2 = 2(0.72)5.78,$$

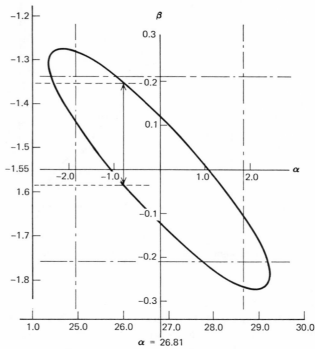

Figure 15.5.3. Boundary of the 95% confidence region for α and β.

the equation of an ellipse centered at $\hat{\alpha} = 26.81$ and $\hat{\beta} = -1.55$. The region is plotted in Figure 15.5.3 along with the superimposed individual confidence intervals for α and β. Naively, the experimenter might feel that the postulated values $\alpha = 26$ and $\beta = -1.7$ were both admissible upon viewing, separately, the *individual* confidence limits for α and β. However these postulated values are outside the *simultaneous* confidence region for α and β and are *not* simultaneously admissible in the light of the data.

Further, if it is known that $\alpha = 26.0$, then the 95% confidence interval for β, given this knowledge, is found by determining the end points of the line segment bounded by the confidence ellipse (15.5.1) when $\alpha = 26$. This means that we must return to (15.5.1) and set $\alpha = 26$. Solving for β yields the 95% confidence interval for β, given $\alpha = 26$, as $(-1.574, -1.357)$. This is illustrated in Figure 15.5.3.

15.6 INFERENCE ABOUT ρ

It frequently happens that we wish to find confidence intervals and make tests of hypotheses concerning the parameter ρ, the correlation coefficient in the population (see Section 6.7(e)). Suppose (x_i, y_i), $i = 1, \ldots, n$ are n independent observations on the bivariate random variable (x, y), where (x, y) has the bivariate normal distribution (see Appendix III). Then, the sampling distribution of the *sample correlation coefficient r*, where r is defined by

(15.6.1)

$$r = \sum_{i=1}^{n} (x_i - \bar{x})(y_i - \bar{y}) \Big/ \sqrt{\sum_{i=1}^{n} (x_i - \bar{x})^2 \sum_{i=1}^{n} (y_i - \bar{y})^2} = \frac{S(XY)}{\sqrt{S(X^2)S(Y^2)}}$$

is very complicated for general values of ρ. We emphasize that here both components of each observation are random, and the joint distribution of both random components is the bivariate normal distribution. This is unlike the cases discussed so far in this chapter, where x_i are assumed to be fixed constants chosen before the y_i are observed "at x_i."

Now for the case when $\rho = 0$, it is known (see for example Cramer (1961)) that the distribution of r is such that the quantity

(15.6.2)

$$\frac{r}{\sqrt{1 - r^2}} \sqrt{n - 2}$$

has the Student t distribution with $(n - 2)$ degrees of freedom. Hence, for example, to test the hypothesis

$$H_0: \quad \rho = 0$$

against the alternatives

$$H_1: \quad \rho \neq 0,$$

at the $100\gamma\%$ level of significance, we use (15.6.2) as our test statistic and reject H_0 in favor of H_1 if the observed value of

(15.6.3)
$$\frac{|r|\sqrt{n-2}}{\sqrt{1-r^2}} > t_{n-2;\gamma/2}.$$

Example 15.6(a). Bennett and Franklin (1954) give the following example. Thirty-six specimens of ball clay were analyzed for carbon content by two different methods. Let

$$x = \text{carbon determined by combustion}$$
$$y = \text{carbon determined by "rational analysis,"}$$

where both x and y are measured in percent.

It turned out that the values of (x_i, y_i), $i = 1, \ldots, 36$ for the 36 specimens gave the following results:

$$n = 36 \qquad \sum x_i = 69.25 \qquad \sum y_i = 102.71$$
$$\sum x_i^2 = 354.0245 \qquad \sum y_i^2 = 826.6842$$
$$\sum x_i y_i = 510.8425.$$

From these we find

$$S(X^2) = \sum (x_i - \bar{x})^2 = 220.8144$$
$$S(Y^2) = \sum (y_i - \bar{y})^2 = 553.6469$$

and

$$S(XY) = \sum (x_i - \bar{x})(y_i - \bar{y}) = \sum x_i y_i - (\sum x_i)(\sum y_i)/n$$
$$= 510.8425 - (69.25)(102.71)/36$$
$$= 510.8425 - 197.5741$$
$$= 313.2684.$$

Hence

$$r = \frac{313.2684}{\sqrt{(220.8144)(553.6469)}} = .9127.$$

Thus, the observed value of the expression in (15.6.3) is

$$\frac{.9127}{\sqrt{1-(.9127)^2}} \sqrt{34} = 13.025,$$

a highly significant result at either the 5% or 1% level of significance, and so we must reject the hypothesis of $\rho = 0$.

Thus in our example it is highly unlikely that the population correlation coefficient ρ could be zero, in which case we might wish to proceed to find a confidence interval for ρ. Now it can be shown that for large n, if $(x_1, y_1), \ldots, (x_n, y_n)$ can be regarded as a random sample from a *bivariate normal distribution* having correlation coefficient ρ, then the quantity

$$(15.6.4) \qquad z = (\tfrac{1}{2}) \log_e \frac{1 + r}{1 - r}$$

is approximately normally distributed with mean

$$(15.6.5) \qquad (\tfrac{1}{2}) \log_e \frac{1 + \rho}{1 - \rho}$$

and variance $1/(n - 3)$. F. N. David has prepared charts (see pp. 518 and 519) for finding 95% and 99% confidence limits for ρ. Consulting these charts and entering at $r = .9127$, the value of r in Example 15.6(a), and interpolating "by eye" between the $n = 25$ and $n = 50$ curves, we find that $(.82, .94)$ is an approximate 95% confidence interval for ρ, and $(.75, .97)$ is an approximate 99% confidence interval for ρ.

We can proceed alternatively as follows. Using (15.6.4) and (15.6.5) we see that an approximate $100(1 - \gamma)\%$ confidence interval for

$$(\tfrac{1}{2}) \log_e \frac{1 + \rho}{1 - \rho}$$

is

$$(15.6.6) \qquad \left[(\tfrac{1}{2}) \log_e \frac{1 + r}{1 - r} \pm z_{\gamma/2} \frac{1}{\sqrt{n - 3}} \right].$$

Consulting Tables II and IV of Appendix VII, we have that, for $1 - \gamma = .95$, the confidence interval for $(\tfrac{1}{2}) \log_e [(1 + \rho)/(1 - \rho)]$ for Example 15.6(a) ($r = .9127, n = 36$) is

$$(15.6.7) \qquad \left(1.544 \pm 1.96 \frac{1}{\sqrt{33}} \right) = (1.203, 1.885).$$

Now using Table IV of Appendix VII again and inversely interpolating, we find that a 95% confidence interval for ρ is

$$(15.6.8) \qquad (.84, .95),$$

compared with $(.82, .94)$ as found previously from the charts. Note that the event the interval (15.6.7) contains $(\tfrac{1}{2}) \log_e [(1 + \rho)/(1 - \rho)]$ implies that the interval (15.6.8) contains ρ, since $(\tfrac{1}{2}) \log_e [(1 + \rho)/(1 - \rho)]$ is increasing in ρ.

15.7 LINEAR REGRESSION PROBLEMS INVOLVING TWO OR MORE INDEPENDENT VARIABLES

Linear regression involving two independent variables can be treated in a manner similar to that presented in Section 15.2. This type of regression, or model fitting, could arise as follows. Suppose a response η is a function of two controlled variables ξ_1 and ξ_2, that is $\eta = f(\xi_1, \xi_2)$. Provided the response is smooth over the experimenter's region of interest of ξ_1 and ξ_2, this response function can often be usefully represented by a first order Taylor's series about a point (ξ_{10}, ξ_{20}) at the center of the region, that is,

$$(15.7.1) \quad \eta \cong f(\xi_{10}, \xi_{20}) + (\xi_1 - \xi_{10}) \frac{\partial f}{\partial \xi_1}\bigg|_{\xi_{10}, \xi_{20}} + (\xi_2 - \xi_{20}) \frac{\partial f}{\partial \xi_2}\bigg|_{\xi_{10}, \xi_{20}}.$$

Evaluating the derivatives $\dfrac{\partial f}{\partial \xi_1}$ and $\dfrac{\partial f}{\partial \xi_2}$ at the point (ξ_{10}, ξ_{20}), and collapsing terms gives the model

$$(15.7.2) \qquad \eta = \alpha + \beta_1 x_1 + \beta_2 x_2,$$

where x_1 and x_2 may be taken to be conveniently coded values of ξ_1 and ξ_2. On many occasions the model (15.7.2) is exactly appropriate.

Suppose now that the observed response y is a random variable which is distributed with mean

$$(15.7.3) \qquad E(y \mid x_1, x_2) = \eta = \alpha + \beta_1 x_1 + \beta_2 x_2$$

and variance σ^2. To estimate $(\alpha, \beta_1, \beta_2)$, we will observe y_i at preselected points (x_{1i}, x_{2i}), $i = 1, \dots, n$. On the basis of this data, we may find least squares estimators of $(\alpha, \beta_1, \beta_2)$, say (a, b_1, b_2), respectively, by minimizing the sum of squares

$$(15.7.4) \qquad Q(\alpha, \beta_1, \beta_2) = \sum (y_i - \alpha - \beta_1 x_{1i} - \beta_2 x_{2i})^2$$

with respect to α, β_1, and β_2.

We emphasize that here the observations constituting the sample are triplets (x_{1i}, x_{2i}, y_i) where x_{1i}, x_{2i}, $i = 1, 2, \dots, n$ are preselected values of the controlled variables x_1 and x_2. This data may then be thought of as a cluster of n points in a three dimensional scatter diagram. Let a, b_1, and b_2 be the values of α, β_1, and β_2 which minimize the sum of squares (15.7.4), that is, which satisfy the three equations $\partial Q / \partial \alpha = 0$, $\partial Q / \partial \beta_1 = 0$, $\partial Q / \partial \beta_2 = 0$. Taking these indicated derivatives, and after some algebraic simplification, we obtain the three normal equations $\left(\sum x_1 \text{ means } \sum_{i=1}^{n} x_{1i}, \text{etc.} \right)$

$$n(a) + \sum x_1(b_1) + \sum x_2(b_2) = \sum y$$

(15.7.5) $$\sum x_1(a) + \sum x_1^2(b_1) + \sum x_1 x_2(b_2) = \sum x_1 y$$

$$\sum x_2(a) + \sum x_1 x_2(b_1) + \sum x_2^2(b_2) = \sum x_2 y.$$

Provided the determinant of the coefficients of these equations does not equal zero our problem reduces to solving these three equations for the three unknowns a, b_1, and b_2.

Dividing the first of the equations by n gives

$$a = \bar{y} - b_1 \bar{x}_1 - b_2 \bar{x}_2$$

where $\bar{y}$, $\bar{x}_1$, and $\bar{x}_2$ are the averages. Placing* this value of a in the remaining two equations gives

(15.7.6) $$b_1 \sum x_{1i}(x_{1i} - \bar{x}_1) + b_2 \sum x_{1i}(x_{2i} - \bar{x}_2) = \sum x_{1i}(y_i - \bar{y}),$$

$$b_1 \sum x_{2i}(x_{1i} - \bar{x}_1) + b_2 \sum x_{2i}(x_{2i} - \bar{x}_2) = \sum x_{2i}(y_i - \bar{y}).$$

These equations may be written as

(15.7.6a) $$b_1 S(X_1^2) + b_2 S(X_1 X_2) = S(X_1 Y),$$

$$b_1 S(X_1 X_2) + b_2 S(X_2^2) = S(X_2 Y),$$

or

(15.7.6b) $$c_{11} b_1 + c_{12} b_2 = c_{1y},$$

$$c_{12} b_1 + c_{22} b_2 = c_{2y}.$$

The quantities $S(X_1^2)$, $S(X_2^2)$, $S(X_1 X_2)$, $S(X_1 Y)$, and $S(X_2 Y)$ or equivalently c_{11}, c_{22}, c_{12}, c_{1y}, c_{2y} are the corrected sums of squares and cross products, that is

$$c_{11} = S(X_1^2) = \sum (x_{1i} - \bar{x}_1)^2 = \sum x_{1i}(x_{1i} - \bar{x}_1) = \sum x_{1i}^2 - (\sum x_{1i})^2 / n$$

$$c_{22} = S(X_2^2) = \sum (x_{2i} - \bar{x}_2)^2 = \sum x_{2i}(x_{2i} - \bar{x}_2) = \sum x_{2i}^2 - (\sum x_{2i})^2 / n$$

$$c_{12} = S(X_1 X_2) = \sum (x_{1i} - \bar{x}_1)(x_{2i} - \bar{x}_2) = \sum x_{1i}(x_{2i} - \bar{x}_2)$$

$$= \sum x_{2i}(x_{1i} - \bar{x}_1) = \sum x_{1i} x_{2i} - (\sum x_{1i})(\sum x_{2i}) / n$$

(15.7.7)

$$c_{1y} = S(X_1 Y) = \sum (x_{1i} - \bar{x}_1)(y_i - \bar{y}) = \sum x_{1i}(y_i - \bar{y})$$

$$= \sum x_{1i} y_i - (\sum x_{1i})(\sum y_i) / n$$

$$c_{2y} = S(X_2 Y) = \sum (x_{2i} - \bar{x}_2)(y_i - \bar{y}) = \sum x_{2i}(y_i - \bar{y})$$

$$= \sum x_{2i} y_i - (\sum x_{2i})(\sum y_i) / n,$$

* The elimination of the constant term a reduces the number of normal equations by one. This step is almost always performed when there are $k \geq 2$ independent variables.

and the corrected sum of squares of the y_i's by

$$(15.7.7a) \qquad S(Y^2) = \sum_{i=1}^{n} (y_i - \bar{y})^2 = \sum y_i^2 - (\sum y_i)^2/n.$$

Solving for the b_i, we find

$$(15.7.8)$$

$$b_1 = \frac{c_{22}}{\Delta} c_{1y} - \frac{c_{12}}{\Delta} c_{2y} = c^{11} c_{1y} + c^{12} c_{2y}$$

$$b_2 = \frac{-c_{12}}{\Delta} c_{1y} + \frac{c_{11}}{\Delta} c_{2y} = c^{12} c_{1y} + c^{22} c_{2y},$$

where we have let $c^{11} = c_{22}/\Delta$, $c^{12} = -c_{12}/\Delta$, and $c^{22} = c_{11}/\Delta$, with $\Delta = c_{11}c_{22} - c_{12}^2$. Considerable savings in arithmetic are clearly possible whenever the x_{1i} and x_{2i} are chosen so that $\sum x_{1i} = \sum x_{2i} = 0$. If, in addition, $\sum x_{1i}x_{2i} = 0$, then $b_1 = \frac{1}{c_{11}} c_{1y}$ and $b_2 = \frac{1}{c_{22}} c_{2y}$.

Using (15.7.3) and (15.7.5) the estimator $\hat{y}_i$ of $E(y_i \mid x_{1i}, x_{2i})$ is given by the fitted model

$$(15.7.9) \qquad \hat{y}_i = a + b_1 x_{1i} + b_2 x_{2i}$$

$$= \bar{y} - b_1 \bar{x}_1 - b_2 \bar{x}_2 + b_1 x_{1i} + b_2 x_{2i}$$

$$= \bar{y} + b_1(x_{1i} - \bar{x}_1) + b_2(x_{2i} - \bar{x}_2),$$

so that the fitted model, in general, is

$$\hat{y} = \bar{y} + b_1(x_1 - \bar{x}_1) + b_2(x_2 - \bar{x}_2).$$

We note in passing that the point $(\bar{x}_1, \bar{x}_2, \bar{y})$ satisfies this equation.

Further, it can be shown that the sum of squares of the deviations $y_i - \hat{y}_i$ is given by (see Problem 15.24)

$$(15.7.10)$$

$$\text{SSD} = \sum (y_i - \hat{y}_i)^2$$

$$= \sum y_i^2 - a \sum y_i - b_1 \sum x_{1i}y_i - b_2 \sum x_{2i}y_i$$

$$= \sum (y_i - \bar{y})^2 - b_1 \sum (x_{1i} - \bar{x}_1)(y_i - \bar{y}) - b_2 \sum (x_{2i} - \bar{x}_2)(y_i - \bar{y})$$

$$= S(Y^2) - b_1 c_{1y} - b_2 c_{2y}.$$

It also may be shown that, if the y_i are independent normal variables $N(\alpha + \beta_1 x_{1i} + \beta_2 x_{2i}, \sigma^2)$, $i = 1, 2, \ldots, n$, SSD/σ^2 is a random variable having the chi-square distribution with $n - 3$ degrees of freedom, and that

$$(15.7.11) \qquad E(\text{SSD}) = (n - 3)\sigma^2.$$

The quantity $\mathrm{SSD}/(n-3)$, which we denote by s_{ey}^2, is thus an unbiased estimator for σ^2.

The computation of the estimate of σ^2 is usually displayed in an "analysis of variance" table as illustrated in Table 15.7.1.

TABLE 15.7.1

Sum of Squares	Degrees of Freedom	Mean Squares
Total corrected sum of squares:		
$S(Y^2) = \sum (y_i - \bar{y})^2$	$n-1$	
Regression sum of squares: $\mathrm{SSR} = \sum b_i c_{iy}$	2	$\mathrm{SSR}/2$
Residual sum of squares:		
$\mathrm{SSD} = S(Y^2) - \sum b_i c_{iy}$	$n-3$	$\mathrm{SSD}/(n-3) = s_{ey}^2$

Under the assumption of normality made above it can be shown that b_1 and b_2 have, respectively, normal distributions $N(\beta_1, c^{11}\sigma^2)$ and $N(\beta_2, c^{22}\sigma^2)$. Further, the estimate a has a normal distribution $N(\alpha, A'\sigma^2)$ where (see Problem 15.25)

$$(15.7.12) \qquad A' = \left(\frac{1}{n} + \sum_{i=1}^{2} c^{ii}\bar{x}_i^2 + 2c^{12}\bar{x}_1\bar{x}_2\right).$$

Also, if we denote $E(y \mid x_{10}, x_{20})$ by η_0, then

$$(15.7.13) \qquad \eta_0 = \alpha + \beta_1 x_{10} + \beta_2 x_{20},$$

and it can be shown that the estimator $\hat{y}_0$ of η_0, given by

$$(15.7.14) \qquad \hat{y}_0 = a + b_1 x_{10} + b_2 x_{20},$$

has the normal distribution $N(\eta_0, A\sigma^2)$, where (see Problem 15.26)

$$(15.7.15) \quad A = \left[\frac{1}{n} + \sum_{i=1}^{2} c^{ii}(x_{i0} - \bar{x}_i)^2 + 2c^{12}(x_{10} - \bar{x}_1)(x_{20} - \bar{x}_2)\right].$$

If σ^2 is not known, as is usually the case, we estimate it by s_{ey}^2, which is a random variable that can be shown to be independent of a, b_1, b_2, and $\hat{y}_0$. It follows from Theorem AVI.7 that the quantitites

$$(15.7.16) \qquad \frac{a - \alpha}{\sqrt{A's_{ey}^2}}, \qquad \frac{b_i - \beta_i}{\sqrt{c^{ii}s_{ey}^2}}, \qquad i = 1, 2, \quad \text{and} \quad \frac{y - \eta_0}{\sqrt{As_{ey}^2}}$$

are all Student t variables with $n-3$ degrees of freedom. From this fact we may proceed to find confidence intervals for α, β_1, β_2, η_0, as in Section 5.4.

It can be shown that a combined (portmanteau) test of the hypothesis H_0: $\beta_1 = 0$ and $\beta_2 = 0$ against the alternative $\beta_1 \neq 0$, $\beta_2 \neq 0$ can be constructed from the analysis of variance table. *If H_0 is true,* then the corrected sum of squares $S(Y^2)$ is a random variable such that $S(Y^2)/\sigma^2$ has a chi-square distribution with $(n-1)$ degrees of freedom, and therefore (see Theorem AVI.5) $s^2 = S(Y^2)/(n-1)$ is an unbiased estimate of σ^2. In Table 15.7.1, $S(Y^2)$ has been partitioned into two independent sum of squares, SSR and SSD with 2 and $(n-3)$ degrees of freedom, respectively. If H_0 is true then the quantities SSR/2 and SSD/$(n-3)$ found in Table 15.7.1 are therefore both independent estimates of σ^2, and their ratio is distributed as $F_{2,n-3}$. (See Theorem AVI.10.) Thus, to test at significance level γ the combined hypothesis H_0: $\beta_1 = 0$, $\beta_2 = 0$, we need only determine whether

$$(15.7.17) \qquad \frac{\text{SSR}/2}{\text{SSD}/(n-3)} = F_{2,n-3} \geq F_{2,n-3;\gamma}.$$

If (15.7.17) holds, we reject H_0, but if $[\text{SSR}/2]/[\text{SSD}/(n-3)] < F_{2,n-3;\gamma}$ we accept H_0. Further, we mention here, without proof that

$(15.7.18)$

$$[c_{11}(b_1 - \beta_1)^2 + 2c_{12}(b_1 - \beta_1)(b_2 - \beta_2) + c_{22}(b_2 - \beta_2)^2]/2s_{ey}^2 = F_{2,n-3}.$$

We may use (15.7.18) to find a $100(1-\gamma)\%$ confidence region for β_1 and β_2, which is, of course, given by

$(15.7.19)$

$$2s_{ey}^2 F_{2,n-3;\gamma} = c_{11}(b_1 - \beta_1)^2 + 2c_{12}(b_1 - \beta_1)(b_2 - \beta_2) + c_{22}(b_2 - \beta_2)^2.$$

The confidence region for β_1 and β_2 based on (15.7.19) does not encompass the identical values of β_1 and β_2 provided by considering the separate confidence intervals which would be found using (15.7.16). (This was illustrated for a simpler model in Figure 15.5.3.)

15.8 EXTENSION TO MORE GENERAL REGRESSION PROBLEMS

It should be noted that the model discussed in Section 15.2 is quite general. For example, suppose $E(y \mid t) = \alpha + \beta \sin t$, where t is an independent variable. We can find estimates of α and β, say a and b, by using the methods and formulas of the previous section, with x replaced by $\sin t$, that is, with x_i replaced by $\sin t_i$, $i = 1, \ldots, n$. If $y_1, \ldots, y_n$ are independent random variables having the normal distributions $N(\alpha + \beta \sin t_1, \sigma^2), \ldots,$

$N(\alpha + \beta \sin t_n, \sigma^2)$, we can make tests, find confidence intervals, etc., for α and β, and, of course, $\alpha + \beta \sin t_0$.

Or consider the problem of fitting the exponential curve $u = \gamma e^{\delta v}$ to pairs of measurements (v_i, u_i), $i = 1, \ldots, n$. Taking logarithms, we have

$$(15.8.1) \qquad \log u = \log \gamma + \delta v.$$

If we let $y = \log u$, $x = v$, $\alpha = \log \gamma$, $\beta = \delta$, then (15.6.1) becomes

$$y = \alpha + \beta x,$$

and we can find estimates of α and β by the methods of Section 15.2. Thus a and b of Section 15.2 are estimators for $\log \gamma$ and δ, while e^a is an estimator for γ.

In addition, if we assume that, for fixed x, y has the normal distribution $N(\alpha + \beta x, \sigma^2)$, we can find confidence intervals for α, β, $\alpha + \beta x_0$ (that is, $\log \gamma$, δ, $\log \gamma + \delta v_0$) as before.

Similarly, using the model (15.7.3), we can extend the range of problems that can be attacked by these methods to many situations. For example, the model $E(y \mid t) = \alpha_0 + \beta_1 \cos \pi t + \beta_2 \sin \pi t$, may be analyzed by the method of Section 15.7, by simply letting $x_{1i} = \cos \pi t_i$ and $x_{2i} = \sin \pi t_i$, $i = 1, 2, \ldots, n$, where the t_i are preselected time points at which we observe the response y_i.

We point out that the model (15.7.3) can itself be extended in a straightforward manner to include $k > 2$ independent variables $x_1, \ldots, x_k$, in which case y would have the distribution $N(\alpha + \beta_1 x_1 + \cdots + \beta_k x_k, \sigma^2)$. [See Draper and Smith (1966) for further details including applications.]

Also note that we have not placed any restrictions as to what the variables $x_1, x_2, \ldots, x_k$ can be. For example, they could be powers of a single variable t, in which case $x_j = t^j$, $j = 1, \ldots, k$, and so we would have a polynomial of degree k in t for the regression curve.

Finally, we mention that all topics in this chapter can be analyzed by a Bayesian approach. However for lack of space we simply refer the reader to Lindley (1965) and the references therein.

PROBLEMS

15.1. On prior theoretical grounds, it has been assumed that $E(y \mid x) = \eta = \beta x$. To gain knowledge of β, it is decided to experiment by setting the independent variable to each of the values $x_1, \ldots, x_n$, and to observe the resulting y's, $y_1, \ldots, y_n$ respectively. Determine by the method of least squares an estimate $\hat{\beta}$ of β and hence an estimate of this regression line of y on x (which passes through the origin). Find the expectation and variance of $\hat{\beta}$, assuming that the y_i's are independent, with expectation βx_i, and variance σ^2. If the y_i's are, in addition, normally distributed, state how you would determine confidence intervals for β, and the true response η_0 when $x = x_0$.

15.2. It is decided to measure the resistance of sheets of a certain metal at temperatures (x) of 100°K, 200°K, 300°K, 400°K, and 500°K. The resistances (y) are found to be 4.7, 7.4, 12.4, 16.5, 19.8, respectively. If the regression of y on x is linear, state the normal equations for the constants in the linear model, and solve.

15.3. An experiment is planned in which three observations will be taken at each of four temperatures, 30°, 50°, 70°, and 90°. When the experiment was actually done, the results obtained are those given in Table P15.3. Find, assuming $\eta = E(y \,|\, x) = \alpha + \beta x$, the least squares estimates of α and β, and hence the least squares estimate of η. Construct 95% confidence intervals for α, β, and $\eta_0 = \alpha + \beta x_0$.

TABLE P15.3

x = temperature	30	30	30	50	50	50	70	70	70	90	90	90
y = response	40	45	31	34	28	35	21	29	25	16	21	23

15.4. It is desired to fit a straight line to data found as a result of observing tensile strength y of 10 test pieces of plastic which have undergone baking (at a uniform temperature) for x minutes, where ten values of x were preselected. The data (in coded units) is given in Table P15.4. Repeat the instructions of Problem 15.3 for this set of data.

TABLE P15.4

x	23	35	45	65	75	95	105	125	155	185
y	2	9.8	9.2	26.2	17.1	24.8	43	55.3	38.4	63.3

15.5. An investigation of the (assumed) linear relationship between the load x on a spring and the subsequent length of the spring y has been carried out with the results as given in Table P15.5. Repeat the instructions of Problem 15.3 for this set of data.

TABLE P15.5

x	5	10	15	20	25	30
y	7.25	8.12	8.95	9.90	10.9	11.8

15.6. A study was made on the effect of pressure (x) on the yield (y) of paint made by a certain chemical process. The results (coded units) are given in Table P15.6. Repeat the instructions of Problem 15.3. Let x_0 take the values $-5, -3, -1, 1, 3, 5$, and draw the "confidence belt" for the estimated regression equation.

TABLE P15.6

x	−5	−4	−3	−2	−1	0	1	2	3	4	5
y	6.5	10.3	9.7	12.1	15.7	13.7	14.2	18.0	19.7	18.8	23.4

15.7. The relationship, (assumed) linear, between yield of bourbon (y) and aging time (x) was studied by observing yields y_i from batches that have been allowed to age $x_i = 2i$ years, $i = 1, 2, \ldots, 6$. The results are given in Table P15.7. Repeat the instructions of Problem 15.6 for this set of data.

TABLE P15.7

x	2	4	6	8	10	12
y	3.0	3.4	4.0	4.5	4.4	5.0

15.8. The moisture (x) of the wet mix of a product is considered to have an effect on the density (y) of the finished product. The moisture of the mix was controlled and finished product densities were as shown in Table P15.8. Repeat the instructions of Problem 15.6 for this set of data.

TABLE P15.8

x	5	6	7	8	9	10	11	12	13	14	15	16	17
y	7.6	9.5	9.3	10.3	11.1	12.1	13.3	12.7	13.0	13.8	14.6	14.8	14.7

15.9. The effect of temperature (x in degrees Kelvin) on the color (y, coded units) of a product was investigated. The data is given in Table P15.9. Exhibit the analysis of variance table and use the table to test the hypothesis H_0: $\beta = 0$. If the test rejects H_0, then fit the above data to the model $E(y \mid x) = \alpha + \beta x$. What is the estimate of $E(y \mid x)$ when $x = 145$? Find a 95% confidence interval for $E(y \mid 145)$.

TABLE P15.9

x	100	110	120	140	150	170	180	200	230	260
y	3.5	7.4	7.1	15.6	11.1	14.9	23.5	27.1	22.1	32.9

15.10. Specimens of blood from 10 different animals were analyzed for blood count y (in millimeters cubed) and packed cell volume count x (units of 10^6) with results as given in Table P15.10. Assuming normality, test the hypothesis that the true correlation coefficient ρ between blood count and volume count is zero. If the test is rejected, use the method of Section 15.6 and Table IV, Appendix VII, to find a 95% confidence interval for ρ.

TABLE P15.10

Animal Number	x	y
1	45	6.5
2	42	6.3
3	56	9.5
4	48	7.5
5	42	7.0
6	35	5.9
7	58	9.5
8	40	6.2
9	39	6.6
10	50	8.7

15.11. Repeat the instructions of Problem 15.10 for the data of Table P15.11. The set of data was obtained by measuring the tensile strength (y) in 1000 psi, and Brinell hardness (x) of each of 15 specimens of cold-drawn copper [data from Bowker and Lieberman (1959)].

TABLE P15.11

Specimen Number	x	y
1	104.2	38.9
2	106.1	40.4
3	105.6	39.9
4	106.3	40.8
5	101.7	33.7
6	104.4	39.5
7	102.0	33.0
8	103.8	37.0
9	104.0	37.6
10	101.5	33.2
11	101.9	33.9
12	100.6	29.9
13	104.9	39.5
14	106.2	40.6
15	103.1	35.1

15.12. Repeat the instructions of Problem 15.10 for the data of Table P15.12. This set of data (due to Bullis and Alderton) was obtained by examining the alpha-resin content of six different specimens of hops by taking colorimeter readings (x) and by direct determination of the concentration (y, in mg per 100 ml).

TABLE P15.12

Specimen Number	x	y
1	8	.12
2	50	.71
3	81	1.09
4	102	1.38
5	140	1.95
6	181	2.05

15.13. Test pieces of boiler plate undergo tests at various times during a production process. The measurements made are force applied in tons per square inch (x) at time of removal from the process and the elongation (y) of the test piece. For the results given in Table P15.13 on 10 test pieces, repeat the instructions of Problem 15.10.

TABLE P15.13

Test Piece Number	x	y
1	1.33	27
2	2.68	50
3	3.57	67
4	4.46	83
5	5.35	101
6	6.24	117
7	7.14	134
8	8.93	154
9	9.82	188
10	10.70	206

15.14. Runs were made at various conditions of saturation (x_1) and transi-somers (x_2). The response SCI (to be denoted by y) is given in Table P15.14 for the corresponding conditions of x_1 and x_2. Test the hypothesis that $\beta_1 = \beta_2 = 0$. If this hypothesis is rejected, then fit the model $E(y \mid x_1, x_2) = \alpha + \beta_1 x_1 + \beta_2 x_2$ to this data. Find a 95% confidence interval for $E(y \mid .25, 105)$, assuming normality of the y's. Find a 95% confidence interval for β_1. Also construct a joint 95% confidence region for β_1 and β_2.

TABLE P15.14

x_1	.04	.04	.04	.04	.20	.20	.20	.20	.38	.38	.38	.38
x_2	100	110	120	130	100	110	120	130	100	110	120	130
y	67.1	64.0	44.3	45.1	69.8	58.5	46.3	44.1	74.5	60.7	49.1	47.6

15.15. A study was made of the relationship between skein strength (y, in pounds) of #225 cotton yarn, and mean fiber length (x_1, in .01 inch) and fiber tensile strength (x_2, in 1000 psi). Twenty combinations of x_1 and x_2 values were used, and y observed at each of these combinations. The observations yielded the following results [data from Duncan (1958)]:

$$\bar{y} = 95.4, \qquad \bar{x}_1 = 77.0, \qquad \bar{x}_2 = 75.1,$$
$$S(Y^2) = 2742.8, \qquad S(X_1^2) = 1217.0, \qquad S(X_1 X_2) = 24.9,$$
$$S(X_1 Y) = 1332.6, \qquad S(X_2^2) = 303.8,$$
$$S(X_2 Y) = 335.2.$$

(a) Estimate the values of α, β_1, β_2, for the regression plane $E(y \mid x_1, x_2) = \alpha + \beta_1 x_1 + \beta_2 x_2$ by least squares.

(b) What estimate would you make for y if a piece of yarn is such that $x_1 = 75$ and $x_2 = 70$?

(c) Estimate the variance σ^2 of y.

(d) Find a 95% confidence interval for $E(y \mid x_1 = 75, x_2 = 70)$, assuming normality of the y's.

15.16. Thirteen combinations of values of x_1, the amount of tricalcium aluminate (in percent) used in a mix of cement, and x_2, the amount of tricalcium (in percent) used in the mix, are used to study the effect on the heat involved, y (in calories), during the hardening of the mix. The observations yield the following results [data from Hald (1952)]:

$$\sum x_{1i} = 97, \qquad \sum x_{2i} = 626, \qquad \sum y_i = 1240.5,$$
$$S(X_1^2) = 415.23, \qquad S(X_2^2) = 2905.69, \qquad S(Y^2) = 2715.76,$$
$$S(X_1 X_2) = 251.08, \qquad S(X_1 Y) = 775.96, \qquad S(X_2 Y) = 2292.95.$$

Assuming the usual linear relationship, find the least squares estimators of the coefficients and of the regression line. Determine 95% confidence intervals for $E(y \mid x_1 = 7, x_2 = 50)$; α; β_1; β_2; and determine a 95% joint confidence region for (β_1, β_2).

15.17. A variable y was observed at twelve different combinations of values of controlled variables x_1 and x_2, with the results as shown in Table P15.17. If the relation between y and (x_1, x_2) may be assumed to be linear in the region covered by the choice of the twelve (x_1, x_2) values, find the least squares estimator of $E(y \mid x_1 = 0, x_2 = 0)$, and give a 95% confidence interval for $E(y \mid x_1 = 0, x_2 = 0)$. Also give a 99% confidence region for (β_1, β_2). (Assume that y is normal.)

TABLE P15.17

x_1	-2	-3	1	4	3	-2	-3	-1	-4	5	0	2
x_2	.5	$-.5$	$-.5$	1.5	-2.5	-4.5	2.5	-3.5	.5	1.5	1.5	3.5
y	15	11	17	18	23	11	17	13	14	32	21	24

15.18. A study of a response η to changes in "factors" x_1 and x_2 yielded the results shown in Table P15.18, obtained by observing y at each of the 12 combinations of (x_1, x_2) shown. (x_1, x_2, and y are quoted in coded units)

TABLE P15.18

x_1	0	0	0	1	1	1	1	2	2	5	5	5
x_2	4	0	2	0	2	4	0	2	2	0	2	4
y	197	222	207	209	227	217	222	237	232	255	270	245

Assuming a linear relationship, determine the least squares estimators of the coefficients of the regression plane. Show the analysis of variance table and, assuming y's to be normal, test the joint hypothesis that $\beta_1 = 0$ and $\beta_2 = 0$. Find a joint 95% confidence region for β_1 and β_2. Develop the 95% confidence "band" for $\eta = E(y \mid x_1 = x_{10}, x_2 = x_{20})$. Evaluate it when $x_{10} = 2$, $x_{20} = 2$.

15.19. Experiments have shown that when a clean tungsten surface is heated by laser irradiation using a focused ruby laser, the rate of evaporation of tungsten from the surface is similar to that obtained by more conventional surface heating methods. The following experimental data relates the observed temperature change as a function of laser amplitude. Fit a first-order polynomial to these data.

Laser pulse Amplitude	0.7	0.9	1.1	1.2	1.5	1.7	1.9	2.0	2.1
Surface Temperature, °C	630	740	700	745	1120	1205	1510	1530	1520

15.20. In an investigation of pure copper bars of small diameter, the following shear stress and shear strain data were collected.

Shear strain, %	8.8	9.3	10.4	11.2	12.3	13.0	13.8	14.4	15.8	16.7	17.3
Shear stress, 1000 psi	11.8	11.9	11.4	11.6	12.3	12.7	13.3	13.7	13.8	14.4	14.5

(i) Fit a straight line viewing the shear strain data as the independent variable, and the shear stress data as the response variate.

(ii) Refit the line with the roles of shear strain and shear stress reversed.

(iii) Why do the two fitted models disagree?

15.21. In a study of internal combustion engines, the following data were observed, associating the net work provided by a cylinder as a function of the fuel fraction:

Fuel fraction	0.15	0.20	0.25	0.30	0.40	0.50	0.60	0.70	0.80	0.90	1.00
Net work, BTU/lbs of air	120	165	204	238	296	373	410, 403, 420, 408	462, 455, 464	520, 518, 525	580	600

Fit a straight line to the data, and test the hypothesis that the true slope is zero.

15.22. Sulfur dioxide can be removed from flue gases at low temperatures (approximately 600°F) through the use of a dry absorbent alkalized alumina. The absorbent, when spent, is later regenerated in a separate process, with elemental sulfur produced as a by-product. In one series of experiments, the following data were obtained, relating the removal of sulfur dioxide as a function of the heights of the absorber tower, under fixed operating conditions:

Height of absorber (ft.)	4.5	8.0	12.0	16.0	20.0	24.0	26.0
Removal of SO_2 (%)	7.2	21.1	28.3	33.0	42.0	54.0	63.8

(a) What is the best estimate of the amount of sulfur dioxide to be removed in a tower 30 feet high?

(b) What height tower is required to remove 95% of the SO_2?

15.23. Estimate the acceleration α of a body disturbed from rest by a constant applied force. The approximate model is $V = \alpha t$ where V is the velocity at time t.

t, seconds	1	2	3	4	5	6	7	8
Velocity V, ft/sec	34.2	57.6	94.3	121.0	146.4	175.2	212.8	247

15.24. Show that SSD is as given in (15.7.10), that is, show that SSD $= \Sigma(y_i - \hat{y}_i)^2$ can be written as

$$\sum y_i^2 - a \sum y_i - b_1 \sum x_{1i} y_i - b_2 \sum x_{2i} y_2$$

or as

$$\sum (y_i - \bar{y})^2 - b_1 \sum (x_{1i} - \bar{x})(y_i - \bar{y}) - b_2 \sum (x_{2i} - \bar{x}_2)(y_i - \bar{y}).$$

15.25. Show that $a = \bar{y} - b_1 \bar{x}_1 - b_2 \bar{x}_2$ has variance $\sigma^2 A'$, where A' is given by (15.7.12).

15.26. Show that $\hat{y}_0 = a + b_1 x_{10} + b_2 x_{20}$ can be written as $\hat{y}_0 = \bar{y} + b_1(x_{10} - \bar{x}_1) + b_2(x_{20} - \bar{x}_2)$. Hence, show that the variance of $\hat{y}_0$ is $\sigma^2 A$, where A is as given in (15.7.15).

CHAPTER 16

Analysis of Variance

16.1 INTRODUCTION

In many experiments the main objective is to determine the effect of various factors (sometimes called "treatments") on some response variable y of basic or primary interest. For instance, in the study of abrasion resistance of a certain type of rubber, it may be important to determine the effect of chlorinating agents on such resistance. Or in the study of strength of synthetic yarn it may be important to determine the effects of viscosity of the material in molten form, rate of extrusion, and other factors on the strength y of the yarn. In making experiments such as these, it is important to design them carefully in terms of numbers of trials as well as choices of levels of the various factors involved.

Experiments designed and analyzed in accordance with certain principles to be discussed in this chapter often make it possible to arrive at clearer and more trustworthy inferences about effects of factors, with a minimum of computational complexity. The principles of experimental design and methods of statistical analysis of experimental results which will be considered in this chapter are commonly referred to as *analysis of variance* methods. This is a vast subject and we can only scratch the surface by considering several of the simpler experimental designs and their statistical analyses. The reader interested in pursuing a study of this subject beyond what is given in this chapter should consult Bennett and Franklin (1954), Kempthorne (1952), Cochran and Cox (1957), and other books devoted to the subject.

16.2 ONE-WAY EXPERIMENTAL LAYOUTS

In Sections 10.4 and 10.5 we discussed the problem of testing hypotheses about the difference of means of two normal populations; on the basis of samples from these populations, we saw that if the two population variances are unknown but could be assumed to be equal, the statistic used for making the test is a Student t statistic involving the two sample means and the two sample variances.

Now suppose we have an experiment involving samples from three or more populations. How do we test a hypothesis concerning the means of the populations from which the samples are assumed to have been drawn? We shall consider this question in this section.

Suppose $A_1, \ldots, A_s$ are levels of some factor A and that we wish to study the effects of $A_1, \ldots, A_s$ on some response variable y of primary or basic interest. We set up an experiment in which n_1 observations are made on y when level A_1 is present, n_2 observations are made on y when level A_2 is present, $\ldots$, and finally n_s observations when level A_s is present. Commonly, the levels of A are called "treatments," there being s treatments in this *experimental design*.

More precisely, we shall consider the following simple model for such an experiment: Suppose $(y_{11}, y_{21}, \ldots, y_{n_1 1})$ is a sample of size n_1 from a population having the normal distribution $N(\mu + \delta_1, \sigma^2)$, $(y_{12}, y_{22}, \ldots, y_{n_2 2})$ is a sample of size n_2 from $N(\mu + \delta_2, \sigma^2)$, $\ldots$, and $(y_{1s}, y_{2s}, \ldots, y_{n_s s})$ is a sample of size n_s from $N(\mu + \delta_s, \sigma^2)$, the samples all being independent and where $(\mu, \delta_1, \ldots, \delta_s, \sigma^2)$ are unknown parameters, with $\delta_1, \ldots, \delta_s$ satisfying the condition

$$(16.2.1) \qquad n_1 \delta_1 + \cdots + n_s \delta_s = 0.$$

The parameters $\delta_1, \ldots, \delta_s$ are referred to as differential effects or simply effects due to $A_1, \ldots, A_s$, respectively. The parameter μ is sometimes called the overall (population) mean and σ^2 is the common variance of the y_{ij}, $i = 1, \ldots, n_j, j = 1, \ldots, s$. The samples may be arranged as shown in Table 16.2.1.

Let $\bar{y}_{.j}$ be the mean of the jth sample, and $\bar{y}$ the mean of all samples pooled together, that is,

$$(16.2.2) \qquad \bar{y}_{.j} = \frac{1}{n_j} \sum_{i=1}^{n_j} y_{ij}, \qquad \bar{y} = \frac{1}{n} \sum \sum y_{ij} = \frac{1}{n} \sum_{j=1}^{s} n_j \bar{y}_{.j},$$

where

$$(16.2.3) \qquad n = n_1 + \cdots + n_s$$

and where $\Sigma \Sigma$ stands for $\sum_{j=1}^{s} \sum_{i=1}^{n_j}$. It can be seen that the total variation $S(Y^2)$ of all observations in the pooled sample can be written in two ways. First,

$$(16.2.4) \qquad S(Y^2) = \sum \sum (y_{ij} - \bar{y})^2 = \sum \sum y_{ij}^2 - n\bar{y}^2.$$

TABLE 16.2.1

ONE-WAY EXPERIMENTAL LAYOUT

Factor-A Levels

A_1	A_2	$\cdots$	A_j	$\cdots$	A_s
y_{11}	y_{12}		y_{1j}		y_{1s}
y_{21}	y_{22}		y_{2j}		y_{2s}
$\cdot$	$\cdot$		$\cdot$		$\cdot$
$\cdot$	$\cdot$		$\cdot$		$\cdot$
$\cdot$	$\cdot$		$\cdot$		$\cdot$
$y_{n_1 1}$			y_{ij}		$y_{n_s s}$
	$y_{n_2 2}$		$\cdot$		
			$\cdot$		
			$y_{n_j j}$		
			$\cdot$		
			$\cdot$		
$\displaystyle\sum_{i=1}^{n_1} y_{i1} = T_1$	$\displaystyle\sum_{i=1}^{n_2} y_{i2} = T_2$	$\cdots$	$\displaystyle\sum_{i=1}^{n_j} y_{ij} = T_j$	$\cdots$	$\displaystyle\sum_{i=1}^{n_s} y_{is} = T_s$

Indeed $S(Y^2)$ is often called the corrected sum of squares with the implication that we have subtracted the *correction term* or *correction factor* $n\bar{y}^2$ from the sum of squares of all the observations.

We may also write $S(Y^2)$ as follows:

$$(16.2.4a) \quad S(Y^2) = \sum\sum (y_{ij} - \bar{y})^2 = \sum\sum [(y_{ij} - \bar{y}_{.j}) + (\bar{y}_{.j} - \bar{y})]^2$$
$$= \sum\sum (y_{ij} - \bar{y}_{.j})^2 + \sum\sum (\bar{y}_{.j} - \bar{y})^2,$$

since the sum of the cross products vanishes, that is,

$$(16.2.5) \qquad \sum\sum (y_{ij} - \bar{y}_{.j})(\bar{y}_{.j} - \bar{y}) = 0.$$

In the second line of (16.2.4a), let us define

$$(16.2.6) \qquad \begin{aligned} S_E &= \sum\sum (y_{ij} - \bar{y}_{.j})^2, \\ S_A &= \sum\sum (\bar{y}_{.j} - \bar{y})^2 = \sum_{j=1}^{s} n_j(\bar{y}_{.j} - \bar{y})^2. \end{aligned}$$

Thus we have

$$S(Y^2) = S_E + S_A,$$

which means that we have broken down the total variation $S(Y^2)$ into two parts; (1) S_A, which reflects the variation between samples, that is, variation due to the various levels of A, and is usually referred to as the between samples sum of squares (of deviations); and (2) S_E, which reflects variation within samples and is usually called the within samples sum of squares.

To look at this situation differently, suppose we denote $y_{ij} - \mu - \delta_j$ by ϵ_{ij}. Then we may write y_{ij} in the following way

$$(16.2.7) \qquad\qquad y_{ij} = \mu + \delta_j + \epsilon_{ij},$$

where, as before, $i = 1, \ldots, n_j, j = 1, \ldots, s$, and the δ_j satisfy (16.2.1), and where the ϵ_{ij}, of course, are n independent $N(0, \sigma^2)$ variables. Hence, y_{ij} is normal and has expectation $\mu + \delta_j$ as before. Replacing the y_{ij} in Table 16.2.1 by the right-hand side of (16.2.7), we see that

$$(16.2.8) \qquad\qquad \begin{aligned} \bar{y}_{.j} &= \mu + \delta_j + \bar{\epsilon}_{.j}, \\ \bar{y} &= \mu + \bar{\epsilon}. \end{aligned}$$

We may now write S_A and S_E as

$$(16.2.9) \qquad\qquad \begin{aligned} S_E &= \sum \sum (\epsilon_{ij} - \bar{\epsilon}_{.j})^2, \\ S_A &= \sum_{j=1}^{s} n_j (\delta_j + \bar{\epsilon}_{.j} - \bar{\epsilon})^2. \end{aligned}$$

Note that S_E depends only on the random variables ϵ_{ij} which are $N(0, \sigma^2)$ variables, and S_A depends on the parameters $\delta_1, \ldots, \delta_s$, as well as on the ϵ_{ij}. In fact, the reader may verify

$$(16.2.10) \qquad\qquad \begin{aligned} E(S_E) &= \sigma^2 \sum_{j=1}^{s} (n_j - 1) = \sigma^2(n - s), \\ E(S_A) &= \sigma^2(s - 1) + \sum_{j=1}^{s} n_j \delta_j^2. \end{aligned}$$

Note that the expectation of S_E does not depend on $\delta_1, \ldots, \delta_s$.

Now it can be shown that, if $\delta_j = 0, j = 1, \ldots, s$, then, since the ϵ_{ij} are normally distributed, the two quantities

$$(16.2.11) \qquad\qquad \frac{S_E}{\sigma^2} \quad \text{and} \quad \frac{S_A}{\sigma^2}$$

are independent random variables, having chi-square distributions with $n - s$ and $s - 1$ degrees of freedom, respectively. Hence it follows from Theorem AVI.10 that the ratio

$$(16.2.12) \qquad\qquad \frac{S_A/(s - 1)\sigma^2}{S_E/(n - s)\sigma^2} = \frac{S_A/(s - 1)}{S_E/(n - s)}$$

has the F distribution with $(s - 1, n - s)$ degrees of freedom.

If we wish to test the null hypothesis of zero effects due to treatments $A_1, \ldots, A_s$ of the A factor, that is,

$$(16.2.13) \qquad\qquad \begin{aligned} H_0&: \quad \delta_1 = \cdots = \delta_s = 0, \\ &\text{against alternatives} \\ H_1&: \quad \delta_1, \ldots, \delta_s \text{ are not all } 0, \end{aligned}$$

such a test may be performed by using the ratio given in (16.2.12). More precisely, we reject H_0 at the $100\alpha\%$ level of significance if

$$(16.2.14) \qquad \frac{S_A/(s-1)}{S_E/(n-s)} > F_{s-1, n-s; \alpha}$$

or accept H_0 if the inequality runs the other way.

If we reject H_0, which means that we reject the hypothesis that the effects $\delta_1, \ldots, \delta_s$ due to $A_1, \ldots, A_s$ are not all zero, we may then proceed to estimate the effects $\delta_1, \ldots, \delta_s$. From (16.2.8) the reader will note that

$$(16.2.15) \qquad \begin{aligned} E(\bar{y}) &= \mu, \\ E(\bar{y}_{.j}) &= \mu + \delta_j, \end{aligned}$$

that is, $\hat{\mu} = \bar{y}$ and $\hat{\delta}_j = \bar{y}_{.j} - \bar{y}$ are unbiased estimators for μ and δ_j, respectively. Furthermore, note that

$$(16.2.16) \qquad n_1 \hat{\delta}_1 + \cdots + n_s \hat{\delta}_s = 0.$$

Of course, from (16.2.10) we see that $\widehat{\sigma^2} = S_E/(n-s)$ is an unbiased estimator for σ^2.

We may collect all of the constituents of our analysis into what is called an *analysis of variance table*, as shown in Table 16.2.2.

TABLE 16.2.2

ANALYSIS OF VARIANCE TABLE FOR A ONE-WAY EXPERIMENTAL LAYOUT

Source of Variation	Sum of Squares	Degrees of Freedom	Mean Square	Expected Mean Square	F-Ratio Test
Between samples	S_A	$s-1$	$\dfrac{S_A}{s-1}$	$\sigma^2 + \dfrac{1}{s-1}\sum_{j=1}^{s} n_j \delta_j^2$	For (16.2.13), use
Within samples	S_E	$n-s$	$\dfrac{S_E}{n-s}$	σ^2	$\dfrac{(n-s)S_A}{(s-1)S_E}$
Total	$S(Y^2)$	$n-1$			

Estimators for $\delta_1, \ldots, \delta_s$ if H_0 is rejected:

$$\hat{\delta}_1 = \bar{y}_{.1} - \bar{y}, \ldots, \hat{\delta}_s = \bar{y}_{.s} - \bar{y}.$$

Estimator for σ^2:

$$\widehat{\sigma^2} = \frac{S_E}{n-s}.$$

Estimator for μ:

$$\hat{\mu} = \bar{y}.$$

A convenient computational scheme for finding the relevant sums of squares in Table 16.2.2 can be given as follows. Denote by T_j and SS_j, the total, and sum of squares respectively of the observations in the jth class, that is,

$$(16.2.17) \qquad T_j = \sum_{i=1}^{n_j} y_{ij} \quad \text{and} \quad SS_j = \sum_{i=1}^{n_j} y_{ij}^2.$$

We denote the grand total by

$$(16.2.17a) \qquad\qquad G = \sum_j T_j.$$

Then we may write the sum of squares entries of the analysis of variance Table 16.2.2 as

$$S_A = \sum_{j=1}^{s} \frac{T_j^2}{n_j} - \frac{G^2}{n}$$

$$(16.2.18) \qquad S_E = S(Y^2) - S_A$$

$$S(Y^2) = \sum_{j=1}^{s} SS_j - \frac{G^2}{n}.$$

Example 16.2(a). Four thermometers labeled 1, 2, 3, 4 were used to make determinations y of the melting point of hydroquinine in degrees centigrade, with the following results.

1	2	3	4
174.0	173.0	171.5	173.5
173.0	172.0	171.0	171.0
173.5		173.0	
173.0			

Suppose we wish to test at the 5% level of significance the hypothesis that there is no significant variation in the means of the melting points as determined by the four thermometers. (The factor in this experiment is thermometer, the levels are the four different thermometers, and their differential effects are δ_1, δ_2, δ_3, and δ_4 respectively.)

We recall from Chapter 5 that, if we code the observations by merely subtracting a constant from each observation, the sums of squares of deviations are not affected. Hence we will analyze values of the variable $y - 170$. This gives the values of $y - 170$ and $(y - 170)^2$ respectively as shown in Table 16.2.3 and 16.2.4.

TABLE 16.2.3

1	2	3	4	Total
4.0	3.0	1.5	3.5	
3.0	2.0	1.0	1.0	
3.5		3.0		
3.0				
13.5	5.0	5.5	4.5	28.5

TABLE 16.2.4

1	2	3	4	Total
16.00	9.00	2.25	12.25	
9.00	4.00	1.00	1.00	
12.25		9.00		
9.00				
46.25	13.00	12.25	13.25	84.75

By using (16.2.18), we find

$$S(Y^2) = (4.0)^2 + (3.0)^2 + \cdots + (1.0)^2 - \frac{(28.5)^2}{11}$$

$$= 84.75 - 73.8409 = 10.9091.$$

We also have

$$S_A = \sum_{j=1}^{4} \frac{T_j^2}{n_j} - \frac{G^2}{n}$$

$$= \frac{(13.5)^2}{4} + \frac{(5)^2}{2} + \frac{(4.5)^2}{3} - \frac{(28.5)^2}{11}$$

$$= 78.2708 - 73.8409$$

$$= 4.4299.$$

Furthermore,

$$S_E = S(Y^2) - S_A = \sum_{j=1}^{4} SS_j - \sum_{j=1}^{4} \frac{T_j^2}{n_j}$$

$$= 46.25 + 13.00 + 12.25 + 13.25 - (78.2708)$$

$$= 6.4792.$$

The analysis of variance table for Example 16.2(a) is shown in Table 16.2.5.

TABLE 16.2.5

ANALYSIS OF VARIANCE TABLE FOR EXAMPLE 16.2(a)

Source of Variation	Sum of Squares	Degrees of Freedom	Mean Square	F-Test
Between thermometers	4.4299	$s - 1 = 3$	1.4766	$\dfrac{1.4766}{.9256} = 1.595$
Within samples	6.4792	$n - s = 7$	.9256	
Total	10.9091	$n - 1 = 10$		

The upper 5% point of $F_{3,7}$ is $F_{3,7;.05} = 4.347$. The observed value is 1.595 and so we accept the hypothesis and conclude that there are no significant differences (at the 5% level) between the thermometers.

16.3 TWO-WAY EXPERIMENTAL LAYOUTS

(a) One Observation Per Cell

Now let us consider the problem of designing an experiment to study the effects of two factors A and B on a response variable y, where factor A has r levels $A_1, \ldots, A_r$, and factor B has s levels $B_1, \ldots, B_s$.

For each possible value of $i(i = 1, \ldots, r)$ and $j(j = 1, \ldots, s)$, let y_{ij} be an observation on y when levels A_i and B_j are present. We will then have rs observations $y_{11}, \ldots, y_{rs}$. A fairly simple mathematical model for this type of experimental design is as follows:

We shall assume that these rs observations y_{ij} are independent random variables having normal distributions

$$(16.3.1) \qquad\qquad N(\mu + \delta_{i.} + \delta_{.j}, \sigma^2)$$

respectively, $i = 1, \ldots, r$, $j = 1, \ldots, s$, where $\delta_{1.}, \ldots, \delta_{r.}$ are *effects* due to levels $A_1, \ldots, A_r$ of factor A, and $\delta_{.1}, \ldots, \delta_{.s}$ are *effects* due to levels $B_1, \ldots, B_s$ of factor B which satisfy the conditions

$$\delta_{1.} + \cdots + \delta_{r.} = 0,$$
$$(16.3.2)$$
$$\delta_{.1} + \cdots + \delta_{.s} = 0.$$

The assumption that the mean of y_{ij} is $\mu + \delta_{i.} + \delta_{.j}$ states that we are considering the effects of A_i and B_j on the response variable y to be additive.

TABLE 16.3.1

THE OBSERVATIONS y_{ij} OF A TWO-WAY EXPERIMENTAL LAYOUT

Factor-B levels

		B_1	B_2	$\cdots$	B_j	$\cdots$	B_s	Row Means
	A_1	y_{11}	y_{12}	$\cdots$	y_{1j}	$\cdots$	y_{1s}	$\bar{y}_{1.}$
	A_2	y_{21}	y_{22}	$\cdots$	y_{2j}	$\cdots$	y_{2s}	$\bar{y}_{2.}$
	.	.	.		.		.	.
Factor-A	.	.	.		.		.	.
Levels	A_i	y_{i1}	y_{i2}	$\cdots$	y_{ij}	$\cdots$	y_{is}	$\bar{y}_{i.}$
	.	.	.		.		.	.
	.	.	.		.		.	.
	A_r	y_{r1}	y_{r2}	$\cdots$	y_{rj}	$\cdots$	y_{rs}	$\bar{y}_{r.}$
Column Means		$\bar{y}_{.1}$	$\bar{y}_{.2}$	$\cdots$	$\bar{y}_{.j}$	$\cdots$	$\bar{y}_{.s}$	$\bar{y}$

If we think of the y_{ij} as being written out in a rectangular (or two-way) array, with r rows and s columns as shown in Table 16.3.1, then we let $\bar{y}_{i.}$ be the mean of the y_{ij} in the ith row (that is, the mean of all observations made with treatment A_i of the A factor present), $\bar{y}_{.j}$ be the mean of the y_{ij} in the jth column (that is, the mean of all observations made with treatment B_j of the B factor present), and $\bar{y}$ be the mean of all y_{ij} (sometimes called the grand mean). The intersection of the ith row and the jth column in the array is called the ijth cell. In the present case we are considering only one observation per cell.

Paralleling the development in Section 16.2, let us examine the total variation of all the observations in this two-way array. The total variation is

(16.3.3)

$$S(Y^2) = \sum_{i=1}^{r} \sum_{j=1}^{s} (y_{ij} - \bar{y})^2$$
$$= \sum \sum [(y_{ij} - \bar{y}_{i.} - \bar{y}_{.j} + \bar{y}) + (\bar{y}_{i.} - \bar{y}) + (\bar{y}_{.j} - \bar{y})]^2$$
$$= \sum \sum (y_{ij} - \bar{y}_{i.} - \bar{y}_{.j} + \bar{y})^2 + \sum \sum (\bar{y}_{i.} - \bar{y})^2 + \sum \sum (\bar{y}_{.j} - \bar{y})^2$$
$$= S_E + S_A + S_B, \quad \text{say,}$$

since, as the reader may verify, the sum of the cross products vanish, that is,

$$\sum \sum (y_{ij} - \bar{y}_{i.} - \bar{y}_{.j} + \bar{y})(\bar{y}_{i.} - \bar{y}) = \sum \sum (y_{ij} - \bar{y}_{i.} - \bar{y}_{.j} + \bar{y})(\bar{y}_{.j} - \bar{y})$$
$$= \sum \sum (\bar{y}_{i.} - \bar{y})(\bar{y}_{.j} - \bar{y}) = 0.$$

Here, of course $\Sigma\Sigma$ stands for $\displaystyle\sum_{i=1}^{r} \sum_{j=1}^{s}$,

$$\bar{y}_{i.} = \frac{1}{s} \sum_{j=1}^{s} y_{ij}, \qquad \bar{y}_{.j} = \frac{1}{r} \sum_{i=1}^{r} y_{ij},$$

and

$$\bar{y} = \frac{1}{rs} \sum \sum y_{ij} = \frac{1}{r} \sum_{i=1}^{r} \bar{y}_{i.} = \frac{1}{s} \sum_{j=1}^{s} \bar{y}_{.j}.$$

Thus, we have partitioned the total variation $S(Y^2)$ into three components S_E, S_A, and S_B [see (16.3.3)]. S_E is called the error sum of squares and represents variation left in the y_{ij} after removing A-treatment effects and B-treatment effects. S_A is called the A-treatment sum of squares and represents variation in the y_{ij} due to A-treatment effects. Similarly, S_B is called the B-treatment sum of squares and represents variation in the y_{ij} due to B-treatment effects.

Now suppose we denote $y_{ij} - \mu - \delta_{i.} - \delta_{.j}$ by ϵ_{ij}. Then we can write

(16.3.4) $y_{ij} = \mu + \delta_{i.} + \delta_{.j} + \epsilon_{ij}, \qquad i = 1, \ldots, r, \quad j = 1, \ldots, s,$

and where $\delta_{i.}$ and $\delta_{.j}$ satisfy (16.3.2) and the ϵ_{ij}, of course, are independent

$N(0, \sigma^2)$ variables. It is to be noted that (16.3.4) implies

$$\bar{y}_{.j} = \mu + \delta_{.j} + \bar{\epsilon}_{.j},$$
(16.3.5)
$$\bar{y}_{i.} = \mu + \delta_{i.} + \bar{\epsilon}_{i.},$$
$$\bar{y} = \mu + \bar{\epsilon},$$

so that S_E, S_A, and S_B may be written as

$$S_E = \sum \sum (\epsilon_{ij} - \bar{\epsilon}_{i.} - \bar{\epsilon}_{.j} + \bar{\epsilon})^2,$$

(16.3.6) $\quad S_A = \sum \sum (\delta_{i.} + \bar{\epsilon}_{i.} - \bar{\epsilon})^2 = s \sum_{i=1}^{r} (\delta_{i.} + \bar{\epsilon}_{i.} - \bar{\epsilon})^2,$

$$S_B = \sum \sum (\delta_{.j} + \bar{\epsilon}_{.j} - \bar{\epsilon})^2 = r \sum_{j=1}^{s} (\delta_{.j} + \bar{\epsilon}_{.j} - \bar{\epsilon})^2.$$

The expectations of these quantities are given by

$$E(S_E) = \sigma^2 (r - 1)(s - 1),$$

(16.3.7) $\quad E(S_A) = \sigma^2 (r - 1) + s \sum_{i=1}^{r} \delta_{i.}^2,$

$$E(S_B) = \sigma^2 (s - 1) + r \sum_{j=1}^{s} \delta_{.j}^2.$$

Now if the $\delta_{i.} = 0$, for all i, it can be shown that as the ϵ_{ij} are normally distributed, the quantities

(16.3.8) $$\frac{S_E}{\sigma^2} \quad \text{and} \quad \frac{S_A}{\sigma^2},$$

are independent random variables having chi-square distributions with $(r - 1)(s - 1)$ and $r - 1$, degrees of freedom respectively. Hence, if the $\delta_{i.}$ are all zero, and no matter what values the $\delta_{.j}$'s have, the ratio (see Appendix VI)

(16.3.9) $$\frac{S_A/(r - 1)\sigma^2}{S_E/[(r - 1)(s - 1)\sigma^2]} = \frac{(s - 1)S_A}{S_E}$$

has the F distribution with $r - 1$, $(r - 1)(s - 1)$ degrees of freedom. Hence, to test the hypothesis of zero A-factor effects, that is,

$$H_0: \quad \delta_1. = \cdots = \delta_r. = 0$$

(16.3.10) $\qquad\qquad\qquad$ against the alternatives

$$H_1: \quad \delta_1., \ldots, \delta_r. \text{ not all zero,}$$

we may use the ratio (16.3.9) and reject H_0 at the $100\alpha\%$ level of significance if

(16.3.11) $$\frac{(s-1)S_A}{S_E} > F_{(r-1),(r-1)(s-1);\alpha}.$$

If we reject H_0, we may proceed to estimate the $\delta_{i.}$ as follows. From (16.3.5) we note that

(16.3.12) $$E(\bar{y}_{i.}) = \mu + \delta_{i.},$$
$$E(\bar{y}) = \mu.$$

Hence $\hat{\delta}_{i.} = \bar{y}_{i.} - \bar{y}$ is an unbiased estimator for $\delta_{i.}$.

TABLE 16.3.2

ANALYSIS OF VARIANCE TABLE FOR A TWO-WAY EXPERIMENTAL LAYOUT

Source of Variation	Sum of Squares	Degrees of Freedom	Mean Square	Expected Mean Square	F-Ratio Test
Due to factor A	S_A	$r-1$	$\dfrac{S_A}{r-1}$	$\sigma^2 + \dfrac{s}{r-1}\sum\limits_{i=1}^{r}\delta_{i.}^2$	For H_0, use (16.3.11)
Due to factor B	S_B	$s-1$	$\dfrac{S_B}{s-1}$	$\sigma^2 + \dfrac{r}{s-1}\sum\limits_{j=1}^{s}\delta_{.j}^2$	For H_0', use (16.3.14)
Error	S_E	$(r-1)(s-1)$	$\dfrac{S_E}{(r-1)(s-1)}$	σ^2	
Total	$S(Y^2)$	$rs-1$			

Estimators for $\delta_{1.}, \ldots, \delta_{r.}$ if H_0 is rejected:
$$\hat{\delta}_{1.} = \bar{y}_{1.} - \bar{y}, \ldots, \hat{\delta}_{r.} = \bar{y}_{r.} - \bar{y}.$$

Estimators for $\delta_{.1}, \ldots, \delta_{.s}$ if H_0' is rejected:
$$\hat{\delta}_{.1} = \bar{y}_{.1} - \bar{y}, \ldots, \hat{\delta}_{.s} = \bar{y}_{.s} - \bar{y}.$$

Estimator for σ^2
$$\widehat{\sigma^2} = \frac{S_E}{(r-1)(s-1)}.$$

Estimator of μ:
$$\hat{\mu} = \bar{y}$$

Similarly, if we wish to test

(16.3.13)

$H_0': \quad \delta_{.1} = \cdots = \delta_{.s} = 0,$ against $H_1': \quad \delta_{.1}, \ldots, \delta_{.s}$ not all zero,

the test procedure is to reject H_0' (at significance level α) if

(16.3.14) $$\frac{(r-1)S_B}{S_E} > F_{(s-1),(r-1)(s-1);\alpha}.$$

If H_0' is rejected, the quantities $\hat{\delta}_{.j} = \bar{y}_{.j} - \bar{y}, j = 1, \ldots, s$ are unbiased estimators for $\delta_{.j}$. Furthermore, no matter what the values of the $\delta_{i.}$ and $\delta_{.j}$ may be, $_{\hat{\delta}^2} = S_E / [(r - 1)(s - 1)]$ is an unbiased estimator for σ^2.

We can arrange the constituents involved in the analysis of results of a two-way experimental layout, in an analysis of variance table as shown in Table 16.3.2.

To compute the sum of squares entries of Table 16.3.2, the following computational scheme is used. We let

(16.3.15)
$$T_{i.} = \sum_{j=1}^{s} y_{ij}, \qquad T_{.j} = \sum_{i=1}^{r} y_{ij}, \qquad G = \sum_{i=1}^{r} \sum_{j=1}^{s} y_{ij}$$
$$SS_{i.} = \sum_{j=1}^{s} y_{ij}^2, \qquad SS_{.j} = \sum_{i=1}^{r} y_{ij}^2,$$

and

(16.3.16)
$$G = \sum_{i=1}^{r} T_{i.} = \sum_{j=1}^{s} T_{.j} = \sum_{i=1}^{r} \sum_{j=1}^{s} y_{ij},$$
$$SS = \sum_{i=1}^{r} SS_{i.} = \sum_{j=1}^{s} SS_{.j} = \sum_{i=1}^{r} \sum_{j=1}^{s} y_{ij}^2.$$

Then it can be verified that

(16.3.17)
$$S_A = \sum_{i=1}^{r} \frac{T_{i.}^2}{s} - \frac{G^2}{rs},$$
$$S_B = \sum_{j=1}^{s} \frac{T_{.j}^2}{r} - \frac{G^2}{rs},$$
$$S_E = SS - \sum_{i=1}^{r} \frac{T_{i.}^2}{s} - \sum_{j=1}^{s} \frac{T_{.j}^2}{r} + \frac{G^2}{rs}$$
$$= S(Y^2) - (S_A + S_B),$$

where

$$S(Y^2) = SS - \frac{G^2}{rs}.$$

Example 16.3(a). An experiment was performed for wear-testing four materials, B_1, B_2, B_3, and B_4. A test piece of each material was taken from each of three positions, A_1, A_2, and A_3, of a testing machine. The loss in weight (due to wear), say y, was measured on each piece of material in milligrams. The values of $y - 230$ [data from Davies (1954)] appear in Table 16.3.3, and values of $(y - 230)^2$ are given in Table 16.3.4. It is desired to test whether there are significant differences due to materials and machine positions.

TABLE 16.3.3

VALUES OF $y - 230$ (IN MG)

	B_1	B_2	B_3	B_4	$T_{i.}$
A_1	11	-35	5	4	-15
A_2	40	11	43	6	100
A_3	44	-12	0	-3	29
$T_{.j}$	95	-36	48	7	114

TABLE 16.3.4

VALUES OF $(y - 230)^2$

	B_1	B_2	B_3	B_4	$SS_{i.}$
A_1	121	1225	25	16	1387
A_2	1600	121	1849	36	3606
A_3	1936	144	0	9	2089
$SS_{.j}$	3657	1490	1874	61	7082

Consulting (16.3.17) and Tables 16.3.3 and 16.3.4, we have that

$$S(Y^2) = SS - \frac{G^2}{12} = 7082 - \frac{(114)^2}{12} = 5999,$$

$$S_A = \sum_{i=1}^{3} \frac{T_{i.}^2}{4} - \frac{(114)^2}{12} = \frac{(-15)^2 + (100)^2 + (29)^2}{4} - \frac{(114)^2}{12}$$

$$= 1683.50,$$

$$S_B = \sum_{j=1}^{4} \frac{T_{.j}^2}{3} - \frac{(114)^2}{12} = \frac{95^2 + (-36)^2 + 48^2 + 7^2}{3} - \frac{(114)^2}{12}$$

$$= 3141.67,$$

and

$$S_E = 5999 - (1683.50 + 3141.67)$$
$$= 1173.83.$$

The analysis of variance table for Example 16.3(a) is given in Table 16.3.5. The upper 5% points of the $F_{2,6}$ and $F_{3,6}$ variables are 5.143 and 4.757 respectively. Consulting the last column of Table 16.3.5, we note that we accept the hypothesis of zero "position" effects, but reject the hypothesis of zero materials effects. That is, we conclude that there is a significant difference due to the materials, but not due to positions. It is left to the reader to estimate the $\delta_{.j}$.

TABLE 16.3.5

ANALYSIS OF VARIANCE FOR WEAR-TESTING EXPERIMENT

Source of Variation	Sums of Squares	Degrees of Freedom	Mean Square	F Test
Between positions	1683.50	2	841.75	$\dfrac{841.75}{195.64} = 4.303$
Between materials	3141.67	3	1047.22	$\dfrac{1047.22}{195.64} = 5.353$
Error	1173.83	6	195.64	
Total	5999	11		

(b) Replicated Two-Way Classification

Referring to Example 16.3(a), it might have been envisaged that the wear of the materials is affected not only by type of material and position on machine, but also by the specific combination of material and position on machine. That is, in general, we might expect y_{ij} to be such that

$$(16.3.18) \qquad E(y_{ij}) = \mu + \delta_{i.} + \delta_{.j} + \lambda_{ij},$$

where λ_{ij} is a differential effect unique to the treatment combination (i, j), called the *ij interaction effect*. We assume $\sum_{i=1}^{r} \lambda_{ij} = 0$ for each j, $\sum_{j=1}^{s} \lambda_{ij} = 0$ for each i, and, as before, $\sum_{i} \delta_{i.} = \sum_{j} \delta_{.j} = 0$. Thus the y_{ij} depends not only on the (differential) effects $\delta_{i.}$ and $\delta_{.j}$ but also on another effect λ_{ij}, the (differential) interaction effect between position on machine A_i and material B_j. Note that in (16.3.18) additivity of the effects $\delta_{i.}$, $\delta_{.j}$, and λ_{ij} is assumed.

Interaction effects are often present and, in the absence of prior information on the way the two factors of the experiment interact, the experimenter might wish to use an experimental design which will give information on λ_{ij}.

A fairly simple way to accomplish this is to run a two-way experimental layout with more than one observation per cell. That is, for each combination of the ith level of factor A and the jth level of factor B, we will have t observations y_{ijk}, where $k = 1, \ldots, t$. Repeating an experiment for such a fixed combination of i and j is called *replication* of the experiment for the ijth cell. If we were to repeat the entire two-way experimental layout (having one observation in each cell) t times, we would have a t replicate of the layout.

If we denote $y_{ijk} - \mu - \delta_{i.} - \delta_{.j} - \lambda_{ij}$ by ϵ_{ijk}, then the model for our experiment can be stated as follows:

$$y_{ijk} = \mu + \delta_{i.} + \delta_{.j} + \lambda_{ij} + \epsilon_{ijk},$$

(16.3.19) $\qquad i = 1, \ldots, r, \quad j = 1, \ldots, s, \quad k = 1, \ldots, t,$

$$\sum_i \lambda_{ij} = \sum_j \lambda_{ij} = \sum_i \delta_{i.} = \sum_j \delta_{.j} = 0,$$

where the ϵ_{ijk} are rst independent $N(0, \sigma^2)$ variables. The observations y_{ijk} for this experimental situation can be exhibited as in Table 16.3.6.

TABLE 16.3.6

THE OBSERVATIONS y_{ijk} OF A REPLICATED TWO-WAY EXPERIMENTAL LAYOUT

		Factor-B Levels				
		B_1	$\cdots$	B_j	$\cdots$	B_s	Row Means
	A_1	y_{111} $\cdot$ $\cdot$ $\cdot$ y_{11t}	$\cdots$	y_{1j1} $\cdot$ $\cdot$ $\cdot$ y_{1jt}	$\cdots$	y_{1s1} $\cdot$ $\cdot$ $\cdot$ y_{1st}	$\bar{y}_{1..}$
		$\cdot$ $\cdot$ $\cdot$		$\cdot$ $\cdot$ $\cdot$		$\cdot$ $\cdot$ $\cdot$	$\cdot$ $\cdot$ $\cdot$
Factor-A Levels	A_i	y_{i11} $\cdot$ $\cdot$ $\cdot$ y_{i1t}	$\cdots$	y_{ij1} $\cdot$ $\cdot$ $\cdot$ y_{ijt}	$\cdots$	y_{is1} $\cdot$ $\cdot$ $\cdot$ y_{ist}	$\bar{y}_{i..}$
		$\cdot$ $\cdot$ $\cdot$		$\cdot$ $\cdot$ $\cdot$		$\cdot$ $\cdot$ $\cdot$	$\cdot$ $\cdot$ $\cdot$
	A_r	y_{r11} $\cdot$ y_{r1t}	$\cdots$	y_{rj1} $\cdot$ y_{rjt}	$\cdots$	y_{rs1} $\cdot$ y_{rst}	$\bar{y}_{r..}$
	Column Means	$\bar{y}_{.1.}$	$\cdots$	$\bar{y}_{.j.}$	$\cdots$	$\bar{y}_{.s.}$	$\bar{y}_{...} = \bar{y}$

(Note that each cell has t observations.)

Now let

$$\bar{y}_{ij.} = \frac{1}{t} \sum_{k=1}^{t} y_{ijk} \quad \text{(cell means)},$$

$$\bar{y}_{i..} = \frac{1}{st} \sum_{j=1}^{s} \sum_{k=1}^{t} y_{ijk} \quad \text{(row means)},$$

(16.3.20)

$$\bar{y}_{.j.} = \frac{1}{rt} \sum_{i=1}^{r} \sum_{k=1}^{t} y_{ijk} \quad \text{(column means)},$$

$$\bar{y}_{...} = \frac{1}{rst} \sum_{i=1}^{r} \sum_{j=1}^{s} \sum_{k=1}^{t} y_{ijk} \quad \text{(grand mean)}.$$

The row means, column means, and grand mean are exhibited in Table 16.3.6. We denote the grand mean $\bar{y}_{...}$ by $\bar{y}$. From (16.3.19), we note that the means defined above take the form

(16.3.21)
$$\begin{aligned}
\bar{y}_{ij.} &= \mu + \delta_{i.} + \delta_{.j} + \lambda_{ij} + \bar{\epsilon}_{ij.}, \\
\bar{y}_{i..} &= \mu + \delta_{i.} && + \bar{\epsilon}_{i..}, \\
\bar{y}_{.j.} &= \mu && + \delta_{.j} && + \bar{\epsilon}_{.j.}, \\
\bar{y} &= \mu && && + \bar{\epsilon}_{....}
\end{aligned}$$

The total variation of the combined sample of rst observations is

$$\begin{aligned}
(16.3.22) \quad S(Y^2) &= \sum_{i=1}^{r} \sum_{j=1}^{s} \sum_{k=1}^{t} (y_{ijk} - \bar{y})^2 \\
&= \sum\sum\sum [(y_{ijk} - \bar{y}_{ij.}) + (\bar{y}_{ij.} - \bar{y}_{i..} - \bar{y}_{.j.} + \bar{y}) \\
&\quad + (\bar{y}_{i..} - \bar{y}) + (\bar{y}_{.j.} - \bar{y})]^2 \\
&= \sum\sum\sum (y_{ijk} - \bar{y}_{ij.})^2 + \sum\sum\sum (\bar{y}_{ij.} - \bar{y}_{i..} - \bar{y}_{.j.} + \bar{y})^2 \\
&\quad + \sum\sum\sum (\bar{y}_{i..} - \bar{y})^2 + \sum\sum\sum (\bar{y}_{.j.} - \bar{y})^2 \\
&= S_E + S_I + S_A + S_B, \quad \text{say},
\end{aligned}$$

since all sums of cross products vanish. $\left(\sum_{i=1}^{r} \sum_{j=1}^{s} \sum_{k=1}^{t} \text{ is denoted by } \sum\sum\sum. \right)$

Referring to (16.3.19) and (16.3.20), note that we may write

(16.3.23)
$$\begin{aligned}
S_E &= \sum\sum\sum (\epsilon_{ijk} - \bar{\epsilon}_{ij.})^2, \\
S_I &= \sum\sum\sum (\lambda_{ij} + \bar{\epsilon}_{ij.} - \bar{\epsilon}_{i..} - \bar{\epsilon}_{.j.} + \bar{\epsilon}_{...})^2, \\
S_A &= \sum\sum\sum (\delta_{i.} + \bar{\epsilon}_{i..} - \bar{\epsilon}_{...})^2 = st \sum_{i} (\delta_{i.} + \bar{\epsilon}_{i..} - \bar{\epsilon}_{...})^2, \\
S_B &= \sum\sum\sum (\delta_{.j} + \bar{\epsilon}_{.j.} - \bar{\epsilon}_{...})^2 = rt \sum_{j} (\delta_{.j} + \bar{\epsilon}_{.j.} - \bar{\epsilon}_{...})^2.
\end{aligned}$$

That is to say, we have partitioned the total variation $S(Y^2)$ into four parts: (1) S_E, the *error sum of squares* representing variation in the y_{ijk} after removing A-factor effects, B-factor effects, and interaction effects; (2) S_I, the *interaction sum of squares* representing variation in the y_{ijk} due to the interaction effects; (3) S_A, the *A-factor sum of squares*, representing variation in the y_{ijk} due to the *A-factor effects;* (4) S_B the *B-factor sum of squares*, representing variation in the y_{ijk} due to the B-factor effects. (See the expected mean square column of Table 16.3.7.)

Now suppose we make a test of

(16.3.24)
$$H_0'': \quad \lambda_{11} = \cdots = \lambda_{rs} = 0, \qquad \text{against alternatives}$$
$$H_1'': \quad \lambda_{11}, \lambda_{12}, \ldots, \lambda_{rs} \quad \text{are not all zero.}$$

It can be shown that under H_0''—the ϵ_{ijk} are independent $N(0, \sigma^2)$ variables—the quantities

(16.3.25)
$$\frac{S_I}{\sigma^2} \quad \text{and} \quad \frac{S_E}{\sigma^2}$$

are independent and distributed as chi-square variables with $(r-1)(s-1)$ and $rs(t-1)$ degrees of freedom, respectively. Hence, the ratio

(16.3.26)
$$\frac{S_I/(r-1)(s-1)\sigma^2}{S_E/rs(t-1)\sigma^2} = \frac{rs(t-1)}{(r-1)(s-1)} \frac{S_I}{S_E}$$

under H_0'', is an F variable with $(r-1)(s-1)$, $rs(t-1)$ degrees of freedom. Hence, we reject H_0'' at the $100\alpha\%$ significance level if

(16.3.27)
$$\frac{rs(t-1)}{(r-1)(s-1)} \frac{S_I}{S_E} > F_{(r-1)(s-1),rs(t-1);\alpha},$$

and accept H_0'' if the inequality runs the other way. If we do reject H_0'', we would then estimate the λ_{ij}. Referring to (16.3.21), we see that

(16.3.28)
$$E(\bar{y}_{ij.} - \bar{y}_{i..} - \bar{y}_{.j.} + \bar{y}) = \lambda_{ij},$$

since $E(\bar{\epsilon}_{ij.}) = E(\bar{\epsilon}_{i..}) = E(\bar{\epsilon}_{.j.}) = E(\bar{\epsilon}_{...}) = 0$. Hence $\hat{\lambda}_{ij} = \bar{y}_{ij.} - \bar{y}_{i..} - \bar{y}_{.j.} + \bar{y}$ is an unbiased estimator of λ_{ij}. The reader should verify that $\sum_i \hat{\lambda}_{ij} = \sum_j \hat{\lambda}_{ij} = 0$.

If H_0 is the hypothesis that $\delta_{1.} = \cdots = \delta_{r.} = 0$, and if H_0' is the hypothesis that $\delta_{.1} = \cdots = \delta_{.s} = 0$, tests of H_0 and H_0' are based on the F ratios $[rs(t-1)S_A]/[(r-1)S_E]$ and $[rs(t-1)S_B]/[(s-1)S_E]$, which

TABLE 16.3.7

ANALYSIS OF VARIANCE TABLE FOR A REPLICATED TWO-WAY EXPERIMENTAL LAYOUT

Source of Variation	Degrees of Freedom	Sum of Squares	Mean Square	Expected Mean Square	F-Ratio Tests
Due to A factor	$r-1$	S_A	$\dfrac{S_A}{r-1}$	$\sigma^2 + \dfrac{st}{r-1}\displaystyle\sum_{i=1}^{r}\delta_{i.}^2$	$H_0: \delta_{1.} = \cdots \delta_{r.} = 0$, use $\dfrac{rs(t-1)S_A}{(r-1)S_E}$.
Due to B factor	$s-1$	S_B	$\dfrac{S_B}{s-1}$	$\sigma^2 + \dfrac{rt}{s-1}\displaystyle\sum_{j=1}^{s}\delta_{.j}^2$	$H_0': \delta_{.1} = \cdots \delta_{.s} = 0$, use $\dfrac{rs(t-1)S_B}{(s-1)S_E}$.
Interaction	$(r-1)(s-1)$	S_I	$\dfrac{S_I}{(r-1)(s-1)}$	$\sigma^2 + \dfrac{t}{(r-1)(s-1)}\displaystyle\sum_i\sum_j \lambda_{ij}^2$	$H_0'': \lambda_{11} = \cdots \lambda_{rs} = 0$, use $\dfrac{rs(t-1)S_I}{(r-1)(s-1)S_E}$.
Error	$rs(t-1)$	S_E	$\dfrac{S_E}{rs(t-1)}$	σ^2	
Total	$\overline{rst-1}$	$\overline{S(Y^2)}$			

Estimators for $\delta_{1.}, \ldots, \delta_{r.}$ if H_0 is rejected:

$$\delta_{1.} = \bar{y}_{1..} - \bar{y}, \ldots, \delta_{r.} = \bar{y}_{r..} - \bar{y}.$$

Estimators for $\delta_{.1}, \ldots, \delta_{.s}$ if H_0' is rejected:

$$\delta_{.1} = \bar{y}_{.1.} - \bar{y}, \ldots, \delta_{.s} = \bar{y}_{.s.} - \bar{y}.$$

Estimators for $\lambda_{11}, \ldots, \lambda_{rs}$ if H_0'' is rejected:

$$\hat{\lambda}_{ij} = \bar{y}_{ij.} - \bar{y}_{i..} - \bar{y}_{.j.} + \bar{y}, \quad i = 1, \ldots, r; \quad j = 1, \ldots, s.$$

Estimator for μ:

$$\hat{\mu} = \bar{y}.$$

Estimator for σ^2:

$$\widehat{\sigma^2} = \frac{S_E}{rs(t-1)}.$$

are $F_{(r-1),rs(t-1)}$ and $F_{(s-1),rs(t-1)}$ random variables respectively. The tests for H_0, H_0', and H_0'' are summarized in Table 16.3.7.

A convenient computational scheme can be set up as follows: Let

$$T_{ij.} = \sum_{k=1}^{t} y_{ijk} \qquad \text{(cell sums)},$$

$$T_{i..} = \sum_{j=1}^{s} \sum_{k=1}^{t} y_{ijk} \qquad \text{(row sums)},$$

$$T_{.j.} = \sum_{i=1}^{r} \sum_{k=1}^{t} y_{ijk} \qquad \text{(column sums)},$$

$$G = \sum_{i=1}^{r} \sum_{j=1}^{s} \sum_{k=1}^{t} y_{ijk} \qquad \text{(grand sum)},$$

$$SS = \sum_{i=1}^{r} \sum_{j=1}^{s} \sum_{k=1}^{t} y_{ijk}^2.$$

Using these quantities, we find that

$$S_A = \sum_{i=1}^{r} \frac{T_{i..}^2}{st} - \frac{G^2}{rst},$$

$$S_B = \sum_{j=1}^{s} \frac{T_{.j.}^2}{rt} - \frac{G^2}{rst},$$

(16.3.29)
$$S_I = \sum_{i=1}^{r} \sum_{j=1}^{s} \frac{T_{ij.}^2}{t} - \frac{G^2}{rst} - S_A - S_B,$$

$$S_E = SS - \sum_{i=1}^{r} \sum_{j=1}^{s} \frac{T_{ij.}^2}{t}$$

$$S(Y^2) = SS - \frac{G^2}{rst}.$$

Note that we can compute S_I by subtraction as follows:

$$S_I = S(Y^2) - (S_A + S_B + S_E).$$

Example 16.3(b). Duncan (1958) quotes the following example. Three analysts, A_1, A_2, A_3, each makes two determinations of the melting point of hydroquinine (in degrees centigrade) with each of four different thermometers B_1, B_2, B_3, B_4. Each reading minus 172°C is given in Table 16.3.8.

TABLE 16.3.8

Thermometer

Analyst	B_1	B_2	B_3	B_4	$T_{i..}$
A_1	2.0 (3.5) 1.5	1.0 (2.5) 1.5	$-.5$ (.0) .5	1.5 (3.0) 1.5	9.0
A_2	1.0 (2.0) 1.0	.0 (1.0) 1.0	-1.0 (-1.0) .0	-1.0 (-1.0) .0	1.0
A_3	1.5 (2.5) 1.0	1.0 (2.5) 1.5	1.0 (2.0) 1.0	.5 (1.5) 1.0	8.5
$T_{.j.}$	8.0	6.0	1.0	3.5	18.5

The entries in parentheses are the cell sums $T_{ij.}$. Here, $r = 3$, $s = 4$, $t = 2$. Using (16.3.29), we find that

$$S_A = \frac{9^2 + 1^2 + 8.5^2}{8} - \frac{(18.5)^2}{24} = 19.281 - 14.260 = 5.021,$$

$$S_B = \frac{8^2 + 6^2 + 1^2 + 3.5^2}{6} - \frac{(18.5)^2}{24} = 18.875 - 14.260 = 4.615,$$

$$S_E = 2^2 + 1.5^2 + \cdots + .5^2 + 1^2 - \frac{3.5^2 + 2.5^2 + \cdots + 1.5^2}{2}$$

$$= 29.250 - 26.625 = 2.625,$$

$$S(Y^2) = 2^2 + \cdots + .5^2 + 1^2 - \frac{18.5^2}{24} = 29.250 - 14.260$$

$$= 14.990,$$

and, hence,

$$S_I = 14.990 - (5.021 + 4.615 + 2.625)$$
$$= 14.990 - 12.261 = 2.729.$$

The numerical results for the analysis of variance table for this example are as follows:

Source of Variation	Degrees of Freedom	Sums of Squares	Mean Square
Between analysts	2	5.021	2.51
Between thermometers	3	4.615	1.54
Interaction	6	2.729	.45
Error	12	2.625	.22
Total	23	14.990	

To test the hypothesis that effects due to analysts are zero, we use

$$F_{2,12} = \frac{2.51}{.22} = 11.410,$$

and since $F_{2,12;.05} = 3.885$, we reject this hypothesis and conclude that there are significant effects due to analysts. Furthermore, to test the hypothesis of zero effects due to thermometers, we use

$$F_{3,12} = \frac{1.54}{.22} = 7.000,$$

and since $F_{3,12;.05} = 3.490$, we reject this hypothesis, thus concluding that there are significant effects due to thermometers. However, to test the hypothesis of zero interaction effects between analysts and thermometers, we use

$$F_{6,12} = \frac{.45}{.22} = 2.040;$$

but since

$$F_{6,12;.05} = 2.996,$$

we conclude that interaction effects are not significantly different from zero. It is left to the reader to compute the estimates $\delta_{i.}$ and $\delta_{.j}$ of the effects $\delta_{i.}$ and $\delta_{.j}$ respectively.

16.4 LATIN SQUARE LAYOUTS

In some experimental situations we have two factors, say A and B, which are known to have additive effects on a response variable y, and we wish to test whether an additional factor C has an effect on y; if it does have an effect, we assume it is also an additive effect. A fairly simple experimental layout for such a situation is provided by a Latin square design.

More precisely, suppose we have r levels $A_1, \ldots, A_r$ of factor A, r levels $B_1, \ldots, B_r$ of factor B, and r levels $C_1, \ldots, C_r$ of factor C. If we set up a square array with r rows and r columns, $A_1, \ldots, A_r$ denoting rows and $B_1, \ldots, B_r$ denoting columns, then we say that the cell formed by the ith row and jth column has *levels A_i and B_j present.* There are r^2 cells in the entire square array. Let us place C_1 in r of the cells, C_2 in r of the remaining cells, $\ldots$, and C_r in the last r cells in such a way that C_1 occurs once and only once in each row and each column, C_2 occurs once and only once in each row and in each column, and so on, for $C_3, \ldots, C_r$, and so that every cell in the array has exactly one of the letters $C_1, \ldots, C_r$. Such an arrangement of $A_1, \ldots, A_r, B_1, \ldots, B_r, C_1, \ldots, C_r$ is called an

$r \times r$ *Latin square design.* An example of a 5×5 Latin square is shown in Table 16.4.1.

TABLE 16.4.1

EXAMPLE OF A 5×5 LATIN SQUARE EXPERIMENTAL DESIGN

Factor B Levels

		B_1	B_2	B_3	B_4	B_5
	A_1	C_1	C_2	C_3	C_4	C_5
Factor A	A_2	C_4	C_5	C_1	C_2	C_3
Levels	A_3	C_5	C_1	C_2	C_3	C_4
	A_4	C_2	C_3	C_4	C_5	C_1
	A_5	C_3	C_4	C_5	C_1	C_2

Let us consider an arbitrary cell, where at the intersection of the ith row and jth column of the array which contains C_t the levels of the A, B, and C factors present are A_i, B_j, and C_t, say. Now suppose y_{ij} is the experimental value of our response variable y determined for the cell described above. We shall assume that y_{ij} is a random variable having the normal distribution

$$(16.4.1) \qquad N(\mu + \delta_{i.} + \delta_{.j} + \delta_t, \sigma^2),$$

where $\delta_{1.}, \ldots, \delta_{r.}$ are effects due to levels $A_1, \ldots, A_r$ of the A factor, $\delta_{.1}, \ldots, \delta_{.r}$ are effects due to levels $B_1, \ldots, B_r$ of the B factor, and $\delta_1, \ldots, \delta_r$ are the effects due to levels $C_1, \ldots, C_r$ of the C factor, and satisfying the conditions:

$$(16.4.2) \qquad \begin{aligned} \delta_{1.} + \cdots + \delta_{r.} &= 0, \\ \delta_{.1} + \cdots + \delta_{.r} &= 0, \\ \delta_1 + \cdots + \delta_r &= 0. \end{aligned}$$

Note particularly, that we are assuming all interaction effects to be zero.

We now define

$$(16.4.3) \qquad \begin{aligned} \bar{y}_{i.} &= \frac{1}{r} \sum_{j=1}^{r} y_{ij}, \qquad \bar{y}_t = \frac{1}{r} \sum_t y_{ij}, \\ \bar{y}_{.j} &= \frac{1}{r} \sum_{i=1}^{r} y_{ij}, \qquad \bar{y} = \frac{1}{r^2} \sum\sum y_{ij}, \end{aligned}$$

where $\sum_t$ denotes summation over all cells in the Latin square array in which C_t occurs, $t = 1, \ldots, r$, while $\Sigma\Sigma$ denotes $\sum_{i=1}^{r} \sum_{j=1}^{r}$.

For the Latin square, the total variation is

(16.4.4) $S(Y^2) = \sum\sum (y_{ij} - \bar{y})^2$

$= \sum\sum [(y_{ij} - \bar{y}_{i.} - \bar{y}_{.j} - \bar{y}_t + 2\bar{y}) + (\bar{y}_{i.} - \bar{y})$

$+ (\bar{y}_{.j} - \bar{y}) + (\bar{y}_t - \bar{y})]^2$

$= \sum\sum (y_{ij} - \bar{y}_{i.} - \bar{y}_{.j} - \bar{y}_t + 2\bar{y})^2$

$+ r \sum_{i=1}^{r} (\bar{y}_{i.} - \bar{y})^2 + r \sum_{j=1}^{r}(\bar{y}_{.j} - \bar{y})^2 + r \sum_{t=1}^{r}(\bar{y}_t - \bar{y})^2$

$= S_E + S_A + S_B + S_C,$

since the sums of the cross products vanish. Denoting $y_{ij} - \mu - \delta_{i.} - \delta_{.j} - \delta_t$ by ϵ_{ij}, we can write

(16.4.5) $y_{ij} = \mu + \delta_{i.} + \delta_{.j} + \delta_t + \epsilon_{ij},$

where the ϵ_{ij} are r^2 independent $N(0, \sigma^2)$ variables. Substituting y_{ij} as given in (16.4.5) into (16.4.4), we obtain

$$S_E = \sum_{i=1}^{r} \sum_{j=1}^{r}(\epsilon_{ij} - \bar{\epsilon}_{i.} - \bar{\epsilon}_{.j} - \bar{\epsilon}_t + 2\bar{\epsilon})^2$$

$$S_A = r \sum_{i=1}^{r} (\delta_{i.} + \bar{\epsilon}_{i.} - \bar{\epsilon})^2,$$

(16.4.6)

$$S_B = r \sum_{j=1}^{r} (\delta_{.j} + \bar{\epsilon}_{.j} - \bar{\epsilon})^2,$$

$$S_C = r \sum_{t=1}^{r} (\delta_t + \bar{\epsilon}_t - \bar{\epsilon})^2.$$

Proceeding as in the previous sections of this chapter, we summarize the analysis of variance in Table 16.4.2.

We should note that for a three-factor situation, we could use a full three-way experimental layout requiring r^3 observations y_{ijk}, where y_{ijk} is the observation yielded when the experiment is performed in the presence of A_i, B_j, and C_k and, of course, $i, j,$ and k can range over $1, \ldots, r$. Our Latin square experimental design is a particular selection of r^2 of the r^3 cells in such a full three-way layout. The use of the Latin square thus effects an economy by requiring r^2 observations instead of r^3.

TABLE 16.4.2

ANALYSIS OF VARIANCE TABLE FOR AN $r \times r$ LATIN SQUARE

Source of Variation	Sum of Squares	Degrees of Freedom	Mean Square	F-Ratio Tests
Due to A factor	S_A	$r - 1$	$\dfrac{S_A}{r - 1}$	$\dfrac{(r - 2)S_A}{S_E}$
Due to B factor	S_B	$r - 1$	$\dfrac{S_B}{r - 1}$	$\dfrac{(r - 2)S_B}{S_E}$
Due to C factor	S_C	$r - 1$	$\dfrac{S_C}{r - 1}$	$\dfrac{(r - 2)S_C}{S_E}$
Error	S_E	$(r - 1)(r - 2)$	$\dfrac{S_E}{(r - 1)(r - 2)}$	
Total	$S(Y^2)$	$r^2 - 1$		

Estimators for $\delta_{1.}, \ldots, \delta_{r.}$ if $H_0 : \delta_{1.} = \cdots = \delta_{r.} = 0$ is rejected:

$$\hat{\delta}_{1.} = \bar{y}_{1.} - \bar{y}, \ldots, \hat{\delta}_{r.} = \bar{y}_{r.} - \bar{y}.$$

Estimators for $\delta_{.1}, \ldots, \delta_{.r}$ if $H_0' : \delta_{.1} = \cdots = \delta_{.r} = 0$ is rejected:

$$\hat{\delta}_{.1} = \bar{y}_{.1} - \bar{y}, \ldots, \hat{\delta}_{.r} = \bar{y}_{.r} - \bar{y}.$$

Estimators for $\delta_1, \ldots, \delta_r$ if $H_0'' : \delta_1 = \cdots = \delta_r = 0$ is rejected:

$$\hat{\delta}_1 = \bar{y}_1 - \bar{y}, \ldots, \hat{\delta}_r = \bar{y}_r - \bar{y}.$$

Estimator for σ^2:

$$\widehat{\sigma^2} = \frac{S_E}{(r - 1)(r - 2)}.$$

Estimator for μ:

$$\hat{\mu} = \bar{y}.$$

A convenient computational scheme for the sums of squares in Table 16.4.2 can be set up as follows. Let

$$T_{i.} = \sum_{j=1}^{r} y_{ij},$$

$$T_{.j} = \sum_{i=1}^{r} y_{ij},$$

(16.4.7)
$$T_t = \sum_{t} y_{ij},$$

$$G = \sum_{i=1}^{r} \sum_{j=1}^{r} y_{ij},$$

$$SS = \sum_{i=1}^{r} \sum_{j=1}^{r} y_{ij}^2,$$

where as before $\sum_{t}$ denotes summation over all cells in the Latin square array in which C_t occurs.

The reader should then verify that [see (16.4.4)]

$$S_A = \sum_{i=1}^{r} \frac{T_{i.}^2}{r} - \frac{G^2}{r^2},$$

$$S_B = \sum_{j=1}^{r} \frac{T_{.j}^2}{r} - \frac{G^2}{r^2},$$

(16.4.8)

$$S_C = \sum_{t=1}^{r} \frac{T_t^2}{r} - \frac{G^2}{r^2},$$

$$S_E = SS - \sum_{i=1}^{r} \frac{T_{i.}^2}{r} - \sum_{j=1}^{r} \frac{T_{.j}^2}{r} - \sum_{t=1}^{r} \frac{T_t^2}{r} + 2\frac{G^2}{r^2}$$

$$= S(Y^2) - (S_A + S_B + S_C),$$

$$S(Y^2) = SS - \frac{G^2}{r^2}.$$

Example 16.4(a). A radioactive counting rate experiment was performed on four specimens of radium C_1, C_2, C_3, C_4. The four different specimens were subjected to a counter with four shielding methods B_1, B_2, B_3, and B_4 in various orders A_1, A_2, A_3, and A_4, so that the entire experiment was conducted according to a Latin square design. The Latin square design adopted and the observations obtained together with certain calculations, are shown in Table 16.4.3.

TABLE 16.4.3

LATIN SQUARE EXPERIMENTAL DESIGN AND OBSERVATIONS IN
A RADIO ACTIVE COUNTING-RATE EXPERIMENT

Shielding Method

		B_1	B_2	B_3	B_4	Row Sums
	A_1	(C_1) 26.46	(C_3) 29.61	(C_2) 27.82	(C_4) 29.15	$T_{1.} = 113.04$
	A_2	(C_2) 27.58	(C_4) 29.52	(C_1) 26.48	(C_3) 29.13	$T_{2.} = 112.71$
Order	A_3	(C_3) 29.54	(C_1) 27.00	(C_4) 29.31	(C_2) 27.90	$T_{3.} = 113.75$
	A_4	(C_4) 29.15	(C_2) 28.03	(C_3) 29.53	(C_1) 26.51	$T_{4.} = 113.22$
	Column sums	$T_{.1} = \overline{112.73}$	$T_{.2} = \overline{114.16}$	$T_{.3} = \overline{113.14}$	$T_{.4} = \overline{112.69}$	$G = \overline{452.72}$

Specimens

(C_1)	(C_2)	(C_3)	(C_4)	
26.46	27.58	29.54	29.15	
27.00	28.03	29.61	29.52	
26.48	27.82	29.53	29.31	
26.51	27.90	29.13	29.15	
$T_1 = \overline{106.45}$	$T_2 = \overline{111.33}$	$T_3 = \overline{117.81}$	$T_4 = \overline{117.13}$	$G = 452.72$

The sum of squares of the y_{ij}, that is, $SS = \Sigma\Sigma\, y_{ij}^2 = 12{,}831.6948$. Using (16.4.8), we find

$$S_A = \frac{(113.04)^2}{4} + \frac{(112.71)^2}{4} + \frac{(113.75)^2}{4} + \frac{(113.22)^2}{4} - \frac{(452.72)^2}{16}$$

$$= .14,$$

$$S_B = \frac{(112.73)^2}{4} + \frac{(114.16)^2}{4} + \frac{(113.14)^2}{4} + \frac{(112.69)^2}{4} - \frac{(452.72)^2}{16}$$

$$= .35,$$

$$S_C = \frac{(106.45)^2}{4} + \frac{(111.33)^2}{4} + \frac{(117.81)^2}{4} + \frac{(117.13)^2}{4} - \frac{(452.72)^2}{16}$$

$$= 21.44,$$

and

$$S(Y^2) = 12{,}831.6948 - \frac{(452.72)^2}{16} = 21.98.$$

Hence

$$S_E = 21.98 - (.14 + .35 + 21.44) = .05.$$

We summarize our numerical results in the following table:

Source of Variation	Degrees of Freedom	Sum of Squares	Mean Square	F-Ratio Test
Due to orders	3	.14	.047	5.6
Due to shielding methods	3	.35	.117	14.0
Due to specimens	3	21.44	7.147	857.6
Error	6	.05	.0083	
Total	15	21.98		

The 5% and 1% significance points of $F_{3,6}$ are 4.757 and 9.780 respectively. Hence order effects are barely significant at the 5% level but not at the 1% level. Shielding method effects and specimen effects are highly significant. It is left to the reader to compute the estimates of the $\delta_{\cdot j}$ and δ_t.

PROBLEMS

16.1. Nine determinations of the ratio of iodine to silver in four different silver preparations were made with the results shown in Table P16.1:

TABLE P16.1

Preparation A	Preparation B	Preparation C	Preparation D
1.176422	1.176441	1.176429	1.176449
1.176425	1.176441	1.176420	1.176450
		1.176437	

(a) Make an analysis of variance of these data.

(b) Test the null hypothesis H_0 that effects due to the different preparations are all zero. If H_0 is rejected, estimate the four preparation effects.

16.2. Determinations were made of the yield of a chemical using three catalytic methods I, II, and III with the results shown in Table P16.2 [from Fraser (1958)]:

TABLE P16.2

Method I	Method II	Method III
47.2	50.1	49.1
49.8	49.3	53.2
48.5	51.5	51.2
48.7	50.9	52.8
		52.3

Test the null hypothesis H_0 that effects due to the different catalytic methods are all zero. If H_0 is rejected, estimate the effects.

16.3. Table P16.3 shows measurements made by Heyl of the gravitational constant G for balls made from gold, platinum, and glass:

TABLE P16.3

Gold	Platinum	Glass
6.683	6.661	6.678
6.681	6.661	6.671
6.676	6.667	6.675
6.678	6.667	6.672
6.679	6.664	6.674
6.672		

Test the hypothesis that gold, platinum, and glass all have the same gravitational constant. If they do not, estimate the effects due the three materials.

16.4. Among the classrooms in the public schools of a given city there are 12 different lighting techniques. Each of these techniques is supposed to provide the same level of illumination. To determine whether or not the illumination they provide is uniform, the data shown in Table P16.4 were complied from a random sample of four lighting techniques. The classrooms are known to be homogeneous and hence can be discounted as a possible source of variability. Observations are in foot-candles on the desk surface.

TABLE P16.4

Lighting Techniques

1	2	3	4
31	31	34	37
38	34	35	34
38	27	39	27
33	27	35	32
31	29	30	26

Test the hypothesis of zero "lighting technique" effects. If there are significant effects, estimate them.

16.5. Total solids (in percent) were determined in each of six batches of wet brewer's yeast a, b, c, d, e, f by each of three analysts A, B, C, with the results shown in Table P16.5.

TABLE P16.5

Batch

Analyst	a	b	c	d	e	f
A	20.1	14.7	13.0	17.8	16.0	14.9
B	20.0	14.9	13.0	17.7	16.2	15.1
C	20.2	14.8	13.1	17.9	16.1	15.0

(a) Make an analysis of variance table of these data.

(b) Test the null hypothesis H_0 that effects due to analysts are all zero. If H_0 is rejected, estimate the effects due to the three analysts.

(c) Test the null hypothesis H_0' that effects due to batches are all zero. If H_0' is rejected estimate the effects.

(d) Comment on the outcome of this experiment.

16.6. An experiment was run to determine the effect of three types of oil, a, b, c, on the wear of four kinds of piston rings A, B, C, D. The measure of wear was taken as the logarithm of loss of piston-ring weight (in grams times 100) in a 12-hour test run. The results of the experiment are shown in Table P16.6.

TABLE P16.6

Piston Ring Type	Oil Type		
	a	b	c
A	1.782	1.568	1.570
B	1.306	1.223	1.240
C	1.149	1.029	1.068
D	1.025	1.919	1.982

Perform an anaylsis of these data similar to that requested in Problem 16.5(a), (b), (c), (d).

16.7. During the manufacture of sheets of building material the permeability (in units we need not describe here) was determined for a sheet from each of

TABLE P16.7

Day

Machine	1	2	3	4	5	6	7	8	9
A	1.404	1.447	1.914	1.887	1.772	1.665	1.918	1.845	1.540
B	1.306	1.241	1.506	1.673	1.227	1.404	1.229	1.583	1.636
C	1.932	1.426	1.382	1.721	1.320	1.633	1.328	1.689	1.703

machines A, B, and C on each of nine days, with the results shown in Table P16.7 [from Hald (1952)]. Perform an analysis of the data similar to that requested in Problem 16.5(a), (b), (c), and (d).

16.8. An experiment was performed to study the effect of plate temperature and filament lighting on transconductance of a certain type of tube. Two levels of plate temperature (550 °F and 600 °F) and four levels of filament lighting current L_1, L_2, L_3, and L_4 were used, and three replicates were made for each combination of plate temperature and filament current. The transconductance measurements are shown in Table P16.8. [from Bowker and Lieberman (1959)] Perform a complete anaylsis of variance of these data in accordance with the procedures used in Example 16.3(b).

TABLE P16.8

Plate Temperature	Filament Lighting Current			
	L_1	L_2	L_3	L_4
	3774	4710	4176	4540
T_1 (550 °F)	4364	4180	4140	4530
	4374	4514	4398	3964
	4216	3828	4122	4484
T_2 (600 °F)	4524	4170	4280	4332
	4136	4180	4226	4390

16.9. Measurement of "filling time" in minutes for specimens of cloths A, B and C taken from machines 1, 2, . . . , 9 in a certain plant gave the results shown in Table P16.9 (original measurement—78.00 minutes). Perform an analysis of variance of these data in accordance with the procedures used in Example 16.3(b).

TABLE P16.9

Machines

Cloth	1	2	3	4	5	6	7	8	9
A	18.76	20.69	19.77	19.85	22.28	20.39	24.31	22.90	19.28
	21.18	23.20	23.94	18.92	20.45	21.80	26.29	25.42	22.04
B	21.28	16.85	20.75	18.72	18.97	19.52	20.08	21.27	15.67
	19.10	20.16	21.49	16.14	20.31	21.27	19.36	17.82	18.84
C	21.74	22.68	21.90	20.28	19.89	21.12	23.02	27.48	18.70
	18.99	23.59	18.61	18.71	18.36	18.59	18.85	22.95	23.39

16.10. In a study of gasoline consumption by city buses, four vehicles, A, B, C, D were tested. In the first run of the day over a specified course, a particular assignment of drivers a, b, c, d was used. In the next run the drivers were re-assigned to the vehicles, and so on, for all four runs, as shown in the Latin square design of Table P16.10. (The variable measured was miles per gallon):

TABLE P16.10

Run No.	Vehicle			
	A	B	C	D
1	9.44(a)	9.83(b)	9.02(c)	9.68(d)
2	9.61(b)	9.22(d)	9.39(a)	8.76(c)
3	9.06(d)	9.02(c)	9.88(b)	8.88(a)
4	8.71(c)	9.02(a)	9.23(d)	9.73(b)

(a) Make an analysis of variance table from these data.

(b) Test (at the 5% significance level) the null hypothesis H_0'' that effects due to drivers are zero. If H_0'' is rejected, estimate the four driver effects.

(c) Perform an analysis on vehicle effects similar to that done in (b) for driver effects.

16.11. An experiment was conducted to study preconditioning of leather on the rate of abrasion of leather. A large square piece of leather was cut into 36 smaller squares which were subjected to a uniform abrasion test at 6 levels of humidity A, B, C, D, E, F in a Latin square arrangement. The rows and squares of the Latin square correspond to the two dimensions of the original large square of leather. The loss of leather due to abrasion was measured in grams. The results are shown in Table P16.11 [from Bennett and Franklin (1954)]:

TABLE P16.11

Rows	Columns					
	1	2	3	4	5	6
a	C 7.38	D 5.39	F 5.03	B 5.50	E 5.01	A 6.79
b	B 7.15	A 8.16	E 4.96	D 5.78	C 6.24	F 5.06
c	D 6.75	F 5.64	C 6.34	E 5.31	A 7.81	B 8.05
d	A 8.05	C 6.45	B 6.31	F 5.46	D 6.05	E 5.51
e	F 5.65	E 5.44	A 7.27	C 6.54	B 7.03	D 5.96
f	E 6.00	B 6.55	D 5.93	A 8.02	F 5.80	C 6.61

(a) Make an analysis of variance table from these data.

(b) Test (at the 5% significance level) the null hypothesis H_0'' that effects due to humidity preconditioning are zero. If H_0'' is rejected, estimate the effects due to the six humidity levels.

16.12. In a 6 × 6 Latin square experimental design, the sums of squares corresponding to the various sources of variation are shown in Table P16.12. Fill out the missing columns and test the null hypothesis that (1) row effects, (2) column effects, and (3) treatment effects are all zero.

TABLE P16.12

Source	Sums of Squares	Degrees of Freedom	Mean Square	Test
Between rows	58.1			
Between columns	78.6			
Between treatments	81.3			
Error	14.2			
Total	232.2			

16.13. Referring to the analysis-of-variance situation of Section 16.3a (see Table 16.3.2), give a procedure for testing the hypothesis

$$\delta_1. = \cdots \delta_r. = \delta_{.1} = \cdots = \delta_{.s} = 0$$

which uses only *one* test.

16.14. Referring to the analysis-of-variance situation of Section 16.3b (see Table 16.3.7), give a procedure for testing the hypothesis

$$\delta_1. = \cdots = \delta_r. = \delta_{.1} = \cdots = \delta_{.s} = \lambda_{11} = \cdots = \lambda_{rs} = 0$$

which uses only *one* test.

16.15. In a preliminary experiment on the percentage vitamin content at different age levels, the results shown in Table P16.15 were obtained:

TABLE P16.15

PERCENT VITAMIN CONTENT OF REPLICATE
SAMPLES AT DIFFERENT AGE LEVELS

Age, days	Vitamin Content, %
0	36, 29, 30, 32
3	47, 11, 50, 55
7	24, 57, 57, 59
10	21, 59, 55, 50
14	56, 54, 51, 33
17	53, 31, 59, 59
21	73, 72, 42, 64, 62, 64

Test the hypothesis that the effects due to age are zero. If the hypothesis is rejected, estimate the age effects.

16.16. An experiment on flies is laid out in 7 blocks of 3 plots each. The treatments were sprays containing 4, 8, and 16 units of the active ingredient designed to kill adult flies as they emerged from the breeding medium. The blocks comprised seven sources of the medium. Number of adult flies found in cases set over the plots are shown in Table P16.16

TABLE P16.16

Block	T_1	T_2	T_3
1	445	414	247
2	113	127	147
3	122	206	138
4	227	78	148
5	132	172	356
6	31	45	29
7	177	103	63

Test the hypothesis H_0 of no treatment effect and the hypothesis H_0' of no block effect.

CHAPTER 17

Some Experimental Designs and Their Analyses

This chapter briefly discusses the *factorial* and *response surface* designs. These designs find widespread applications in situations of the following type.

Suppose a *response*, $\eta = E(y)$, is an unknown function of variables $\xi_1, \xi_2, \ldots, \xi_k$, which are under the control of the experimenter, that is,

$$\eta = f(\xi_1, \xi_2, \ldots, \xi_k).$$

Now the objective of the experimenter may be to *screen* the k controlled variables in order to discover subsets whose effects upon the response are greatest, or the objective may be to provide a *graduating* function, or map, of the function over selected ranges of the controlled variables. To meet these objectives, the factorial and response surface designs are frequently employed. These designs are a natural outgrowth of the "one-way experimental layouts for s treatments each containing n observations" discussed in Chapter 16. However, for both the factorial and response surface designs, the treatment levels are now selected in unique ways in order to meet the objectives of interest to the experimenter.

Whenever the number of treatments and/or the number of required experiments becomes large, the usual consequence is that unwanted sources of variability are introduced into the experimental environment. For example, several different testing machines rather than one may have to be used in order to complete a program of experiments within a certain time period. Differences in performance of the machines then adds to the overall variability of the experiments. On other occasions it may not be possible to complete all the required experiments within a single day, variability associated with successive days having effects upon the mean performance of a series of experiments. The contributions of sources of variability, such as test machines and/or days, can often be kept from enlarging the experimental variance by *blocking* the experimental design. In brief, most experimental programs are conceived as containing two types of variables, both influencing the response: (i) the experimental or *treatment* variables, and (ii) the variability or *blocking* variables. Designs containing both treatments and blocks are analogous to the *two-way classification* layouts described in Chapter 16. The Latin square designs are employed when there are two separate sources of blocking variables (for example, test machines as well as days) in addition to the treatment variables.

The subject of experimental designs and their associated analyses is a very extensive one. In this chapter we show something of the vast variety of these designs and, for a few of these designs, some of the variety of data interpretations open to the statistical analyst.

17.1 THE FACTORIAL DESIGNS

By a *factorial design* we mean a design that is constructed by taking all combinations of l_1 levels (or versions) of factor A, with the l_2 levels of factor B, with the l_3 levels of factor $C, \ldots$, and, finally, the l_k levels of factor k. The complete factorial design then contains a total of $t = l_1 \times l_2 \times l_3 \times \cdots \times l_k$ "treatments" with r_j observations per treatment. (Usually, but not necessarily, a fixed number of *replicate* observations say r, are taken on each treatment, providing a total of $N = r \times l_1 \times l_2 \times l_3 \times \cdots \times l_k$ experiments.) A single observation, y_{ij}, $i = 1$, $2, \ldots, r_j$ and $j = 1, 2, \ldots, t$, is recorded for each experiment. If $r_j = r$ for all j, then we say that we have replicated each treatment r number of times. Ideally the entire program of experiments is run in a random sequence.

For example, an experimenter may wish to study the effects of temperature at three levels, two types of catalyst, and four pump speeds upon the yield of a particular hydrocarbon in a fluid bed reactor. The factorial design needed here would then consist of $3 \times 2 \times 4 = 24$ treatments. To provide a measure of experimental error each treatment might be performed twice to give a total of $N = 2 \times 2 \times 3 \times 4 = 48$ experiments. The program is then run in random order.

A primary concern of the experimenter will rest in comparisons between the various treatment averages. In the above example he may wish to compare the mean performance of two catalysts, or to determine whether there is a meaningful linear or quadratic trend associating the process yield with the various temperature levels, at the various pumping speeds. To estimate these and other effects we can use certain statistics, called *contrasts*, which we now define formally.

Consider a collection of N observations y_u, $u = 1, 2, \ldots, N$. A *contrast* is a linear combination of the observations of the form $\sum_{u=1}^{N} d_u y_u$, subject to the constraint that $\Sigma d_u = 0$. Similarly we may define contrasts between t treatment averages $\bar{y}_j$, $j = 1, 2, \ldots, t$ each based on r observations, by

$$(17.1.1) \qquad \sum_{j=1}^{t} c_j \bar{y}_j, \qquad \text{where} \qquad \sum_{j=1}^{t} c_j = 0.$$

Two contrasts $\sum_{u=1}^{N} d_u y_u$ and $\sum_{u=1}^{N} d_u' y_u$ are *orthogonal* whenever $\sum_u d_u d_u' = 0$. Occasionally, contrasts may be required between t treatment averages, $\bar{y}_j$ each based on a different number of observations r_j, $j = 1, 2, \ldots, t$. Under these circumstances the contrast must be of the form

$$(17.1.1a) \qquad \sum c_j r_j \bar{y}_j,$$

and subject to the constraint that $\Sigma c_j r_j = 0$. Given the constraints on the constants d_u, it is possible to construct $v = (N - 1)$ orthogonal contrasts. There is no limit to the number of sets of $(N - 1)$ orthogonal contrasts that can be constructed from N observations, except for the trivial case where $N = 2$.

Now, given a set of $N - 1$ orthogonal contrasts, $L_v = \sum_{u=1}^{N} d_{uv} y_u$, $v = 1, 2, \ldots, N - 1$, it can be shown that

$$S(Y^2) = \sum_{u=1}^{N} (y_u - \bar{y})^2 = \sum_{v=1}^{N-1} \left[\left(\frac{L_v^2}{\sum d_{uv}^2} \right) \right].$$

Thus, the $(N - 1)$ degrees of freedom associated with the corrected sum of squares in an analysis of variance table can be partitioned into $(N - 1)$ separate additive components, where each single degree of freedom component, $(L_v^2 / \Sigma d_{uv}^2)$, is associated with a single orthogonal contrast L_v. Further, under the assumption that the errors of observation are independent $N(0, \sigma^2)$ random variables, the statistic

$$(17.1.2) \qquad \frac{\sum d_u y_u - E[\sum d_u y_u]}{\sqrt{(\sum d_u^2) s^2}} = t_v$$

may be used to test hypotheses concerning the expected value of a contrast. Here, as previously, t_v is the Student t variable with v degrees of freedom, and s^2 is the usual unbiased estimator of σ^2. If we wish to test the hypothesis that $E[\sum d_u y_u] = 0$, at significance level α, we would thus reject the hypothesis if the observed value of t_v is such that

$$(17.1.2a) \qquad |t_v| = |\textstyle\sum d_u y_u|/\sqrt{(\sum d_u^2)s^2} > t_{v;\alpha/2},$$

and accept the hypothesis otherwise. Now since $t_v^2 = F_{1,v}$, we see that this test is equivalent to computing the *ratio* of the contrast mean square to the residual mean square in the analysis of variance table and referring this ratio to $F_{1,v;\alpha}$.

The limits of the $100(1 - \alpha)\%$ confidence interval estimate for $E[\sum d_u y_u]$ are given by

$$(17.1.3) \qquad \textstyle\sum d_u y_u \pm t_{v;\alpha/2}\sqrt{(\sum d_u^2)s^2},$$

where v is the number of degrees of freedom associated with s^2.

Often in a designed experiment, each of the t treatment averages $\bar{y}_j$ is based on the same number of observations r. Thus, individual degree of freedom *treatment* contrasts $\sum_{j=1}^{t} c_j \bar{y}_j$ can be constructed [see (17.1.1)]. Again, for treatment averages with the same number r of observations, two treatment contrasts, $\sum_{j=1}^{t} c_j \bar{y}_j$ and $\sum_{j=1}^{t} c'_j \bar{y}_j$, are orthogonal if and only if $\sum_{j=1}^{t} c_j c'_j = 0$, and for any set of treatment averages one may construct $(t - 1)$ orthogonal treatment contrasts. Further, the treatment sum of squares S_A [see (16.2.6) or (16.2.18)] can be partitioned into $(t - 1)$ individual degree of freedom components associated with each orthogonal contrast. Each contrast sum of squares is given by

$$(17.1.4) \qquad [\textstyle\sum c_j (r_j \bar{y}_j)]^2 / \sum r_j c_j^2 \qquad \text{or} \qquad r\frac{(\sum c_j \bar{y}_j)^2}{\sum c_j^2}$$

when all r_j are equal.

The 2^k Factorial Design

To illustrate the above discussion consider a 2^k *factorial design* consisting of the $N = 2^k$ treatments formed from all possible combinations of two versions of each of the k factors under the control of the experimenter. For example, if one of the factors is temperature, then the two versions might be the lower and upper temperatures $100°C$ and $120°C$. Or if the factor is qualitative, the two versions might be catalyst A and catalyst B or, as another example, the presence of catalyst A and the absence of catalyst A. These designs are frequently called the *two level* factorials without regard to the qualitative or quantitative nature of the controlled factors.

For $k = 3$ factors, the eight treatments of the 2^3 factorial design are those displayed, using three equivalent notations, in Table 17.1.1

TABLE 17.1.1

THE 2^3 FACTORIAL DESIGN IN "YATES' ORDER"
IN THREE NOTATIONS

Experiment Number	Notation 1 A B C	Notation 2 1 2 3	Notation 3 x_1 x_2 x_3
(i)	1	0 0 0	− − −
(ii)	a	1 0 0	+ − −
(iii)	b	0 1 0	− + −
(iv)	ab	1 1 0	+ + −
(v)	c	0 0 1	− − +
(vi)	ac	1 0 1	+ − +
(vii)	bc	0 1 1	− + +
(viii)	abc	1 1 1	+ + +

In the first notation, the three factors are labeled A, B, and C, and the two levels of each factor indicated by the presence or absence of the associated lower-case letter. The symbol 1 represents that treatment in which all the factors appear at their "low" level. In the second notation, the factors are identified as 1, 2, and 3 and their two levels by 0 and 1, respectively. Finally, in the third notation the factors are identified as x_1, x_2, and x_3, and their levels by plus and minus signs. The plus and minus notation provides a convenient geometric representation of the design, the eight settings $(\pm 1, \pm 1, \pm 1)$ defining the vertices of a cube in the three space (x_1, x_2, x_3) of the factors, the center of the cube located at the origin of the x_1, x_2, x_3 coordinate system. The design is illustrated in Figure 17.1.1(b).

In all three cases the array of treatments, called the *design matrix*, has been written down in standard order, or "Yates' order," in honor of one of the early proponents of these designs. The experimental program is, of course, run in random order.

The design matrices for the 2^2 and the 2^4 factorials are displayed in Table 17.1.2. These designs when viewed geometrically are the 2^2 vertices of a square and the 2^4 vertices of the tesseract or cube in four space, as illustrated in Figures 17.1.1(a) and 17.1.1(c). The extension for $k > 4$ should be obvious.

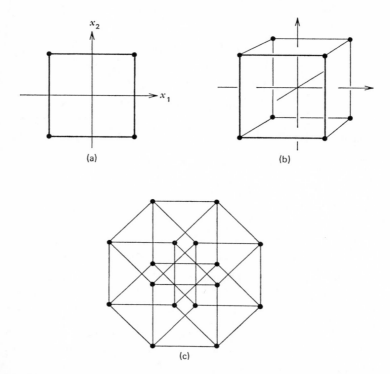

Figure 17.1.1. Geometric displays of the 2^2, 2^3, and 2^4 factorial designs. (a) The 2^2 factorial. (b) The 2^3 factorial. (c) The 2^4 factorial (although one cannot "see" the tesseract, it's projection to fewer dimensions can be visualized).

For the two-way experimental layout, the postulated model [see (16.3.4)], is

$$y_{ij} = \mu + \delta_{i.} + \delta_{.j} + \epsilon_{ij},$$

where the $\delta_{i.}$ are (usually) defined to be the block effects, the $\delta_{.j}$ the treatment effects and, further, where $\sum_i \delta_{i.} = 0$ and $\sum_j \delta_{.j} = 0$. Thus, if there are t treatments, then $t - 1$ of the $\delta_{.j}$ can be independently estimated from the data. For the two-level factorial designs, each of the treatment effects

TABLE 17.1.2

THE DESIGN MATRICES FOR THE 2^2 AND 2^4 FACTORIALS
IN YATES ORDER

				x_1	x_2	x_3	x_4
	x_1	x_2		$-$	$-$	$-$	$-$
				$+$	$-$	$-$	$-$
2^2	$-$	$-$	2^4	$-$	$+$	$-$	$-$
Factorial	$+$	$-$	Factorial	$+$	$+$	$-$	$-$
	$-$	$+$		$-$	$-$	$+$	$-$
	$+$	$+$		$+$	$-$	$+$	$-$
				$-$	$+$	$+$	$-$
				$+$	$+$	$+$	$-$
				$-$	$-$	$-$	$+$
				$+$	$-$	$-$	$+$
				$-$	$+$	$-$	$+$
				$+$	$+$	$-$	$+$
				$-$	$-$	$+$	$+$
				$+$	$-$	$+$	$+$
				$-$	$+$	$+$	$+$
				$+$	$+$	$+$	$+$

$\delta_{.j}$ is redefined and replaced by an exactly equivalent additive "factorial" effect consisting of, for the 2^k factorial design,

$$k \quad \text{main effects}$$

$$k(k-1)/2 \quad \text{two factor interaction effects}$$

(17.1.5) $k(k-1)(k-2)/2 \cdot 3$ three factor interaction effects
.
.
.

$$k(k-1)(k-2)\cdots(k-h+1)/h! \quad h\text{-factor interaction effects,}$$
.
.
.

and a single k-factor interaction effect, for a total of $2^k - 1$ factorial effects. These are mutually orthogonal contrasts and they may be estimated by setting out the plus and minus signs for each factorial effect. For example, suppose we illustrate for the case of the 2^3 factorial. Consulting Table 17.1.3, we see that the last column of the table is used simply to tabulate the total of all the observations taken at each of the eight treatments identified by the ($\mp$, $\mp$, $\mp$) signs in the first three columns. Now, for example, the main effect for factor {2} may be estimated by using the arrangement of the ($\mp$) in the column headed {2} and the column of treatment totals, as follows:

$$(17.1.6) \quad \left\{\begin{array}{l}\text{estimate of main effect} \\ \text{of factor } \{2\}\end{array}\right\}$$

$$= \frac{1}{r2^{3-1}}[-T_1 - T_2 + T_3 + T_4 - T_5 - T_6 + T_7 + T_8].$$

The factor $\frac{1}{2^{3-1}}$, or, in general, $\frac{1}{2^{k-1}}$, is the reciprocal of the number of plus signs in a column, while the factor r in the denominator of this expression equals the common number of observations that make up each treatment total. Hence (17.1.6) equals $\bar{y}_+ - \bar{y}_-$ for factor {2}, where by $\bar{y}_+$ we mean the average of all the observations recorded when factor {2} was at its plus level, and $\bar{y}_-$ the corresponding average of observations taken at the minus level of factor {2}.

TABLE 17.1.3

CONTRAST COEFFICIENTS FOR THE 2^3 FACTORIAL DESIGN

Row	{1}	{2}	{3}	{12}	{13}	{23}	{123}	Treatment Totals
1	−	−	−	+	+	+	−	T_1
2	+	−	−	−	−	+	+	T_2
3	−	+	−	−	+	−	+	T_3
4	+	+	−	+	−	−	−	T_4
5	−	−	+	+	−	−	+	T_5
6	+	−	+	−	+	−	−	T_6
7	−	+	+	−	−	+	−	T_7
8	+	+	+	+	+	+	+	T_8

In general, the first k columns of signs in a *contrast coefficient table* such as Table 17.1.3 are identical to those given in the design matrix for the 2^k factorial design, and these columns may be used as bases for estimates of the k main effects. The remaining columns of signs are used to estimate "interaction" effects and are generated from the row-wise product of the signs in the first k columns. For example, to obtain the entry in row three of the column labeled {12}, we have that the entry in the column {1}, row three, is −, and the entry in the column {2}, row three is +, and hence the entry for row three in column {12} is $(+)(-) = -$. The generalization of this table for the 2^k factorial designs is straightforward. The estimate of the $ij \cdots k$ interaction effect is proportional to the sum of the products of the corresponding $\{ij \cdots \}k$ elements and treatment totals $\Sigma \{ij \cdots k\}T$ and is given by

(17.1.7)   estimated $ij \cdots k$ effect $= \dfrac{1}{r2^{k-1}} (\Sigma \{ij \cdots k\}T)$,

where $\{ij \cdots k\}$ stands for the 2^k elements of plus and minus signs in the appropriate column. Thus, in the example above, the estimate of the $\{12\}$ or $x_1 x_2$ interaction effect would be

(17.1.8) $\begin{Bmatrix} \text{estimated } x_1 x_2 \\ \text{interaction effect} \end{Bmatrix}$

$$= \dfrac{1}{(r)(4)} [+T_1 - T_2 - T_3 + T_4 + T_5 - T_6 - T_7 + T_8].$$

The associated single degree of freedom component of the treatment sum of squares is, from (17.1.4) and (17.1.7),

(17.1.9) $\begin{Bmatrix} \text{sum of squares} \\ \text{for } ij \cdots k \text{ effect} \end{Bmatrix}$

$$= [\Sigma \{ij \cdots k\}T]^2/r2^k = r2^{k-2} \text{ (estimate of effect)}^2.$$

The reader may check for himself that the $(k - 1)$ estimates are all orthogonal contrasts, and that each contrast is the difference between two averages of $r2^{k-1}$ observations each. Further the variance for *each* estimate of the several effects is

(17.1.10) $V[\text{estimate of effect}] = v = \left(\dfrac{1}{r2^{k-1}} + \dfrac{1}{r2^{k-1}} \right) \sigma^2 = \dfrac{1}{r2^{k-2}} \sigma^2.$

We will denote an estimate of (17.1.10) by $\hat{v}$, that is,

(17.1.10a) $\hat{v} = \dfrac{1}{r2^{k-2}} \widehat{\sigma^2}$,

where $\widehat{\sigma^2}$ is the estimate of σ^2 obtained by dividing the appropriate error sum of squares by its associated degrees of freedom. (See the example of the next section.)

A test of hypothesis about any effect is provided by using the fact that the following is distributed as Student's t, namely, we have

(17.1.11) $t_v = \dfrac{(\text{estimate of effect}) - E(\text{estimate of effect})}{\sqrt{\hat{v}}}.$

An interval estimate for any effect is given by

(17.1.12) $(\text{estimate of effect}) \pm t_{v;\alpha/2} \sqrt{\hat{v}}.$

TABLE 17.2.1
DATA FOR 2^2 FACTORIAL IN FIVE REPLICATES

Treatment	Studied Factors Type of Dyestuff	Studied Factors Preparation Method	Design Variables x_1	Design Variables x_2	Responses y_i; (loss of reflectance) Blocking Variable: Days (i)	(ii)	(iii)	(iv)	(v)	Treatment Totals T_i	Treatment Averages $\bar{y}_i$
(i)	Old	Standard	−	−	23	21	23	21	22	110	22.0
(ii)	New	Standard	+	−	30	28	31	28	23	140	28.0
(iii)	Old	Alternative	−	+	12	8	5	9	6	40	8.0
(iv)	New	Alternative	+	+	23	19	17	14	17	90	18.0
Day totals T_j					88	76	76	72	68	$G = 380$	
Day averages $\bar{y}_j$					22.0	19.0	19.0	18.0	17.0		$\bar{\bar{y}} = 19$

17.2 AN EXAMPLE OF A 2^2 FACTORIAL REPLICATED FIVE TIMES

An experimenter wishes to determine the effects of a new dyestuff and alternative methods for preparing fabric upon fabric color fastness. The following 2^2 factorial design was employed in which the plus and minus levels of x_1 correspond to the new dyestuff and standard dyestuff, respectively, and the plus and minus signs of factor x_2 represent the alternative and standard methods for preparing the fabric, respectively. The measured response y (here modified and coded for convenience) was the loss in reflectance of dyed fabric after twenty hours of exposure to a carbon-arc lamp, in keeping with standards given in AATCC Method 16A-1864 (American Association of Textile Chemists and Colorists). To provide estimates of treatment effects with sufficient precision it was decided to replicate the experimental program five times. Before running the experiment it was also felt that day-to-day variability might inflate the variance of the observations and so, to protect the treatment comparisons against this unwanted source of variability, the 2^2 factorial program was randomly run once on each of five days. The 2^2 factorial design, and recorded responses, are given in Table 17.2.1.

A test of the hypothesis that there are no treatment effects is carried out by first postulating the model, [see (16.3.4)]:

$$y_{ij} = \mu + \delta_{i.} + \delta_{.j} + \epsilon_{ij},$$

where the y_{ij} are the recorded observed reflectance measurements, $i = 1, \ldots, 4$, $j = 1, \ldots, 5$, and where the $\delta_{i.}$ are here defined to be the day effects, $\delta_{.j}$ the effects due to treatments, and the ϵ_{ij} are independent

TABLE 17.2.2

Source of Variation	Sum of Squares	Degrees of Freedom	Mean Square	Expected Mean Squares	F Ratio
Total $S(Y^2)$	1176	19			
SSq Treatments	1060	3	353.3	$\sigma^2 + \frac{5}{3}\sum_{j=1}^{5} \delta_{.j}^2$	$F_{3,12} = 70.64$
SSq Days	56	4	14.0	$\sigma^2 + \frac{4}{4}\sum_{i=1}^{4} \delta_{i.}^2$	$F_{4,12} = 2.80$
Error SSq	60	12	$5.0 = s^2$	σ^2	
Pooled Error SSq	116	16	$11.0 = s^2$ (see below)		$F_{3,16} = 32.12$

$N(0, \sigma^2)$. We then use the analysis of variance, (see Table 16.3.2) to estimate σ^2, the variance of the observations, and finally to test the hypothesis that the treatment effects $\delta_{.j}$ are all equal to zero. The analysis of variance is given in Table 17.2.2. (SSq is a commonly used abbreviation for "sum of squares.")

$$\text{totals } S(Y^2) = \sum y^2 - \frac{G^2}{r2^k}$$

$$= 23^2 + 30^2 + \cdots + 14^2 - (380)^2/20$$

$$= 8396 - 7220 = 1176$$

$$\text{treatments SSq} = \frac{1}{r} \sum T_i^2 - \frac{G^2}{r2^k}$$

$$= \tfrac{1}{5}(110^2 + 140^2 + 40^2 + 90^2) - (380)^2/20$$

$$= 8280 - 7220 = 1060$$

$$\text{days SSq} = \frac{1}{2^k} \sum T_j^2 - \frac{G^2}{r2^k}$$

$$= \tfrac{1}{4}(88^2 + 76^2 + \cdots + 68^2) - (380)^2/20$$

$$= 7276 - 7220 = 56.$$

To test the hypothesis that the treatment effects $\delta_{.j}$ are all zero, we note that, under this hypothesis, both the treatment mean square and the error mean square, are independent estimates of σ^2, and hence their ratio will be distributed as F with degrees of freedom (3, 12). The observed ratio of $F_{3,12} = 353.3/5.0 = 70.04$; upon reference to Table VI of Appendix VII, this is found to be a most unusual value, since the Prob $\{F_{3,12} \geq 7.22\} = .005$ (here, as elsewhere in this chapter, the significance level of a test is .05). Hence the hypothesis, all $\delta_{.j} = 0$, is rejected. We now turn to making additional inferences about $\delta_{.j}$.

Employing (17.1.9) we can separate the three degrees of freedom for treatments into individual orthogonal one degree of freedom contrasts associated with (i) the main effect of x_1: changing dyestuff; (ii) the main effect of x_2: changing fabric preparation; and (iii) a measure of the dyestuff-preparation interaction effect $x_1 x_2$. The computation of these contrasts [see (17.1.7)], or, equivalently, the estimation of these effects, is displayed in Table 17.2.3, along with the 95% confidence limits for the true value of the effects, based on the estimate of the variance of an effect $\hat{v} = [1/(5)(1)] s^2 = [1/(5)(1)] (5) = 1.0$ [see (17.1.10a) and the analysis of variance table 17.2.2].

TABLE 17.2.3

CONTRAST COEFFICIENTS FOR FACTORIAL EFFECTS

Contrast Coefficients			Treatment Totals
{1}	{2}	{12}	T
−	−	+	110
+	−	−	140
−	+	−	40
+	+	+	90

Estimated Effects

$$\left\{ \begin{matrix} x_1 \text{ (dyestuff)} \\ \text{effect} \end{matrix} \right\} = \frac{1}{(52)} [-110 + 140 - 40 + 90]$$

$$= \frac{230}{10} - \frac{150}{10} = 8.0$$

$$\left\{ \begin{matrix} x_2 \text{ (preparation)} \\ \text{effect} \end{matrix} \right\} = \frac{1}{5(2)} [-110 - 140 + 40 + 90]$$

$$= \frac{130}{10} - \frac{250}{10} = -12.0$$

$$\left\{ \begin{matrix} x_1 x_2 \text{ (interaction)} \\ \text{effect} \end{matrix} \right\} = \frac{1}{5(2)} [+110 - 140 - 40 + 90]$$

$$= \frac{200}{10} - \frac{180}{10} = 2.0$$

Here $\hat{v} = \frac{1}{5(1)} s^2 = \frac{1}{5(1)} (5) = 1.0$ [see (17.1.10a), and the analysis of variance table 17.2.2]. The associated single degree of freedom portions of the treatment sum of squares are given in Table 17.2.4 [see (17.1.9)].

TABLE 17.2.4

PARTITIONING TREATMENT SUM OF SQUARES
INTO INDIVIDUAL DEGREE OF FREEDOM COMPONENTS

Source	Sum of Squares	Degrees of Freedom	Mean Squares
x_1 (dyestuff) effect $= 5(2°)(8.0)^2 = 320$	320	1	320
x_2 (preparation) effect $= 5(2°)(-12.0)^2 = 720$	720	1	720
$x_1 x_2$ (interaction) effect $= 5(2°)(2.0)^2 = 20$	20	1	20
Total treatment sum of squares $= 1060$	1060	3	353.3

A test of the hypothesis that the effect of changing dyestuff is zero, that is, that the x_1 effect is zero, is provided by the ratio of the dyestuff and error mean squares. Thus the relevant observed F is

$$F_{1,12} = \frac{320}{5.0} = 64.0,$$

a very rare $F_{1,12}$, since Prob $\{F_{1,12} \geq 4.74\} = .05$. Thus we reject the hypothesis that the x_1 effect is zero. An equivalent test is provided by using the t-statistic [see (17.1.11)],

$$t_{12} = \frac{8.0 - \text{zero}}{\sqrt{\dfrac{1}{5(1)}(5.0)}} = 8.0,$$

and we note, once again, that $F_{1,v} = (t_v)^2$. Similarly, the hypothesis that there is no effect due to factor x_2 (preparation) must be rejected, since $F_{1,12} = 720/5.0 = 144.0$. However, there is no strong evidence that an interaction effect between the type of dyestuff, and method of preparation exists since the corresponding test of the null hypothesis "interaction effect $= 0$" gives an observed $F_{1,12} = \frac{20}{5} = 4.0$, and Prob $\{F_{1,12} > 4\} >$ .05. The 95% confidence limits for each individual effect are [see (17.1.12)]

$$\text{estimate of effect} \pm 2.179 \sqrt{\tfrac{1}{5}(5.0)} = \text{estimate of effect} \pm 2.2$$

since the estimate of the variance of an effect, $\hat{v}$

$$= \frac{1}{(5)2^{2-2}} \hat{\sigma}^2 = \tfrac{1}{5}s^2 = \tfrac{1}{5}(5.0).$$

In summation, on the basis of the evidence provided by the 2^2 factorial, the effect of changing to the new dyestuff will be to increase the reflectance, and hence the color fastness of the dyed fabric, by 8.0 ± 2.2 units, whereas changing to the new method for preparing the fabric prior to dying has a deleterious effect of -12.0 ± 2.2 units. Changing to the new dyestuff and continuing with the standard mode of preparation of fabric is thus strongly suggested by these data.

The analysis of these data need not end with the investigation of the effects of the treatments. The experimenter could also test the hypothesis that the day to day effects $\delta_{i.}$ were zero. Investigating the variation in color fastness due to days is not the primary objective of the experimenter, but the design and associated analysis of variance table provides a ready opportunity to perform this test of hypothesis. The corresponding F ratio, $F_{4,5} = 14.0/5.0 = 2.8$ is not a rare event at the 5% level of significance, since the Prob $\{F_{4,12} \geq 3.26\} = .05$. However, it may be considered a "rare" event at the 10% level, since the Prob $\{F_{4,12} \geq 2.48\} = 0.10$. A variety of interpretations are now open to the experimenter. He could decide, on the basis of the 5% test, that no day effects existed and that blocking the experiment by days was an unnecessary nuisance. He might then *pool* the sums of squares and degrees of freedom for days and error and produce the pooled estimate of variance $s^2 = 116/16 = 11.0$ with sixteen degrees of freedom as noted in the last line of Table 17.2.2. We note that the hypothesis that the treatment effects are zero is still rejected, for the new observed $F_{3,16} = 353.3/11.0 = 32.12$ is still an extraordinarily rare event, since the Prob $\{F_{3,16} \geq 6.30\} = .005$. Or, the experimenter might decide, on the basis of the 10% test, that blocking to eliminate day to day effects had been worthwhile, since real day to day effects were detected and hence, used only the error mean square $s^2 = 5.0$ in the analysis. Alternatively he might still pool the day to day error contributions to see whether the treatment effects were detectably nonzero in the presence of this additional acknowledged source of variance due to days. The fact that they are, in this example, enhances the experimenter's ability to make statements about the treatments across future days.*

The reader may have noticed that there seems to be a trend across days, the daily average diminishing gradually in time. Since days are equally spaced in time, one may easily construct orthogonal single degree of freedom contrasts, and associated sum of squares, reflecting the day-to-day variability that could be assigned to a linear, or quadratic trend across days. The necessary sets of constant coefficients for the linear and quadratic trends are displayed in Table 17.2.5.

TABLE 17.2.5

	Successive Daily Averages 22.0 19.0 19.0 18.0 17.0					Contrast $\sum c_i \bar{y}_{i.}$	$SSq = \dfrac{r(\sum c_i \bar{y}_i)^2}{\sum c_i^2}$
Linear effect contrast coefficients, c_i:	-2	-1	0	1	2	11	$4(11)^2/10 = 48.40$
Quadratic effect contrast coefficients, c_i':	2	-1	-2	-1	2	3	$4(3)^2/14 = 2.57$

The sum of squares and degrees of freedom for days may now be partitioned as illustrated in Table 17.2.6.

---

* The experimenter here assumes that there is no day by treatment interaction. If, before running the experiments, he had thought such interactions likely, the design would have to be modified to permit estimation of these effects. This could be accomplished by repeating the treatments within each day (see Section 16.3.6).

TABLE 17.2.6
PARTITIONING BLOCK SUM OF SQUARES

	SSq	Degrees of Freedom	Mean Square	
Linear day effect	48.40	1	48.40	$F_{1.12} = 9.68$
Quadratic day effect	2.57	1	2.57	$F_{1.12} < 1$
Other effects (by subtraction)	5.03	2	2.51	$F_{2.12} < 1$
Total sum of squares for days	56.00	4	14.0	

The test of the hypothesis that no linear trend exists (that is, that the linear contrasts amongst the true treatment means equals zero against the alternative that the linear contrast is not zero) is given by the ratio of the linear effect mean square divided by the error mean square, $s^2 = 5.0$ as found in Table 17.2.2. Thus the observed $F_{1.12} = 48.40/5.0 = 9.68$, and since Prob $\{F_{1.12} \geq 4.74\} = .05$, the observed $F_{1.12}$ is a rare event. We can thus reject the hypothesis that no linear trend exists, which means that we accept the hypothesis that a linear trend does exist. We note that this apparent trend accounts for almost all the variation between days. In this instance, one assignable cause was the wearing down of the electrodes in the carbon arc lamp used in the instrument for measuring reflectance.

The ease and richness of the present analysis is due in very large part to the experimental design.

17.3   UNREPLICATED 2^k FACTORIAL DESIGNS

For moderate values of k, $k \geq 4$, the total number of treatments specified by a 2^k factorial design quickly becomes large and experimenters often become unwilling to repeat, i.e., replicate the experimental program. When replication is ruled out no estimate of σ^2 is available, that is, none that can be constructed from replicated observations. One compromise is to partially replicate the full design, the choice of the repeated treatments being carefully selected. However, an estimate of σ^2 can be constructed whenever the number of variables k is large. To explain, when working with k factors, it is simply unlikely that all the $(2^k - 1)$ factorial effects will, in fact, be large. Under the assumption that the response being investigated changes smoothly over the range of the factors being varied, it becomes unlikely that high-order interaction effects exist. When they exist, the magnitude of such effects is usually small relative to the main effects of the lower order interactions. In circumstances where these assumptions seems reasonable, the $[(2^k - 1) - k - k(k - 1)/2]$ degrees of freedom available for the estimation of the three-factor and high-order inter-action effects are employed, instead, to provide an estimate of the variance. In other cases, it could become clear, after experiments have been per-formed, that $h = 1, 2, \ldots$, where $h < k$, of the factors have only *very small or no effects* on the response when compared to the effects of the remaining $(k - h)$ factors. When this occurs, the experimenter often declares the h factors (over the ranges studied) as having zero effects, or as having effects whose magnitudes can not be distinguished from the contributions of the random errors. The program then becomes a 2^{k-h} factorial replicated 2^h times providing $(2^h - 1)2^{k-h}$ degrees of freedom for estimating σ^2.

The 2^k factorial designs can be viewed as "equal opportunity" designs since they permit the orthogonal estimation of all $(2^k - 1)$ factorial effects, each effect estimated with minimum variance $\sigma^2/2^{k-2}$. Now the experi-menter often is anxious to determine which subset of these $(2^k - 1)$ candidate effects has, in fact, the largest influence on the response. We say that the experimenter wishes to *screen* the many effects to discover the important ones. When the factorial design is not replicated, this search becomes difficult, since the experimenter will have to determine which of the *estimated* effects are due to the experimental error, and which are reflections of real and hopefully large effects. To help identify the real effects, the estimates of the effects may be plotted on normal probability paper.* Since linear combinations of random variables are statistics tending to have a normal distribution (see Appendix IV), those estimates that have values which are primarily due to the errors of observation should have the appearance of events from a normal distribution.

Following Section 13.7, the $(2^k - 1)$ estimates of the factorial effects are first ordered, say $e_{(1)}, \ldots, e_{(2^k-1)}$, and these ordered values $e_{(i)}$ are plotted against $P_i = (i - .5)/(2^k - 1)$, $i = 1, \ldots, 2^k - 1$, on normal probability paper. Now if the effects have true value zero, then the ordered estimated effects $e_{(i)}$, will, when plotted against P_i, tend to fall along a "straight line," confirming the hypothesis that these estimates are due solely to errors. However, the estimates of largest magnitude will, *if they reflect real effects*, lie off the straight line. (The student is reminded that the straight line is determined by the points in the inner quartiles, see (Section 13.7)). Having identified these large estimates as distinguish-able from estimates that may be assumed to be manifestations of error only, the experimenter may use some, or all, of the degrees of freedom associated with the "error-like" effects and contrive an estimate of σ^2.

---

* Or perhaps more resourcefully on "half normal paper" as described by Daniel (1959).

TABLE 17.3.1

Run Number	Factor and Levels				Equivalent Design Levels				Alternative Design	y_i Observations
	Temperature °C	Speed, 1000 rpm	Catalyst mols	Pressure, 100 psig	x_1	x_2	x_3	x_4	A B C D	
1	30	1.0	.6	7	−	−	−	−	1	62
2	32	1.0	.6	7	+	−	−	−	a	88
3	30	1.2	.6	7	−	+	−	−	b	63
4	32	1.2	.6	7	+	+	−	−	ab	83
5	30	1.0	1.0	7	−	−	+	−	c	88
6	32	1.0	1.0	7	+	−	+	−	ac	80
7	30	1.2	1.0	7	−	+	+	−	bc	99
8	32	1.2	1.0	7	+	+	+	−	abc	92
9	30	1.0	.6	10	−	−	−	+	d	65
10	32	1.0	.6	10	+	−	−	+	ad	123
11	30	1.2	.6	10	−	+	−	+	bd	65
12	32	1.2	.6	10	+	+	−	+	abd	121
13	30	1.0	1.0	10	−	−	+	+	cd	97
14	32	1.0	1.0	10	+	−	+	+	acd	105
15	30	1.2	1.0	10	−	+	+	+	bcd	92
16	32	1.2	1.0	10	+	+	+	+	abcd	117

TABLE 17.3.2

TABLE OF CONTRAST COEFFICIENTS FOR A 2^4 FACTORIAL DESIGN

Main Effects				Two-Factor Interactions						Three-Factor				Four-Factor	Response
{1}	{2}	{3}	{4}	{12}	{13}	{14}	{23}	{24}	{34}	{123}	{124}	{134}	{234}	{1234}	y
−	−	−	−	+	+	+	+	+	+	−	−	−	−	+	62
+	−	−	−	−	−	−	+	+	+	+	+	+	−	−	88
−	+	−	−	−	+	+	−	−	+	+	+	−	+	−	63
+	+	−	−	+	−	−	−	−	+	−	−	+	+	+	83
−	−	+	−	+	−	+	−	+	−	+	−	+	+	−	88
+	−	+	−	−	+	−	−	+	−	−	+	−	+	+	80
−	+	+	−	−	−	+	+	−	−	−	+	+	−	+	99
+	+	+	−	+	+	−	+	−	−	+	−	−	−	−	92
−	−	−	+	+	+	−	+	−	−	−	+	+	+	−	65
+	−	−	+	−	−	+	+	−	−	+	−	−	+	+	123
−	+	−	+	−	+	−	−	+	−	+	−	+	−	+	65
+	+	−	+	+	−	+	−	+	−	−	+	−	−	−	121
−	−	+	+	+	−	−	−	−	+	+	+	−	−	+	97
+	−	+	+	−	+	+	−	−	+	−	−	+	−	−	105
−	+	+	+	−	−	−	+	+	+	−	−	−	+	−	92
+	+	+	+	+	+	+	+	+	+	+	+	+	+	+	117

We illustrate this method for constructing an estimate of σ^2 from an unreplicated factorial experiment in the following example.

> **Example 17.3(a).** An experimenter is interested in studying the effects of the $k = 4$ factors: temperature, speed of agitation, catalyst concentration, and pressure, upon the yield of a chemical process, and performed experiments comprising a 2^4 factorial design. The experiments were performed in a random sequence and Table 17.3.1 displays the actual factor settings, the experimental design in two alternative notations, and the responses y_i, $i = 1, 2, \ldots, 16$.

The table of contrast coefficients required for estimating the fifteen factorial effects are displayed in Table 17.3.2.

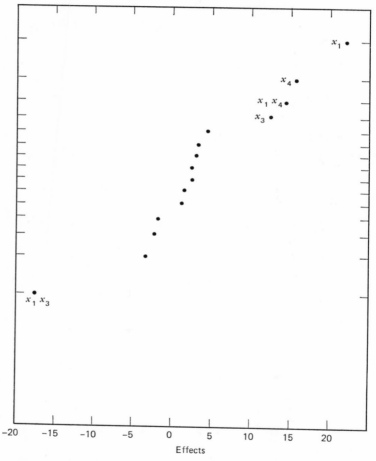

Figure 17.3.1. Ordered effects for the 2^4 factorial design of Table 17.2.1.

The estimated factorial effects are displayed in Table 17.3.3 [see (17.1.7)].

TABLE 17.3.3

Main Effects		Estimates, using Table 17.3.2, of:					
		Two-Factor Interactions		Three-Factor Interactions		Four-Factor Interactions	
x_1	22.25	x_1x_2	1.25	$x_1x_2x_3$	3.25	$x_1x_2x_3x_4$	1.50
x_2	3.00	x_1x_3	−17.75	$x_1x_2x_4$	2.50		
x_3	12.50	x_1x_4	14.50	$x_1x_3x_4$	2.50		
x_4	16.25	x_2x_3	4.50	$x_2x_3x_4$	−2.25		
		x_2x_4	−1.75				
		x_3x_4	−3.25				

Computing the estimates using Table 17.3.2 is long and tedious. An alternative and more rapid estimation technique is provided by Yates' algorithm, illustrated for this example in the Appendix to this chapter (see page 461).

Now let us consider the problem of obtaining an estimate of σ^2. As a preliminary to determining an estimate of σ^2, the ordered effects (see table below) are plotted on normal paper, as in Figure 17.3.1. On viewing this plot (and remembering that extreme points should have little weight in orientating the fitted line on Normal probability paper), we note that the smallest estimates (those near zero) lie reasonably along a straight line, while the largest estimates are far off the line. This suggests strongly that these large estimates are not acting in a matter compatible with the suggestion that they are due to random errors.

Order	Effects	Estimates	$(i - \frac{1}{2})/15$
$i = 15$	1	22.25	.9667
14	4	16.25	.9000
13	14	14.50	.8333
12	3	12.50	.7667
11	23	4.50	.7000
10	123	3.25	.6333
9	2	3.00	.5667
8	124	2.50	.5000
7	134	2.50	.4333
6	1234	1.50	.3667
5	12	1.25	.3000
4	24	−1.75	.2333
3	234	−2.25	.1667
2	34	−3.25	.1000
1	13	−17.75	.0333

One immediate interpretation of the data open to the experimenter is that variable x_2 (speed of agitation) has no influence on the response since none of the large effects involves x_2. Under this assumption, the program becomes a replicated 2^3 factorial in variables x_1, x_3, and x_4. An estimate of σ^2 may be found, then, from the corresponding analysis of variance table, based on this simplifying assumption, as given in Table 17.3.4.

TABLE 17.3.4

Source	Sum of Squares	Degrees of Freedom	Mean Square	F Ratios
$S(Y^2)$	6062	15		
x_1 (temperature) effect	1980.25	1	1980.25	$F_{1,8} = 68.3$
x_3 (catalyst) effect	625.00	1	625.00	$F_{1,8} = 21.6$
x_4 (pressure) effect	1056.25	1	1056.25	$F_{1,8} = 36.4$
x_1x_3 interaction	1260.25	1	1260.25	$F_{1,8} = 43.5$
x_1x_4 interaction	841.00	1	841.00	$F_{1,8} = 29.0$
x_3x_4 interaction	42.25	1	42.25	$F_{1,8} = 1.5$
$x_1x_3x_4$ interaction	25.00	1	25.00	$F_{1,8} = .9$
Residual sum of squares	232.00	8	$29.00 = s^2$	

Now the sum of squares contribution of the main effect of temperature (x_1) is given by [see (17.1.9)]: $2(2)(22.25)^2 = 1980.25$. (The contribution to the sum of squares is also quickly obtained using Yates' algorithm.) Using the estimate $s^2 = 29.00$, based on eight degrees of freedom, the three main effects, temperature (x_1), catalyst concentration (x_2), and pressure (x_3), are clearly significant, as well as the temperature–catalyst and temperature–pressure interactions, since Prob $\{F_{1,8} \geq 5.32\} = .05$. The 95% confidence limits for the effects are (see (17.1.10))

$$\text{(estimate of the effect)} \pm t_{8;.025}\sqrt{(\tfrac{1}{8} + \tfrac{1}{8})s^2}$$

(17.3.1) $$= \text{(estimate of the effect)} \pm 2.306 \times \sqrt{29.00/4}$$

$$= \text{(estimate of the effect)} \pm 6.21.$$

An alternative procedure for obtaining an estimate of σ^2 is to assume the effects of the four three-factor interactions and the single four-factor interaction to be nil, and to pool together the sums of squares associated with these high-order interactions. Consulting Table 17.3.3, we see that the sum of squares for these effects is

(17.3.2) $2(2)[(3.25)^2 + (2.50)^2 + (2.50)^2 + (-2.25)^2 + (1.50)^2] = 121.5.$

Thus an estimate of σ^2 based now on five degrees of freedom is $s^2 =$

$121.5/5 = 24.3$. Inferences concerning which effects were important would not be materially changed had the experimenter contrived the estimate of σ^2 in this fashion. In this example, as in the previous example of the replicated 2^2 factorial design (Section 17.2), analyses can vary.

17.4 BLOCKING THE 2^k FACTORIAL DESIGNS

It is never possible to conduct experiments in an environment wherein all sources of variability are eliminated. However, it is often possible to control some sources of variability and thereby establish environments within which the experimental variability is decreased. This is accomplished by partitioning the design into subsets or *blocks* of experiments, the experimental environment within each block being held as constant as possible. Separating a 2^k factorial design into two blocks can only be done by sacrificing a degree of freedom ordinarily employed to estimate a factorial effect or contrast. The factorial contrast of least interest to the experimenter is usually the contrast associated with the kth-order, that is, the highest order interaction effect. This contrast is employed for blocking, the plus and minus signs of the $\{ij \cdots k\}$ factorial contrast being used to separate the 2^k treatments into two blocks of 2^{k-1} treatments each. The treatments are, of course, run randomly within each block.

When we have two blocks of 2^{k-1} treatments each, which may be replicated r times, $r = 1, 2, \ldots$, then the estimate of the kth-order interaction effect, $\dfrac{1}{r2^{k-1}} (\Sigma \{ij \cdots k\}T)$, has an expected value of $(ij \cdots k)$ effect $+ \delta$, where δ is the block effect. Whenever two effects are estimated by an identical linear statistic, *or* if an estimate has an expected value equal to some combination of effects, then the effects are said to be *confounded*, or the estimate is said to have an *alias* structure. Thus, we will say that in this instance the $ij \cdots k$ interaction effect and block effect are confounded. The appropriate mathematical model when using a blocked design is a variation on the one-way classification model, $y_{ij} = \mu + \delta_j + \epsilon_{ij}$, $i = 1, 2, \ldots, r$; $j = 1, 2, \ldots, 2^k$, where the $(2^k - 1)$ degrees of freedom for treatment effects are partitioned into $2^k - 2$ degrees of freedom for factorial effects, and a single degree of freedom for an effect (usually the highest order interaction) confounded with the block effect. The 2^3 factorial, partitioned into two blocks of four runs each, is displayed in Table 17.4.1.

It is, in fact, possible to partition the 2^k designs into 2^{k-1} blocks of two runs each. In this case there will be $(2^{k-1} - 1)$ degrees of freedom assignable to blocks leaving 2^{k-1} degrees of freedom for estimating factorial effects. The nature of the confounding pattern, or alias structure, for such

TABLE 17.4.1

BLOCKING A 2^3 FACTORIAL DESIGN

Partitioning the 2^3 Design into Two Blocks

Yates Run Number	x_1	x_2	x_3	Blocking Contrast $x_1x_2x_3$
1	−	−	−	−
2	+	−	−	+
3	−	+	−	+
4	+	+	−	−
5	−	−	+	+
6	+	−	+	−
7	−	+	+	−
8	+	+	+	+

The 2^3 Factorial in Two Blocks

Yates Number	Block 1 $x_1x_2x_3 = +$			Yates Number	Block 2 $x_1x_2x_3 = -$		
	x_1	x_2	x_3		x_1	x_2	x_3
2	+	−	−	1	−	−	−
3	−	+	−	4	+	+	−
5	−	−	+	6	+	−	+
8	+	+	+	7	−	+	+

TABLE 17.4.2

THE 2^3 AND 2^4 FACTORIAL DESIGNS IN BLOCKS OF TWO TREATMENTS

Block		x_1	x_2	x_3
	(i)	−	−	−
		+	+	+
	(ii)	+	−	−
		−	+	+
	(iii)	−	+	−
		+	−	+
	(iv)	+	+	−
		−	−	+

Block		x_1	x_2	x_3	x_4
	(i)	−	−	−	−
		+	+	+	+
	(ii)	+	−	−	−
		−	+	+	+
	(iii)	−	+	−	−
		+	−	+	+
	(iv)	+	+	−	−
		−	−	+	+
	(v)	−	−	+	−
		+	+	−	+
	(vi)	+	−	+	−
		−	+	−	+
	(vii)	−	+	+	−
		+	−	−	+
	(viii)	+	+	+	−
		−	−	−	+

a design is beyond the scope of this discussion (see Box and Hunter (1961) for details). However, the resulting designs are of considerable interest. The 2^3 and 2^4 factorial designs, partitioned into blocks of two treatments each, are displayed in Table 17.4.2. These designs are structured so that the main effects estimates are not affected at all (that is, not confounded) by differences between the blocks. (The designs are sometimes termed *main effect clear* designs.) The reader will note that the pairs of treatments comprising the blocks are "complementary" or "foldover" pairs. Thus, differences between treatments *within* a block are unaffected by the block means. The 2^{k-1} differences determined from within the 2^{k-1} blocks supply all the necessary information for estimating the k main effects, clear of the block effects.

> *Example 17.4(a).* A replicated 2^3 factorial in four blocks of four treatments each.
> A study was performed to determine the effects of texturing on the breaking tenacity of artificial fiber. The process variables selected for study were x_1:spindle speed, x_2:temperature of plates, and x_3:amount of twist. A 2^3 factorial design in $r = 2$ replicates was chosen. Further, since only four experiments could be run during a single day, each 2^3 factorial design was partitioned into two blocks of four runs each, and the program completed on four separate days. The experiments were randomly run within each day. The 2^3 design was partitioned as illustrated in Table 17.4.3, that is, the contrast for the three-factor interaction effect was used to block the design. The results are displayed in Table 17.4.3.

To estimate the factorial effects one may use either the table of contrast coefficients for the 2^3 factorial (Table 17.1.3), or Yates' algorithm (see page 461). The estimated effects are given in Table 17.4.3. The three-factor interaction effect is not included since this contrast was used by the experimenter to identify the blocks. The block sum of squares and the single degree of freedom contributions of the factorial effects [see (17.1.9)] are entered in the analysis of variance table. The error sum of squares is obtained by subtraction. The estimate of the experimental error variance is $s^2 = .9940$ with six degrees of freedom. The confidence limits for the factorial effects [see (17.1.12)] are given by

$$(\text{estimate of effect}) \pm 2.447\sqrt{\frac{1}{2(4)}(.9940)} = (\text{estimate of effect}) \pm .86.$$

It seems clear from this analysis that all three factors have detectable effects upon breaking tenacity, the role of spindle speed (x_1) demonstrating itself primarily through its interaction with the amount of twist (x_3). Both temperature (x_2) and twist (x_3) have detectable main effects.

TABLE 17.4.3

REPLICATED 2^3 IN FOUR BLOCKS

	Day 1				Day 2				Day 3				Day 4		
x_1	x_2	x_3	y	x_1	x_2	x_3	y	x_1	x_2	x_3	y	x_1	x_2	x_3	y
+	−	−	19.8	−	−	−	18.8	+	−	−	18.4	−	+	+	19.5
+	+	+	13.6	+	+	−	18.0	+	+	+	14.8	+	+	−	15.0
−	−	+	22.7	−	+	+	20.6	−	+	−	12.8	−	−	−	17.7
−	+	−	11.8	+	−	+	19.0	−	−	+	23.8	+	−	+	19.8
			67.9				76.4				69.8				72.0

Grand Total
$\Sigma y = 286.1$
$\Sigma y^2 = 5292.79$

Standard Order $x_1 \; x_2 \; x_3$	Treatment Totals	Estimated Effects		95% Confidence Limits
− − −	36.5	Grand Average =	17.88	±.61
+ − −	38.2	x_1 effect =	−.58	
− + −	24.6	x_2 effect =	−2.12	
+ + −	33.0	x_3 effect =	1.34	±.86
− − +	46.5	$x_1 x_2$ effect =	.17	
+ − +	38.8	$x_1 x_3$ effect =	−1.84	
− + +	40.1	$x_2 x_3$ effect =	.02	
+ + +	28.4			

Analysis of Variance

	Source	SSq	Degrees of Freedom	
	$S(Y^2)$	176.9644	15	
	Block SSq	10.0269	3	
	x_1	5.4056	1	
Effects	x_2	71.8256	1	
	x_3	28.8906	1	
	$x_1 x_2$	.4556	1	
	$x_1 x_3$	54.3906	1	
	$x_2 x_3$	.0056	1	
	Error	5.9639	6	.9940 = s^2

$F_{1,6;0.05} = 5.99$.

17.5 THE TWO-LEVEL FRACTIONAL FACTORIAL DESIGNS

Whenever the number of variables k becomes large, the number of treatments required by the 2^k factorial designs becomes burdensomely large. When completed, the data from a 2^k design provides, in addition to estimates of the k main effects and the $k(k-1)/2$ two-factor interactions, estimates of all the three-factor and higher order effects. However, it is often the case that the three-factor and higher order effects can be assumed *a priori* to be zero, or at least to be small relative to the lower order effects. When this is true, only a fraction of the 2^k design need be employed. This discussion will be restricted to the one-half replicate designs, that is, the so-called 2^{k-1} fractional factorials. The general fractional factorial design is beyond the scope of this book, but we refer the interested reader to Box and Hunter (1961).

The 2^{k-1} fractional factorial designs may be constructed by first partitioning the 2^k factorial into two blocks of 2^{k-1} runs each, using the highest order interaction contrast. Each block is then a 2^{k-1} design. For example, to construct the 2^{4-1} design, one begins (as demonstrated in Table 17.5.1) by writing down the full 2^4 factorial design, and then partitioning the design into 2 blocks of eight runs each, the $x_1x_2x_3x_4$ four-factor interaction contrast vector being used.

TABLE 17.5.1

THE TWO 1/2 REPLICATES OF THE 2^4 FACTORIAL.

GENERATOR CONTRAST: $x_1x_2x_3x_4$

2^4 FACTORIAL					2^{4-1} (Design A)				2^{4-1} (Design B)			
x_1	x_2	x_3	x_4	$x_1x_2x_3x_4$	x_1	x_2	x_3	x_4	x_1	x_2	x_3	x_4
−	−	−	−	+	−	−	−	−	+	−	−	−
+	−	−	−	−	+	+	−	−	−	+	−	−
−	+	−	−	−	+	−	+	−	−	−	+	−
+	+	−	−	+	−	+	+	−	+	+	+	−
−	−	+	−	−	+	−	−	+	−	−	−	+
+	−	+	−	+	−	+	−	+	+	+	−	+
−	+	+	−	+	−	−	+	+	+	−	+	+
+	+	+	−	−	+	+	+	+	−	+	+	+
−	−	−	+	−	Generator: $+x_1x_2x_3x_4$				Generator: $-x_1x_2x_3x_4$			
+	−	−	+	+	Defining				Defining			
−	+	−	+	+	Relation $I + 1234$				Relation $I - 1234$			
+	+	−	+	−								
−	−	+	+	+								
+	−	+	+	−								
−	+	+	+	−								
+	+	+	+	+								

The 2^{4-1} design, called Design A in Table 17.5.1, consists of the eight runs of the 2^4 factorial which contain a plus sign in the four-factor interaction contrast vector $x_1x_2x_3x_4$, while Design B consists of the eight runs possessing a minus sign in this vector. Since we are using the $x_1x_2x_3x_4$ column in this way, the *generators* of these fractional factorials are said to be $+x_1x_2x_3x_4$ and $-x_1x_2x_3x_4$, respectively.

With only 2^{k-1} treatments it is obviously impossible to estimate all the $(2^k - 1)$ individual effects in the factorial model. However, 2^{k-1} orthogonal contrasts can be determined and it is important to identify the confounded factorial effects estimated by these statistics. The confounding pattern, or alias structure, is best explained by example. In Table 17.5.2, we see the 2^{4-1} fractional factorial design with *generator* $+x_1x_2x_3x_4$ along with a set of corresponding observations. The design has been listed in Yates' order with respect to variables x_1, x_2, and x_3. The design was run in random order.

TABLE 17.5.2

A 2^{4-1} FRACTIONAL FACTORIAL DESIGN

Design				Observations	Factorial Effect Contrast Coefficients			
x_1	x_2	x_3	x_4	y	{1}	{234}	{12}	{34}
−	−	−	−	8.4	−	−	+	+
+	−	−	+	15.7	+	+	−	−
−	+	−	+	12.6	−	−	−	−
+	+	−	−	15.4	+	+	+	+
−	−	+	+	5.2	−	−	+	+
+	−	+	−	6.7	+	+	−	−
−	+	+	−	4.1	−	−	−	−
+	+	+	+	11.3	+	+	+	+

To estimate the x_1 effect we have [see (17.1.7)], $\frac{1}{4}(\Sigma\, y\{1\}) = 4.70$. To estimate the $x_2x_3x_4$ effect, we find that the contrast coefficients {234} are identical to those already used to estimate the x_1 effect, that is $\{1\} = \{234\}$. Hence the x_1 and $x_2x_3x_4$ effects are confounded. In fact it can be shown that $E[\frac{1}{4}(\Sigma\, y\{1\})] = (x_1$ effect $+ x_2x_3x_4$ effect). Similarly the contrast coefficients for the x_1x_2 interaction effect are identical to those for the x_3x_4 interaction, and so we write $\{12\} = \{34\}$. Thus $E[\frac{1}{4}(\Sigma\, y\{12\})] = (x_1x_2$ effect $+ x_3x_4$ effect) and these two interactions are confounded. The reader may satisfy himself that, in this example, the four main effects are each confounded with a single three-factor interaction, and that the six two-factor interactions are confounded in pairs.

A convenient method for determining the confounding pattern for 2^{k-1} fractional factorials, and hence the expected value of the $2^{k-1} - 1$ orthogonal factorial contrasts, is provided by the design's *defining relation*. In the above example the design generator is $+x_1 x_2 x_3 x_4$, or the "word," $+\textbf{1234}$. The defining relation for the designs is then the "sentence" $\textbf{I} + \textbf{1234}$, where the symbol $\textbf{I}$ is called the identity. (In general, for the 2^{k-p} fractional factorials there will be p generators and the defining relation will be a sentence containing 2^p words.) Multiplying through the defining relation by, say, the symbol $\textbf{1}$ gives $\textbf{I 1} + \textbf{1}^2\textbf{234}$. We now adopt the rule that *any* symbol which appears to an even power converts to the identity $\textbf{I}$, thus giving $\textbf{1} + \textbf{I234}$ or simply, dropping the identity $\textbf{I}$, $\textbf{1} + \textbf{234}$. Similarly, multiplying the defining relation by $\textbf{12}$ gives $\textbf{12} + \textbf{34}$. If the generator of the design had been $-x_1 x_2 x_3 x_4$, the defining relation would be $\textbf{I} - \textbf{1234}$. The expected value of any factorial contrast, $\dfrac{1}{r2^{k-p-1}} \Sigma\, y\{ij \cdots k\}$, may be determined by multiplying the defining relation by the corresponding word symbol, $ij \cdots k$, that identifies the contrast. For the full 2^k factorial designs, the defining relation is simply $\textbf{I}$.

The analysis of a 2^{k-p} fractional factorial is accomplished by initially considering the data as having been provided by a 2^h full factorial design, where $h = k - p$ is some convenient subset of the k variables. The $(2^h - 1)$ factorial effects are then estimated using Yates' algorithm, or a table of contrast coefficients. Using only the h variables, each factorial effect is then labeled with its *naive* name or word. The expected value of each estimate is then determined by multiplying the defining relations of the design respectively by its naive word. The assumption is usually made that three-factor and higher order effects may be ignored. This assumption simplifies the confounding pattern, or alias structure.

> **Example 17.5(a).** A development laboratory is attempting to improve performance of a packaging machine. Five components (each consisting of a small metal arm of unique shape) have been redesigned, and the objective of the experiments is to determine whether changing one or more of the components will have a salutary effect on the response: the crease retention of the packaging paper. The one-half replicate of the 2^5 factorial design was employed where, in the design matrix, the minus sign was reserved for the standard component, and the plus sign for the redesigned component. The 2^{5-1} design with generator $-x_1 x_2 x_3 x_4 x_5$ was chosen since it contained the standard condition $(-, -, -, -, -)$. (Usually the fraction chosen is determined by flipping a coin.) The sixteen runs were performed in a random sequence. The data, arrayed for our convenience in Yates' order, on variables x_1, x_2, x_3, and x_4 are displayed in Table 17.5.3.

TABLE 17.5.3

2^{5-1} Fractional Factorial

Design Generator $-x_1x_2x_3x_4x_5$

Defining Relation I $- 12345$

x_1	x_2	x_3	x_4	x_5	y	Yates' Algorithm				Naive Estimates		Identification of Estimates
−	−	−	−	−	4.01	7.10	13.88	31.33	64.56	$\bar{y} =$	4.0350	
+	−	−	−	+	3.09	6.78	17.45	33.23	zero	$\hat{1} =$	zero	1 − 2345 → **1**
−	+	−	−	+	3.23	9.10	13.81	−.25	−1.02	$\hat{2} =$	−.1275	2 −1345 → **2**
+	+	−	−	−	3.55	8.35	19.42	.25	6.26	$\widehat{12} =$	.7825	12 − 345 → **12**
−	−	+	−	+	4.93	6.87	−.60	−1.07	9.18	$\hat{3} =$	1.1475	3 − 1245 → **3**
+	−	+	−	−	4.17	6.94	.35	.05	.78	$\widehat{13} =$	.0975	13 − 245 → **13**
−	+	+	−	−	3.62	9.72	.21	3.11	−.52	$\widehat{23} =$	−.0650	23 − 145 → **23**
+	+	+	−	+	4.73	9.70	.04	3.15	.68	$\widehat{123} =$	.0850	123 − 45 → **−45**
−	−	−	+	+	3.77	−.92	−.32	3.57	1.90	$\hat{4} =$	.2375	4 − 1235 → **4**
+	−	−	+	−	3.10	.32	−.75	5.61	.50	$\widehat{14} =$	.0625	14 − 235 → **14**
−	+	−	+	−	3.03	−.76	.07	.95	1.12	$\widehat{24} =$	.1400	24 − 135 → **24**
+	+	−	+	+	3.91	1.11	−.02	−.17	.04	$\widehat{124} =$	.0050	124 − 35 → **−35**
−	−	+	+	−	5.25	−.67	1.24	−.43	2.04	$\widehat{34} =$	.2550	34 − 125 → **34**
+	−	+	+	+	4.47	.88	1.87	−.09	−1.12	$\widehat{134} =$	−.1400	134 − 25 → **25**
−	+	+	+	+	4.44	−.78	1.55	.63	.34	$\widehat{234} =$	.0425	234 − 15 → **−15**
+	+	+	+	−	5.26	.82	1.60	.05	−.58	$\widehat{1234} =$	−.0725	1234 − 5 → **−5**

$\Sigma y = 64.56$

$\Sigma y^2 = 269.0512$ (Check: 4304.8192)

Identification of Effects	Estimates	Rank Order i	$(i - \frac{1}{2})/15$
3-1245	1.1475	15	.9667
12-345	.7825	14	.9000
34-125	.2550	13	.8333
4-1235	.2375	12	.7667
24-135	.1400	11	.7000
13-245	.0975	10	.6333
123-45	.0850	9	.5667
14-235	.0625	8	.5000
234-15	.0425	7	.4333
124-35	.0050	6	.3667
1-2345	zero	5	.3000
23-145	−.0650	4	.2333
1234-5	−.0725	3	.1667
2-1345	−.1275	2	.1000
134-25	−.1400	1	.0333

Fifteen orthogonal factorial contrasts can now be estimated from the sixteen observations. This is quickly accomplished using Yates' algorithm on four of the five factors, as illustrated for factors 1, 2, 3, and 4 in Table 17.5.3. The initial or naive identification for the contrasts is also listed. The expected value for each estimate is determined from the design defining relation $I - 12345$. Assuming now that three-factor and higher order interaction effects are zero, we obtain orthogonal estimates of the $k = 5$ main effects and of each of the $5(4)/2 = 10$ two-factor interaction effects. When using the defining relation care must be taken in affixing the proper sign to the estimated effects. For example, the estimated main effect of x_5 equals .0725, and the estimated x_1x_5 interaction effect is −.0425. The reader can check for himself that the contrasts estimating these effects are

x_5 effect:

$$\tfrac{1}{8}(-4.01 + 3.09 + 3.23 + \cdots + 4.47 + 4.44 - 5.26) = .0725$$

x_1x_5 effect:

$$\tfrac{1}{8}(4.01 + 3.09 - 3.23 - \cdots + 4.47 - 4.44 - 5.26) = -.0425.$$

The normal plot of the estimates of the effects, Figure 17.5.1, suggests that only the x_3 and x_1x_2 effects are clearly distinguishable from estimates that might reasonably be viewed as linear combinations of normal independent, mean zero variables. Assuming now that only factors 1, 2, and

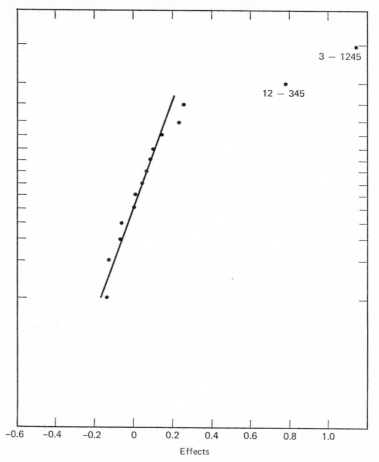

Figure 17.5.1. Normal plot of ordered estimates of effects (Table 17.5.3).

3 have effects upon the response, the 2^{5-1} fractional factorial design becomes a replicated 2^3 design in these variables.

The usefulness of the fractional factorial designs for screening is now apparent. Before the experiment, $k = 5$ candidate variables were thought possibly to have large effects. The experimental strategy has located a particular subset of three variables. There were, of course, $\binom{5}{3} = 10$ possible subsets of three. The 2^{5-1} design becomes a 2^3 factorial, replicated, in any selected subset of three variables. Similarly, if four of the original variables had been found to have important effects, the 2^{5-1} design would become a 2^4 factorial. There are, of course, five such possibilities. If two of the variables prove important, the design would have collapsed into one of the $\binom{5}{2} = 10$ possible 2^2 factorials, replicated four times. Finally if only one variable is important, each of the five main effects is separately estimable.

17.6 RESPONSE SURFACES

It is often necessary to "explore" an unknown response function $\eta = f(\xi_1, \xi_2, \ldots, \xi_k)$ over an experimental region defined by acceptable ranges of k controlled variables $\xi_1, \xi_2, \ldots, \xi_k$. If the function is continuous over the region, it may be usefully approximated by a first-order Taylor's series about $\boldsymbol{\xi}_0 = (\xi_{10}, \xi_{20}, \ldots, \xi_{k0})$, a selected point within the region. Thus we would have

$$(17.6.1) \quad \eta \simeq f(\xi_{10}, \xi_{20}, \ldots, \xi_{k0}) + (\xi_1 - \xi_{10}) \left.\frac{\partial f}{\partial \xi_1}\right|_{\xi_0}$$
$$+ (\xi_2 - \xi_{20}) \left.\frac{\partial f}{\partial \xi_2}\right|_{\xi_0} + \cdots + (\xi_k - \xi_{k0}) \left.\frac{\partial f}{\partial \xi_k}\right|_{\xi_0}$$

where each of the derivatives is evaluated at the point $\boldsymbol{\xi}_0$. The model may then be rewritten as

$$(17.6.2) \quad \eta \simeq \gamma_0 + \gamma_1 \xi_1 + \gamma_2 \xi_2 + \cdots + \gamma_k \xi_k = \gamma_0 + \sum_{i=1}^{k} \gamma_i \xi_i,$$

where

$$(17.6.3) \quad \gamma_0 = f(\xi_{10}, \xi_{20}, \ldots, \xi_{k0}) - \xi_{10} \left.\frac{\partial f}{\partial \xi_1}\right|_{\xi_0}$$
$$- \xi_{20} \left.\frac{\partial f}{\partial \xi_2}\right|_{\xi_0} - \cdots - \xi_{k0} \left.\frac{\partial f}{\partial \xi_k}\right|_{\xi_0}$$

with

$$(17.6.4) \quad \gamma_i = \left.\frac{\partial f}{\partial \xi_i}\right|_{\xi_0}$$

(these are called first-order coefficients.) The model of (17.6.1) may be viewed as a first-order approximating polynomial. If this first-order model should prove inappropriate to represent the response function, a second-order Taylor's series is often employed. The second-order polynomial approximation is:

$$(17.6.5) \quad \eta = \gamma_0 + \sum_{i=1}^{k} \gamma_i \xi_i + \sum_{i=1}^{k} \gamma_{ii} \xi_i^2 + \sum_{i<j} \gamma_{ij} \xi_i \xi_j$$

where γ_{ii} is the quadratic coefficient, and γ_{ij} the cross product or two-factor interaction coefficients between variable ξ_i and ξ_j. Together, the k quadratic and $k(k-1)/2$ cross product coefficients comprise the second-order portion of the polynomial model.

In discussing both the experimental designs, and the analysis of the data, one can replace the $u = 1, 2, \ldots, N$ settings of the variables ξ_{iu} by standardized variables x_{iu} defined by

$$(17.6.6) \qquad x_{iu} = \frac{\xi_{iu} - \xi_{i0}}{c_i},$$

where ξ_{i0} is the midpoint of the experimental region with respect to ξ_{iu}, and where c_i is some convenient scale factor chosen so that $\Sigma\, x_{iu}^2 = N$, approximately. The first-order model may now be written

$$(17.6.7) \quad \eta = \beta_0 + \beta_1 x_1 + \beta_2 x_2 + \cdots + \beta_k x_k = \beta_0 + \sum_i \beta_i x_i,$$

and the second-order model

$$(17.6.8) \qquad \eta = \beta_0 + \sum_i \beta_i x_i + \sum_i \beta_{ii} x_i^2 + \sum\sum_{i<j} \beta_{ij} x_i x_j,$$

where β_0 is a constant, β_i are the k first-order coefficients, β_{ii} are the k quadratic coefficients, and β_{ij} are the $k(k-1)/2$ cross product or interaction coefficients for the models written in terms of the x_i.

Given that observed responses y_u at fixed settings of the x_{iu} are normally and independently distributed random variables, with common variance σ^2, then the maximum likelihood estimates of the coefficients are those which minimize [see (15.7.4)]

$$(17.6.9) \qquad Q = \sum_{u=1}^{N} (y_u - \eta_u)^2$$

where, of course, $\eta_u = \beta_0 + \Sigma\,\beta_i x_{iu} + \Sigma\,\beta_{ii} x_{iu}^2 + \Sigma\Sigma\,\beta_{ij} x_{iu} x_{ju}$. Before discussing the least squares estimation procedure for these models, and some of the associated experimental designs, it is helpful to adopt the following notation:

$$(17.6.10) \quad \sum x_{iu} = [i], \qquad \sum x_{iu}^2 = [ii], \qquad \sum x_{iu} x_{ju} = [ij],$$
$$\sum x_{iu}^2 x_{ju} = [iij], \quad \text{etc.},$$

where in every case, the summation is over $u = 1, 2, \ldots, N$. The quantities in the square brackets when divided by N are termed the *moments* of the design.

In general, the $k + 1$ normal equations associated with the first-order model (17.6.7) are given by

$$Nb_0 + [1]b_1 + [2]b_2 + \cdots + [k]b_k = \sum y_u$$

$$[1]b_0 + [11]b_1 + [12]b_2 + \cdots + [1k]b_k = \sum x_{1u}y_u$$

$$[2]b_0 + [12]b_1 + [22]b_2 + \cdots + [2k]b_k = \sum x_{2u}y_u$$

(17.6.11)
$$\vdots \qquad \vdots \qquad \vdots \qquad\qquad \vdots \qquad \vdots$$

$$[k]b_0 + [1k]b_1 + [2k]b_2 + \cdots + [kk]b_k = \sum x_{ku}y_u.$$

If we now select the settings of the ith controlled variable ξ_{iu} so that $\xi_{i0} = \bar{\xi}_i$, where $\bar{\xi}_i = \sum \xi_{iu}/N$ is the average setting, then $[i] = 0$. Further if the settings ξ_{iu} and ξ_{ju} are symmetrically chosen, it is usually possible to have $[ij] = 0$ for all i and j. Thus, with a design that has these moments, namely, $[i] = 0$ and $[ij] = 0$, the solution for the $k + 1$ coefficients becomes simply

(17.6.12) $b_0 = \sum y_u/N$ and $b_i = \sum x_{iu}y_u/[ii]$, $i = 1, 2, \ldots, k$.

The fitted *first-order model* is then

(17.6.13) $\hat{y} = b_0 + b_1x_1 + b_2x_2 + \cdots + b_kx_k = b_0 + \sum_i b_ix_i$.

The variances of the coefficients are

$$V(b_0) = \sigma^2/N, \qquad V(b_i) = \sigma^2/[ii],$$

with all covariances equal to zero. The sum of squares of residuals $(y_u - \hat{y}_u)$, often called the sum of squares of deviations (SSD), equals

(17.6.14) $\text{SSD} = \sum (y_u - \hat{y}_u)^2$
$$= S(Y^2) - b_1 \sum x_{1u}y_u - b_2 \sum x_{2u}y_u - \cdots - b_k \sum x_{ku}y_u,$$

where $S(Y^2)$, the total corrected sum of squares, equals

$$\sum y_u^2 - b_0 \sum y_u = \sum y_u^2 - N\bar{y}^2.$$

Given that the model is appropriate, we have that $E(\text{SSD}) = (N - k - 1)\sigma^2$, and hence the estimate of σ^2 is given by $s^2 = (\text{SSD})/(N - k - 1)$ with $N - k - 1$ degrees of freedom. The associated analysis of variance is given in Table 17.6.1.

TABLE 17.6.1

ANALYSIS OF VARIANCE TABLE FOR FIRST-ORDER MODEL

IN k VARIABLES: $\eta = \beta_0 + \sum_{i=1}^{k} \beta_i x_i$

Source	Degrees of Freedom
Total corrected sum of squares $S(Y^2) = \sum y_u - b_0 \sum y_u$	$N - 1$
$SSb_1 = b_1 \sum x_{1u} y_u$	1
$SSb_2 = b_2 \sum x_{2u} y_u$	1
Regression sum of squares $\cdot$ $\cdot$ $\cdot$	$\cdot$ $\cdot$ $\cdot$
$SSb_k = b_k \sum x_{ku} y_u$	1

Residual sum of squares

$$\text{SSD} = S(Y^2) - \sum_{i=1}^{k} b_i \sum x_{iu} y_u \qquad N - k - 1 \qquad s^2 = (\text{SSD})/(N - k - 1).$$

Individual tests of the hypothesis H_i: $\beta_i = 0$ against the alternative H_i': $\beta_i \neq 0$ are provided by using the fact that the ratio

(17.6.15) $$F = b_i \sum x_{iu} y_u / s^2$$

is distributed as $F_{1,(N-k-1)}$ if H_i is true. The identical test may be performed using Student's t. We have that

(17.6.16) $$t_{N-k-1} = \frac{b_i - \beta}{\sqrt{s^2/[ii]}} = \frac{b_i}{\sqrt{s^2/[ii]}},$$

if H_i is true. From (17.6.12), we have then, that $t_{N-k-1}^2 = [ii] b_i^2/s^2 = b_i (\sum x_{iu} y_u)/s^2 = F_{1,N-k-1}$.

A test of the composite hypothesis that $\beta_i = 0$ for all k variables can be based on the fact that

(17.6.17) $$\frac{\sum_i b_i \sum_u x_{iu} y_u / k}{s^2} = F_{k,N-k-1},$$

if the hypothesis, H_0: all $\beta_i = 0$, is true. If the observed $F_{k,N-k-1}$ is greater than $F_{k,N-k-1;\alpha}$ we would reject the hypothesis at significance level α, etc.

When using polynomial functions it is important that the approximation to the unknown function be checked by performing various *lack of fit* tests. We may do one sort of checking by using the residuals $(y_u - \hat{y}_u)$. For example, the histogram of the residuals should appear normal in shape. If the model used is adequate, then no patterns or runs should be apparent amongst the residuals. Another important check on the adequacy of the model is available whenever portions of the experimental design have been replicated. In such cases, an estimate of σ^2 can be constructed in the usual manner at each replicated point, and these estimates pooled to give an estimate of σ^2, say s^2 based on ν degrees of freedom. It is therefore possible to partition the residual sum of squares into two portions, one due to intrinsic variability alone as represented by s^2, based on ν degrees of freedom, and a remainder representing the failure of the fitted responses to estimate the true response, that is a measure of $\sum \{E(\hat{y}_u) - E(y_u)\}^2$, based upon $N - k - 1 - \nu$ degrees of freedom. The quantity $\sum_u \{E(\hat{y}_u) - E(y_u)\}^2$ is called the lack of fit sum of squares, and if the model postulated is correct, has value zero. We usually tabulate the relevant sums of squares in a table such as Table 17.6.2.

TABLE 17.6.2

PARTITIONING THE RESIDUAL SUM OF SQUARES

Source	Sum of Squares	Degrees of Freedom	
Residual sum of squares	SSD	$N - k - 1$	
Lack of fit sum of squares	SSL = SSD − SSE	$N - k - 1 - \nu$	$\text{SSL}/(N - k - 1 - \nu)$
Error sum of squares	SSE	ν	$\text{SSE}/\nu = s^2, \quad E(s^2) = \sigma^2$

The sum of squares SSE may be found by computing the total within sum of squares, that is, computing the within sum of squares at each replicated point, and summing [see Example 17.6(a)]. Of course, SSD may be found, as usual, by taking $\text{SSD} = S(Y^2) - \sum_{i=1}^{k} b_i(\sum x_{iu}y_u)$. Now it is known that the distribution of the ratio of the lack of fit mean square, say with ℓ degrees of freedom, to the mean square for error, say with ν degrees of freedom, is *not* the $F_{\ell,\nu}$ distribution. In practice, the observed ratio is deemed significant at the α level if the observed ratio is greater than $2F_{\ell,\nu;\alpha}$. It turns out that the ratio of the mean squares

$$[\text{SSL}/(N - k - 1 - \nu)]/[\text{SSE}/\nu]$$

is approximately distributed as $F_{N-k-1-\nu,\nu}$. If the observed value of F is less than the critical value $F_{N-k-1-\nu,\nu;.05}$ the fitted model may be declared adequate to represent the response function. However, quite acceptable fitted models are found in practice which give a lack of fit F ratio *two* or *three times* larger than the critical F. Hence, the lack of fit F ratio should only be taken as a *signal* of nonadequacy.

Good experimental designs for fitting first order polynomial models are the simplex, the 2^k factorial, or the 2^{k-p} fractional factorials. (The simplex is a regular figure having $k + 1$ vertices in k-dimensional space. Examples are the vertices of the equilateral triangle, and the tetrahedron as illustrated in Table 17.6.3.)

TABLE 17.6.3
DESIGN MATRICES FOR $k = 2, 3$ USEFUL FOR
FITTING FIRST-ORDER MODELS

x_1	x_2	x_1	x_2	x_1	x_2	x_3	x_1	x_2	x_3
.87	−.50	−1	−1	−1	−1	1	−1	−1	−1
−.87	.50	1	−1	1	−1	−1	1	−1	−1
0	1.00	−1	1	−1	1	−1	−1	1	−1
0	0	1	1	1	1	1	1	1	−1
.	.	0	0	0	0	0	−1	−1	1
.	.	.	.	.	.	.	1	−1	1
.	.	.	.	.	.	.	−1	1	1
0	0	.	.	.	.	.	1	1	1
Equilateral		0	0	0	0	0	0	0	0
triangle with		Square (or 2^2		Tetrahedron			.	.	.
center points		factorial)		(or simplex,			.	.	.
		with center		or 2^{3-1} factorial)			.	.	.
		points		with center			0	0	0
				points			Cube (or 2^3		
							factorial)		
							with center		
							points		

Example 17.6(a). In a pilot plant an experimenter was interested in determining how the time and temperature conditions of a clave affected the buildup of an unwanted by-product in a chemical process. Theoretical explanations were available, but for the purposes at hand, it was simpler merely to explore the region of interest in time and temperature by a series of experiments, and to fit an approximating first-order mathematical model. To provide a measure of experimental error, the 2^2 factorial portion of the design was repeated, and the *center point* replicated four times. The entire sequence of twelve runs was performed in random order. The settings of the variables: time and temperature, the associated design matrix in the standardized variables x_1 and x_2, and the recorded responses are displayed in Table 17.6.4.

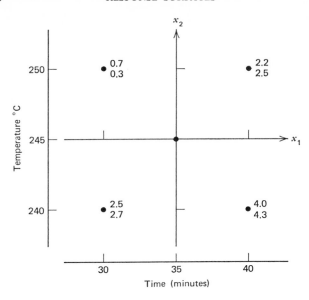

TABLE 17.6.4

| Variables | | | | |
Time (min)	Temperature °C	x_1	x_2	Response y
30	240	−1	−1	2.5
40	240	1	−1	4.0
30	250	−1	1	.7
40	250	1	1	2.2
35	245	0	0	2.5
35	245	0	0	2.3
30	240	−1	−1	2.7
40	240	1	−1	4.3
30	250	−1	1	.3
40	250	1	1	2.5
35	245	0	0	3.0
35	245	0	0	2.4

The first-order model is $\eta = \beta_0 + \beta_1 x_1 + \beta_2 x_2$, and the observations y are taken to be independent $N(\eta, \sigma^2)$. The design employed is a 2^2 factorial with center points, replicated. For this design we have

$$[1] = 0, \quad [2] = 0, \quad [11] = 8, \quad [22] = 8, \quad [12] = 0.$$

Thus, following (17.6.12) we have

$$b_0 = \sum y/N \qquad = 29.4/12 = 2.450$$

(17.6.18) $$b_1 = (\sum x_1 y)/[11] = 6.8/8 = .850$$

$$b_2 = (\sum x_2 y)/[22] = -7.8/8 = -.975,$$

giving the fitted model

(17.6.19) $$\hat{y} = 2.450 + .850x_1 - .975x_2.$$

The associated analysis of variance table is displayed in Table 17.6.5.

TABLE 17.6.5

ANALYSIS OF VARIANCE FOR A FITTED FIRST-ORDER MODEL

Source		SSq	Degrees of Freedom	Mean Squares	
$S(Y^2) = \sum y^2 - b_0 \sum y$		13.9700	11		
$SSb_1 =$	$b_1 \sum x_1 y$	5.7800	1	5.7800	$F_{1,9} = 88.92$
$SSb_2 =$	$b_2 \sum x_2 y$	7.6050	1	7.6050	$F_{1,9} = 117.00$
Residual (SSD)		.5850	9	.0650	
Lack of fit (SSL)		.1050	2	.0525	$F_{2,7} < 1$, nonsignificant
Error SSq (SSE)		.4800	7	$.0686 = s^2$	

To obtain the entry for the error sum of squares SSE, we first draw up Table 17.6.6, from Table 17.6.4.

TABLE 17.6.6

Treatments	$(-1, -1)$	$(-1, 1)$	$(1, -1)$	$(1, 1)$	$(0, 0)$
	2.5	.7	4.0	2.2	2.5, 3.0
	2.7	.3	4.3	2.5	2.3, 2.4
Sums	5.2	1.0	8.3	4.7	10.2
Means	2.6	.5	4.15	2.35	2.55

The total within sum of squares is easily obtained. For the four treatments with the paired observations we have for the within sum of squares $\sum d_i^2/2$ where d_i is the difference of the ith pair, $i = 1, 2, 3, 4$. For the single treatment with four observations we compute $\sum (y_i - \bar{y})^2$. Thus, the total within sum of squares is

$$[(2.5 - 2.7)^2/2 + (.7 - .3)^2/2 + \cdots + (2.2 - 2.5)^2/2]$$
$$+ [(2.5 - 2.55)^2 + (3.0 - 2.55)^2 + \cdots + (2.4 - 2.55)^2] = .4800.$$

Of course, we may also obtain the error sum of squares quickly, using the above data, as follows. The reader may recall that a table such as the above results when considering a one-way analysis of variance situation. The model would be $y_{ij} = \eta + \tau_j + \epsilon_{ij}$ [see Table 16.2.2 and relations (16.2.18)].

Here, the treatment effects $\tau_1, \ldots, \tau_5$ are identified with the design settings $(-1, -1)$, $(-1, 1)$, $(1, -1)$, $(1, 1)$, $(0, 0)$, respectively. The relations (16.2.18) lead to the Table 17.6.7.

TABLE 17.6.7

Total SSq	$S(Y^2) = \Sigma\, y^2 - N\bar{y}^2$	13.9700	11
Treatment SSq	$\Sigma\, \dfrac{T_j^2}{n_j} - \dfrac{(\Sigma\, T_j)^2}{12}$	13.4900	4
Error SSq	(by subtraction)	0.4800	7

$$\left(\text{Treatment SSq} = \frac{(5.2)^2}{2} + \frac{(1.0)^2}{2} + \frac{(8.3)^2}{2} + \frac{(4.7)^2}{2} + \frac{(10.2)^2}{4} - \frac{(29.4)^2}{12}\right.$$

$$\left. = 13.4900\right)$$

As a partial check of the two analysis of variance tables 17.6.5 and 17.6.7 one notes that the treatment sum of squares with four degrees of freedom equals the total of the sum of squares due to b_1, b_2 and the lack of fit sum of squares.

A test of the hypothesis that the model is adequate to represent the unknown response function is given by the ratio of mean squares $F_{2.7} = 0.0525/0.0686 = 0.765$, which is not statistically significant. This information (coupled with a plot of the residuals $y_u - \hat{y}_u$) leads to the conclusion that the model appears to be adequate. The residual mean square is thus used to provide the estimate of variance, $s^2 = 0.0650$ with $\nu = 9$ degrees of freedom. The test of hypothesis that $\beta_1 = 0$ and that $\beta_2 = 0$ both produce very large observed $F_{1.9}$ values and the two hypotheses are rejected. The individual 95% confidence interval limits are, respectively,

$$b_1 \pm t_{9;.025}\sqrt{s^2/[11]} = \quad .850 \pm 2.26\sqrt{.0650/8} = \quad .850 \pm .090$$

$$b_2 \pm t_{9;.025}\sqrt{s^2/[22]} = -.975 \pm 2.26\sqrt{.0650/8} = -.975 \pm .090.$$

The fitted equation may now be employed to map the unknown response function over the experimental region. For example, the *contour* for the predicted response $\hat{y} = 3.0$ is given by substituting in (17.6.19) to give $3.000 = 2.450 + .850x_1 - .975x_2$, the equation of a straight line in the coordinate system of x_1 and x_2. The contours for $\hat{y} = 1, 2, 3,$ and 4 are

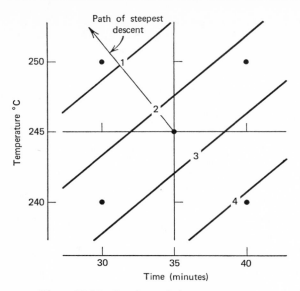

Figure 17.6.1. Contours of planar response.

plotted in Figure 17.6.1. These are the contours of a plane surface. Progress to a lower response (the response y is an unwanted by-product) can be most quickly explored by performing experiments along the path of steepest descent, that is, along a path perpendicular to the contour lines originating at the center of the experimental region. In Figure 17.6.1 we show a path of steepest ascent originating at the center of the experimental region.

17.7 FITTING A SECOND-ORDER MODEL

In order to fit the second-order model in k variables,

$$(17.7.1) \qquad \eta = \beta_0 + \sum_{i=1}^{k} \beta_i x_i + \sum_{i=1}^{k} \beta_{ii} x_i^2 + \sum_{i<j}^{k} \beta_{ij} x_i x_j,$$

a total of $\binom{k+2}{2}$ coefficients must be estimated. Further, in order to estimate quadratic coefficients, a minimum of three levels of each of the variable ξ_i must be used. This would seem to suggest that the 3^k factorial designs would be useful for securing data for the purpose of second-order model fitting. However, many other appropriate designs exist, with valuable properties, requiring fewer experimental points.

Some of the most frequently used designs for $k = 2$ and 3 variables are displayed in Figure 17.7.1.

k = 2 variables
The 3² factorial

x_1	x_2
−	−
−	0
−	+
0	−
0	0
0	+
+	−
+	0
+	+

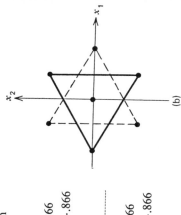

(a)

The Hexagon Design

	x_1	x_2
Block I	1.00	0
	−.50	$\sqrt{.75} = .866$
	−.50	$-\sqrt{.75} = -.866$
	0	0
	0	0
Block II	−1.00	0
	.50	$\sqrt{.75} = .866$
	.50	$-\sqrt{.75} = -.866$
	0	0
	0	0

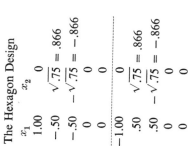

(b)

The Octagon Design

	x_1	x_2
Block I	−1	−1
	1	−1
	−1	1
	1	1
	0	0
	0	0
Block II	$\sqrt{2}$	0
	$-\sqrt{2}$	0
	0	$\sqrt{2}$
	0	$-\sqrt{2}$
	0	0
	0	0

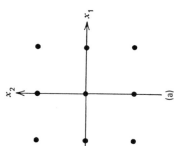

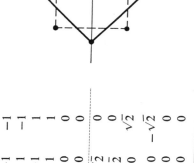

(c)

Figure 17.7.1. Second-order designs for k = 2 variables.

The Central Composite Design

	x_1	x_2	x_3
Block I	-1	-1	1
	1	-1	-1
	-1	1	-1
	1	1	1
	0	0	0
	0	0	0
Block II	-1	-1	-1
	1	-1	1
	-1	1	1
	1	1	-1
	0	0	0
	0	0	0
Block III	1.63	0	0
	-1.63	0	0
	0	-1.63	0
	0	1.63	0
	0	0	-1.63
	0	0	1.63
	0	0	0
	0	0	0

The Cuboctahedron Design

x_1	x_2	x_3
-1	-1	0
1	-1	0
-1	1	0
1	1	0
-1	0	-1
1	0	-1
-1	0	1
1	0	1
0	-1	-1
0	1	-1
0	-1	1
0	1	1
0	0	0
.	.	.
.	.	.
.	.	.
0	0	0

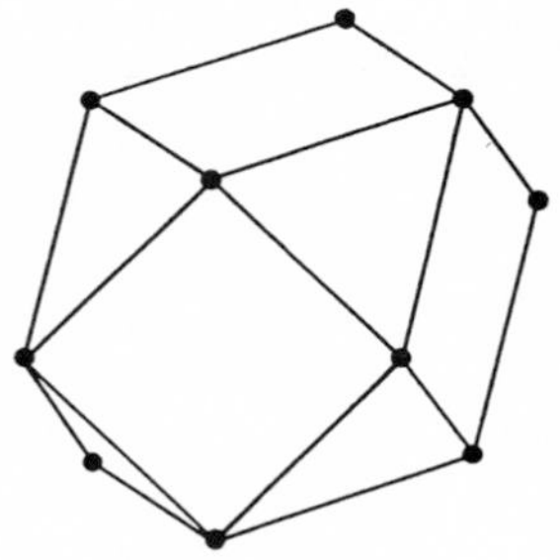

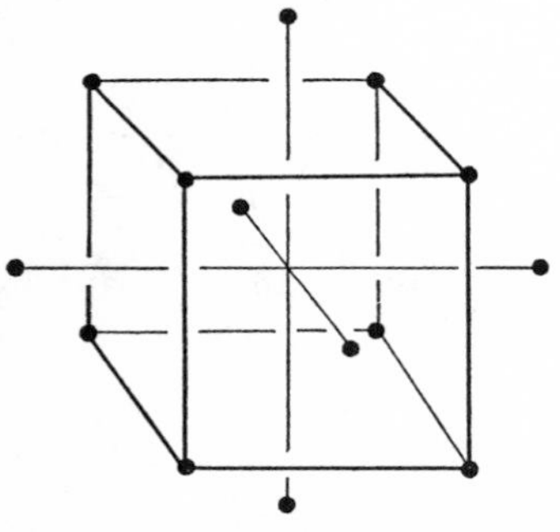

Figure 17.7.2. Second-order designs for $k = 3$ variables.

For $k = 2$ the second-order model is $\eta = \beta_0 + \beta_1 x_1 + \beta_2 x_2 + \beta_{11} x_1^2 + \beta_{22} x_2^2 + \beta_{12} x_1 x_2$. The normal equations associated with this model are:

(17.7.1)
$$N b_0 + [1] b_1 + [2] b_2 + [11] b_{11} + [22] b_{22} + [12] b_{12} = \sum y_u$$
$$[1] b_0 + [11] b_1 + [12] b_2 + [111] b_{11} + [122] b_{22} + [112] b_{12} = \sum x_{1u} y_u$$
$$[2] b_0 + [12] b_1 + [22] b_2 + [112] b_{11} + [222] b_{22} + [122] b_{12} = \sum x_{2u} y_u$$

$[11]b_0 + [111]b_1 + [112]b_2 + [1111]b_{11} + [1122]b_{22} + [1112]b_{12} = \sum x_{1u}^2 y_u$

$[22]b_0 + [122]b_1 + [222]b_2 + [1122]b_{11} + [2222]b_{22} + [1222]b_{12} = \sum x_{2u}^2 y_u$

$[12]b_0 + [112]b_1 + [122]b_2 + [1112]b_{11} + [1222]b_{22} + [1122]b_{12} = \sum x_{1u} x_{2u} y_u.$

The generalization of these normal equations for $k > 2$ should be obvious. A characteristic of the symmetrical designs displayed in Figure 17.7.1 and 17.7.2 is that the mixed second moments $[ij]$, all the odd moments $[i]$, $[iii]$, $[iij]$, and all the fourth-order moments of the form $[iiij]$ are all zero, leading to a simplification of the normal equations. For these designs we have directly

$$(17.7.2) \quad b_i = \sum x_{iu} y_u / [ii] \quad \text{and} \quad b_{ij} = \sum x_{iu} x_{ju} y_u / [iijj], \qquad i \neq j \neq 0.$$

The k second-order coefficients b_{ii}, along with b_0 are readily obtained by ordinary algebra.

The analysis of variance for this case is displayed in Table 17.7.1.

TABLE 17.7.1

ANALYSIS OF VARIANCE TABLE FOR A SECOND-ORDER MODEL
IN k VARIABLES

$$\eta = \beta_0 + \sum_{i=1}^{k} \beta_i x_i + \sum_{i=1}^{k} \beta_{ii} x_i^2 + \sum_{i<j}^{k} \beta_{ij} x_i x_j$$

Total corrected SSq	$= \Sigma y^2 - (\Sigma y)^2/N = S(Y^2)$	$N - 1$
First-order coefficients	$SSq(b_i) = \sum_i^k b_i \left(\sum_{u=1}^{N} x_{iu} y_u \right)$	k
Second-order coefficients	$SSq(b_{ii}) = \sum_{i=1}^{k} b_{ii} \left(\sum_{u=1}^{N} x_{iu}^2 y_u \right) + b_0 \Sigma y - (\Sigma y)^2/N$	k
	$SSq(b_{ij}) = \sum_{i<j}^{k} b_{ij} \left(\sum_{u=1}^{N} x_{iu} x_{ju} y_u \right)$	$k(k-1)/2$

Residual SSq	$SSD = S(Y^2) - SSq(b_i) - SSq(b_{ii}) - SSq(b_{ij})$	$N - \binom{k+2}{2}$
Lack of fit SSq	$= SSD - SSE$	$N - \binom{k+2}{2} - \nu$
Error SSq	$= SSE$	ν
	$s_r^2 = SSD \Big/ \left[N - \binom{k+2}{2} \right]$	$s_e^2 = SSE/\nu$
	$= $ estimate of σ^2 based on residual sum of squares.	$= $ estimate of σ^2 based on error sum of squares.

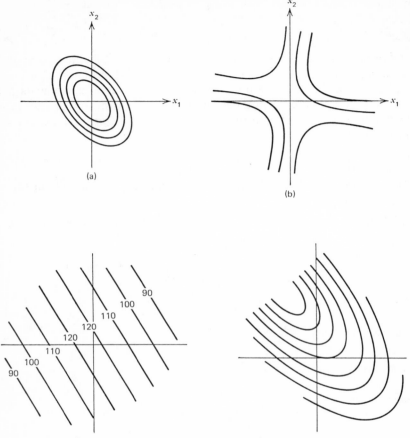

Figure 17.7.3. Illustrative contours provided by fitted second-order models.   (a) Concentric elliptical contours.   (b) Concentric hyperbolic contours.

Whenever replicated runs can be used to provide an error sum of squares with v degrees of freedom, the residual sum of squares SSD may be partitioned, as in Table 17.7.1 above, into two parts:

$$SSD = SSE + \text{lack of fit sum of squares}.$$

The value of SSE may be computed in two alternative ways—see Section 17.6. The lack of fit and error mean squares provides a measure of the adequacy of the fitted model.

The fitted model is

(17.7.3) $\hat{y} = b_0 + b_1 x_1 + b_2 x_2 + b_{11} x_1^2 + b_{22} x_2^2 + b_{12} x_1 x_2.$

If the model is accepted as adequate to represent the unknown response function, (17.7.3) can be used to provide an estimated, that is approximate, map of the response over the experimental region. The contours of the fitted surface are obtained by setting $\hat{y}$ equal to selected specific values of the response and plotting the resulting second degree equations in the coordinate system of x_1 and x_2. The contours will be concentric ellipses, or hyperbolas, or even straight lines, as illustrated in Figure 17.7.3.

> **Example 17.7(a).** In a study to determine the optimum conditions for the growth of large crystals of ZnS of great purity, two factors were varied: the temperature of the melt, and the rate of withdrawal of the crucible in which the crystal was grown. The experimenters began their investigation by employing a replicated simplex design with repeated center points. The factor settings, the settings of the experimental design variables, and a recorded response (here coded) are displayed in Table 17.7.2.

TABLE 17.7.2

Factor Settings		Design Variables		Response	Computations
Temperature °C	Rate (inches/day)	x_1	x_2	y	
1920	1.00	1.00	0	7.2, 6.9	$N = 10, \quad \Sigma y = 101.5$
1890	1.05	−.50	.866	9.3, 9.6	$\Sigma y^2 = 1065.73$
1890	.95	−.50	−.866	10.4, 9.8	$\Sigma x_1 y = -5.4500$
1900	1.00	0	0	12.3, 11.7	$\Sigma x_2 y = -1.1258$
				12.2, 12.1	$[11] = [22] = 3.0$

The object of the above experimental program was to see if a plane would be an acceptable approximation to the response function, so that a *path of steepest ascent*, determined in the space of the factors, would lead to a region of higher responses. The data of Example 17.7(a) are plotted in Figure 17.7.4 and it is obvious that a plane would be inadequate to represent the response function. The observed response at the center of the experimental region appears to be larger than the average response at the peripheral design points, an immediate signal that the response surface is curved. The hypothesis that the surface is nonplanar can be verified by fitting the first-order model and then testing to see whether the fitted model is adequate to represent the data. Fitting the model $\eta = \beta_0 + \beta_1 x_1 + \beta_2 x_2$ to these data gives

$$b_0 = \frac{\Sigma y}{N} = \frac{101.5}{12} = 8.4584; \qquad b_1 = \frac{\Sigma x_1 y}{[11]} = \frac{-5.45}{3.0} = -1.8167;$$

$$b_2 = \frac{\Sigma x_2 y}{[22]} = \frac{-1.1258}{3.0} = -0.3753,$$

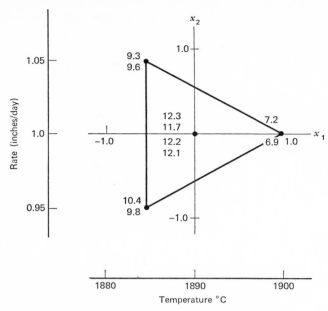

Figure 17.7.4. Simplex design for fitting a first-order model.

so that $\hat{y} = 8.46 - 1.82x_1 - 0.38x_2$. The corresponding analysis of variance is displayed in Table 17.7.3.

TABLE 17.7.3

$S(Y^2) = \Sigma y^2 - b_0 \Sigma y$		35.5050	9
$SSb_1 =$	$b_1 \Sigma x_1 y$	9.9010	1
$SSb_2 =$	$b_2 \Sigma x_2 y$	.4225	1

Residual SSD		25.1815	7	
Lack of fit		24.7040	1	24.7040
Error SSq		.4775	6	$.0796 = s^2$

Although the ratio of the lack of fit mean square to the error mean square is only approximately distributed as a $F_{1,6}$ variable, it is obvious that the observed ratio $F = \dfrac{24.7040}{.0796} = 310$ is highly unusual.

An alternative test, and for this design an exactly equivalent test, of the hypothesis that no curvature exists is provided by comparing the observations at the center of the design against those on the periphery of the design, by constructing a relevant contrast.

Now recall from Section 1 that a contrast is a linear combination of observations, with the constants in the linear combinations summing to zero. Denote the observations taken at $(0, 0)$, $(1, 0)$, $(-.05, .866)$, $(-.05, -.866)$ by y_{0j}, y_{1j}, y_{2j}, y_{3j}, respectively. To compare the observations taken at $(0, 0)$, the center of the design, with the observations taken at the periphery, namely, at $(1, 0)$, $(-.05, .866)$, $(-.05, -.866)$, we will use the contrast

$$3[y_{01} + y_{02} + y_{03} + y_{04}] - 2[y_{11} + y_{12}] - 2[y_{21} + y_{22}] - 2[y_{31} + y_{32}].$$

(Note that the constants in this contrast do sum to zero.) Alternatively we may write this as

$$3(4)\bar{y}_0 - 2(2)\bar{y}_1 - 2(2)\bar{y}_2 - 2(2)\bar{y}_3,$$

which is the form $\sum_{j=0}^{3} c_j d_j \bar{y}_j$, with d_j the number of observations at each treatment. That is,

$$c_0 = 3 \quad c_1 = -2 \quad c_2 = -2 \quad c_3 = -2$$
$$d_0 = 4 \quad d_1 = 2 \quad d_2 = 2 \quad d_3 = 2,$$

and we note, again, that $\sum_{j=0}^{3} c_j d_j = 0$. Inserting the actual observations from Table 17.7.2, or the information below, we find that $\sum c_j d_j \bar{y}_j = 38.50$.

	$\bar{y}_0$	$\bar{y}_1$	$\bar{y}_2$	$\bar{y}_3$
	12.075	7.050	9.450	10.100; $\sum c_j(d_j\bar{y}_j) = 38.50$
$d_j(c_j)$	4(3)	2(−2)	2(−2)	2(−2)

The corresponding single degree of freedom sum of squares is

$$(17.7.4) \quad [\sum c_j(d_j\bar{y}_j)]^2 / \sum d_j c_j^2 = (38.50)^2/[4(9) + 2(4) + 2(4) + 2(4)]$$
$$= (38.50)^2/60 = 24.7042,$$

which, except for rounding errors, equals the lack of fit sum of squares. Note that if $d_j = r$ for all j, then the left-hand side of (17.7.4) reduces to (17.1.4). The hypothesis that the contrast effect is zero, that is, that no curvature exists, is thus rejected, for the observed $F_{1,6} = 24.7042/.0796 = 310$, is obviously significantly large.

On review of these data the experimenters decided to form a hexagon design by adding a second replicated simplex with two centerpoints, as illustrated in Figure 17.7.5. The factor settings, levels of the design variables, and observed responses are displayed in Table 17.7.4.

TABLE 17.7.4

Factor Settings		Design Variables		Response	Computations
Temperature °C	Rate (inches/day)	x_1	x_2	y	
1920	1.00	1.000	0	7.2, 6.9	$N = 20$, $\Sigma y = 200.8$
1890	1.05	-.500	.866	9.3, 9.6	$\Sigma y^2 = 2118.48$
1890	.95	-.500	-.866	10.4, 9.8	$\Sigma x_1 y = -3.5000$, $\Sigma x_2 y = -10.5652$
1900	1.00	0	0	12.3, 11.7	$\Sigma x_1^2 y = 48.1000$ $\Sigma x_2^2 y = 55.5000$
1900	1.00	0	0	12.2, 12.1	$\Sigma x_1 x_2 y = -4.1568$
1880	1.00	-1.000	0	7.7, 7.8	Design moments
1910	1.05	.500	.866	6.2, 5.8	$[11] = [22] = 6.0$
1910	.95	.500	-.866	11.3, 11.6	$[1111] = [2222] = 4.50$
1900	1.00	0	0	11.8, 12.4	$[1122] = 1.5$
1900	1.00	0	0	12.7, 12.0	All other moments zero.

Treatments Totals	(0,0) 97.2	(1,0) 14.1	(-.5, .866) 18.9	(-.5, -.866) 21.2	(-1, 0) 15.1	(.5, .866) 12.0	(.5, -.866) 22.9

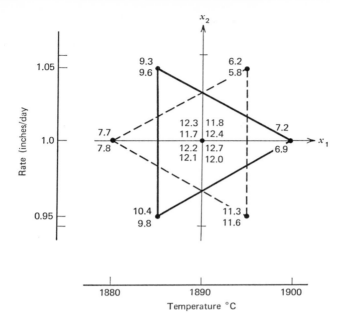

Figure 17.7.5. Hexagon design in two blocks for fitting a second-order model.

The proposed second-order model is

$$(17.7.5) \quad \eta = \beta_0 + \beta_1 x_1 + \beta_2 x_2 + \beta_{11} x_1^2 + \beta_{22} x_2^2 + \beta_{12} x_1 x_2.$$

The corresponding normal equations are

$$
\begin{array}{llll}
20b_0 & & + 6.00b_{11} + 6.00b_{22} & = \ \ \ 200.8000 \\
& + 6.0b_1 & & = \ \ \ {-3.5000} \\
(17.7.6) & \quad + 6.0b_2 & & = \ \ \ {-10.5652} \\
6.00b_0 & & + 4.50b_{11} + 1.50b_{22} & = \ \ \ 48.1000 \\
6.00b_0 & & + 1.50b_{11} + 4.50b_{22} & = \ \ \ 55.5000 \\
& & + 1.50b_{12} & = \ \ \ {-4.1568}
\end{array}
$$

The estimates b_1, b_2, and b_{12} are obtained directly:

$$(17.7.7) \qquad b_1 = \frac{-3.5000}{6.0} = -0.5833; \qquad b_2 = \frac{-10.5652}{6.0} = -1.7609;$$

$$b_{12} = \frac{-4.1568}{1.50} = -2.7712.$$

The remaining normal equations are

$$20.0b_0 + 6.0b_{11} + 6.0b_{22} = 200.8000$$

(17.7.8) $\qquad 6.0b_0 + 4.5b_{11} + 1.5b_{22} = \quad 48.1000$

$$6.0b_0 + 1.5b_{11} + 4.5b_{22} = \quad 55.5000,$$

which, on solving, gives

(17.7.9) $\quad b_0 = 12.1500, \qquad b_{11} = -4.7500, \qquad b_{22} = -2.2833.$

(All computations have been carried out to extra decimal places to reduce the role of rounding errors.) The fitted model is then

(17.7.10) $\quad \hat{y} = 12.15 - 0.58x_1 - 1.76x_2 - 4.75x_1^2 - 2.28x_2^2 - 2.77x_1x_2.$

The associated analysis of variance is displayed in Table 17.7.5.

TABLE 17.7.5

Source	SSq	Degrees of Freedom	Mean Square	Ratio
$S(Y^2) = \Sigma y^2 - (\Sigma y)^2/N =$	102.4480	19		
First-order coefficients $\begin{cases} b_1 \Sigma x_1 y = \;\;2.0417 \\ b_2 \Sigma x_2 y = 18.6039 \end{cases}$	20.6456	2		
Second-order coefficients $\begin{cases} b_{11}, b_{22}: \; 68.4898 \\ b_{12} \Sigma x_1 x_2 y = 11.5193 \end{cases}$	80.0091	3	26.6697	$F_{3,14} = 208$
Residual SSq $= \Sigma (y - \hat{y})^2$	1.7883	14	0.1277	
Lack of fit SSq	0.6483	1	0.6483	$F_{1,13} = 7.40$
Error SSq	1.1400	13	$0.0876 = s^2$	

Using Table 17.7.1 we have

Sum of Squares due quadratic effects $b_{11},\ b_{22} = -4.7500(48.1000)$
$$- 2.2833(55.5000) + 12.15(200.8000) - 2016.0320 = 68.4898;$$

error sum of squares

$$= S(Y^2) - \text{treatment SSq} = 102.4480$$

$$- \left[\frac{(97.2)^2}{8} + \frac{(14.1)^2}{2} + \frac{(18.9)^2}{2} + \cdots + \frac{(22.9)^2}{2} - 2016.0320 \right]$$

$$= 102.4480 - 101.3080 = 1.1400.$$

The ratio of the lack of fit mean square to the error mean square is less than twice the critical value of $F_{1,13;.05} = 4.67$ and the fitted second-order model is declared adequate to represent the unknown function. We recall now that the lack of fit for F should, usually, be two or three times the critical value before the model is declared completely inadequate to represent the data. Here the decision is taken to continue with the second-order polynomial model, recognizing that this fitted model may not be the best model that can be proposed but, for the purposes of empirical approximation, it can be useful.

Using the residual sum of squares and degrees of freedom, the new estimate of the variance is $s^2 = 1.7883/14 = .1277$ with fourteen degrees of freedom. A test of the hypothesis that all second-order contributions are zero is provided by the test $F_{3,14} = 26.6697/.1277 = 208$. Since Prob $\{F_{3,14} \geq 3.34\} = .05$, the hypothesis is rejected.

The fitted second-order model can now be used to determine the approximate contours of the response function. Thus, setting $\hat{y} = 10$ into (17.7.10) gives

$$10 = 12.15 - 0.58x_1 - 1.76x_2 - 4.75x_1^2 - 2.28x_2^2 - 2.77x_1x_2,$$

the equation of an ellipse. The contours of the estimated response are shown in Figure 17.7.6.

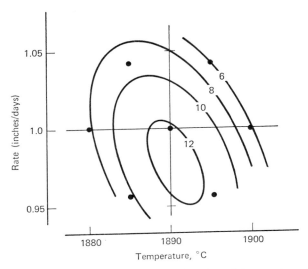

Figure 17.7.6. Contours of crystal purity as a function of temperature and rate of growth.

17.8   RESPONSE SURFACE METHODS FOR $k > 2$

When the response η is a function of two variables ξ_1, ξ_2, first and second-order polynomial models can provide useful approximating maps of the function over the studied ranges of the variables. Such maps are provided by plotting the contour lines obtained from the fitted model. When there are $k = 3$ variables ξ_1, ξ_2, ξ_3, useful maps of η can be constructed using contour *surfaces*.

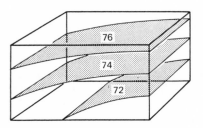

Figure 17.8.1. Contour surfaces of temperature in a room.

An everyday realization of contour surfaces is provided by first considering all the points located within the space of a room, each point identified by three coordinates. At each point let the response be the recorded temperature. In most rooms where the air has not been disturbed, high temperature readings will be found near the ceiling, and low temperature readings along the floor. The result may be viewed schematically, as in Figure 17.8.1, where contour *surfaces* are used to display the different observed temperatures. The temperature at any chosen point in the room can be quickly determined by interpolation between the adjacent contour surfaces. Figure 17.8.1 is a display of a four-dimensional system comprised of the temperature and the three spatial dimensions.

The first-order model

(17.8.1) $$\eta = \beta_0 + \beta_1 x_1 + \beta_2 x_2 + \beta_3 x_3$$

may be used to describe the contours of a planar response η in the three space of x_1, x_2, and x_3. The fitted model $\hat{y} = b_0 + \sum_{i=1}^{3} b_i x_i$ provides, for each value of $\hat{y}$, the equation of a plane in the space of x_1, x_2, x_3, as illustrated in Figure 17.8.2. Here the contours are planar surfaces, analogous to the contour straight lines used to map a plane in the two space of ξ_1, ξ_2, as illustrated earlier in Figure 17.6.1. Useful experimental designs for fitting the first-order model are the simplex with center point, or the 2^3 factorial with center point. With the simplex design a measure of

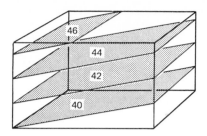

Figure 17.8.2. The contours of a hyper*plane* determined from a first-order polynomial model.

lack of fit of the planar model is provided by the contrast comparing the average response $\bar{y}_0$ at the center against the overall average response. If this contrast is significantly large, the curvature of the response surface is evident. For the 2^3 factorial with center point additional degrees of freedom for lack of fit are provided by the estimated interactions.

The second-order model

$$(17.8.2) \quad \eta = \beta_0 + \beta_1 x_1 + \beta_2 x_2 + \beta_3 x_3 + \beta_{11} x_1^2 + \beta_{22} x_2^2 + \beta_{33} x_3^2$$
$$+ \beta_{12} x_1 x_2 + \beta_{13} x_1 x_3 + \beta_{23} x_2 x_3$$

may be used to determine the contours of a nonplanar response η in the x_1, x_2, x_3 space. For each value for y, the fitted model $\hat{y} = b_0 + \Sigma b_i x_i + \Sigma b_{ii} x_i^2 + \sum_{i<j} \sum b_{ij} x_i x_j$ provides a second-order surface in the form of a

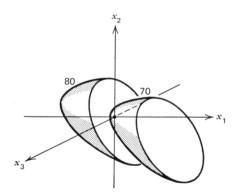

Figure 17.8.3. The contour surfaces of a nonplanar surface obtained from a second-order polynomial model.

sphere, ellipsoid, paraboloid, hyperboloid of one or two sheets, or cylinder. The actual form of the contour surface will, obviously, depend on the signs and magnitude of the various second-order coefficients. An example is shown in Figure 17.8.3 where, over the region of interest to the experimenter, the contour surfaces of the response are segments of concentric ellipsoids.

Useful experimental designs for estimating all the coefficients in a second-order model for $k = 3$ are the central composite design (which may be put together sequentially) and the cuboctahedron with center point. Both designs are illustrated in Figure 17.7.2. Of course, the 3^3 factorial design may also be employed. However, this design requires many more experiments than either of the central composite or cuboctahedron, and has other minor disadvantages. In general the 3^k and larger (4^k, 5^k, etc.) factorial designs can almost always be replaced by designs with both fewer experiments and additional advantageous properties.

For $k \geq 3$, the corresponding first and second-order models are

$$(17.8.3) \qquad \eta = \beta_0 + \sum_i^k \beta_i x_i$$

and

$$(17.8.4) \qquad \eta = \beta_0 + \sum_i^k \beta_i x_i + \sum_i^k \beta_{ii} x_i^2 + \sum\sum_{i<j}^k \beta_{ij} x_i x_j.$$

The experimental designs used for fitting these models are the multivariate extensions of the simplex, the 2^k and 2^{k-p} fractional designs with center points, and, for the second-order models, multivariate extensions of the central composite. Another class of second-order designs are the Box-Behnken designs, which are extensions of the three-level cuboctahedron design to higher dimensionality. Many of the Box-Behnken designs find their origins in the design matrices for the balanced incomplete block designs.

Once a second-order model has been fitted, the identification of the fitted contour surfaces and the interpretation of the fitted response function is enhanced by a *canonical* analysis of the fitted equation. Every second-order polynomial model of the form (17.8.4) can, by a proper location and rotation of the coordinate system, be reexpressed in its canonical form,

$$(17.8.5) \qquad \hat{y} = B_0 + \sum B_{ii} X_i^2,$$

where the coefficients B_0 and B_{ii} are certain linear combinations of the fitted coefficients, and the $X_i = \sum a_i x_i$ are the coordinate axes of the new system. The details for performing a canonical analysis of a second-order polynomial function can be found in text books on geometry.

Example 17.8(a). To study the effects of three variables on the cutting tool life: ξ_1, cutting tool speed (feet per minute); ξ_2, the feed rate (inches per revolution), and ξ_3, the depth of cut (inches), a central composite design was employed. This study is fully reported in Wu (1964). The settings of the three controlled variables ξ_1, ξ_2, ξ_3, the equivalent settings of the design variables x_1, x_2, and x_3, and the observed response y (measured tool wear) are listed in Table 17.8.1. It was decided to fit a second-order model to the response function. The design employed was a central composite design consisting of a 2^3 factorial with the center point replicated four times, plus a twice-replicated *star* design.

TABLE 17.8.1

THE DESIGN MATRIX, AND RECORDED RESPONSES FOR
A $k = 3$, CENTRAL COMPOSITE DESIGN

ξ_1 Speed: fpm	ξ_2 Feed: ipm	ξ_3 Depth: in	x_1	x_2	x_3	Tool Life $y = \ln$ (min)
330	.010	.049	−1	−1	−1	5.08
700	.010	.049	1	−1	−1	3.61
330	.022	.049	−1	1	−1	5.11
700	.022	.049	1	1	−1	3.30
330	.010	.100	−1	−1	1	5.15
700	.010	.100	1	−1	1	3.56
330	.022	.100	−1	1	1	4.79
700	.022	.100	1	1	1	2.89
480	.015	.070	0	0	0	4.19
480	.015	.070	0	0	0	4.42
480	.015	.070	0	0	0	4.26
480	.015	.070	0	0	0	4.41
226	.015	.070	−2	0	0	5.48
1020	.015	.070	2	0	0	2.64
480	.0072	.070	0	−2	0	4.70
480	.034	.070	0	2	0	3.99
480	.015	.034	0	0	−2	4.60
480	.015	.143	0	0	2	4.25
226	.015	.070	−2	0	0	5.41
1020	.015	.070	2	0	0	2.71
480	.0072	.070	0	−2	0	4.53
480	.034	.070	0	2	0	3.74
480	.015	.034	0	0	−2	4.66
480	.015	.143	0	0	2	4.17

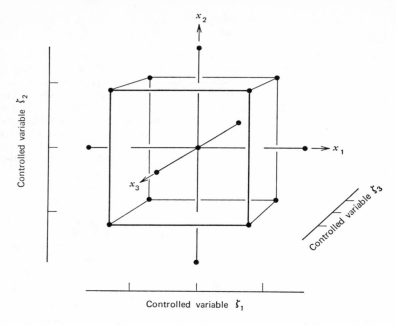

Figure 17.8.4. A central composite design, $k = 3$.

The normal equations provided by the model and the design are:

$$
\begin{aligned}
24b_0 && + 24b_{11} + 24b_{22} + 24b_{33} &&&= 101.65 \\
& 24b_1 &&&&= -17.85 \\
&& 24b_2 &&&= -4.31 \\
(17.8.6) && 24b_3 &&&= -2.39 \\
24b_0 && + 72b_{11} + 8b_{22} + 8b_{33} &&&= 98.45 \\
24b_0 && + 8b_{11} + 72b_{22} + 8b_{22} &&&= 101.33 \\
24b_0 && + 8b_{11} + 8b_{22} + 72b_{33} &&&= 104.21 \\
&& 8b_{12} &&&= -0.65 \\
&& 8b_{13} &&&= -0.21 \\
&& 8b_{23} &&&= -0.75
\end{aligned}
$$

The first-order coefficients b_i, and cross product coefficients b_{ij} may be solved directly. The coefficients of the fitted second-order model, and the corresponding analysis of variance table are displayed in Table 17.8.2.

TABLE 17.8.2

Estimated Coefficients	Analysis of Variance for the Fitted Second-Order Model			
$b_0 = \quad 4.29541$	$\Sigma\, y^2$	445.478900	24	
$b_1 = -\; .74375$	SSq b_0	430.530104	1	
$b_2 = -\; .17958$	First-order terms	14.287923	3	
$b_3 = -\; .09958$	Quadratic terms	.294491 $\Big\}$ .423129	6	0.705215
$b_{11} = -\; .06484$	Cross products	.128638		
$b_{22} = -\; .02000$	Residual SSq	.237744	14	0.01698
$b_{33} = \quad .02500$	Lack of fit	.141944	5	0.02839
$b_{12} = -\; .08125$	Error SSq	.095800	9	0.01064
$b_{13} = -\; .02625$		$F_{9.5} = 2.67$ not significant		
$b_{23} = -\; .09375$				

Nine degrees of freedom are available for estimating σ^2 independent of the choice of model. Three of these degrees of freedom are provided by the four repeated center points, and the remainder from the six paired axial points. The error sum of squares, subtracted from the residual sum of squares, gives the lack of fit sum of squares as displayed in Table 17.8.2. The lack of fit mean square does not seem unduly large when compared to the error mean square using the lack of fit F ratio, and the postulated second-order model is declared adequate to represent the response function. The residual mean square, $s^2 = .01698$, with fourteen degrees of freedom may now be used as the estimate of σ^2. The mean square for the second-order terms, .705215 with six degrees of freedom is also an estimate of σ^2 provided the composite hypothesis that all second-order coefficients in the model are zero is true. The corresponding observed F ratio equals $F_{6.14} = .705215/.01698 = 41.53$, leads to the rejection of the hypothesis, and the decision is taken to include all second-order terms in the model.

By substituting values for $\hat{y}$ in the fitted model,

$$\hat{y} = 4.295 - .744x_1 - .179x_2 - .100x_3 - .065x_1^2 - .020x_2^2$$
$$+ .025x_3^2 - .081x_1x_2 - .026x_1x_3 - .094x_2x_3,$$

we may determine contour surfaces in the usual manner.

17.9 APPENDIX. YATES' ALGORITHM FOR THE 2^k FACTORIAL

Yates' Algorithm is a method for greatly speeding the computations required in analyzing a two level factorial or fractional factorial design.

To employ the algorithm it is essential that the 2^k factorial design be written down in standard order, the first column of the design array x_1, consisting of alternating minus and plus signs, the second column x_2, alternating pairs of minus and plus signs, the third column x_3, alternating groups of four minus and plus signs, and so on until the kth column, which contains 2^{k-1} minus signs followed by the same number of plus signs. A minus sign is the first element in each column. The 2^4 factorial in standard (or Yates') order is shown in Table 17.3.1. The observation y_i recorded for each of the $i = 1, 2, \ldots, 2^k$ treatments (or the total T_i of the observations for each treatment) is also recorded. The observations are now grouped into separate pairs, and by a series, first of additions of observations in pairs and then of subtractions of observations in pairs, 2^k entries forming a second column of data are constructed. This new column is, in turn used to construct a third column, and the same procedure is repeated until k columns have been formed. Specifically one proceeds as follows:

(i) The pairs of entries are algebraically summed, the sum of each successive pair providing successively the first 2^{k-1} entries in the new column.

(ii) The pairs of entries are successively algebraically summed after changing the sign of the first entry in each of the pairs. These new successive sums provide the final 2^{k-1} entries in the new column.

The top entry in the final column will be the total of all the observations. Dividing by the total number of observations $N = r2^k$, where r is the common number of replicates of the 2^k treatments, gives the grand average $\bar{\bar{y}}$.

(iii) The estimated effects are given by the remaining $2^k - 1$ entries in the final column of the algorithm divided by $N/2 = r2^{k-1}$.

(iv) The estimated effects are identified by noting the plus signs in the treatment identification on the same line in the design array.

(v) The individual degree of freedom sum of squares for each of the $2^k - 1$ estimated effects are given by squaring the entries in the last column and dividing by N. Note that the first element, squared and divided by N, is the correction factor.

To check the computations, let $S = \Sigma\, T^2$ be the sum of squares of the treatment totals. Then the sum of squares of the entries in the kth column will equal $2^k S$.

The student should work through the following example (Table 17.9.1)

TABLE 17.9.1

Yates' Run Number	Notation 1	x_1	x_2	x_3	x_4	Observations	(i)	(ii)	(iii)	(iv)	Estimated Effects	Identification of Effects		Effect Contribution to Treatment SSq
1	1	−	−	−	−	62	150	296	655	1440	$\bar{y} = 1440/16$ $= 90.0$			
2	a	+	−	−	−	88	146	359	785	178	22.25	A	x_1	(1980.25)
3	b	−	+	−	−	63	168	374	31	24	3.00	B	x_2	36.00
4	ab	+	+	−	−	83	191	411	147	10	1.25	AB	$x_1 x_2$	6.25
5	c	−	−	+	−	88	188	46	19	100	12.50	C	x_3	625.00
6	ac	+	−	+	−	80	186	−15	5	−142	−17.75	AC	$x_1 x_3$	1260.25
7	bc	−	+	+	−	99	202	114	−5	36	4.50	BC	$x_2 x_3$	81.00
8	abc	+	+	+	−	92	209	33	15	26	3.25	ABC	$x_1 x_2 x_3$	42.25
9	d	−	−	−	+	65	26	−4	63	130	16.25	D	x_4	1056.25
10	ad	+	−	−	+	123	20	23	37	116	14.50	AD	$x_1 x_4$	841.00
11	bd	−	+	−	+	65	−8	−2	−61	−14	−1.75	BD	$x_2 x_4$	12.25
12	abd	+	+	−	+	121	−7	7	−81	20	2.50	ABD	$x_1 x_2 x_4$	25.00
13	cd	−	−	+	+	97	58	−6	27	−26	−3.25	CD	$x_3 x_4$	42.25
14	acd	+	−	+	+	105	56	1	9	20	2.50	ACD	$x_1 x_3 x_4$	25.00
15	bcd	−	+	+	+	92	8	−2	7	−18	−2.25	BCD	$x_2 x_3 x_4$	20.25
16	abcd	+	+	+	+	117	25	17	19	12	1.50	$ABCD$	$x_1 x_2 x_3 x_4$	9.00

Total 1440

Sum of squares $S = 135662$; Sum of squares col. (iv): $2{,}170{,}592 = 2^4 (135662)$: checks

PROBLEMS

17.1. Penicillin production requires a fermentation step that must be done in batches. One difficulty in production of successive batches is the nutrient, corn steep liquor, which varies. A study was begun to determine whether changes in temperature and pH might increase the penicillin yields for a new set of fermenters. A 2^2 factorial design was employed, a new batch of corn steep liquor employed for each set of four runs. The following results were obtained:

Design		Pencillin Yields				
(pH)	(Temperature)		Corn Steep Liquor Batches			
x_1	x_2	1	2	3	4	5
-1	-1	40	35	28	27	33
1	-1	95	80	94	76	83
-1	1	66	50	48	45	61
1	1	124	98	105	96	100

(i) Write out a suitable analysis of variance model.

(ii) Test the hypothesis that there are no treatment differences.

(iii) Test the hypothesis that there are no difference between the corn steep liquor batches.

(iv) Determine the pH and temperature effects.

(v) Separate individual degree of freedom sums of squares for these effects in the analysis of variance table.

(vi) Estimate σ^2 and make a 95% interval statement for the pH $\times$ Temperature interaction effect.

17.2. In the study of the flooding capacity of a pulse column, a 2^3 factorial design was employed in which two pulse amplitudes, two frequencies of pulsation, and two levels of flow ratio were varied. The following results were obtained:

Amplitude	Frequency	Flow Ratio	Capacity	
-1	-1	-1	179	184
1	-1	-1	330	338
-1	1	-1	280	297
1	1	-1	300	312
-1	-1	1	185	187
1	-1	1	288	304
-1	1	1	251	271
1	1	1	193	198

Estimate the factorial effects.

Estimate the variance of the observations.

Make a 95% interval statement for each of the various effects.

Can any of the variables, amplitude, frequency or flow ratio, be considered to be unimportant in affecting the response over the experimental region?

Consider the second column of observations as having been randomly run by a different operator. Reestimate the variance and determine whether there is a statistically significant operator-to-operator difference.

17.3. To determine the best production characteristics for a new method of manufacturing adiponitile (ADN), considerable development work was required on the purification system. The following replicated 2^3 factorial design was performed to study the effects of x_1: ADN feed rate; x_2: solvent to feed ratio, and x_3: temperature, on the response variable ADN purity. Each replicate represented one week's work. The trials were performed randomly within each week.

| | | | Response y: Replicates | | |
| | | | | | |
x_1	x_2	x_3	1	2	3
−	−	−	2.58	2.66	2.74
+	−	−	3.04	2.96	3.26
−	+	−	2.81	2.63	3.07
+	+	−	3.12	3.17	3.28
−	−	+	2.45	2.49	2.65
+	−	+	2.65	2.62	2.81
−	+	+	2.45	2.54	2.67
+	+	+	2.74	2.72	3.00

(i) Using Yates' algorithm estimate the factorial effects.

(ii) Construct an appropriate analysis of variance table.

(iii) Comment on the hypothesis that there are no statistically significant differences between the replicates.

(iv) Test the hypothesis that the contrast replicate 3 vs. replicates 1 and 2 equals zero.

(v) Make a 95% interval statement for the effect of ADN feed rate, that is, for the x_1 effect.

17.4. A study to determine whether modest changes in four critical dimensions in an automobile carburetor would change the horsepower produced by a standard six-cylinder engine, employed the following 2^4 factorial design.

| Dimensions | | | | Response |
A	B	C	D	y
−	−	−	−	14.8
+	−	−	−	24.8
−	+	−	−	12.3
+	+	−	−	20.1
−	−	+	−	13.8
+	−	+	−	22.3
−	+	+	−	12.0
+	+	+	−	20.0
−	−	−	+	16.3
+	−	−	+	23.7
−	+	−	+	13.5
+	+	−	+	19.4
−	−	+	+	11.3
+	−	+	+	23.6
−	+	+	+	11.2
+	+	+	+	21.8

(i) Which dimension is the most critical in influencing the response?

(ii) Assuming $\sigma^2 = 4$, make a 95% interval statement for the effect of the most important dimension.

(iii) Construct a normal probability plot of the estimated effects.

(iv) Using the analysis of variance table, obtain an estimate of σ^2.

17.5. In the manufacture of chemical products by electrolysis, the product yields depend upon x_1: current density; x_2: cathode configuration, and x_3: the flow rate of the catholyte. The following 2^3 factorial, blocked into groups of four experiments to eliminate day-to-day effects, was run to study the effects of these variables on process yields y.

Block 1				Block 2				Block 3				Block 4			
x_1	x_2	x_3	y	x_1	x_2	x_3	y	x_1	x_2	x_3	y	x_1	x_2	x_3	y
−	−	−	14.6	−	−	+	19.3	−	−	+	21.3	−	−	−	21.7
+	−	+	17.4	+	−	−	16.4	+	−	−	19.5	+	−	+	22.5
−	+	+	17.4	−	+	−	15.2	−	+	−	17.6	−	+	+	24.0
+	+	−	10.2	+	+	+	16.0	+	+	+	18.3	+	+	−	16.5

(i) Using Yates' algorithm, estimate the effects of the controlled variables.

(ii) Using an analysis of variance table, estimate σ^2.

(iii) Is there a statistically significant linear time trend between the four blocks?

17.6. Microminiature integrated circuits are manufactured, in part, by depositing thin films of dielectric material in predesigned patterns. Prior to mounting the film, the substrate must be prepared. In a study to determine the best operating condition for preparation, four variables were studied. The

Variables	(−)	(+)
x_1, substrate temperature, °C	250,	300
x_2, vacuum in chamber 1, mm of Hg	1×10^{-5},	1.5×10^{-5}
x_3, vacuum in chamber 2, mm of Hg	1×10^{-5},	1.5×10^{-5}
x_4, pattern type	Type A,	Type B

Block I					Block II				
x_1	x_2	x_3	x_4	y	x_1	x_2	x_3	x_4	y
−	−	−	−	3.43	−	−	−	+	3.62
+	−	−	+	4.04	+	−	−	−	4.17
−	+	−	+	3.57	−	+	−	−	3.48
+	+	−	−	3.86	+	+	−	+	4.18
−	−	+	+	4.09	−	−	+	−	3.27
+	−	+	−	3.27	+	−	+	+	4.35
−	+	+	−	3.15	−	+	+	+	4.09
+	+	+	+	4.20	+	+	+	−	3.52

experiment was blocked into two blocks of eight runs each, each block corresponding to a day. The design, and the response (thickness of deposited SiO film in 1000Å) are shown in the table at the bottom of p. 466.

(i) Employing Yates' algorithm, estimate the effects and determine, using normal probability paper, the statistically significant effects.
(ii) Construct the analysis of variance table and estimate σ^2.
(iii) Test the hypothesis that there is no difference between the block means.

17.7 (continuation of 17.6). Reanalyze only the data provided by Block I of the previous problem. What is the generator of this fractional factorial design?

17.8. In a study to determine the compressive strength of cylinders of concrete, five variables—type of sand, type of cement, amount of water, time to mix, and time in mold—were simultaneously studied using a 2^{5-1} fractional factorial design, with generator -12345. The (coded) responses are displayed in the random sequence in which the experiments were performed. (i) Determine the effect having the greatest influence on the compressive strength. (ii) Construct an estimate of the variance σ^2. (iii) Make a 95% interval statement for the effects.

x_1	x_2	x_3	x_4	x_5	y
−	+	+	−	−	10.2
+	−	−	+	−	17.5
−	−	+	+	−	13.0
−	+	+	+	+	17.3
−	−	−	−	−	13.4
−	−	+	−	+	20.2
+	−	+	−	−	18.1
+	+	−	+	+	15.7
+	+	+	+	−	15.1
−	+	−	+	−	10.6
+	−	+	+	+	19.1
−	+	−	−	+	16.9
+	+	−	−	−	14.8
−	−	−	+	+	19.5
+	+	+	−	+	15.7
+	−	−	−	+	19.2

17.9. Using the data of Problem 17.1:

(i) Fit a first-order model to the response function.
(ii) Assuming the postulated model is appropriate, plot the contours of the best fitting plane.
(iii) Plot the path of steepest ascent.

17.10. In a study of the breaking strength of concrete cylinders, it was decided to map breaking strength as a function of x_1: hours in mold, and x_2: age at test. The octagon design was used and the following data were obtained.

Run Number	x_1	x_2	y
1	$-\sqrt{2}$	0	77
2	0	0	95
3	0	0	94
4	-1	-1	84
5	1	1	95
6	0	0	93
7	0	$-\sqrt{2}$	76
8	-1	1	79
9	0	0	96
10	0	$\sqrt{2}$	88
11	1	-1	78
12	$\sqrt{2}$	0	84

(i) Plot the design points and, at each point, record the corresponding observations.

(ii) Determine whether a first-order model appears adequate to represent the response function.

(iii) Now fit a second-order model to the data.

(iv) To each of the observations in runs 1, 3, 7, 9, 10, and 12, add 5.0 and refit the second-order model. What difference do you observe between the two fitted second-order models?

17.11. Given the following hexagonal design with center points, and the recorded observations below:

(i) Plot the design coordinates and at each point record the corresponding observations.

(ii) Fit a first-order model.

(iii) Does the model adequately represent the data?

(iv) Are both variables x_1 and x_2 necessary to explain the data?

(v) Sketch the fitted response function.

Run Number	x_1	x_2	y
1	0	0	4.2
2	$-.500$	.866	3.7
3	0	0	4.4
4	.500	$-.866$	4.8
5	1	0	7.4
6	0	0	3.9
7	$-.500$	$-.866$	1.8
8	-1	0	1.1
9	.500	.866	6.8
10	0	0	4.2

17.12. For the following experimental design and data,
   (i) Plot the design and at each point record the corresponding observations.
   (ii) Consider only the first eight runs and analyze as a 2^2 factorial design
Does a first-order model adequately represent these data?
   (iii) Using all the data, fit a second-order model.

Run Number	x_1	x_2	y
1	−1	−1	45.9
2	−1	−1	53.3
3	−1	1	57.5
4	−1	1	58.8
5	1	−1	60.6
6	1	−1	58.0
7	1	1	58.6
8	1	1	52.6
9	0	0	56.9
10	2	0	55.4
11	−2	0	46.9
12	0	2	57.5
13	0	−2	55.0
14	0	0	58.9
15	0	0	50.3

17.13. A 3^2 factorial design was employed in a study to determine the best
freezing conditions for orange juice, the response measured being the percent
remaining natural vitamin B after eight weeks. The two variables are x_1: depth
of freeze in °C, and x_2: rate of freeze.

x_1	x_2	y
0	0	81.5
1	1	75.8
0	1	80.2
0	−1	79.2
−1	1	76.0
1	1	74.3
0	0	81.3
1	−1	80.1
1	0	79.1
−1	−1	70.2
−1	0	75.2
−1	−1	71.7
−1	1	76.2
1	−1	81.0

   (i) Plot the design and at each point record the corresponding observations.
   (ii) Fit a second-order model.
   (iii) Determine whether the model adequately represents the data.

Appendices

APPENDIX I. UNITY OF THE AREA UNDER THE NORMAL PROBABILITY DENSITY FUNCTION

In Chapter 7, we stated without proof that $\int_{-\infty}^{\infty} f(x) \, dx = 1$, where

$$f(x) = \frac{1}{\sigma\sqrt{2\pi}} \exp\left[-\frac{1}{2}\left(\frac{x - \mu}{\sigma}\right)^2\right], \qquad -\infty < x < \infty.$$

We now prove this statement.
Let

$$I = \int_{-\infty}^{\infty} \frac{1}{\sigma\sqrt{2\pi}} \exp\left[-\frac{1}{2}\left(\frac{x - \mu}{\sigma}\right)^2\right] dx.$$

We shall show that $I^2 = 1$. Since $f(x)$ is positive for all values of x, proving $I^2 = 1$ will imply that $I = 1$.

Now

$$I^2 = \left\{\int_{-\infty}^{\infty} \frac{1}{\sigma\sqrt{2\pi}} \exp\left[-\frac{1}{2}\left(\frac{x - \mu}{\sigma}\right)^2\right] dx\right\}$$

$$\times \left\{\int_{-\infty}^{\infty} \frac{1}{\sigma\sqrt{2\pi}} \exp\left[-\frac{1}{2}\left(\frac{y - \mu}{\sigma}\right)^2\right] dy\right\}$$

$$= \int_{-\infty}^{\infty} \int_{-\infty}^{\infty} \frac{1}{\sigma\sqrt{2\pi}} \exp\left[-\frac{1}{2}\left(\frac{x - \mu}{\sigma}\right)^2\right] \frac{1}{\sigma\sqrt{2\pi}} \exp\left[-\frac{1}{2}\left(\frac{y - \mu}{\sigma}\right)^2\right] dx \, dy.$$

Let $(x - \mu)/\sigma = u$, and $(y - \mu)/\sigma = v$.
Hence

$$I^2 = \int_{-\infty}^{\infty} \int_{-\infty}^{\infty} \frac{1}{2\pi} \exp\left[-\tfrac{1}{2}(u^2 + v^2)\right] du \, dv.$$

Let $u = r \cos \theta$, $v = r \sin \theta$. The Jacobian of this transformation is

$$\frac{\partial(u, v)}{\partial(r, \theta)} = \begin{vmatrix} \dfrac{\partial u}{\partial r} & \dfrac{\partial u}{\partial \theta} \\ \dfrac{\partial v}{\partial r} & \dfrac{\partial v}{\partial \theta} \end{vmatrix} = r.$$

Hence

$$I^2 = \frac{1}{2\pi} \int_0^{2\pi} \int_0^{\infty} e^{-r^2/2} r \, dr \, d\theta$$

$$= \frac{1}{2\pi} \int_0^{2\pi} (-e^{-r^2/2}|_0^{\infty}) \, d\theta$$

$$= \frac{1}{2\pi} \int_0^{2\pi} d\theta = 1,$$

hence $I = 1$.

APPENDIX II. THE MOMENT GENERATING FUNCTION OF THE NORMAL DISTRIBUTION

From (6.10.2) the moment generating function of a random variable x having the normal distribution $N(\mu, \sigma^2)$ is

$$M_x(t) = E(e^{tx}) = \int_{-\infty}^{\infty} \frac{1}{\sigma\sqrt{2\pi}} e^{tx} \exp\left[-\frac{1}{2}\left(\frac{x-\mu}{\sigma}\right)^2\right] dx.$$

Let $(x - \mu)/\sigma = z$, that is, $x = \mu + \sigma z$. Then

$$M_x(t) = e^{\mu t} \int_{-\infty}^{\infty} \frac{1}{\sqrt{2\pi}} e^{\sigma t z} e^{-z^2/2} \, dz$$

$$= e^{\mu t} \int_{-\infty}^{\infty} \frac{1}{\sqrt{2\pi}} \exp\left[-\tfrac{1}{2}(z^2 - 2\sigma t z + \sigma^2 t^2 - \sigma^2 t^2)\right] dz$$

$$= e^{\mu t} e^{\sigma^2 t^2/2} \int_{-\infty}^{\infty} \frac{1}{\sqrt{2\pi}} e^{-(z-\sigma t)^2/2} \, dz,$$

and letting $z - \sigma t = w$, we have

$$M_x(t) = e^{\mu t} e^{\sigma^2 t^2/2} \int_{-\infty}^{\infty} \frac{1}{\sqrt{2\pi}} e^{-w^2/2} \, dw$$

$$= e^{\mu t + \sigma^2 t^2/2},$$

since the value of the integral is unity (see Appendix I).

Differentiating, we have

$$M_x'(t) = (\mu + \sigma^2 t)e^{\mu t + \sigma^2 t^2/2},$$

$$M_x''(t) = \sigma^2 e^{\mu t + \sigma^2 t^2/2} + (\mu + \sigma^2 t)^2 e^{\mu t + \sigma^2 t^2/2},$$

and hence $M_x'(0) = \mu$, $M_x''(0) = \sigma^2 + \mu^2$, that is, $E(x) = \mu$ and $E(x^2) = \sigma^2 + \mu^2$. We thus have another proof (see Chapter 7) that

$$E(x) = \mu,$$

$$\mathrm{Var}\,(x) = E(x^2) - [E(x)]^2 = \sigma^2 + \mu^2 - \mu^2.$$

$$= \sigma^2.$$

Note that the moment generating function of the normal distribution has in its exponent the mean of the distribution as the coefficient of t, and the variance of the distribution as the coefficient of $t^2/2$.

Summarizing, we have the following:

THEOREM AII.1.   *If x is an $N(\mu, \sigma^2)$ random variable, its moment generating function $M_x(t)$ is given by*

$$M_x(t) = e^{\mu t + \sigma^2 t^2/2}.$$

If $x_1, \ldots, x_n$ are n independent random variables having normal distributions $N(\mu_1, \sigma_1^2), \ldots, N(\mu_n, \sigma_n^2)$ respectively, the moment generating function $M_L(t)$ of L, where

$$L = c_1 x_1 + \cdots + c_n x_n$$

and where c_i are arbitrary constants, is

$$M_L(t) = E(e^{t(c_1 x_1 + \cdots + c_n x_n)}) = \prod_{i=1}^{n} E(e^{t c_i x_i})$$

$$= \exp\left[t \sum \mu_i c_i + \frac{t^2}{2} \sum c_i^2 \sigma_i^2 \right],$$

which is the moment generating function of a random variable having the normal distribution $N(\Sigma c_i \mu_i, \Sigma c_i^2 \sigma_i^2)$.

We summarize by stating the following:

THEOREM AII.2.   *If $(x_1, \ldots, x_n)$ are n independent random variables having distributions $N(\mu_1, \sigma_1^2), \ldots, N(\mu_n, \sigma_n^2)$ respectively, then the random variable $L = \sum_{i=1}^{n} c_i x_i$ (that is, a linear combination of independent normal variables) is normally distributed with mean and variance given by $\sum_{i=1}^{n} c_i \mu_i$ and $\sum_{i=1}^{n} c_i^2 \sigma_i^2$ respectively.*

If the x_i have identical distributions, namely, $N(\mu, \sigma^2)$, that is, if $(x_1, \ldots, x_n)$ is a sample of n independent observations from $N(\mu, \sigma^2)$, then

$$E(L) = \mu \sum c_i,$$
$$\text{Var}(L) = \sigma^2 \sum c_i^2.$$

It is left to the reader as an exercise to show that the moment generating function of $z = (x - \mu)/\sigma$ is $e^{t^2/2}$ and, hence, that z has the normal distribution $N(0, 1)$. As a further exercise, he should show that the moment generating function of $L = ax + b$ is

$$e^{(a\mu + b)t + a^2 \sigma^2 t^2/2},$$

that is, L has the $N(a\mu + b, a^2 \sigma^2)$ distribution.

APPENDIX III. THE BIVARIATE NORMAL
DISTRIBUTION

One of the most important bivariate or two-dimensional continuous distributions, and one which is useful as a model for many situations in which we take pairs of measurements, is the *normal bivariate distribution*. If we denote the pair of random variables by (w_1, w_2), then the p.d.f. of the bivariate normal distribution is given by

$$(AIII.1) \quad f(w_1, w_2) = \frac{1}{2\pi\sigma_1\sigma_2\sqrt{1 - \rho^2}} \exp\left\{ - \frac{1}{2(1 - \rho^2)} \right.$$
$$\left. \times \left[\left(\frac{w_1 - \mu_1}{\sigma_1}\right)^2 - \frac{2\rho}{\sigma_1\sigma_2}(w_1 - \mu_1)(w_2 - \mu_2) + \left(\frac{w_2 - \mu_1}{\sigma_2}\right)^2 \right] \right\},$$

for any point (w_1, w_2) in the w_1, w_2 plane.

Note that $f(w_1, w_2)$ is specified by the five parameters $\mu_1, \mu_2, \sigma_1, \sigma_2$, and ρ. It can be verified that (in the notation of Chapter 6) the marginal p.d.f.'s of w_1 and w_2 are, respectively

$$(AIII.2) \qquad f_1(w_1) = \frac{1}{\sqrt{2\pi}\sigma_1} \exp\left\{ -\left[\frac{w_1 - \mu_1}{\sigma_1}\right]^2 \bigg/ 2 \right\}$$

and

$$(AIII.3) \qquad f_2(w_2) = \frac{1}{\sqrt{2\pi}\sigma_2} \exp\left\{ -\left[\frac{w_2 - \mu_2}{\sigma_2}\right]^2 \bigg/ 2 \right\}.$$

Hence, the marginal distributions of w_1 and w_2 are $N(\mu_1, \sigma_1^2)$ and $N(\mu_2, \sigma_2^2)$, respectively. From (AIII.2) and (AIII.3) we see that

$$(AIII.4) \qquad \begin{aligned} E(w_1) &= \mu_1 \\ E(w_2) &= \mu_2, \end{aligned}$$

and that

$$(AIII.5) \qquad \begin{aligned} \text{Var}(w_1) &= \sigma_1^2 \\ \text{Var}(w_2) &= \sigma_2^2. \end{aligned}$$

It can also be verified that the covariance between w_1 and w_2 is given by

$$(AIII.6) \qquad \sigma_{12} = E[(w_1 - \mu_1)(w_2 - \mu_2)] = \rho\sigma_1\sigma_2.$$

Now recall from (6.7.17) that the correlation coefficient between w_1 and w_2 is defined as $\sigma_{12}/\sigma_1\sigma_2$. Hence, the parameter ρ appearing in the density

(AIII.1) is, from (AIII.6), the correlation coefficient between w_1 and w_2. If w_1 and w_2 are independent, then of course, $\rho = 0$. If we are given the fact that w_1 and w_2 are bivariate normal, and if $\rho = 0$, then substituting in (AIII.1), we find

$$(\text{AIII.7}) \quad f(w_1, w_2) = \frac{1}{2\pi\sigma_1\sigma_2} \exp\left\{-\left[\left(\frac{w_1 - \mu_1}{\sigma_1}\right)^2 + \left(\frac{w_2 - \mu_2}{\sigma_2}\right)^2\right]/2\right\}$$

$$= \frac{1}{\sqrt{2\pi}\sigma_1} \exp\left\{-\left[\frac{w_1 - \mu_1}{\sigma_1}\right]^2/2\right\}$$

$$\times \frac{1}{\sqrt{2\pi}\sigma_2} \exp\left\{-\left[\frac{w_2 - \mu_2}{\sigma_2}\right]^2/2\right\} = f_1(w_1)f_2(w_2).$$

That is, if $\rho = 0$, the joint density of (w_1, w_2) factors into a product of their marginals, so that w_1 and w_2 are independent. We summarize this important result in the following theorem.

THEOREM AIII.1. *Let w_1 and w_2 have a bivariate normal distribution with means μ_1 and μ_2, variances σ_1^2 and σ_2^2, and correlation coefficient ρ. Then w_1 and w_2 are statistically independent if and only if $\rho = 0$.*

Now using (AIII.1) and (AIII.2), the reader should verify that in the notation of Section 6.7d, the conditional probability density function of w_2, given w_1, is

$$(\text{AIII.8}) \quad f(w_2 \mid w_1) = \frac{1}{\sqrt{2\pi}\,\sigma_2(1 - \rho^2)^{1/2}}$$

$$\times \exp\left[-\frac{1}{2\sigma_2^2(1 - \rho^2)}\left\{w_2 - \mu_2 - \rho\frac{\sigma_2}{\sigma_1}(w_1 - \mu_1)\right\}^2\right].$$

Thus, w_2, *given* w_1, has a normal probability density function with mean $E(w_2 \mid w_1) = \mu_2 + \rho(\sigma_2/\sigma_1)(w_1 - \mu_1)$, and variance $V(w_2 \mid w_1) = \sigma_2^2(1 - \rho^2)$. Note that the mean of w_2, given w_1, depends on w_1 but the variance does not. If we write $E(w_2 \mid w_1) = \alpha + \beta x$, then $\alpha = \mu_2 - \beta\mu_1$, $\beta = \rho\sigma_2/\sigma_1 = \sigma_{12}/\sigma_1^2$, and further, if we write $V(w_2 \mid w_1) = \sigma^2$, then $\sigma^2 = \sigma_2^2(1 - \rho^2)$. The quantity $\beta = \rho\sigma_2/\sigma_1 = \sigma_{12}/\sigma_1^2$ is called the *regression coefficient* of w_2 on w_1. Note that apart from the factor σ_2/σ_1, which is a factor that takes into account the units of measurement of w_1 and w_2 and the variability of w_1 and w_2, β depends on ρ, so that ρ is a quantity that measures the amount of *linear* dependence of w_2 and w_1, given that w_2 and w_1 are bivariate normal.

It is interesting to compare the above with the regression model used in Chapter 15. There we use the variables (x, y) to stand for (w_1, w_2), and assume that for *fixed* values of x, say $x_1, \ldots, x_n$, the random variables $y_1, \ldots, y_n$ are independently and normally distributed with means $\alpha + \beta x_1, \ldots, \alpha + \beta x_n$, respectively, and variances all equal to σ^2. From the pairs of measurements $(x_1, y_1), \ldots, (x_i, y_i), \ldots, (x_n, y_n)$, we then proceed to find the "least squares" estimators for α and β. Indeed, for $\beta = \sigma_{12}/\sigma_1^2 = \sigma_{xy}/\sigma_x^2$ we found its "least square estimator" to be $b = S(XY)/S(X^2)$, and for $\alpha = \mu_y - \beta\mu_x$, we found that the least squares estimator is $a = \bar{y} - b\bar{x}$ [see (15.2.6)].

APPENDIX IV. THE CENTRAL LIMIT THEOREM

In Section 7.3, it was stated that for large values of n the random variable having the binomial distribution with p.f. given by (3.2.2) has approximately the normal distribution $N[n\theta, n\theta(1 - \theta)]$. In Section 7.4 it was stated that in large samples the sample sum T (or the sample mean $\bar{x}$) has approximately the normal distribution $N(n\mu, n\sigma^2)$ [or $N(\mu, \sigma^2/n)$]. The proof of this statement is provided by the following special case:

CENTRAL LIMIT THEOREM. *Let $(x_1, \ldots, x_n)$ be a random sample of size n from a distribution having mean μ and variance σ^2. Then if $T = x_1 + \cdots + x_n$, the limiting distribution of the ratio*

$$\frac{T - n\mu}{\sqrt{n}\,\sigma} \quad \text{or of} \quad \frac{(\bar{x} - \mu)\sqrt{n}}{\sigma}$$

as $n \to \infty$ is the standard normal distribution $N(0, 1)$.

It is convenient to say that the asymptotic distribution of T for large n is $N(n\mu, n\sigma^2)$ [or of $\bar{x}$ for large n is $N(\mu, \sigma^2/n)$].

To prove this theorem we consider the moment generating function of

$$T' = \frac{x_1 - \mu}{\sqrt{n}} + \frac{x_2 - \mu}{\sqrt{n}} + \cdots + \frac{x_n - \mu}{\sqrt{n}}$$

and show that its limit as $n \to \infty$ is that of $N(0, \sigma^2)$. From Section 6.10 we have that

$$M_{x_i}(t) = 1 + \mu_1' t + \mu_2' \frac{t^2}{2!} + \mu_3' \frac{t^3}{3!} + \cdots$$

and hence

$$M_{x_i - \mu}(t) = 1 + \sigma^2 \frac{t^2}{2!} + \mu_3 \frac{t^3}{3!} + \cdots.$$

Furthermore, we know that

$$M_{(x_i-\mu)/\sqrt{n}}(t) = M_{x_i-\mu}\left(\frac{t}{\sqrt{n}}\right) = 1 + \sigma^2 \frac{t^2}{2!\,n} + \mu_3 \frac{t^3}{3!\,n^{3/2}} + \cdots,$$

which may be put in the form

$$M_{(x_i-\mu)/\sqrt{n}}(t) = 1 + \frac{\sigma^2 t^2 + \delta_n}{2n}$$

where δ_n approaches 0 as $n \to \infty$. Since the x_i are independent, we have

$$M_{T'}(t) = \left(1 + \frac{\sigma^2 t^2 + \delta_n}{2n}\right)^n.$$

As $n \to \infty$, we have

$$\lim_{n \to \infty} M_{T'}(t) = e^{\sigma^2 t^2/2},$$

which is the moment generating function of the normal distribution $N(0, \sigma^2)$ (see Appendix II).

Hence $(T - n\mu)/\sqrt{n}\sigma$ has as its limiting distribution as $n \to \infty$, the standard normal distribution $N(0, 1)$. In arriving at this conclusion we make use of a fundamental theorem which states roughly that, under certain conditions, if a sequence of c.d.f.'s has a limit which is a c.d.f., this c.d.f. is uniquely determined by the limit of the corresponding sequence of moment generating functions.

The shape of the distribution from which we are sampling determines the value of n required before the distribution of T approximates a normal distribution sufficiently closely to enable one to compute probabilities concerning T from its normal approximation with a satisfactory degree of accuracy. However, for some distributions (for instance, even the rectangular or triangular), the approximation to the distribution of T by the normal is very good for n as low as 10. If the population being sampled has a long tail (skewed) in either direction, then the required n is larger.

Example AIV(a). As an example of how we may apply the Central Limit Theorem, we establish the normal approximation to the binomial distribution used in Chapter 7.

Suppose we have a random variable y having sample space with two points, namely 0 and 1, such that

$$P(y = 1) = \theta, \qquad P(y = 0) = 1 - \theta,$$

where $0 < \theta < 1$.

Then,

$$E(y) = 1.\theta + 0(1 - \theta) = \theta,$$

$$E(y^2) = 1^2.\theta + 0^2(1 - \theta) = \theta,$$

hence

$$\text{Var}(y) = E(y - \theta)^2 = E(y^2) - \theta^2 = \theta - \theta^2 = \theta(1 - \theta).$$

Now, if $(y_1, \ldots, y_n)$ is a sample of size n on the random variable y, and if

$$x = y_1 + \cdots + y_n,$$

then, as we have seen in Chapter 3, x is a random variable having the binomial distribution $b(x)$ defined by (3.2.2) with

$$E(x) = E(y_1) + \cdots + E(y_n) = \theta + \cdots + \theta = n\theta,$$

$$\sigma_x^2 = \sigma_{y_1}^2 + \cdots + \sigma_{y_n}^2 = \theta(1 - \theta) + \cdots + \theta(1 - \theta) = n\theta(1 - \theta).$$

Hence x is the sum of n independent random variables all having the same distribution with mean θ and variance $\theta(1 - \theta)$; hence by the central limit theorem $(x - n\theta)/[\sqrt{n\theta(1 - \theta)}]$ is a random variable whose limiting distribution as $n \to \infty$ is the normal distribution $N(0, 1)$.

APPENDIX V. ALGEBRAIC PROOFS OF SOME LEAST SQUARES FORMULAS

In Chapter 15 we developed formulas for the regression line of y on x using methods of the calculus. In this Appendix, we give an algebraic proof of the same formulas concerning this regression line.

Suppose we have n independent observations $(w_1, \ldots, w_n)$ and we wish to minimize the quantity $V(k)$ with respect to k, where

$$V(k) = \sum_{i=1}^{n} (w_i - k)^2.$$

We wish to show that $V(k)$ is minimized if $k = \bar{w}$. We may write

$$V(k) = \sum_{i=1}^{n} [(w_i - \bar{w}) + (\bar{w} - k)]^2$$

(AV.1)
$$= \sum_{i=1}^{n} [(w_i - \bar{w})^2 + (\bar{w} - k)^2 + 2(\bar{w} - k)(w_i - \bar{w})]$$

$$= \sum_{i=1}^{n} (w_i - \bar{w})^2 + n(\bar{w} - k)^2,$$

since $\sum_{i=1}^{n} (w_i - \bar{w}) = \Sigma w_i - n\bar{w} = n\bar{w} - n\bar{w} = 0$. Now $V(k)$ is the sum of two non-negative quantities and obviously has its minimum for fixed $w_1, \ldots, w_n$ if k is put equal to $\bar{w}$, that is, at $k = \bar{w}$.

The regression problem requires the values of α and β which minimize

$$Q(\alpha, \beta) = \sum_{i=1}^{n} (y_i - \alpha - \beta x_i)^2,$$

where $x_1, \ldots, x_n$ are not all equal. If we let $y_i - \beta x_i = w_i$, we have $\bar{y} - \beta \bar{x} = \bar{w}$ and

$$Q(\alpha, \beta) = \sum_{i=1}^{n} (w_i - \alpha)^2.$$

Using the identity (AV.1), we have that

$$Q(\alpha, \beta) = \sum_{i=1}^{n} (w_i - \bar{w})^2 + n(\bar{w} - \alpha)^2,$$

which, of course, is minimized for any fixed value of β by choosing

$$\alpha = \bar{w} = \bar{y} - \beta \bar{x}.$$

Hence for any value of β, $Q(\alpha, \beta)$ is minimized with respect to α for $\alpha = \bar{y} - \beta \bar{x}$, in which case $Q(\alpha, \beta)$ has the value

(AV.2)
$$Q(\bar{y} - \beta \bar{x}, \beta) = \sum_{i=1}^{n} [y_i - (\bar{y} - \beta \bar{x}) - \beta x_i]^2$$
$$= \sum_{i=1}^{n} [(y_i - \bar{y}) - \beta(x_i - \bar{x})]^2.$$

Note that (AV.2) is a function of β alone, which we will denote by $Q(\beta)$. To find the value of β which minimizes $Q(\beta)$ for fixed values of $x_1, \ldots, x_n$ not all equal, we write

$$Q(\beta) = A\beta^2 - 2B\beta + C,$$

which can be written

(AV.3)
$$Q = A\left[\left(\beta - \frac{B}{A} \right)^2 + \left(\frac{AC - B^2}{A^2} \right) \right].$$

Hence $Q(\beta)$ is minimized with respect to β if we choose

(AV.4)
$$\beta = \frac{B}{A}.$$

On expanding (AV.2), $Q(y - \beta \bar{x}, \beta)$ may be written as

$$Q = \beta^2 \left[\sum_{i=1}^{n} (x_i - \bar{x})^2 \right] - 2\beta \sum_{i=1}^{n} (x_i - \bar{x})(y_i - \bar{y}) + \sum_{i=1}^{n} (y_i - \bar{y})^2.$$

[Comparing with $Q(\beta)$ above, we have $A = \sum_{i=1}^{n} (x_i - \bar{x})^2$, $B = \sum_{i=1}^{n} (x_i - \bar{x}) \times (y_i - \bar{y})$, and $C = \sum_{i=1}^{n} (y_i - \bar{y})^2$.] Hence $Q(\bar{y} - \beta \bar{x}, \beta)$ is minimized with respect to β for fixed $x_1, \ldots, x_n$ not all equal [see (AV.4)] if we choose

$$\beta = \frac{\Sigma (x_i - \bar{x})(y_i - \bar{y})}{\Sigma (x_i - \bar{x})^2}.$$

That is, $Q(\alpha, \beta)$ is minimized if we choose

$$\alpha = \bar{y} - \beta\bar{x},$$

and

$$\beta = \frac{\Sigma\,(x_i - \bar{x})(y_i - \bar{y})}{\Sigma\,(x_i - \bar{x})^2}.$$

It is convenient to denote the value of (α, β) which minimizes $Q(\alpha, \beta)$ by (a, b), that is,

$$a = \bar{y} - b\bar{x},$$

$$b = \frac{\Sigma\,(x_i - \bar{x})(y_i - \bar{y})}{\Sigma\,(x_i - \bar{x})^2}.$$

Note that the minimum value of $Q(\alpha, \beta)$ is $Q(a, b)$, where

$$
\begin{aligned}
Q(a, b) &= C - \frac{B^2}{A} \\
&= \sum_{i=1}^{n}(y_i - \bar{y})^2 - \frac{\left[\displaystyle\sum_{i=1}^{n}(x_i - \bar{x})(y_i - \bar{y})\right]^2}{\Sigma\,(x_i - \bar{x})^2} \\
&= \sum_{i=1}^{n}(y_i - \bar{y})^2 - b^2\sum_{i=1}^{n}(x_i - \bar{x})^2 \\
&= \sum_{i=1}^{n}(y_i - \bar{y})^2 - r^2\sum(y_i - \bar{y})^2 \\
&= \sum_{i=1}^{n}(y_i - \bar{y})^2(1 - r^2),
\end{aligned}
$$

and where r is the sample correlation coefficient given by

$$r = \frac{\Sigma\,(x_i - \bar{x})(y_i - \bar{y})}{\sqrt{\Sigma\,(x_i - \bar{x})^2}\sqrt{\Sigma\,(y_i - \bar{y})^2}} = b\,\frac{\sqrt{\Sigma\,(x_i - \bar{x})^2}}{\sqrt{\Sigma\,(y_i - \bar{y})^2}}.$$

Now since Q is positive, we have that $\Sigma\,(y_i - \bar{y})^2(1 - r^2) \geq 0$, that is,

$$-1 \leq r \leq 1.$$

[Compare this result with (AIII.8).]

APPENDIX VI. SAMPLING FROM A NORMAL DISTRIBUTION

In Section 6.9 we have defined what we mean by a random sample $(x_1, \ldots, x_n)$ from a probability density function $f(x)$, and we have stated that the sum T and mean $\bar{x}$ of a sample from a normal distribution have normal distributions themselves. In this Appendix we shall present some theorems about various sampling distributions which arise when sampling from a normal distribution and which are used at various places in Chapters 8, 9, 10, 11, 12, 15, and 16.

(a) Distribution of Mean of a Sample From a Normal Distribution

If $(x_1, \ldots, x_n)$ is a sample of size n from $N(\mu, \sigma^2)$, then by definition $(x_1, \ldots, x_n)$ are independent random variables all having mean μ and variance σ^2, and hence the following theorem is a corollary of Theorem AII.2.

THEOREM AVI.1.   *If $(x_1, \ldots, x_n)$ is a sample of size n from $N(\mu, \sigma^2)$, then the sample mean $\bar{x}$ has the normal distribution $N(\mu, \sigma^2/n)$, while the sample sum T has the normal distribution $N(n\mu, n\sigma^2)$.*

To see that $\bar{x}$ has the distribution $N(\mu, \sigma^2/n)$, we let $c_1 = \cdots = c_n = 1/n$, $\mu_1 = \cdots = \mu_n = \mu$, $\sigma_1^2 = \cdots = \sigma_n^2 = \sigma^2$ in Theorem AII.2, in which case L becomes $\bar{x}$ whose moment generating function is

$$M_{\bar{x}}(t) = e^{t\mu + t^2\sigma^2/2n},$$

which is the moment generating function of $N(\mu, \sigma^2/n)$, and hence $\bar{x}$ has the distribution $N(\mu, \sigma^2/n)$.

To see that T has the distribution $N(n\mu, n\sigma^2)$, we let $\mu_1 = \cdots = \mu_n = \mu$, $\sigma_1^2 = \cdots = \sigma_n^2 = \sigma^2$, and $c_1 = \cdots = c_n = 1$ in Theorem AII.2, and thus find

$$M_T(t) = e^{tn\mu + t^2 n\sigma^2/2},$$

which is the moment generating function of $N(n\mu, n\sigma^2)$, and hence T has the normal distribution $N(n\mu, n\sigma^2)$.

Now suppose $\bar{x}_1$ and $\bar{x}_2$ are means of independent samples of size n_1 and n_2 from the normal distributions $N(\mu_1, \sigma_1^2)$ and $N(\mu_2, \sigma_2^2)$ respectively. Then $\bar{x}_1$ and $\bar{x}_2$ are $N(\mu_1, \sigma_1^2/n_1)$ and $N(\mu_2, \sigma_2^2/n_2)$ random variables respectively and are independent.

If we consider the linear combination $\bar{x}_1 - \bar{x}_2$ of the two independent random variables $\bar{x}_1$ and $\bar{x}_2$, having normal distributions $N(\mu_1, \sigma_1^2/n_1)$ and $N(\mu_2, \sigma_2^2/n_2)$ respectively, we find from Theorem AII.2 that $\bar{x}_1 - \bar{x}_2$ is an $N(\mu_1 - \mu_2, \sigma_1^2/n_1 + \sigma_2^2/n_2)$ random variable. Summarizing, we have the following theorem:

THEOREM AVI.2.   *Let $\bar{x}_1$ and $\bar{x}_2$ be means of independent samples of sizes n_1 and n_2 from normal distributions $N(\mu_1, \sigma_1^2)$ and $N(\mu_2, \sigma_2^2)$ respectively. Then $\bar{x}_1 - \bar{x}_2$ has the normal distribution $N(\mu_1 - \mu_2, \sigma_1^2/n_1 + \sigma_2^2/n_2)$.*

(b) The Chi-Square Distribution

Suppose $(x_1, \ldots, x_n)$ is a sample of size n from a normal distribution $N(0, 1)$, and let us consider the probability that

(AVI.1) $\chi^2 < x_1^2 + \cdots + x_n^2 < \chi^2 + h,$ for $h > 0.$

We have

(AVI.2) $\quad P(\chi^2 < x_1^2 + \cdots + x_n^2 < \chi^2 + h)$

$$= \left(\frac{1}{2\pi}\right)^{n/2} \int \cdots \int_S \exp\left(-\frac{1}{2}\sum_{i=1}^{n} x_i^2\right) dx_1 \cdots dx_n,$$

where $h > 0$. The desired probability is that contained in the n-dimensional spherical shell S (of volume ΔV) whose outer radius is $\sqrt{\chi^2 + h}$ and whose inner radius is $\sqrt{\chi^2}$. The value of $\exp\left(-\frac{1}{2}\sum_1^{n} x_i^2\right)$ is smallest in S for points $(x_1, \ldots, x_n)$ on the outer sphere, and largest for points $(x_1, \ldots, x_n)$ on the inner sphere. At points on the outer sphere

$$e^{-\Sigma x_i^2/2} = e^{-(\chi^2+h)/2},$$

and at points on the inner sphere

$$e^{-\Sigma x_i^2/2} = e^{-\chi^2/2}.$$

Therefore, we can write

(AVI.3) $\quad \left(\frac{1}{2\pi}\right)^{n/2} e^{-(\chi^2+h)/2} \Delta V < P(\chi^2 < x_1^2 + \cdots + x_n^2 < \chi^2 + h)$

$$< \left(\frac{1}{2\pi}\right)^{n/2} e^{-\chi^2/2} \Delta V,$$

where ΔV is the volume of the spherical shell S. Using the fact that the volume of an n-dimensional sphere of radius r is Kr^n, where K is a constant depending on n, we see that

$$\Delta V = K[(\chi^2 + h)^{n/2} - (\chi^2)^{n/2}].$$

Dividing (AVI.3) by h and taking limits as $h \to 0$, and using the fact that $\lim_{h \to 0} \dfrac{\Delta V}{h} = \dfrac{n}{2}(\chi^2)^{(n/2)-1}$, $\lim_{h \to 0} e^{-h/2} = 1$, we find that

$$C(\chi^2)^{n/2-1}e^{-\chi^2/2} \leq \lim_{h \to 0} \frac{P(\chi^2 < x_1^2 + \cdots + x_n^2 < \chi^2 + h)}{h}$$

$$\leq C(\chi^2)^{n/2-1}e^{-\chi^2/2},$$

where C is constant.

Therefore the p.e. of the random variable $x_1^2 + \cdots + x_n^2$, which we denote by χ^2, is

(AVI.4) $\qquad\qquad\qquad C(\chi^2)^{n/2-1}e^{-\chi^2/2}\, d\chi^2$

for $\chi^2 > 0$, and $0\, d\chi^2$ otherwise.

Since

$$C \int_0^\infty (\chi^2)^{n/2-1} e^{-\chi^2/2} \, d\chi^2 = 1,$$

it will be found by successive integration by parts that

$$C = \frac{1}{2^{n/2} \Gamma(n/2)},$$

where

$$\Gamma\left(\frac{n}{2}\right) = \left(\frac{n-2}{2}\right)\left(\frac{n-4}{2}\right) \cdots (1) \qquad \text{if } n \text{ is even,}$$

$$\Gamma\left(\frac{n}{2}\right) = \left(\frac{n-2}{2}\right)\left(\frac{n-4}{2}\right) \cdots \frac{1}{2}\sqrt{\pi} \qquad \text{if } n \text{ is odd.}$$

The function $\Gamma(m)$ is called the *gamma function* of m and satisfies the difference equation $\Gamma(m+1) = m\Gamma(m)$. The p.d.f. contained in (AVI.4) is called the chi-square p.d.f. with n degrees of freedom. We summarize by the following theorem:

THEOREM AVI.3.   *If $(x_1, \ldots, x_n)$ is a sample of size n from the normal distribution $N(0, 1)$, the random variable $x_1^2 + \cdots + x_n^2$, denoted by χ^2, has the probability element*

(AVI.5) $$\frac{(\chi^2/2)^{n/2-1}}{2\Gamma(n/2)} e^{-\chi^2/2} \, d\chi^2,$$

for $\chi^2 > 0$, and $0 \, d\chi^2$ otherwise.

A random variable χ^2 having the chi-square distribution (AVI.5) with n degrees of freedom is often denoted by χ_n^2 and is called a chi-square variable with n degrees of freedom.

We now state the following:

COROLLARY.   *If $(x_1, \ldots, x_n)$ is a sample from the normal distribution $N(\mu, \sigma^2)$, the random variable*

$$\frac{1}{\sigma^2} \sum_1^n (x_i - \mu)^2$$

has the chi-square distribution with n degrees of freedom.

For if $(x_1, \ldots, x_n)$ is a sample from $N(\mu, \sigma^2)$,

$$[(x_1 - \mu)/\sigma, \ldots, (x_n - \mu)/\sigma]$$

is a sample from $N(0, 1)$. Thus, it follows from Theorem AVI.3 that $\sum_1^n \left(\frac{x_i - \mu}{\sigma}\right)^2$, that is, $\frac{1}{\sigma^2} \sum_1^n (x_i - \mu)^2$, has the chi-square distribution with n degrees of freedom.

An important property of the chi-square distribution may be stated in the following theorem:

THEOREM AVI.4.   *If $\chi^2_{n_1}$ and $\chi^2_{n_2}$ are independent random variables having chi-square distributions with n_1 and n_2 degrees of freedom respectively, then $\chi^2_{n_1} + \chi^2_{n_2}$ is a random variable having the chi-square distribution with $n_1 + n_2$ degrees of freedom.*

The proof of this theorem is quite evident if we consider two independent samples $(x_1, \ldots, x_{n_1})$ and $(x'_1, \ldots, x'_{n_2})$ from the normal distribution $N(0, 1)$.   It follows from Theorem AVI.3 that $x_1^2 + \cdots + x_{n_1}^2$ and $x_1'^2 + \cdots + x_{n_2}'^2$ are independent random variables having chi-square distributions with n_1 and n_2 degrees of freedom respectively, and also that $(x_1^2 + \cdots + x_{n_1}^2) + (x_1'^2 + \cdots + x_{n_2}'^2)$ has the chi-square distribution with $n_1 + n_2$ degrees of freedom.

Now, since the integral of the expression in (AVI.5) is 1 over the interval $(0, \infty)$, it is seen that

$$\int_0^\infty \left(\frac{\chi^2}{2}\right)^{n/2-1} e^{-\chi^2/2} \, d\chi^2 = 2\Gamma\left(\frac{n}{2}\right).$$

Replacing n by $n + 2$ on both sides and then multiplying both sides by $[2\Gamma(n/2)]^{-1}$, we then see that

$$E\left(\frac{\chi^2}{2}\right) = \frac{\Gamma(n/2 + 1)}{\Gamma(n/2)} = \frac{n}{2}.$$

Hence

$$E(\chi_n^2) = E(\chi^2) = n,$$

that is, the expectation of χ_n^2 is n.   The graphs of the p.d.f. of χ_n^2 for several values of n are sketched in Figure AVI.1.

Values of $\chi^2_{n;\alpha}$ for which

$$P(\chi_n^2 > \chi^2_{n;\alpha}) = \alpha$$

have been tabulated for various values of n and α in Table III.   We now state two other important theorems without proof concerning sampling from normal distributions.   Proofs can be found in books in advanced statistics [for example, Cramer (1946) or Wilks (1962)].

THEOREM AVI.5.   *Let $(x_1, \ldots, x_n)$ be a sample of size n from the normal distribution $N(\mu, \sigma^2)$, and let $s^2 = [1/(n-1)] \sum_{i=1}^{n} (x_i - \bar{x})^2$ and $\bar{x} = (1/n) \sum_{i=1}^{n} x_i$. Then $[(n-1)s^2]/\sigma^2$ has the chi-square distribution with $n-1$ degrees of freedom.*

Considering the joint distribution of $\bar{x}$ and s, we have another important theorem:

THEOREM AVI.6.   *If* $(x_1, \ldots, x_n)$ *is a sample from the normal distribution* $N(\mu, \sigma^2)$, $\bar{x}$ *and* $[(n - 1)s^2]/\sigma^2$ *are independent random variables having the normal distribution* $N(\mu, \sigma^2/n)$ *and the chi-square distribution with* $n - 1$ *degrees of freedom respectively.*

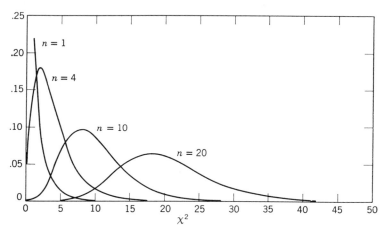

Figure AVI.1. Graphs of probability density functions of χ^2 for 1, 4, 10, and 20 degrees of freedom.

(c) The Student *t* Distribution

One of the most important distributions in statistics in problems involving small samples from normal distributions can be stated in the following theorem

THEOREM AVI.7.   *If u and v are independent random variables having the normal distribution* $N(0, 1)$ *and the chi-square distribution with m degrees of freedom respectively, then the random variable*

$$t = \frac{u}{\sqrt{v/m}}$$

has the following probability element

(AVI.6) $$\frac{\Gamma[\frac{1}{2}(m + 1)]}{\sqrt{m\pi}\ \Gamma(\frac{1}{2}m)}\left(1 + \frac{t^2}{m}\right)^{-(m+1)/2} dt$$

on the interval $(-\infty, +\infty)$.

To prove this we note that, since u and v are independent, (u, v) has probability element

(AVI.7) $$\frac{1}{\sqrt{2\pi}} e^{-u^2/2} \frac{\left(\frac{v}{2}\right)^{m/2-1} e^{-v/2}}{2\Gamma(m/2)} \, du \, dv$$

for $u > 0$ and $-\infty < v < +\infty$. Applying to (AVI.7) the transformation

$$u = \frac{t\sqrt{w}}{\sqrt{m}},$$

$$v = w,$$

for which the Jacobian is $\sqrt{w/m}$, we obtain

(AVI.8) $$\frac{1}{\sqrt{2\pi m}\, 2\Gamma(m/2)} \left(\frac{w}{2}\right)^{[(m+1)/2]-1} \exp\left[-\frac{1}{2} w\left(1 + \frac{t^2}{m}\right)\right] dw \, dt.$$

Taking the marginal distribution with respect to w, that is, integrating with respect to w for a fixed t, we obtain (AVI.6).

It can be shown that the mean and variance of t are given by

$$E(t) = 0,$$

$$\text{Var}\,(t) = \frac{m}{m-2},$$

the variance being finite only for $m > 2$.

The distribution having the probability element (AVI.6) is called the *Student t distribution with m degrees of freedom*. A random variable having (AVI.6) as its probability element is called a *Student t variable with m degrees of freedom* and is often denoted by t_m. It will be noted that the p.d.f. of t is *symmetric* about zero. Its graph looks something like that of the p.d.f. of the normal distribution $N(0, 1)$. In fact, it can be proved that as $m \to \infty$ the probability element (AVI.6) has the probability element of $N(0, 1)$ as its limit.

Values of t_α for which

$$P(t_m > t_{m;\alpha}) = \alpha$$

have been tabulated in Table V for various values of m and α. Because of the symmetry about zero, note that $t_{m;\alpha} = -t_{m;1-\alpha}$.

As an immediate application of Theorem AVI.7, we have the following theorem:

THEOREM AVI.8.   *If* $(x_1, \ldots, x_n)$ *is a sample from the normal distribution* $N(\mu, \sigma^2)$, *the random variable*

$$t = \frac{(\bar{x} - \mu)\sqrt{n}}{s}$$

has the Student t distribution with $n - 1$ *degrees of freedom.*

To establish this theorem, we recall from Theorem AVI.6 that the random variable

$$u = \frac{(\bar{x} - \mu)\sqrt{n}}{\sigma}$$

has the normal distribution $N(0, 1)$ and

$$v = \frac{(n - 1)s^2}{\sigma^2}$$

has the chi-square distribution with $n - 1$ degrees of freedom. Furthermore, u and v are independent. Therefore by Theorem AVI.7,

$$t = \frac{u}{\sqrt{v/(n - 1)}} = \frac{(\bar{x} - \mu)\sqrt{n}}{s}$$

has the Student t distribution with $n - 1$ degrees of freedom.

We also state the following important theorem:

THEOREM AVI.9.   *Let $\bar{x}_1$ and $\bar{x}_2$ be means and s_1^2 and s_2^2 be variances of independent samples of sizes n_1 and n_2 from normal distributions $N(\mu_1, \sigma^2)$ and $N(\mu_2, \sigma^2)$. Then the random variable*

$$t = \frac{(\bar{x}_1 - \bar{x}_2) - (\mu_1 - \mu_2)}{s_w\sqrt{1/n_1 + 1/n_2}},$$

where $s_w^2 = [(n_1 - 1)s_1^2 + (n_2 - 1)s_2^2]/[n_1 + n_2 - 2]$, has the Student t distribution with $n_1 + n_2 - 2$ degrees of freedom.

To establish this theorem we note that, if in Theorem AVI.2 we put $\sigma_1^2 = \sigma_2^2 = \sigma^2$,

$$u = \frac{(\bar{x}_1 - \bar{x}_2) - (\mu_1 - \mu_2)}{\sigma\sqrt{1/n_1 + 1/n_2}}$$

has the normal distribution $N(0, 1)$. Furthermore,

$$v = \frac{(n_1 - 1)s_1^2}{\sigma^2} + \frac{(n_2 - 1)s_2^2}{\sigma^2},$$

being the sum of two independent random variables having chi-square distributions with $n_1 - 1$ and $n_2 - 1$ degrees of freedom respectively, has a chi-square distribution with $(n_1 - 1) + (n_2 - 1) = n_1 + n_2 - 2$ degrees of freedom.

Therefore by Theorem AVI.7,

$$t = \frac{u}{\sqrt{v/(n_1 + n_2 - 2)}} = \frac{(\bar{x}_1 - \bar{x}_2) - (\mu_1 - \mu_2)}{s_w\sqrt{1/n_1 + 1/n_2}}$$

has the Student t distribution with $n_1 + n_2 - 2$ degrees of freedom.

(d) The Snedecor F Distribution

Another important distribution in the theory of sampling from normal distributions can be stated as follows:

THEOREM AVI.10.	*Let v_1 and v_2 be two independent random variables having chi-square distributions with m_1 and m_2 degrees of freedom respectively. Then the random variable*

$$(AVI.9) \qquad\qquad F = \frac{v_1/m_1}{v_2/m_2}$$

has probability element

$$(AVI.10) \quad \frac{\Gamma[(m_1 + m_2)/2]}{\Gamma(m_1/2)\Gamma(m_2/2)}\left(\frac{m_1}{m_2}\right)^{m_1/2} F^{(m_1/2)-1}\left(1 + \frac{m_1 F}{m_2}\right)^{-(m_1+m_2)/2} dF,$$

where $F > 0$. (F is sometimes called the variance ratio or the Snedecor F variable, and is often denoted by F_{m_1, m_2}.)

To prove this theorem we begin with the probability element of (v_1, v_2), where v_1 and v_2 are independent random variables having chi-square distributions with m_1 and m_2 degrees of freedom respectively. The joint probability element is

$$(AVI.11) \qquad\qquad \frac{(v_1/2)^{(m_1/2)-1}(v_2/2)^{(m_2/2)-1}}{4\Gamma(m_1/2)\Gamma(m_2/2)} e^{-(v_1+v_2)/2} \, dv_1 \, dv_2.$$

Let us make the transformation

$$v_1 = \frac{m_1}{m_2} GF,$$

$$v_2 = G,$$

of which the Jacobian is $(m_1/m_2)G$. The probability element (AVI.10) becomes

$$\frac{(m_1/m_2)^{m_1/2}(G/2)^{(m_1+m_2)/2-1}}{2\Gamma(m_1/2)\Gamma(m_2/2)} F^{m_1/2-1} \exp\left[-\frac{1}{2}G\left(1 + \frac{m_1}{m_2}F\right)\right] dG \, dF.$$

Integrating this expression with respect to G from 0 to $+\infty$ for a fixed F gives (AVI.10). The distribution having the probability element (AVI.10) is known as the *Snedecor F distribution* or *variance ratio distribution* with m_1 and m_2 degrees of freedom.

Values of $F_{m_1, m_2; \alpha} = F_\alpha$ for which

$$P(F_{m_1, m_2} > F_{m_1; m_2; \alpha}) = \alpha$$

have been tabulated for various values of m_1 and m_2 and α and are given in Table VI. Some p.d.f.'s of F are graphed in Figure AVI.2.

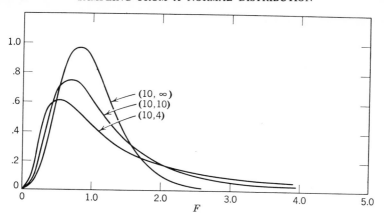

Figure AVI.2. Graphs of probability density functions of F for $(m_1, m_2) = (10, \infty)$, $(10, 10)$, and $(10, 4)$.

It can be verified that the mean value of F_{m_1, m_2} is $m_2/(m_2 - 2)$ which is finite only if $m_2 > 2$. The variance of F is

$$[2m_2^2(m_1 + m_2 - 2)]/[m_1(m_2 - 2)^2(m_2 - 4)]$$

which is finite only if $m_2 > 4$. Note that $t_m^2 = (v_1/1)/(v_2/m)$, where v_1 is a chi-square variable with one degree of freedom and v_2 is an independent chi-square variable with m degrees of freedom. Hence, $t_m^2 = F_{1,m}$.

APPENDIX VII.   TABLES

TABLE I

5,000 RANDOM DIGITS*

00000	10097 32533	76520 13586	34673 54876	80959 09117	39292 74945
00001	37542 04805	64894 74296	24805 24037	20636 10402	00822 91665
00002	08422 68953	19645 09303	23209 02560	15953 34764	35080 33606
00003	99019 02529	09376 70715	38311 31165	88676 74397	04436 27659
00004	12807 99970	80157 36147	64032 36653	98951 16877	12171 76833
00005	66065 74717	34072 76850	36697 36170	65813 39885	11199 29170
00006	31060 10805	45571 82406	35303 42614	86799 07439	23403 09732
00007	85269 77602	02051 65692	68665 74818	73053 85247	18623 88579
00008	63573 32135	05325 47048	90553 57548	28468 28709	83491 25624
00009	73796 45753	03529 64778	35808 34282	60935 20344	35273 88435
00010	98520 17767	14905 68607	22109 40558	60970 93433	50500 73998
00011	11805 05431	39808 27732	50725 68248	29405 24201	52775 67851
00012	83452 99634	06288 98083	13746 70078	18475 40610	68711 77817
00013	88685 40200	86507 58401	36766 67951	90364 76493	29609 11062
00014	99594 67348	87517 64969	91826 08928	93785 61368	23478 34113
00015	65481 17674	17468 50950	58047 76974	73039 57186	40218 16544
00016	80124 35635	17727 08015	45318 22374	21115 78253	14385 53763
00017	74350 99817	77402 77214	43236 00210	45521 64237	96286 02655
00018	69916 26803	66252 29148	36936 87203	76621 13990	94400 56418
00019	09893 20505	14225 68514	46427 56788	96297 78822	54382 14598
00020	91499 14523	68479 27686	46162 83554	94750 89923	37089 20048
00021	80336 94598	26940 36858	70297 34135	53140 33340	42050 82341
00022	44104 81949	85157 47954	32979 26575	57600 40881	22222 06413
00023	12550 73742	11100 02040	12860 74697	96644 89439	28707 25815
00024	63606 49329	16505 34484	40219 52563	43651 77082	07207 31790
00025	61196 90446	26457 47774	51924 33729	65394 59593	42582 60527
00026	15474 45266	95270 79953	59367 83848	82396 10118	33211 59466
00027	94557 28573	67897 54387	54622 44431	91190 42592	92927 45973
00028	42481 16213	97344 08721	16868 48767	03071 12059	25701 46670
00029	23523 78317	73208 89837	68935 91416	26252 29663	05522 82562
00030	04493 52494	75246 33824	45862 51025	61962 79335	65337 12472
00031	00549 97654	64051 88159	96119 63896	54692 82391	23287 29529
00032	35963 15307	26898 09354	33351 35462	77974 50024	90103 39333
00033	59808 08391	45427 26842	83609 49700	13021 24892	78565 20106
00034	46058 85236	01390 92286	77281 44077	93910 83647	70617 42941
00035	32179 00597	87379 25241	05567 07007	86743 17157	85394 11838
00036	69234 61406	20117 45204	15956 60000	18743 92423	97118 96338
00037	19565' 41430	01758 75379	40419 21585	66674 36806	84962 85207
00038	45155 14938	19476 07246	43667 94543	59047 90033	20826 69541
00039	94864 31994	36168 10851	34888 81553	01540 35456	05014 51176

APPENDICES

TABLE I (*continued*)

00040	98086 24826	45240 28404	44999 08896	39094 73407	35441 31880
00041	33185 16232	41941 50949	89435 48581	88695 41994	37548 73043
00042	80951 00406	96382 70774	20151 23387	25016 25298	94624 61171
00043	79752 49140	71961 28296	69861 02591	74852 20539	00387 59579
00044	18633 32537	98145 06571	31010 24674	05455 61427	77938 91936
00045	74029 43902	77557 32270	97790 17119	52527 58021	80814 51748
00046	54178 45611	80993 37143	05335 12969	56127 19255	36040 90324
00047	11664 49883	52079 84827	59381 71539	09973 33440	88461 23356
00048	48324 77928	31249 64710	02295 36870	32307 57546	15020 09994
00049	69074 94138	87637 91976	35584 04401	10518 21615	01848 76938
00050	09188 20097	32825 39527	04220 86304	83389 87374	64278 58044
00051	90045 85497	51981 50654	94938 81997	91870 76150	68476 64659
00052	73189 50207	47677 26269	62290 64464	27124 67018	41361 82760
00053	75768 76490	20971 87749	90429 12272	95375 05871	93823 43178
00054	54016 44056	66281 31003	00682 27398	20714 53295	07706 17813
00055	08358 69910	78542 42785	13661 58873	04618 97553	31223 08420
00056	28306 03264	81333 10591	40510 07893	32604 60475	94119 01840
00057	53840 86233	81594 13628	51215 90290	28466 68795	77762 20791
00058	91757 53741	61613 62269	50263 90212	55781 76514	83483 47055
00059	89415 92694	00397 58391	12607 17646	48949 72306	94541 37408
00060	77513 03820	86864 29901	68414 82774	51908 13980	72893 55507
00061	19502 37174	69979 20288	55210 29773	74287 75251	65344 67415
00062	21818 59313	93278 81757	05686 73156	07082 85046	31853 38452
00063	51474 66499	68107 23621	94049 91345	42836 09191	08007 45449
00064	99559 68331	62535 24170	69777 12830	74819 78142	43860 72834
00065	33713 48007	93584 72869	51926 64721	58303 29822	93174 93972
00066	85274 86893	11303 22970	28834 34137	73515 90400	71148 43643
00067	84133 89640	44035 52166	73852 70091	61222 60561	62327 18423
00068	56732 16234	17395 96131	10123 91622	85496 57560	81604 18880
00069	65138 56806	87648 85261	34313 65861	45875 21069	85644 47277
00070	38001 02176	81719 11711	71602 92937	74219 64049	65584 49698
00071	37402 96397	01304 77586	56271 10086	47324 62605	40030 37438
00072	97125 40348	87083 31417	21815 39250	75237 62047	15501 29578
00073	21826 41134	47143 34072	64638 85902	49139 06441	03856 54552
00074	73135 42742	95719 09035	85794 74296	08789 88156	64691 19202
00075	07638 77929	03061 18072	96207 44156	23821 99538	04713 66994
00076	60528 83441	07954 19814	59175 20695	05533 52139	61212 06455
00077	83596 35655	06958 92983	05128 09719	77433 53783	92301 50498
00078	10850 62746	99599 10507	13499 06319	53075 71839	06410 19362
00079	39820 98952	43622 63147	64421 80814	43800 09351	31024 73167

TABLE I (*continued*)

00080	59580 06478	75569 78800	88835 54486	23768 06156	04111 08408
00081	38508 07341	23793 48763	90822 97022	17719 04207	95954 49953
00082	30692 70668	94688 16127	56196 80091	82067 63400	05462 69200
00083	65443 95659	18288 27437	49632 24041	08337 65676	96299 90836
00084	27267 50264	13192 72294	07477 44606	17985 48911	97341 30358
00085	91307 06991	19072 24210	36699 53728	28825 35793	28976 66252
00086	68434 94688	84473 13622	62126 98408	12843 82590	09815 93146
00087	48908 15877	54745 24591	35700 04754	83824 52692	54130 55160
00088	06913 45197	42672 78601	11883 09528	63011 98901	14974 40344
00089	10455 16019	14210 33712	91342 37821	88325 80851	43667 70883
00090	12883 97343	65027 61184	04285 01392	17974 15077	90712 26769
00091	21778 30976	38807 36961	31649 42096	63281 02023	08816 47449
00092	19523 59515	65122 59659	86283 68258	69572 13798	16435 91529
00093	67245 52670	35583 16563	79246 86686	76463 34222	26655 90802
00094	60584 47377	07500 37992	45134 26529	26760 83637	41326 44344
00095	53853 41377	36066 94850	58838 73859	49364 73331	96240 43642
00096	24637 38736	74384 89342	52623 07992	12369 18601	03742 83873
00097	83080 12451	38992 22815	07759 51777	97377 27585	51972 37867
00098	16444 24334	36151 99073	27493 70939	85130 32552	54846 54759
00099	60790 18157	57178 65762	11161 78576	45819 52979	65130 04860

* From *A Million Random Digits*, by the RAND Corporation (1955), The Free Press. Reproduced by permission of the publishers.

TABLE II

Values of the Standardized Normal Distribution Function $\Phi(z)$*†

z	.00	.01	.02	.03	.04	.05	.06	.07	.08	.09
.0	.5000	.5040	.5080	.5120	.5160	.5199	.5239	.5279	.5319	.5359
.1	.5398	.5438	.5478	.5517	.5557	.5596	.5636	.5675	.5714	.5753
.2	.5793	.5832	.5871	.5910	.5948	.5987	.6026	.6064	.6103	.6141
.3	.6179	.6217	.6255	.6293	.6331	.6368	.6406	.6443	.6480	.6517
.4	.6554	.6591	.6628	.6664	.6700	.6736	.6772	.6808	.6844	.6879
.5	.6915	.6950	.6985	.7019	.7054	.7088	.7123	.7157	.7190	.7224
.6	.7257	.7291	.7324	.7357	.7389	.7422	.7454	.7486	.7517	.7549
.7	.7580	.7611	.7642	.7673	.7703	.7734	.7764	.7794	.7823	.7852
.8	.7881	.7910	.7939	.7967	.7995	.8023	.8051	.8078	.8106	.8133
.9	.8159	.8186	.8212	.8238	.8264	.8289	.8315	.8340	.8365	.8389
1.0	.8413	.8438	.8461	.8485	.8508	.8531	.8554	.8577	.8599	.8621
1.1	.8643	.8665	.8686	.8708	.8729	.8749	.8770	.8790	.8810	.8830
1.2	.8849	.8869	.8888	.8907	.8925	.8944	.8962	.8980	.8997	.90147
1.3	.90320	.90490	.90658	.90824	.90988	.91149	.91309	.91466	.91621	.91774
1.4	.91924	.92073	.92220	.92364	.92507	.92647	.92785	.92922	.93056	.93189
1.5	.93319	.93448	.93574	.93699	.93822	.93943	.94062	.94179	.94295	.94408
1.6	.94520	.94630	.94738	.94845	.94950	.95053	.95154	.95254	.95352	.95449
1.7	.95543	.95637	.95728	.95818	.95907	.95994	.96080	.96164	.96246	.96327
1.8	.96407	.96485	.96562	.96638	.96712	.96784	.96856	.96926	.96995	.97062
1.9	.97128	.97193	.97257	.97320	.97381	.97441	.97500	.97558	.97615	.97670

z										
2.0	.97725	.97778	.97831	.97882	.97932	.97982	.98030	.98077	.98124	.98169
2.1	.98214	.98257	.98300	.98341	.98382	.98422	.98461	.98500	.98537	.98574
2.2	.98610	.98645	.98679	.98713	.98745	.98778	.98809	.98840	.98870	.98899
2.3	.98928	.98956	.98983	$.9^20097$	$.9^20358$	$.9^20613$	$.9^20863$	$.9^21106$	$.9^21344$	$.9^21576$
2.4	$.9^21802$	$.9^22024$	$.9^22240$	$.9^22451$	$.9^22656$	$.9^22857$	$.9^23053$	$.9^23244$	$.9^23431$	$.9^23613$
2.5	$.9^23790$	$.9^23963$	$.9^24132$	$.9^24297$	$.9^24457$	$.9^24614$	$.9^24766$	$.9^24915$	$.9^25060$	$.9^25201$
2.6	$.9^25339$	$.9^25473$	$.9^25604$	$.9^25731$	$.9^25855$	$.9^25975$	$.9^26093$	$.9^26207$	$.9^26319$	$.9^26427$
2.7	$.9^26533$	$.9^26636$	$.9^26736$	$.9^26833$	$.9^26928$	$.9^27020$	$.9^27110$	$.9^27197$	$.9^27282$	$.9^27365$
2.8	$.9^27445$	$.9^27523$	$.9^27599$	$.9^27673$	$.9^27744$	$.9^27814$	$.9^27882$	$.9^27948$	$.9^28012$	$.9^28074$
2.9	$.9^28134$	$.9^28193$	$.9^28250$	$.9^28305$	$.9^28359$	$.9^28411$	$.9^28462$	$.9^28511$	$.9^28559$	$.9^28605$
3.0	$.9^28650$	$.9^28694$	$.9^28736$	$.9^28777$	$.9^28817$	$.9^28856$	$.9^28893$	$.9^28930$	$.9^28965$	$.9^28999$
3.1	$.9^30324$	$.9^30646$	$.9^30957$	$.9^31260$	$.9^31553$	$.9^31836$	$.9^32112$	$.9^32378$	$.9^32636$	$.9^32886$
3.2	$.9^33129$	$.9^33363$	$.9^33590$	$.9^33810$	$.9^34024$	$.9^34230$	$.9^34429$	$.9^34623$	$.9^34810$	$.9^34991$
3.3	$.9^35166$	$.9^35335$	$.9^35499$	$.9^35658$	$.9^35811$	$.9^35959$	$.9^36103$	$.9^36242$	$.9^36376$	$.9^36505$
3.4	$.9^36631$	$.9^36752$	$.9^36869$	$.9^36982$	$.9^37091$	$.9^37197$	$.9^37299$	$.9^37398$	$.9^37493$	$.9^37585$
3.5	$.9^37674$	$.9^37759$	$.9^37842$	$.9^37922$	$.9^37999$	$.9^38074$	$.9^38146$	$.9^38215$	$.9^38282$	$.9^38347$
3.6	$.9^38409$	$.9^38469$	$.9^38527$	$.9^38583$	$.9^38637$	$.9^38689$	$.9^38739$	$.9^38787$	$.9^38834$	$.9^38879$
3.7	$.9^38922$	$.9^38964$	$.9^40039$	$.9^40426$	$.9^40799$	$.9^41158$	$.9^41504$	$.9^41838$	$.9^42159$	$.9^42468$
3.8	$.9^42765$	$.9^43052$	$.9^43327$	$.9^43593$	$.9^43848$	$.9^44094$	$.9^44331$	$.9^44558$	$.9^44777$	$.9^44988$
3.9	$.9^45190$	$.9^45385$	$.9^45573$	$.9^45753$	$.9^45926$	$.9^46092$	$.9^46253$	$.9^46406$	$.9^46554$	$.9^46696$
4.0	$.9^46833$	$.9^46964$	$.9^47090$	$.9^47211$	$.9^47327$	$.9^47439$	$.9^47546$	$.9^47649$	$.9^47748$	$.9^47843$

$$* \; \Phi(z) = \frac{1}{\sqrt{2\pi}} \int_{-\infty}^{z} e^{-x^2/2}\, dx$$

† From *Statistical Tables and Formulas*, by A. Hald, John Wiley and Sons, New York (1952); reproduced by permission of Professor A. Hald and the publishers.

TABLE III

Percentage points of the χ^2_m Distribution*†

m \ α	.995	.990	.975	.950	.050	.025	.010	.005
1	392704×10^{-10}	157088×10^{-9}	982069×10^{-9}	393214×10^{-8}	3.84146	5.02389	6.63490	7.87944
2	.0100251	.0201007	.0506356	.102587	5.99147	7.37776	9.21034	10.5966
3	.0717212	.114832	.215795	.351846	7.81473	9.34840	11.3449	12.8381
4	.206990	.297110	.484419	.710721	9.48773	11.1433	13.2767	14.8602
5	.411740	.554300	.831211	1.145476	11.0705	12.8325	15.0863	16.7496
6	.675727	.872085	1.237347	1.63539	12.5916	14.4494	16.8119	18.5476
7	.989265	1.239043	1.68987	2.16735	14.0671	16.0128	18.4753	20.2777
8	1.344419	1.646482	2.17973	2.73264	15.5073	17.5346	20.0902	21.9550
9	1.734926	2.087912	2.70039	3.32511	16.9190	19.0228	21.6660	23.5893
10	2.15585	2.55821	3.24697	3.94030	18.3070	20.4831	23.2093	25.1882
11	2.60321	3.05347	3.81575	4.57481	19.6751	21.9200	24.7250	26.7569
12	3.07382	3.57056	4.40379	5.22603	21.0261	23.3367	26.2170	28.2995
13	3.56503	4.10691	5.00874	5.89186	22.3621	24.7356	27.6883	29.8194
14	4.07468	4.66043	5.62872	6.57063	23.6848	26.1190	29.1413	31.3193
15	4.60094	5.22935	6.26214	7.26094	24.9958	27.4884	30.5779	32.8013
16	5.14224	5.81221	6.90766	7.96164	26.2962	28.8454	31.9999	34.2672
17	5.69724	6.40776	7.56418	8.67176	27.5871	30.1910	33.4087	35.7185
18	6.26481	7.01491	8.23075	9.39046	28.8693	31.5264	34.8053	37.1564
19	6.84398	7.63273	8.90655	10.1170	30.1435	32.8523	36.1908	38.5822

m								
20	7.43386	8.26040	9.59083	10.8508	31.4104	34.1696	37.5662	39.9968
21	8.03366	8.89720	10.28293	11.5913	32.6705	35.4789	38.9321	41.4010
22	8.64272	9.54249	10.9823	12.3380	33.9244	36.7807	40.2894	42.7956
23	9.26042	10.19567	11.6885	13.0905	35.1725	38.0757	41.6384	44.1813
24	9.88623	10.8564	12.4011	13.8484	36.4151	39.3641	42.9798	45.5585
25	10.5197	11.5240	13.1197	14.6114	37.6525	40.6465	44.3141	46.9278
26	11.1603	12.1981	13.8439	15.3791	38.8852	41.9232	45.6417	48.2899
27	11.8076	12.8786	14.5733	16.1513	40.1133	43.1944	46.9630	49.6449
28	12.4613	13.5648	15.3079	16.9279	41.3372	44.4607	48.2782	50.9933
29	13.1211	14.2565	16.0471	17.7083	42.5569	45.7222	49.5879	52.3356
30	13.7867	14.9535	16.7908	18.4926	43.7729	46.9792	50.8922	53.6720
40	20.7065	22.1643	24.4331	26.5093	55.7585	59.3417	63.6907	66.7659
50	27.9907	29.7067	32.3574	34.7642	67.5048	71.4202	76.1539	79.4900
60	35.5346	37.4848	40.4817	43.1879	79.0819	83.2976	88.3794	91.9517
70	43.2752	45.4418	48.7576	51.7393	90.5312	95.0231	100.425	104.215
80	51.1720	53.5400	57.1532	60.3915	101.879	106.629	112.329	116.321
90	59.1963	61.7541	65.6466	69.1260	113.145	118.136	124.116	128.299
100	67.3276	70.0648	74.2219	77.9295	124.342	129.561	135.807	140.169

* That is, values of $\chi^2_{m;\alpha}$, where m represents degrees of freedom and

$$\int_0^{\chi^2_{m;\alpha}} \frac{1}{2\Gamma(m/2)} \left(\frac{\chi^2}{2}\right)^{(m/2)-1} e^{-\chi^2/2} \, d\chi^2 = 1 - \alpha.$$

For $m < 100$, linear interpolation is adequate. For $m > 100$, $\sqrt{2\chi^2_m}$ is approximately normally distributed with mean $\sqrt{2m-1}$ and unit variance, so that percentage points may be obtained from Table II.

† From Biometrika Tables for Statisticians, Vol. 1 (2nd edition), Cambridge University Press (1958); edited by E. S. Pearson and H. O. Hartley; reproduced by permission of the publishers.

TABLE IV

$$\text{Values of the Function } g = \frac{1}{2} \ln \frac{1 + r}{1 - r}$$

r	g	r	g	r	g	r	g	r	g	r	g
.00	$.000_{20}$	.40	$.424_{24}$	.80	1.099_{28}	.940	1.738_{9}	.960	1.946_{13}	.980	2.298_{25}
.02	$.020_{20}$	.42	$.448_{24}$	.81	$.127_{30}$	.941	$.747_{9}$	.961	$.959_{13}$	.981	$.323_{28}$
.04	$.040_{20}$	.44	$.472_{25}$	.82	$.157_{31}$	.942	756_{8}	.962	$.972_{14}$	.982	$.351_{29}$
.06	$.060_{20}$	.46	$.497_{26}$	.83	$.188_{33}$	.943	$.764_{10}$	.963	1.986_{14}	.983	$.380_{30}$
.08	$.080_{20}$	.48	$.523_{26}$	.84	$.221_{35}$	.944	$.774_{9}$	.964	2.000_{14}	.984	$.410_{33}$
.10	$.100_{21}$	.50	$.549_{27}$	.85	1.256_{37}	.945	1.783_{9}	.965	2.014_{15}	.985	2.443_{34}
.12	$.121_{20}$	.52	$.576_{28}$	.86	$.293_{40}$	.946	$.792_{10}$	.966	$.029_{15}$	.986	$.477_{38}$
.14	$.141_{20}$	.54	$.604_{29}$	.87	$.333_{43}$	.947	$.802_{10}$	.967	$.044_{16}$	.987	$.515_{40}$
.16	$.161_{21}$	.56	$.633_{29}$	.88	$.376_{46}$	.948	$.812_{10}$	.968	$.060_{16}$	.988	$.555_{44}$
.18	$.182_{21}$	.58	$.662_{31}$	.89	$.422_{50}$	.949	$.822_{10}$	.969	$.076_{16}$	.989	$.599_{48}$
.20	$.203_{21}$	.60	$.693_{32}$	.90	1.472_{56}	.950	1.832_{10}	.970	2.092_{18}	.990	2.647_{53}
.22	$.224_{21}$	.62	$.725_{33}$	.91	$.528_{61}$	.951	$.842_{11}$	.971	$.110_{17}$	.991	$.700_{59}$
.24	$.245_{21}$	.64	$.758_{35}$	.92	$.589_{69}$	.952	$.853_{10}$	.972	$.127_{19}$	.992	$.759_{67}$
.26	$.266_{22}$	.66	$.793_{36}$	.93	$.658_{80}$	.953	$.863_{11}$	.973	$.146_{19}$	.993	$.826_{77}$
.28	$.288_{22}$	.68	$.829_{38}$	.94	.738	.954	$.874_{12}$	.974	$.165_{20}$	.994	.903
.30	$.310_{22}$	.70	$.867_{41}$	.95	1.832	.955	1.886_{11}	.975	2.185_{20}	.995	2.994
.32	$.332_{22}$	.72	$.908_{42}$	.96	1.946	.956	$.897_{12}$	.976	$.205_{22}$	.996	3.106
.34	$.354_{23}$	.74	$.950_{46}$	.97	2.092	.957	$.909_{12}$	.977	$.227_{22}$	.997	.250
.36	$.377_{23}$	.76	$.996_{49}$	.98	2.298	.958	$.921_{12}$	.978	$.249_{24}$	.998	.453
.38	$.400_{24}$	.78	1.045_{54}	.99	2.647	.959	$.933_{13}$	.979	$.273_{25}$	.999	3.800
.40	.424	.80	1.099	1.00	∞	.960	1.946	.980	2.298	1.000	∞

(see next columns)

From *Cambridge Elementary Statistical Tables*, by D. V. Lindley and J. C. P. Miller, Cambridge University Press (1958); reproduced by permission of the publishers.

TABLE V

m \ α	.25	.1	.05	.025	.01	.005
1	1.000	3.078	6.314	12.706	31.821	63.657
2	.816	1.886	2.920	4.303	6.965	9.925
3	.765	1.638	2.353	3.182	4.541	5.841
4	.741	1.533	2.132	2.776	3.747	4.604
5	.727	1.476	2.015	2.571	3.365	4.032
6	.718	1.440	1.943	2.447	3.143	3.707
7	.711	1.415	1.895	2.365	2.998	3.499
8	.706	1.397	1.860	2.306	2.896	3.355
9	.703	1.383	1.833	2.262	2.821	3.250
10	.700	1.372	1.812	2.228	2.764	3.169
11	.697	1.363	1.796	2.201	2.718	3.106
12	.695	1.356	1.782	2.179	2.681	3.055
13	.694	1.350	1.771	2.160	2.650	3.012
14	.692	1.345	1.761	2.145	2.624	2.977
15	.691	1.341	1.753	2.131	2.602	2.947
16	.690	1.337	1.746	2.120	2.583	2.921
17	.689	1.333	1.740	2.110	2.567	2.898
18	.688	1.330	1.734	2.101	2.552	2.878
19	.688	1.328	1.729	2.093	2.539	2.861
20	.687	1.325	1.725	2.086	2.528	2.845
21	.686	1.323	1.721	2.080	2.518	2.831
22	.686	1.321	1.717	2.074	2.508	2.819
23	.685	1.319	1.714	2.069	2.500	2.807
24	.685	1.318	1.711	2.064	2.492	2.797
25	.684	1.316	1.708	2.060	2.485	2.787
26	.684	1.315	1.706	2.056	2.479	2.779
27	.684	1.314	1.703	2.052	2.473	2.771
28	.683	1.313	1.701	2.048	2.467	2.763
29	.683	1.311	1.699	2.045	2.462	2.756
30	.683	1.310	1.697	2.042	2.457	2.750
40	.681	1.303	1.684	2.021	2.423	2.704
60	.679	1.296	1.671	2.000	2.390	2.660
120	.677	1.289	1.658	1.980	2.358	2.617
∞	.674	1.282	1.645	1.960	2.326	2.576

* That is, values of $t_{m;\alpha}$, where m equals degrees of freedom and

$$\int_{-\infty}^{t_{m;\alpha}} \frac{\Gamma[(m+1)/2]}{\sqrt{\pi m}\,\Gamma(m/2)} \left(1 + \frac{t^2}{m}\right)^{-(m+1)/2} dt = 1 - \alpha.$$

† From *Biometrika Tables for Statisticians*, Vol. 1 (2nd edition) Cambridge University Press (1958); edited by E. S. Pearson and H. O. Hartley; reproduced by permission of the publishers.

‡ Where necessary, interpolation should be carried out using the reciprocals of the degrees of freedom, and for this the function $120/m$ is convenient.

TABLE VI PERCENTAGE POINTS

$\alpha = .10$

m_1 / m_2	1	2	3	4	5	6	7	8	9
1	39.864	49.500	53.593	55.833	57.241	58.204	58.906	59.439	59.858
2	8.5263	9.0000	9.1618	9.2434	9.2926	9.3255	9.3491	9.3668	9.3805
3	5.5383	5.4624	5.3908	5.3427	5.3092	5.2847	5.2662	5.2517	5.2400
4	4.5448	4.3246	4.1908	4.1073	4.0506	4.0098	3.9790	3.9549	3.9357
5	4.0604	3.7797	3.6195	3.5202	3.4530	3.4045	3.3679	3.3393	3.3163
6	3.7760	3.4633	3.2888	3.1808	3.1075	3.0546	3.0145	2.9830	2.9577
7	3.5894	3.2574	3.0741	2.9605	2.8833	2.8274	2.7849	2.7516	2.7247
8	3.4579	3.1131	2.9238	2.8064	2.7265	2.6683	2.6241	2.5893	2.5612
9	3.3603	3.0065	2.8129	2.6927	2.6106	2.5509	2.5053	2.4694	2.4403
10	3.2850	2.9245	2.7277	2.6053	2.5216	2.4606	2.4140	2.3772	2.3473
11	3.2252	2.8595	2.6602	2.5362	2.4512	2.3891	2.3416	2.3040	2.2735
12	3.1765	2.8068	2.6055	2.4801	2.3940	2.3310	2.2828	2.2446	2.2135
13	3.1362	2.7632	2.5603	2.4337	2.3467	2.2830	2.2341	2.1953	2.1638
14	3.1022	2.7265	2.5222	2.3947	2.3069	2.2426	2.1931	2.1539	2.1220
15	3.0732	2.6952	2.4898	2.3614	2.2730	2.2081	2.1582	2.1185	2.0862
16	3.0481	2.6682	2.4618	2.3327	2.2438	2.1783	2.1280	2.0880	2.0553
17	3.0262	2.6446	2.4374	2.3077	2.2183	2.1524	2.1017	2.0613	2.0284
18	3.0070	2.6239	2.4160	2.2858	2.1958	2.1296	2.0785	2.0379	2.0047
19	2.9899	2.6056	2.3970	2.2663	2.1760	2.1094	2.0580	2.0171	1.9836
20	2.9747	2.5893	2.3801	2.2489	2.1582	2.0913	2.0397	1.9985	1.9649
21	2.9609	2.5746	2.3649	2.2333	2.1423	2.0751	2.0232	1.9819	1.9480
22	2.9486	2.5613	2.3512	2.2193	2.1279	2.0605	2.0084	1.9668	1.9327
23	2.9374	2.5493	2.3387	2.2065	2.1149	2.0472	1.9949	1.9531	1.9189
24	2.9271	2.5383	2.3274	2.1949	2.1030	2.0351	1.9826	1.9407	1.9063
25	2.9177	2.5283	2.3170	2.1843	2.0922	2.0241	1.9714	1.9292	1.8947
26	2.9091	2.5191	2.3075	2.1745	2.0822	2.0139	1.9610	1.9188	1.8841
27	2.9012	2.5106	2.2987	2.1655	2.0730	2.0045	1.9515	1.9091	1.8743
28	2.8939	2.5028	2.2906	2.1571	2.0645	1.9959	1.9427	1.9001	1.8652
29	2.8871	2.4955	2.2831	2.1494	2.0566	1.9878	1.9345	1.8918	1.8560
30	2.8807	2.4887	2.2761	2.1422	2.0492	1.9803	1.9269	1.8841	1.8498
40	2.8354	2.4404	2.2261	2.0909	1.9968	1.9269	1.8725	1.8289	1.7929
60	2.7914	2.3932	2.1774	2.0410	1.9457	1.8747	1.8194	1.7748	1.7380
120	2.7478	2.3473	2.1300	1.9923	1.8959	1.8238	1.7675	1.7220	1.6843
∞	2.7055	2.3026	2.0838	1.9449	1.8473	1.7741	1.7167	1.6702	1.6315

OF THE F_{m_1, m_2} DISTRIBUTION*†

$$\alpha = .10$$

10	12	15	20	24	30	40	60	120	∞
60.195	60.705	61.220	61.740	62.002	62.265	62.529	62.794	63.061	63.328
9.3916	9.4081	9.4247	9.4413	9.4496	9.4579	9.4663	9.4746	9.4829	9.4913
5.2304	5.2156	5.2003	5.1845	5.1764	5.1681	5.1597	5.1512	5.1425	5.1337
3.9199	3.8955	3.8689	3.8443	3.8310	3.8174	3.8036	3.7896	3.7753	3.7607
3.2974	3.2682	3.2380	3.2067	3.1905	3.1741	3.1573	3.1402	3.1228	3.1050
2.9369	2.9047	2.8712	2.8363	2.8183	2.8000	2.7812	2.7620	2.7423	2.7222
2.7025	2.6681	2.6322	2.5947	2.5753	2.5555	2.5351	2.5142	2.4928	2.4708
2.5380	2.5020	2.4642	2.4246	2.4041	2.3830	2.3614	2.3391	2.3162	2.2926
2.4163	2.3789	2.3396	2.2983	2.2768	2.2547	2.2320	2.2085	2.1843	2.1592
2.3226	2.2841	2.2435	2.2007	2.1784	2.1554	2.1317	2.1072	2.0818	2.0554
2.2482	2.2087	2.1671	2.1230	2.1000	2.0762	2.0516	2.0261	1.9997	1.9721
2.1878	2.1474	2.1049	2.0597	2.0360	2.0115	1.9861	1.9597	1.9323	1.9036
2.1376	2.0966	2.0532	2.0070	1.9827	1.9576	1.9315	1.9043	1.8759	1.8462
2.0954	2.0537	2.0095	1.9625	1.9377	1.9119	1.8852	1.8572	1.8280	1.7973
2.0593	2.0171	1.9722	1.9243	1.8990	1.8728	1.8454	1.8168	1.7867	1.7551
2.0281	1.9854	1.9399	1.8913	1.8656	1.8388	1.8108	1.7816	1.7507	1.7182
2.0009	1.9577	1.9117	1.8624	1.8362	1.8090	1.7805	1.7506	1.7191	1.6856
1.9770	1.9333	1.8868	1.8368	1.8103	1.7827	1.7537	1.7232	1.6910	1.6567
1.9557	1.9117	1.8647	1.8142	1.7873	1.7592	1.7298	1.6988	1.6659	1.6308
1.9367	1.8924	1.8449	1.7938	1.7667	1.7382	1.7083	1.6768	1.6433	1.6074
1.9197	1.8750	1.8272	1.7756	1.7481	1.7193	1.6890	1.6569	1.6228	1.5862
1.9043	1.8593	1.8111	1.7590	1.7312	1.7021	1.6714	1.6389	1.6042	1.5668
1.8903	1.8450	1.7964	1.7439	1.7159	1.6864	1.6554	1.6224	1.5871	1.5490
1.8775	1.8319	1.7831	1.7302	1.7019	1.6721	1.6407	1.6073	1.5715	1.5327
1.8658	1.8200	1.7708	1.7175	1.6890	1.6589	1.6272	1.5934	1.5570	1.5176
1.8550	1.8090	1.7596	1.7059	1.6771	1.6468	1.6147	1.5805	1.5437	1.5036
1.8451	1.7989	1.7492	1.6951	1.6662	1.6356	1.6032	1.5686	1.5313	1.4906
1.8359	1.7895	1.7395	1.6852	1.6560	1.6252	1.5925	1.5575	1.5198	1.4784
1.8274	1.7808	1.7306	1.6759	1.6465	1.6155	1.5825	1.5472	1.5090	1.4670
1.8195	1.7727	1.7223	1.6673	1.6377	1.6065	1.5732	1.5376	1.4989	1.4564
1.7627	1.7146	1.6624	1.6052	1.5741	1.5411	1.5056	1.4672	1.4248	1.3769
1.7070	1.6574	1.6034	1.5435	1.5107	1.4755	1.4373	1.3952	1.3476	1.2915
1.6524	1.6012	1.5450	1.4821	1.4472	1.4094	1.3676	1.3203	1.2646	1.1926
1.5987	1.5458	1.4871	1.4206	1.3832	1.3419	1.2951	1.2400	1.1686	1.0000

APPENDICES

TABLE VI

$\alpha = .05$

m_2 \ m_1	1	2	3	4	5	6	7	8	9
1	161.45	199.50	215.71	224.58	230.16	233.99	236.77	238.88	240.54
2	18.513	19.000	19.164	19.247	19.296	19.330	19.353	19.371	19.385
3	10.128	9.5521	9.2766	9.1172	9.0135	8.9406	8.8868	8.8452	8.8123
4	7.7086	6.9443	6.5914	6.3883	6.2560	6.1631	6.0942	6.0410	5.9988
5	6.6079	5.7861	5.4095	5.1922	5.0503	4.9503	4.8759	4.8183	4.7725
6	5.9874	5.1433	4.7571	4.5337	4.3874	4.2839	4.2066	4.1468	4.0990
7	5.5914	4.7374	4.3468	4.1203	3.9715	3.8660	3.7870	3.7257	3.6767
8	5.3177	4.4590	4.0662	3.8378	3.6875	3.5806	3.5005	3.4381	3.3881
9	5.1174	4.2565	3.8626	3.6331	3.4817	3.3738	3.2927	3.2296	3.1789
10	4.9646	4.1028	3.7083	3.4780	3.3258	3.2172	3.1355	3.0717	3.0204
11	4.8443	3.9823	3.5874	3.3567	3.2039	3.0946	3.0123	2.9480	2.8962
12	4.7472	3.8853	3.4903	3.2592	3.1059	2.9961	2.9134	2.8486	2.7964
13	4.6672	3.8056	3.4105	3.1791	3.0254	2.9153	2.8321	2.7669	2.7144
14	4.6001	3.7389	3.3439	3.1122	2.9582	2.8477	2.7642	2.6987	2.6458
15	4.5431	3.6823	3.2874	3.0556	2.9013	2.7905	2.7066	2.6408	2.5876
16	4.4940	3.6337	3.2389	3.0069	2.8524	2.7413	2.6572	2.5911	2.5377
17	4.4513	3.5915	3.1968	2.9647	2.8100	2.6987	2.6143	2.5480	2.4943
18	4.4139	3.5546	3.1599	2.9277	2.7729	2.6613	2.5767	2.5102	2.4563
19	4.3808	3.5219	3.1274	2.8951	2.7401	2.6283	2.5435	2.4768	2.4227
20	4.3513	3.4928	3.0984	2.8661	2.7109	2.5990	2.5140	2.4471	2.3928
21	4.3248	3.4668	3.0725	2.8401	2.6848	2.5727	2.4876	2.4205	2.3661
22	4.3009	3.4434	3.0491	2.8167	2.6613	2.5491	2.4638	2.3965	2.3419
23	4.2793	3.4221	3.0280	2.7955	2.6400	2.5277	2.4422	2.3748	2.3201
24	4.2597	3.4028	3.0088	2.7763	2.6207	2.5082	2.4226	2.3551	2.3002
25	4.2417	3.3852	2.9912	2.7587	2.6030	2.4904	2.4047	2.3371	2.2821
26	4.2252	3.3690	2.9751	2.7426	2.5868	2.4741	2.3883	2.3205	2.2655
27	4.2100	3.3541	2.9604	2.7278	2.5719	2.4591	2.3732	2.3053	2.2501
28	4.1960	3.3404	2.9467	2.7141	2.5581	2.4453	2.3593	2.2913	2.2360
29	4.1830	3.3277	2.9340	2.7014	2.5454	2.4324	2.3463	2.2782	2.2229
30	4.1709	3.3158	2.9223	2.6896	2.5336	2.4205	2.3343	2.2662	2.2107
40	4.0848	3.2317	2.8387	2.6060	2.4495	2.3359	2.2490	2.1802	2.1240
60	4.0012	3.1504	2.7581	2.5252	2.3683	2.2540	2.1665	2.0970	2.0401
120	3.9201	3.0718	2.6802	2.4472	2.2900	2.1750	2.0867	2.0164	1.9588
∞	3.8415	2.9957	2.6049	2.3719	2.2141	2.0986	2.0096	1.9384	1.8799

(*continued*)

α = .05

10	12	15	20	24	30	40	60	120	∞
241.88	243.91	245.95	248.01	249.05	250.09	251.14	252.20	253.25	254.32
19.396	19.413	19.429	19.446	19.454	19.462	19.471	19.479	19.487	19.496
8.7855	8.7446	8.7029	8.6602	8.6385	8.6166	8.5944	8.5720	8.5494	8.5265
5.9644	5.9117	5.8578	5.8025	5.7744	5.7459	5.7170	5.6878	5.6581	5.6281
4.7351	4.6777	4.6188	4.5581	4.5272	4.4957	4.4638	4.4314	4.3984	4.3650
4.0600	3.9999	3.9381	3.8742	3.8415	3.8082	3.7743	3.7398	3.7047	3.6688
3.6365	3.5747	3.5108	3.4445	3.4105	3.3758	3.3404	3.3043	3.2674	3.2298
3.3472	3.2840	3.2184	3.1503	3.1152	3.0794	3.0428	3.0053	2.9669	2.9276
3.1373	3.0729	3.0061	2.9365	2.9005	2.8637	2.8259	2.7872	2.7475	2.7067
2.9782	2.9130	2.8450	2.7740	2.7372	2.6996	2.6609	2.6211	2.5801	2.5379
2.8536	2.7876	2.7186	2.6464	2.6090	2.5705	2.5309	2.4901	2.4480	2.4045
2.7534	2.6866	2.6169	2.5436	2.5055	2.4663	2.4259	2.3842	2.3410	2.2962
2.6710	2.6037	2.5331	2.4589	2.4202	2.3803	2.3392	2.2966	2.2524	2.2064
2.6021	2.5342	2.4630	2.3879	2.3487	2.3082	2.2664	2.2230	2.1778	2.1307
2.5437	2.4753	2.4035	2.3275	2.2878	2.2468	2.2043	2.1601	2.1141	2.0658
2.4935	2.4247	2.3522	2.2756	2.2354	2.1938	2.1507	2.1058	2.0589	2.0096
2.4499	2.3807	2.3077	2.2304	2.1898	2.1477	2.1040	2.0584	2.0107	1.9604
2.4117	2.3421	2.2686	2.1906	2.1497	2.1071	2.0629	2.0166	1.9681	1.9168
2.3779	2.3080	2.2341	2.1555	2.1141	2.0712	2.0264	1.9796	1.9302	1.8780
2.3479	2.2776	2.2033	2.1242	2.0825	2.0391	1.9938	1.9464	1.8963	1.8432
2.3210	2.2504	2.1757	2.0960	2.0540	2.0102	1.9645	1.9165	1.8657	1.8117
2.2967	2.2258	2.1508	2.0707	2.0283	1.9842	1.9380	1.8895	1.8380	1.7831
2.2747	2.2036	2.1282	2.0476	2.0050	1.9605	1.9139	1.8649	1.8128	1.7570
2.2547	2.1834	2.1077	2.0267	1.9838	1.9390	1.8920	1.8424	1.7897	1.7331
2.2365	2.1649	2.0889	2.0075	1.9643	1.9192	1.8718	1.8217	1.7684	1.7110
2.2197	2.1479	2.0716	1.9898	1.9464	1.9010	1.8533	1.8027	1.7488	1.6906
2.2043	2.1323	2.0558	1.9736	1.9299	1.8842	1.8361	1.7851	1.7307	1.6717
2.1900	2.1179	2.0411	1.9586	1.9147	1.8687	1.8203	1.7689	1.7138	1.6541
2.1768	2.1045	2.0275	1.9446	1.9005	1.8543	1.8055	1.7537	1.6981	1.6377
2.1646	2.0921	2.0148	1.9317	1.8874	1.8409	1.7918	1.7396	1.6835	1.6223
2.0772	2.0035	1.9245	1.8389	1.7929	1.7444	1.6928	1.6373	1.5766	1.5089
1.9926	1.9174	1.8364	1.7480	1.7001	1.6491	1.5943	1.5343	1.4673	1.3893
1.9105	1.8337	1.7505	1.6587	1.6084	1.5543	1.4952	1.4290	1.3519	1.2539
1.8307	1.7522	1.6664	1.5705	1.5173	1.4591	1.3940	1.3180	1.2214	1.0000

TABLE VI

$$\alpha = .025$$

m_1 m_2	1	2	3	4	5	6	7	8	9
1	647.79	799.50	864.16	899.58	921.85	937.11	948.22	956.66	963.28
2	38.506	39.000	39.165	39.248	39.298	39.331	39.355	39.373	39.387
3	17.443	16.044	15.439	15.101	14.885	14.735	14.624	14.540	14.473
4	12.218	10.649	9.9792	9.6045	9.3645	9.1973	9.0741	8.9796	8.9047
5	10.007	8.4336	7.7636	7.3879	7.1464	6.9777	6.8531	6.7572	6.6810
6	8.8131	7.2598	6.5988	6.2272	5.9876	5.8197	5.6955	5.5996	5.5234
7	8.0727	6.5415	5.8898	5.5226	5.2852	5.1186	4.9949	4.8994	4.8232
8	7.5709	6.0595	5.4160	5.0526	4.8173	4.6517	4.5286	4.4332	4.3572
9	7.2093	5.7147	5.0781	4.7181	4.4844	4.3197	4.1971	4.1020	4.0260
10	6.9367	5.4564	4.8256	4.4683	4.2361	4.0721	3.9498	3.8549	3.7790
11	6.7241	5.2559	4.6300	4.2751	4.0440	3.8807	3.7586	3.6638	3.5879
12	6.5538	5.0959	4.4742	4.1212	3.8911	3.7283	3.6065	3.5118	3.4358
13	6.4143	4.9653	4.3472	3.9959	3.7667	3.6043	3.4827	3.3880	3.3120
14	6.2979	4.8567	4.2417	3.8919	3.6634	3.5014	3.3799	3.2853	3.2093
15	6.1995	4.7650	4.1528	3.8043	3.5764	3.4147	3.2934	3.1987	3.1227
16	6.1151	4.6867	4.0768	3.7294	3.5021	3.3406	3.2194	3.1248	3.0488
17	6.0420	4.6189	4.0112	3.6648	3.4379	3.2767	3.1556	3.0610	2.9849
18	5.9781	4.5597	3.9539	3.6083	3.3820	3.2209	3.0999	3.0053	2.9291
19	5.9216	4.5075	3.9034	3.5587	3.3327	3.1718	3.0509	2.9563	2.8800
20	5.8715	4.4613	3.8587	3.5147	3.2891	3.1283	3.0074	2.9128	2.8365
21	5.8266	4.4199	3.8188	3.4754	3.2501	3.0895	2.9686	2.8740	2.7977
22	5.7863	4.3828	3.7829	3.4401	3.2151	3.0546	2.9338	2.8392	2.7628
23	5.7498	4.3492	3.7505	3.4083	3.1835	3.0232	2.9024	2.8077	2.7313
24	5.7167	4.3187	3.7211	3.3794	3.1548	2.9946	2.8738	2.7791	2.7027
25	5.6864	4.2909	3.6943	3.3530	3.1287	2.9685	2.8478	2.7531	2.6766
26	5.6586	4.2655	3.6697	3.3289	3.1048	2.9447	2.8240	2.7293	2.6528
27	5.6331	4.2421	3.6472	3.3067	3.0828	2.9228	2.8021	2.7074	2.6309
28	5.6096	4.2205	3.6264	3.2863	3.0625	2.9027	2.7820	2.6872	2.6106
29	5.5878	4.2006	3.6072	3.2674	3.0438	2.8840	2.7633	2.6686	2.5919
30	5.5675	4.1821	3.5894	3.2499	3.0265	2.8667	2.7460	2.6513	2.5746
40	5.4239	4.0510	3.4633	3.1261	2.9037	2.7444	2.6238	2.5289	2.4519
60	5.2857	3.9253	3.3425	3.0077	2.7863	2.6274	2.5068	2.4117	2.3344
120	5.1524	3.8046	3.2270	2.8943	2.6740	2.5154	2.3948	2.2994	2.2217
∞	5.0239	3.6889	3.1161	2.7858	2.5665	2.4082	2.2875	2.1918	2.1136

(*continued*)

$\alpha = .025$

10	12	15	20	24	30	40	60	120	∞
968.63	976.71	984.87	993.10	997.25	1001.4	1005.6	1009.8	1014.0	1018.3
39.398	39.415	39.431	39.448	39.456	39.465	39.473	39.481	39.490	39.498
14.419	14.337	14.253	14.167	14.124	14.081	14.037	13.992	13.947	13.902
8.8439	8.7512	8.6565	8.5599	8.5109	8.4613	8.4111	8.3604	8.3092	8.2573
6.6192	6.5246	6.4277	6.3285	6.2780	6.2269	6.1751	6.1225	6.0693	6.0153
5.4613	5.3662	5.2687	5.1684	5.1172	5.0652	5.0125	5.9589	4.9045	4.8491
4.7611	4.6658	4.5678	4.4667	4.4150	4.3624	4.3089	4.2544	4.1989	4.1423
4.2951	4.1997	4.1012	3.9995	3.9472	3.8940	3.8398	3.7844	3.7279	3.6702
3.9639	3.8682	3.7694	3.6669	3.6142	3.5604	3.5055	3.4493	3.3918	3.3329
3.7168	3.6209	3.5217	3.4186	3.3654	3.3110	3.2554	3.1984	3.1399	3.0798
3.5257	3.4296	3.3299	3.2261	3.1725	3.1176	3.0613	3.0035	2.9441	2.8828
3.3736	3.2773	3.1772	3.0728	3.0187	2.9633	2.9063	2.8478	2.7874	2.7249
3.2497	3.1532	3.0527	2.9477	2.8932	2.8373	2.7797	2.7204	2.6590	2.5955
3.1469	3.0501	2.9493	2.8437	2.7888	2.7324	2.6742	2.6142	2.5519	2.4872
3.0602	2.9633	2.8621	2.7559	2.7006	2.6437	2.5850	2.5242	2.4611	2.3953
2.9862	2.8890	2.7875	2.6808	2.6252	2.5678	2.5085	2.4471	2.3831	2.3163
2.9222	2.8249	2.7230	2.6158	2.5598	2.5021	2.4422	2.3801	2.3153	2.2474
2.8664	2.7689	2.6667	2.5590	2.5027	2.4445	2.3842	2.3214	2.2558	2.1869
2.8173	2.7196	2.6171	2.5089	2.4523	2.3937	2.3329	2.2695	2.2032	2.1333
2.7737	2.6758	2.5731	2.4645	2.4076	2.3486	2.2873	2.2234	2.1562	2.0853
2.7348	2.6368	2.5338	2.4247	2.3675	2.3082	2.2465	2.1819	2.1141	2.0422
2.6998	2.6017	2.4984	2.3890	2.3315	2.2718	2.2097	2.1446	2.0760	2.0032
2.6682	2.5699	2.4665	2.3567	2.2989	2.2389	2.1763	2.1107	2.0415	1.9677
2.6396	2.5412	2.4374	2.3273	2.2693	2.2090	2.1460	2.0799	2.0099	1.9353
2.6135	2.5149	2.4110	2.3005	2.2422	2.1816	2.1183	2.0517	1.9811	1.9055
2.5895	2.4909	2.3867	2.2759	2.2174	2.1565	2.0928	2.0257	1.9545	1.8781
2.5676	2.4688	2.3644	2.2533	2.1946	2.1334	2.0693	2.0018	1.9299	1.8527
2.5473	2.4484	2.3438	2.2324	2.1735	2.1121	2.0477	1.9796	1.9072	1.8291
2.5286	2.4295	2.3248	2.2131	2.1540	2.0923	2.0276	1.9591	1.8861	1.8072
2.5112	2.4120	2.3072	2.1952	2.1359	2.0739	2.0089	1.9400	1.8664	1.7867
2.3882	2.2882	2.1819	2.0677	2.0069	1.9429	1.8752	1.8028	1.7242	1.6371
2.2702	2.1692	2.0613	1.9445	1.8817	1.8152	1.7440	1.6668	1.5810	1.4822
2.1570	2.0548	1.9450	1.8249	1.7597	1.6899	1.6141	1.5299	1.4327	1.3104
2.0483	1.9447	1.8326	1.7085	1.6402	1.5660	1.4835	1.3883	1.2684	1.0000

TABLE VI

$\alpha = .01$

m_2 \ m_1	1	2	3	4	5	6	7	8	9
1	4052.2	4999.5	5403.3	5624.6	5763.7	5859.0	5928.3	5981.6	6022.5
2	98.503	99.000	99.166	99.249	99.299	99.332	99.356	99.374	99.388
3	34.116	30.817	29.457	28.710	28.237	27.911	27.672	27.489	27.345
4	21.198	18.000	16.694	15.977	15.522	15.207	14.976	14.799	14.659
5	16.258	13.274	12.060	11.392	10.967	10.672	10.456	10.289	10.158
6	13.745	10.925	9.7795	9.1483	8.7459	8.4661	8.2600	8.1016	7.9761
7	12.246	9.5466	8.4513	7.8467	7.4604	7.1914	6.9928	6.8401	6.7188
8	11.259	8.6491	7.5910	7.0060	6.6318	6.3707	6.1776	6.0289	5.9106
9	10.561	8.0215	6.9919	6.4221	6.0569	5.8018	5.6129	5.4671	5.3511
10	10.044	7.5594	6.5523	5.9943	5.6363	5.3858	5.2001	5.0567	4.9424
11	9.6460	7.2057	6.2167	5.6683	5.3160	5.0692	4.8861	4.7445	4.6315
12	9.3302	6.9266	5.9526	5.4119	5.0643	4.8206	4.6395	4.4994	4.3875
13	9.0738	6.7010	5.7394	5.2053	4.8616	4.6204	4.4410	4.3021	4.1911
14	8.8616	6.5149	5.5639	5.0354	4.6950	4.4558	4.2779	4.1399	4.0297
15	8.6831	6.3589	5.4170	4.8932	4.5556	4.3183	4.1415	4.0045	3.8948
16	8.5310	6.2262	5.2922	4.7726	4.4374	4.2016	4.0259	3.8896	3.7804
17	8.3997	6.1121	5.1850	4.6690	4.3359	4.1015	3.9267	3.7910	3.6822
18	8.2854	6.0129	5.0919	4.5790	4.2479	4.0146	3.8406	3.7054	3.5971
19	8.1850	5.9259	5.0103	4.5003	4.1708	3.9386	3.7653	3.6305	3.5225
20	8.0960	5.8489	4.9382	4.4307	4.1027	3.8714	3.6987	3.5644	3.4567
21	8.0166	5.7804	4.8740	4.3688	4.0421	3.8117	3.6396	3.5056	3.3981
22	7.9454	5.7190	4.8166	4.3134	3.9880	3.7583	3.5867	3.4530	3.3458
23	7.8811	5.6637	4.7649	4.2635	3.9392	3.7102	3.5390	3.4057	3.2986
24	7.8229	5.6136	4.7181	4.2184	3.8951	3.6667	3.4959	3.3629	3.2560
25	7.7698	5.5680	4.6755	4.1774	3.8550	3.6272	3.4568	3.3239	3.2172
26	7.7213	5.5263	4.6366	4.1400	3.8183	3.5911	3.4210	3.2884	3.1818
27	7.6767	5.4881	4.6009	4.1056	3.7848	3.5580	3.3882	3.2558	3.1494
28	7.6356	5.4529	4.5681	4.0740	3.7539	3.5276	3.3581	3.2259	3.1195
29	7.5976	5.4205	4.5378	4.0449	3.7254	3.4995	3.3302	3.1982	3.0920
30	7.5625	5.3904	4.5097	4.0179	3.6990	3.4735	3.3045	3.1726	3.0665
40	7.3141	5.1785	4.3126	3.8283	3.5138	3.2910	3.1238	2.9930	2.8876
60	7.0771	4.9774	4.1259	3.6491	3.3389	3.1187	2.9530	2.8233	2.7185
120	6.8510	4.7865	3.9493	3.4796	3.1735	2.9559	2.7918	2.6629	2.5586
∞	6.6349	4.6052	3.7816	3.3192	3.0173	2.8020	2.6393	2.5113	2.4073

(*continued*)

$$\alpha = .01$$

10	12	15	20	24	30	40	60	120	∞
6055.8	6106.3	6157.3	6208.7	6234.6	6260.7	6286.8	6313.0	6339.4	6366.0
99.399	99.416	99.432	99.449	99.458	99.466	99.474	99.483	99.491	99.501
27.229	27.052	26.872	26.690	26.598	26.505	26.411	26.316	26.221	26.125
14.546	14.374	14.198	14.020	13.929	13.838	13.745	13.652	13.558	13.463
10.051	9.8883	9.7222	9.5527	9.4665	9.3793	9.2912	9.2020	9.1118	9.0204
7.8741	7.7183	7.5590	7.3958	7.3127	7.2285	7.1432	7.0568	6.9690	6.8801
6.6201	6.4691	6.3143	6.1554	6.0743	5.9921	5.9084	5.8236	5.7372	5.6495
5.8143	5.6668	5.5151	5.3591	5.2793	5.1981	5.1156	5.0316	4.9460	4.8588
5.2565	5.1114	4.9621	4.8080	4.7290	4.6486	4.5667	4.4831	4.3978	4.3105
4.8492	4.7059	4.5582	4.4054	4.3269	4.2469	4.1653	4.0819	3.9965	3.9090
4.5393	4.3974	4.2509	4.0990	4.0209	3.9411	3.8596	3.7761	3.6904	3.6025
4.2961	4.1553	4.0096	3.8584	3.7805	3.7008	3.6192	3.5355	3.4494	3.3608
4.1003	3.9603	3.8154	3.6646	3.5868	3.5070	3.4253	3.3413	3.2548	3.1654
3.9394	3.8001	3.6557	3.5052	3.4274	3.3476	3.2656	3.1813	3.0942	3.0040
3.8049	3.6662	3.5222	3.3719	3.2940	3.2141	3.1319	3.0471	2.9595	2.8684
3.6909	3.5527	3.4089	3.2588	3.1808	3.1007	3.0182	2.9330	2.8447	2.7528
3.5931	3.4552	3.3117	3.1615	3.0835	3.0032	2.9205	2.8348	2.7459	2.6530
3.5082	3.3706	3.2273	3.0771	2.9990	2.9185	2.8354	2.7493	2.6597	2.5660
3.4338	3.2965	3.1533	3.0031	2.9249	2.8442	2.7608	2.6742	2.5839	2.4893
3.3682	3.2311	3.0880	2.9377	2.8594	2.7785	2.6947	2.6077	2.5168	2.4212
3.3098	3.1729	3.0299	2 8796	2.8011	2.7200	2.6359	2.5484	2.4568	2.3603
3.2576	3.1209	2.9780	2.8274	2.7488	2.6675	2.5831	2.4951	2.4029	2.3055
3.2106	3.0740	2.9311	2.7805	2.7017	2.6202	2.5355	2.4471	2.3542	2.2559
3.1681	3.0316	2.8887	2.7380	2.6591	2.5773	2.4923	2.4035	2.3099	2.2107
3.1294	2.9931	2.8502	2.6993	2.6203	2.5383	2.4530	2.3637	2.2695	2.1694
3.0941	2.9579	2.8150	2.6640	2.5848	2.5026	2.4170	2.3273	2.2325	2.1315
3.0618	2.9256	2.7827	2.6316	2.5522	2.4699	2.3840	2.2938	2.1984	2.0965
3.0320	2.8959	2.7530	2.6017	2.5223	2.4397	2.3535	2.2629	2.1670	2.0642
3.0045	2.8685	2.7256	2.5742	2.4946	2.4118	2.3253	2.2344	2.1378	2.0342
2.9791	2.8431	2.7002	2.5487	2.4689	2.3860	2.2992	2.2079	2.1107	2.0062
2.8005	2.6648	2.5216	2.3689	2.2880	2.2034	2.1142	2.0194	1.9172	1.8047
2.6318	2.4961	2.3523	2.1978	2.1154	2.0285	1.9360	1.8363	1.7263	1.6006
2.4721	2.3363	2.1915	2.0346	1.9500	1.8600	1.7628	1.6557	1.5330	1.3805
2.3209	2.1848	2.0385	1.8783	1.7908	1.6964	1.5923	1.4730	1.3246	1.0000

APPENDICES

TABLE VI

$\alpha = .005$

m_2 \ m_1	1	2	3	4	5	6	7	8	9
1	16211	20000	21615	22500	23056	23437	23715	23925	24091
2	198.50	199.00	199.17	199.25	199.30	199.33	199.36	199.37	199.39
3	55.552	49.799	47.467	46.195	45.392	44.838	44.434	44.126	43.882
4	31.333	26.284	24.259	23.155	22.456	21.975	21.622	21.352	21.139
5	22.785	18.314	16.530	15.556	14.940	14.513	14.200	13.961	13.772
6	18.635	14.544	12.917	12.028	11.464	11.073	10.786	10.566	10.391
7	16.236	12.404	10.882	10.050	9.5221	9.1554	8.8854	8.6781	8.5138
8	14.688	11.042	9.5965	8.8051	8.3018	7.9520	7.6942	7.4960	7.3386
9	13.614	10.107	8.7171	7.9559	7.4711	7.1338	6.8849	6.6933	6.5411
10	12.826	9.4270	8.0807	7.3428	6.8723	6.5446	6.3025	6.1159	5.9676
11	12.226	8.9122	7.6004	6.8809	6.4217	6.1015	5.8648	5.6821	5.5368
12	11.754	8.5096	7.2258	6.5211	6.0711	5.7570	5.5245	5.3451	5.2021
13	11.374	8.1865	6.9257	6.2335	5.7910	5.4819	5.2529	5.0761	4.9351
14	11.060	7.9217	6.6803	5.9984	5.5623	5.2574	5.0313	4.8566	4.7173
15	10.798	7.7008	6.4760	5.8029	5.3721	5.0708	4.8473	4.6743	4.5364
16	10.575	7.5138	6.3034	5.6378	5.2117	4.9134	4.6920	4.5207	4.3838
17	10.384	7.3536	6.1556	5.4967	5.0746	4.7789	4.5594	4.3893	4.2535
18	10.218	7.2148	6.0277	5.3746	4.9560	4.6627	4.4448	4.2759	4.1410
19	10.073	7.0935	5.9161	5.2681	4.8526	4.5614	4.3448	4.1770	4.0428
20	9.9439	6.9865	5.8177	5.1743	4.7616	4.4721	4.2569	4.0900	3.9564
21	9.8295	6.8914	5.7304	5.0911	4.6808	4.3931	4.1789	4.0128	3.8799
22	9.7271	6.8064	5.6524	5.0168	4.6088	4.3225	4.1094	3.9440	3.8116
23	9.6348	6.7300	5.5823	4.9500	4.5441	4.2591	4.0469	3.8822	3.7502
24	9.5513	6.6610	5.5190	4.8898	4.4857	4.2019	3.9905	3.8264	3.6949
25	9.4753	6.5982	5.4615	4.8351	4.4327	4.1500	3.9394	3.7758	3.6447
26	9.4059	6.5409	5.4091	4.7852	4.3844	4.1027	3.8928	3.7297	3.5989
27	9.3423	6.4885	5.3611	4.7396	4.3402	4.0594	3.8501	3.6875	3.5571
28	9.2838	6.4403	5.3170	4.6977	4.2996	4.0197	3.8110	3.6487	3.5186
29	9.2297	6.3958	5.2764	4.6591	4.2622	3.9830	3.7749	3.6130	3.4832
30	9.1797	6.3547	5.2388	4.6233	4.2276	3.9492	3.7416	3.5801	3.4505
40	8.8278	6.0664	4.9759	4.3738	3.9860	3.7129	3.5088	3.3498	3.2220
60	8.4946	5.7950	4.7290	4.1399	3.7600	3.4918	3.2911	3.1344	3.0083
120	8.1790	5.5393	4.4973	3.9207	3.5482	3.2849	3.0874	2.9330	2.8083
∞	7.8794	5.2983	4.2794	3.7151	3.3499	3.0913	2.8968	2.7444	2.6210

(*continued*)

$$\alpha = .005$$

10	12	15	20	24	30	40	60	120	∞
24224	24426	24630	24836	24940	25044	25148	25253	25359	25465
199.40	199.42	199.43	199.45	199.46	199.47	199.47	199.48	199.49	199.51
43.686	43.387	43.085	42.778	42.622	42.466	42.308	42.149	41.989	41.829
20.967	20.705	20.438	20.167	20.030	19.892	19.752	19.611	19.468	19.325
13.618	13.384	13.146	12.903	12.780	12.656	12.530	12.402	12.274	12.144
10.250	10.034	9.8140	9.5888	9.4741	9.3583	9.2408	9.1219	9.0015	8.8793
8.3803	8.1764	7.9678	7.7540	7.6450	7.5345	7.4225	7.3088	7.1933	7.0760
7.2107	7.0149	6.8143	6.6082	6.5029	6.3961	6.2875	6.1772	6.0649	5.9505
6.4171	6.2274	6.0325	5.8318	5.7292	5.6248	5.5186	5.4104	5.3001	5.1875
5.8467	5.6613	5.4707	5.2740	5.1732	5.0705	4.9659	4.8592	4.7501	4.6385
5.4182	5.2363	5.0489	4.8552	4.7557	4.6543	4.5508	4.4450	4.3367	4.2256
5.0855	4.9063	4.7214	4.5299	4.4315	4.3309	4.2282	4.1229	4.0149	3.9039
4.8199	4.6429	4.4600	4.2703	4.1726	4.0727	3.9704	3.8655	3.7577	3.6465
4.6034	4.4281	4.2468	4.0585	3.9614	3.8619	3.7600	3.6553	3.5473	3.4359
4.4236	4.2498	4.0698	3.8826	3.7859	3.6867	3.5850	3.4803	3.3722	3.2602
4.2719	4.0994	3.9205	3.7342	3.6378	3.5388	3.4372	3.3324	3.2240	3.1115
4.1423	3.9709	3.7929	3.6073	3.5112	3.4124	3.3107	3.2058	3.0971	2.9839
4.0305	3.8599	3.6827	3.4977	3.4017	3.3030	3.2014	3.0962	2.9871	2.8732
3.9329	3.7631	3.5866	3.4020	3.3062	3.2075	3.1058	3.0004	2.8908	2.7762
3.8470	3.6779	3.5020	3.3178	3.2220	3.1234	3.0215	2.9159	2.8058	2.6904
3.7709	3.6024	3.4270	3.2431	3.1474	3.0488	2.9467	2.8408	2.7302	2.6140
3.7030	3.5350	3.3600	3.1764	3.0807	2.9821	2.8799	2.7736	2.6625	2.5455
3.6420	3.4745	3.2999	3.1165	3.0208	2.9221	2.8198	2.7132	2.6016	2.4837
3.5870	3.4199	3.2456	3.0624	2.9667	2.8679	2.7654	2.6585	2.5463	2.4276
3.5370	3.3704	3.1963	3.0133	2.9176	2.8187	2.7160	2.6088	2.4960	2.3765
3.4916	3.3252	3.1515	2.9685	2.8728	2.7738	2.6709	2.5633	2.4501	2.3297
3.4499	3.2839	3.1104	2.9275	2.8318	2.7327	2.6296	2.5217	2.4078	2.2867
3.4117	3.2460	3.0727	2.8899	2.7941	2.6949	2.5916	2.4834	2.3689	2.2469
3.3765	3.2111	3.0379	2.8551	2.7594	2.6601	2.5565	2.4479	2.3330	2.2102
3.3440	3.1787	3.0057	2.8230	2.7272	2.6278	2.5241	2.4151	2.2997	2.1760
3.1167	2.9531	2.7811	2.5984	2.5020	2.4015	2.2958	2.1838	2.0635	1.9318
2.9042	2.7419	2.5705	2.3872	2.2898	2.1874	2.0789	1.9622	1.8341	1.6885
2.7052	2.5439	2.3727	2.1881	2.0890	1.9839	1.8709	1.7469	1.6055	1.4311
2.5188	2.3583	2.1868	1.9998	1.8983	1.7891	1.6691	1.5325	1.3637	1.0000

* That is, values of $F_{m_1, m_2}; \alpha$, where (m_1, m_2) is the pair of degrees of freedom in F_{m_1, m_2} and

$$\frac{\Gamma((m_1 + m_2)/2)}{\Gamma(m_1/2)\Gamma(m_2/2)} \left(\frac{m_1}{m_2}\right)^{m_1/2} \int_0^{F_{m_1, m_2}; \alpha} F^{(m_1/2)-1}\left(1 + \frac{m_1}{m_2} F\right)^{-(m_1+m_2)/2} dF = 1 - \alpha.$$

† From "Tables of percentage points of the Inverted Beta (F) Distribution," *Biometrika*, Vol. 33 (1943), pp. 73–88, by Maxine Merrington and Catherine M. Thompson; reproduced by permission of E. S. Pearson. If necessary, interpolation should be carried out using the reciprocals of the degrees of freedom.

TABLE VII
VALUES OF e^{-x}

x	e^{-x}	x	e^{-x}	x	e^{-x}
.00	1.00000	1.70	.18268	3.40	.03337
.10	.90484	1.80	.16530	3.50	.03020
.20	.81873	1.90	.14957	3.60	.02732
.30	.74082	2.00	.13534	3.70	.02472
.40	.67032	2.10	.12246	3.80	.02237
.50	.60653	2.20	.11080	3.90	.02024
.60	.54881	2.30	.10026	4.00	.01832
.70	.49659	2.40	.09072	4.10	.01657
.80	.44933	2.50	.08208	4.20	.01500
.90	.40657	2.60	.07427	4.30	.01357
1.00	.36788	2.70	.06721	4.40	.01228
1.10	.33287	2.80	.06081	4.50	.01111
1.20	.30119	2.90	.05502	4.60	.01005
1.30	.27253	3.00	.04979	4.70	.00910
1.40	.24660	3.10	.04505	4.80	.00823
1.50	.22313	3.20	.04076	4.90	.00745
1.60	.20190	3.30	.03688	5.00	.00674

TABLE VIII

PERCENTAGE POINTS OF THE BEHRENS-FISHER DISTRIBUTION*

γ	ν_2	ν_1 \ ϕ	0°	15°	30°	45°	60°	75°	90°
2.5% points	$\nu_2 = 6$	6	2.447	2.440	2.435	2.435	2.435	2.440	2.447
		8	2.447	2.430	2.398	2.364	2.331	2.310	2.306
		12	2.447	2.423	2.367	2.301	2.239	2.193	2.179
		24	2.447	2.418	2.342	2.247	2.156	2.088	2.064
		∞	2.447	2.413	2.322	2.201	2.082	1.993	1.960
	$\nu_2 = 8$	6	2.306	2.310	2.331	2.364	2.398	2.430	2.447
		8	2.306	2.300	2.294	2.292	2.294	2.300	2.306
		12	2.306	2.292	2.262	2.229	2.201	2.183	2.179
		24	2.306	2.286	2.236	2.175	2.118	2.077	2.064
		∞	2.306	2.281	2.215	2.128	2.044	1.982	1.960
	$\nu_2 = 12$	6	2.179	2.193	2.239	2.301	2.367	2.423	2.447
		8	2.179	2.183	2.201	2.229	2.262	2.292	2.306
		12	2.179	2.175	2.169	2.167	2.169	2.175	2.179
		24	2.179	2.168	2.142	2.112	2.085	2.069	2.064
		∞	2.179	2.163	2.120	2.064	2.011	1.973	1.960
	$\nu_2 = 24$	6	2.064	2.088	2.156	2.247	2.342	2.418	2.447
		8	2.064	2.077	2.118	2.175	2.236	2.286	2.306
		12	2.064	2.069	2.085	2.112	2.142	2.168	2.179
		24	2.064	2.062	2.058	2.056	2.058	2.062	2.064
		∞	2.064	2.056	2.035	2.009	1.983	1.966	1.960
	$\nu_2 = \infty$	6	1.960	1.993	2.082	2.201	2.322	2.413	2.447
		8	1.960	1.982	2.044	2.128	2.215	2.281	2.306
		12	1.960	1.973	2.011	2.064	2.120	2.163	2.179
		24	1.960	1.966	1.983	2.009	2.035	2.056	2.064
		∞	1.960	1.960	1.960	1.960	1.960	1.960	1.960
.5% points	$\nu_2 = 6$	6	3.707	3.654	3.557	3.514	3.557	3.654	3.707
		8	3.707	3.643	3.495	3.363	3.307	3.328	3.355
		12	3.707	3.636	3.453	3.246	3.104	3.053	3.055
		24	3.707	3.631	3.424	3.158	2.938	2.822	2.797
		∞	3.707	3.626	3.402	3.093	2.804	2.627	2.576
	$\nu_2 = 8$	6	3.355	3.328	3.307	3.363	3.495	3.643	3.707
		8	3.355	3.316	3.239	3.206	3.239	3.316	3.355
		12	3.355	3.307	3.192	3.083	3.032	3.039	3.055
		24	3.355	3.301	3.158	2.988	2.862	2.805	2.797
		∞	3.355	3.295	3.132	2.916	2.723	2.608	2.576
	$\nu_2 = 12$	6	3.055	3.053	3.104	3.246	3.453	3.636	3.707
		8	3.055	3.039	3.032	3.083	3.192	3.307	3.355
		12	3.055	3.029	2.978	2.954	2.978	3.029	3.055
		24	3.055	3.020	2.938	2.853	2.803	2.793	2.797
		∞	3.055	3.014	2.909	2.775	2.661	2.595	2.576
	$\nu_2 = 24$	6	2.797	2.822	2.938	3.158	3.424	3.631	3.707
		8	2.797	2.805	2.862	2.988	3.158	3.301	3.355
		12	2.797	2.793	2.803	2.853	2.938	3.020	3.055
		24	2.797	2.785	2.759	2.747	2.759	2.785	2.797
		∞	2.797	2.777	2.726	2.664	2.613	2.585	2.576
	$\nu_2 = \infty$	6	2.576	2.627	2.804	3.093	3.402	3.626	3.707
		8	2.576	2.608	2.723	2.916	3.132	3.295	3.355
		12	2.576	2.595	2.661	2.775	2.909	3.014	3.055
		24	2.576	2.585	2.613	2.664	2.726	2.777	2.797
		∞	2.576	2.576	2.576	2.576	2.576	2.576	2.576

* Table VIII is taken from Table V_1 of Fisher and Yates: *Statistical Tables for Biological, Agricultural and Medical Research*, published by Oliver and Boyd, Edinburgh and by the kind permission of the authors and publishers.

TABLE IX

Percentage Points of the Behrens-Fisher Distribution* when ν_1 is large ($\nu_1 > 60$)

γ	ν_2 \ ϕ	0°	10°	20°	30°	40°	50°	60°	70°	80°	90°
.05	10	1.812	1.808	1.794	1.774	1.749	1.721	1.693	1.668	1.651	1.645
	12	1.782	1.778	1.767	1.751	1.730	1.707	1.684	1.664	1.650	1.645
	15	1.753	1.750	1.741	1.728	1.711	1.693	1.675	1.659	1.649	1.645
	20	1.725	1.722	1.716	1.706	1.694	1.680	1.667	1.656	1.648	1.645
	30	1.697	1.696	1.692	1.685	1.677	1.668	1.659	1.652	1.647	1.645
	60	1.671	1.670	1.668	1.665	1.661	1.656	1.652	1.648	1.646	1.645
	∞	1.645	1.645	1.645	1.645	1.645	1.645	1.645	1.645	1.645	1.645
.025	10	2.228	2.219	2.194	2.157	2.112	2.066	2.024	1.989	1.967	1.960
	12	2.179	2.171	2.151	2.120	2.083	2.046	2.011	1.984	1.966	1.960
	15	2.131	2.126	2.109	2.085	2.056	2.026	1.999	1.978	1.965	1.960
	20	2.086	2.082	2.069	2.051	2.030	2.008	1.989	1.973	1.963	1.960
	30	2.042	2.039	2.031	2.019	2.005	1.991	1.978	1.968	1.962	1.960
	60	2.000	1.999	1.995	1.989	1.982	1.975	1.969	1.964	1.961	1.960
	∞	1.960	1.960	1.960	1.960	1.960	1.960	1.960	1.960	1.960	1.960
.010	10	2.764	2.748	2.704	2.637	2.559	2.481	2.414	2.364	2.335	2.326
	12	2.681	2.668	2.631	2.576	2.513	2.450	2.396	2.356	2.334	2.326
	15	2.602	2.592	2.563	2.520	2.470	2.421	2.379	2.349	2.332	2.326
	20	2.528	2.520	2.498	2.466	2.430	2.394	2.364	2.343	2.330	2.326
	30	2.457	2.452	2.438	2.417	2.393	2.370	2.351	2.337	2.329	2.326
	60	2.390	2.388	2.380	2.370	2.358	2.347	2.338	2.331	2.328	2.326
	∞	2.326	2.326	2.326	2.326	2.326	2.326	2.326	2.326	2.326	2.326

γ	ϕ / ν_2	0°	10°	20°	30°	40°	50°	60°	70°	80°	90°
.005	10	3.169	3.148	3.086	2.993	2.883	2.775	2.684	2.620	2.586	2.576
	12	3.055	3.037	2.985	2.909	2.820	2.733	2.661	2.611	2.584	2.576
	15	2.947	2.932	2.892	2.831	2.762	2.695	2.640	2.603	2.582	2.576
	20	2.845	2.835	2.804	2.760	2.709	2.661	2.622	2.595	2.580	2.576
	30	2.750	2.743	2.723	2.693	2.661	2.630	2.605	2.588	2.579	2.576
	60	2.660	2.657	2.647	2.632	2.616	2.601	2.590	2.582	2.577	2.576
	∞	2.576	2.576	2.576	2.576	2.576	2.576	2.576	2.576	2.576	2.576
.0025	10	3.581	3.553	3.473	3.350	3.203	3.058	2.939	2.859	2.818	2.807
	12	3.429	3.405	3.338	3.237	3.119	3.003	2.910	2.848	2.816	2.807
	15	3.286	3.267	3.214	3.134	3.042	2.954	2.884	2.838	2.814	2.807
	20	3.153	3.139	3.099	3.040	2.974	2.911	2.861	2.829	2.812	2.807
	30	3.030	3.020	2.994	2.955	2.912	2.872	2.841	2.821	2.810	2.807
	60	2.915	2.910	2.897	2.878	2.857	2.838	2.823	2.814	2.809	2.807
	∞	2.807	2.807	2.807	2.807	2.807	2.807	2.807	2.807	2.807	2.807
.001	10	4.144	4.106	3.999	3.832	3.630	3.425	3.259	3.152	3.103	3.090
	12	3.930	3.898	3.809	3.671	3.508	3.347	3.219	3.138	3.100	3.090
	15	3.733	3.708	3.636	3.528	3.401	3.280	3.185	3.126	3.098	3.090
	20	3.552	3.533	3.479	3.399	3.308	3.222	3.156	3.116	3.096	3.090
	30	3.386	3.372	3.336	3.284	3.226	3.172	3.131	3.106	3.094	3.090
	60	3.232	3.225	3.207	3.181	3.153	3.128	3.110	3.098	3.092	3.090
	∞	3.090	3.090	3.090	3.090	3.090	3.090	3.090	3.090	3.090	3.090

* Table IX is taken from Table V_2 of Fisher and Yates: *Statistical Tables for Biological, Agricultural and Medical Research,* published by Oliver and Boyd, Edinburgh and by the kind permission of the authors and publishers.

APPENDIX VIII.   CHARTS

CHART I

The 95% Confidence Limits for the Population Correlation Coefficient ρ, Given the Sample Correlation Coefficient r. The Numbers on the Curves Are the Sample Size.*

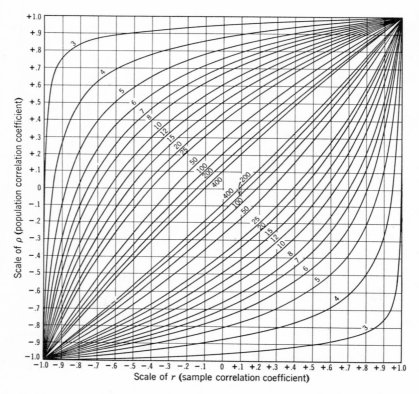

* From *Biometrika Tables for Statisticians*, Vol. 1 (2nd edition), Cambridge University Press (1958); edited by E. S. Pearson and H. O. Hartley; reproduced by permission of the publishers.

CHART II

THE 99% CONFIDENCE LIMITS FOR THE POPULATION CORRELATION COEFFICIENT
ρ, GIVEN THE SAMPLE CORRELATION COEFFICIENT r. THE NUMBERS ON THE
CURVES ARE THE SAMPLE SIZE.*

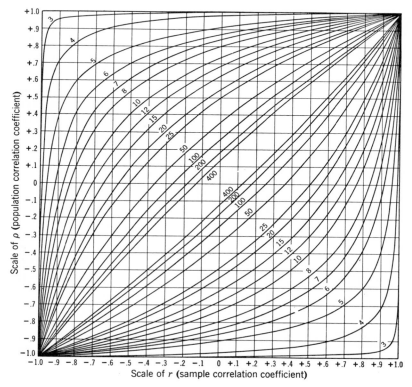

Scale of ρ (population correlation coefficient)

Scale of r (sample correlation coefficient)

* From *Biometrika Tables for Statisticians*, Vol. 1 (2nd edition), Cambridge
University Press (1958); edited by E. S. Pearson and H. O. Hartley; reproduced
by permission of the publishers.

CHART III

THE 95% CONFIDENCE LIMITS FOR THE BINOMIAL PARAMETER θ, GIVEN THE SAMPLE PROPORTION T/n. THE NUMBERS ON THE CURVES DENOTE THE SAMPLE SIZE.*

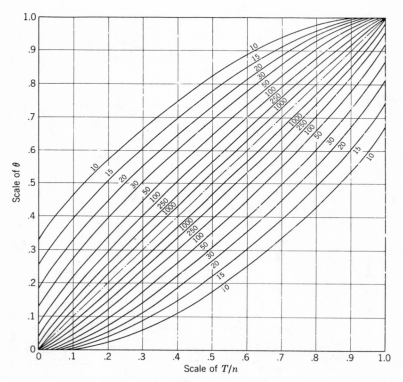

* From E. S. Pearson and C. J. Clopper, "The Use of Confidence or Fiducial Limits Illustrated in the Case of the Binomial," *Biometrika*, **26** (1934), p. 404; reproduced by permission of E. S. Pearson.

CHART III

THE 95% CONFIDENCE LIMITS FOR THE BINOMIAL PARAMETER θ, GIVEN THE SAMPLE PROPORTION T/n. THE NUMBERS ON THE CURVES DENOTE THE SAMPLE SIZE.*

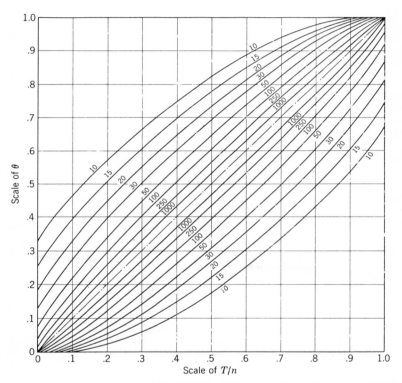

* From E. S. Pearson and C. J. Clopper, "The Use of Confidence or Fiducial Limits Illustrated in the Case of the Binomial," *Biometrika*, **26** (1934), p. 404; reproduced by permission of E. S. Pearson.

CHART II

THE 99% CONFIDENCE LIMITS FOR THE POPULATION CORRELATION COEFFICIENT
ρ, GIVEN THE SAMPLE CORRELATION COEFFICIENT r. THE NUMBERS ON THE
CURVES ARE THE SAMPLE SIZE.*

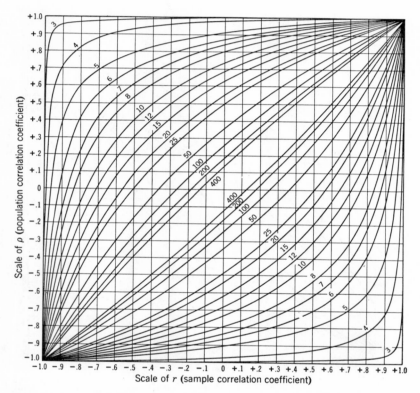

* From *Biometrika Tables for Statisticians*, Vol. 1 (2nd edition), Cambridge
University Press (1958); edited by E. S. Pearson and H. O. Hartley; reproduced
by permission of the publishers.

References

American Society for Testing Materials (1947), *Manual on Presentation of Data*, Philadelphia.

Bayes, Thomas (1763), "An Essay Towards Solving a Problem in the Doctrine of Chances," *Philosophical Transactions of the Royal Society of London*. Reprinted in *Biometrika*, vol. 45 (1958), pp. 296–315.

Bennett, C. A., and N. L. Franklin (1954), *Statistical Analysis in Chemistry and the Chemical Industry*, Wiley.

Bowker, A. H., and G. J. Lieberman (1959), *Engineering Statistics*, Prentice-Hall.

Box, G. E. P. (1954), "The Exploration and Exploitation of Response Surfaces," *Biometrics*, vol. 10, pp. 16–60.

Box, G. E. P., and J. S. Hunter (1957), "Multi-factor Experimental Designs for Exploring Response Surfaces," *Annals of Mathematical Statistics*, vol. 28, pp. 195–241.

Box, G. E. P., and J. S. Hunter (1961), "The 2^{k-p} Fractional Factorial Designs," *Technometrics*, vol. 3, pp. 311–351 and 449–458.

Brownlee, K. A. (1960), *Statistical Theory and Methodology in Science and Engineering*, Wiley.

Bulmer, M. G. (1957), Approximate Confidence Limits for Components of Variance, *Biometrika*, vol. 44, pp. 159–167.

Cochran, W. G., and G. M. Cox (1957), *Experimental Designs* (second edition), Wiley.

Chemical Corps Engineering Agency of the U.S. Army, Manual No. 2 (1953), *Master Sampling Plans for Single, Duplicate, Double and Multiple Sampling*, Army Chemical Center, Md.

Cramer, H. (1946), *Mathematical Methods of Statistics*, Princeton University Press.

Daniel, C. (1959), "Use of Half-Normal Plots in Interpreting Two Level Experiments," *Technometrics*, vol. 1, pp. 311–342.

Davies, O. L. (1954), *Design and Analysis of Industrial Experiments*, Oliver and Boyd.

Dixon, W. J., and F. J. Massey, Jr. (1957), *Introduction to Statistical Analysis*, McGraw-Hill.

Dodge, H. F., and H. G. Romig (1959), *Sampling Inspection Tables* (second edition) Wiley.

Draper, N. R., and H. Smith, Jr. (1966), *Applied Regression Analysis*, Wiley.

Duncan, A. G. (1958), *Quality Control and Industrial Statistics*, R. D. Irwin, Inc.

Feller, William (1957), *An Introduction to Probability Theory and Its Applications* (second edition), Wiley.

Fraser, D. A. S. (1958), *Statistics: An Introduction*, Wiley.

Hald, A. (1952), *Statistical Theory with Engineering Applications*, Wiley.

Hoel, P. G. (1954), *Introduction to Mathematical Statistics* (second edition), Wiley.

Hogg, R. V., and A. T. Craig (1959), *Introduction to Mathematical Statistics*, Macmillan.

Kempthorne, O. (1952), *The Design and Analysis of Experiments*, Wiley.

Kenney, J. F., and E. S. Keeping (1956), *Mathematics of Statistics*, Vols. 1 and 2 (third edition), Van Nostrand.

Lindley, D. V. (1965), *Introduction to Probability and Statistics from a Bayesian Viewpoint. Part II. Inference*, Cambridge University Press.

Mood, A. M. (1950), *Introduction to the Theory of Statistics*, McGraw-Hill.

Murphy, R. B. (1948), Non-Parametric Tolerance Limits, *Annals of Mathematical Statistics*, vol. 19, pp. 581–589.

Parzen, E. (1960), *Modern Probability Theory and Its Applications*, Wiley.

Raiffa, H., and Robert Schlaifer (1961), *Applied Statistical Decision Theory*, Graduate School of Business Administration, Harvard University.

RAND Corporation (1955), *A Million Random Digits with* 100,000 *Normal Deviates*, The Free Press.

Resnikoff, G. J., and G. J. Lieberman (1957), *Tables of the Non-Central t-Distribution*, Stanford University Press.

Siegel, S. (1956), *Nonparametric Statistics for the Behavioral Sciences*, McGraw-Hill.

Somerville, P. N. (1958), Tables for Obtaining Non-Parametric Tolerance Limits, *Annals of Mathematical Statistics*, vol. 29, pp. 599–601.

Statistical Research Group, Columbia University (1948), *Sampling Inspection*, McGraw-Hill.

Steel, R. G. D., and J. H. Torrie (1960), *Principles and Procedures of Statistics*, McGraw-Hill.

United States Department of Defense (1950), *Military Standard. Sampling Procedures and Tables for Inspection by Attributes.* (Military Standard 105A.) Washington: Government Printing Office.

Uspensky, G. V. (1937), *Introduction to Mathematical Probability*, McGraw-Hill.

Wald, A. (1947), *Sequential Analysis*, Wiley.

Wilks, S. S. (1948), *Elementary Statistical Analysis*, Princeton University Press.

Wilks, S. S. (1962), *Mathematical Statistics*, Wiley.

Woodward, R. H., and P. L. Goldsmith (1964), *Cumulative Sum Techniques*, Oliver and Boyd.

Wu, S. M., (1964), Tool Life Testing by Response Surface Methodology, I and II, *Trans. of the ASME, Jour. Engr. Ind.*, vol. 86, p. 105.

Answers to Selected Problems

CHAPTER 2

2.1 (a) 93.9% (b) 6.1% (c) 95.8% (d) 98.9%.

2.3 (b) 119 (c) 126 (d) .9; 93 (e) $\frac{15}{900}$.

2.5 (a) .72 (b) .28 (c) .06.

2.7 (a) $\binom{900}{10} \Big/ \binom{1000}{10}$ (b) $1 - \binom{900}{10} \Big/ \binom{1000}{10}$.

2.9 Sampling without replacement (a) .2 (b) $\frac{2}{9}$ (c) $\frac{2}{9}$ (d) $\frac{17}{45}$; Sampling with replacement (a) .18 (b) .25 (c) .25 (d) .36.

2.13 $\dfrac{P(365, r)}{(365)^r}$.

2.15 (a) $3p_1 - 3p_2 + p_3$ (b) $4p_1 - 6p_2 + 4p_3 - p_4$ (c) $\dfrac{p_4}{p_2}$
(d) P (exactly 1) $= 4[p_1 - 3p_2 + 3p_3 - p_4]$; P (exactly 2) $= 6[p_2 - 2p_3 + p_4]$;
P (exactly 3) $= 4[p_3 - p_4]$.

2.23 .0525.

2.25 .2895.

2.27 $\frac{91}{120}$.

2.29 $\dfrac{\dbinom{90}{1} \Big/ \dbinom{100}{1} \times \dbinom{10}{10} \Big/ \dbinom{99}{10}}{\dfrac{\dbinom{10}{9}\dbinom{90}{1}}{\dbinom{100}{10}} \times \dfrac{1}{\dbinom{90}{1}}} = 1/10.$

523

CHAPTER 3

3.1 $h(x) = \dfrac{\binom{7}{x}\binom{43}{10-x}}{\binom{50}{10}}$ $0 \le X \le 7.$

3.3 P (at least one ace) $= 1 - (\tfrac{5}{6})^5$; P (exactly one ace) $= (\tfrac{5}{6})^5$; P (exactly two aces) $= (\tfrac{5}{2})(\tfrac{1}{6})^2(\tfrac{5}{6})^3.$

3.5 $\dfrac{\binom{12}{0}\binom{36}{13}}{\binom{52}{13}}\,.$

3.7   (a) $\dfrac{\binom{13}{x}\binom{39}{13-x}}{\binom{52}{13}}$ $0 \le x \le 13$, x = number of spades

(b) $\dfrac{\binom{13}{y}\binom{39}{13-y}}{\binom{52}{13}}$ $0 \le y \le 13$, y = number of hearts

(c) $\dfrac{\binom{13}{x}\binom{13}{y}\binom{26}{13-x-y}}{\binom{52}{13}}$ $0 \le x \le 13, 0 \le y \le 13, \;\; 0 \le x+y \le 13$

3.9 $1 - \sum\limits_{x=0}^{3}\binom{10,000}{x}(.000005)^x(.999995)^{10,000-x}$;   Use   Poisson: $\mu = n\theta = $ $.05$; so that we have, approximately, $1 - \sum\limits_{x=0}^{3} e^{-.05}\dfrac{(.05)^x}{x!}\,.$

3.11   (a) Since $\mu = 5 \times 10 \times 4/100 = 2$, we have $1 - \sum\limits_{x=0}^{2}\dfrac{e^{-2}2^x}{x!} = 1 - 5e^{-2}$, (b) e^{-6}.

3.13   (a) $\dfrac{10!}{3!3!4!}\left(\dfrac{1}{4}\right)^3\left(\dfrac{1}{4}\right)^3\left(\dfrac{1}{2}\right)^4$ (b) $\left(\dfrac{1}{2}\right)^{10}.$

3.15 $\left[\dfrac{\binom{10}{4}\binom{20}{x-5}}{\binom{30}{x-1}}\right]\dfrac{6}{30-x+1}$ $5 \le x \le 25.$

3.17   (a) $\binom{m}{x}\theta^x(1-\theta)^{m-x}$ (b) $\binom{n}{y}(1-\theta)^{my}(1-(1-\theta)^m)^{n-y}.$

3.19 $\theta = \frac{1}{9}$.

3.21 $(1 - \theta)^{k-1}\theta$ (b) $\binom{x-1}{k-1}(1-\theta)^{x-k}\theta^{k}$.

3.23 $P_A(x) = \binom{x-1}{3}(.85)^4(.15)^{x-4}$, $x = 4.5$; $P_A(6) = \binom{5}{3}(.85)^4(.15)^2$.

CHAPTER 4

4.1 (a) and (b)

θ	0	$\frac{1}{12}$	$\frac{2}{12}$	$\frac{3}{12}$	$\frac{4}{12}$	$\frac{5}{12}$	$\frac{6}{12}$	$\frac{7}{12}$	$\frac{8}{12}$	$\frac{9}{12}$
$P(A\mid\theta)$	1	$\frac{3}{4}$	$\frac{6}{11}$	$\frac{21}{55}$	$\frac{14}{55}$	$\frac{7}{44}$	$\frac{1}{11}$	$\frac{1}{22}$	$\frac{1}{55}$	$\frac{1}{220}$
$\theta P(A\mid\theta)$	0	$\frac{1}{16}$	$\frac{1}{11}$	$\frac{21}{220}$	$\frac{14}{165}$	$\frac{35}{528}$	$\frac{1}{22}$	$\frac{7}{264}$	$\frac{2}{165}$	$\frac{3}{880}$

θ	$\frac{10}{12}$	$\frac{11}{12}$	$\frac{12}{12}$
$P(A\mid\theta)$	0	0	0
$\theta P(A\mid\theta)$	0	0	0

(c) $AOQL$ is $\frac{21}{220}$ at $\theta^* = \frac{3}{12} = .25$.

4.3 (b) $AOQL = .101$ and occurs at $\theta^* = .24$.

4.5 (c) $AOQL = .067$ at $\theta^* = \frac{1}{6}$ (d) $\theta_1 = .01$ (e) $\theta_2 = .369$.

4.7 (c) $AOQL = .136$ at $\theta^* = .20$ (d) $\theta_1 \simeq .085$ (e) $\theta_2 \simeq .45$.

4.9 (c) $AOQL \simeq .0089$ at $\theta^* \simeq .0229$ (d) $\theta_1 \simeq .0082$ (e) $\theta_2 \simeq .053$.

4.11 (a) $AOQL = .0074$ at $\theta^* = .02$ (d) $\theta_1 = .001$ (e) $\theta_2 = .046$.

4.13 (c) $AOQL = .072$, $\theta^* = .12$ (d) $\theta_1 \simeq .04$ (e) $\theta_2 \simeq .26$.

4.15 (c) $AOQL = .0175$ at $\theta^* = .033$ (d) $\theta_1 \simeq .009$ (e) $\theta_2 \simeq .096$.

CHAPTER 5

5.1 (a) $Q_{25} = 53$; $Q_{50} = Q_{75} = 55$ (b) $\bar{x} = 53.88$, $S_x = 1.81$.

5.3 (b) $\bar{x} = 70.835$, $S_x = 1.81$.

5.5

(a)	(e)	(f)	(g)
.4095 − .4145	.006	6	.006
.4145 − .4195	.034	40	.040
.4195 − .4245	.132	172	.172
.4245 − .4295	.179	351	.351
.4295 − .4345	.218	569	.569
.4345 − .4395	.183	752	.752
.4395 − .4445	.146	898	.898
.4445 − .4495	.069	967	.967
.4495 − .4545	.030	997	.997
.4545 − .4595	.003	1,000	1.000
	1.000		

(b) $\bar{x} \cong .43324$, $s_x^2 \cong (.005)^2 s_u^2$; $s_x \cong .00865$.

5.7 (b) $\bar{m} = .15212$, $S_m = .0022$.

5.9 (b) $\bar{m} = 13.4264$, $S_m = .1131$.

5.11 $\bar{m} = 2.175$; $S_m^2 = 1.7665$.

5.13 $\bar{m} = 6.92$, $S_m^2 = 6.478$.

5.15 $\bar{m} = 115.6$, $S_m^2 = 137$, $S_m = 11.7$.

CHAPTER 6

6.1 (a) $p(x) = \dfrac{(x-1)(x-2)}{240}$, $x = 3, 4, \ldots 10$ (c) $\mu = 8.25$, $\sigma = 1.70$.

6.3 (a) $F(x') = P\{x \le x'\} = \begin{cases} 0, & x' \le 0 \\ \dfrac{x'^2}{R^2}, & 0 < x \le R \\ 1, & x > R \end{cases}$

$f(x) = \dfrac{2x}{R^2}, 0 < x \le R$

$= 0$ otherwise

(b) median $= .707R$ (c) $\mu = \frac{2}{3}R$, var $(x) = \dfrac{R^2}{18}$.

6.5 $E(T) = 350$, var $(T) = 13$, $\sigma(T) = \sqrt{13}$.

6.7 $\mu_T = .045$, $\sigma_T^2 = .00000081$, $\sigma_T = .0009$.

6.9 (a) $p(x_2) = \dbinom{n}{x_2} \theta_2^{x_2}(1-\theta_2)^{n-x_2}$ (b) $E(x_1) = n\theta_1$, $\sigma_{x_1}^2 = n\theta_1(1-\theta_1)$;

$E(x_2) = n\theta_2$, $\sigma_{x_2}^2 = n\theta_2(1-\theta_2)$ (c) Cov $(x_1, x_2) = -n\theta_1\theta_2$.

6.11 $\mu_T = 4.5n$, $\sigma_T^2 = 8.25n$.

6.13 $n \cong 2340.6$.

6.15 $K = 1,112,430$.

6.17 $E(x) = 1$, Var $(x) = 1$.

6.27 $p(y_i) = \theta(1-\theta)^{y_i-1}, y_i = 1, 2, \ldots$; $p(x) = \dbinom{x-1}{k-1}\theta^k(1-\theta)^{x-k}, x = k, k+1, \ldots$; $E(x) = \dfrac{k}{\theta}$, var $(x) = \dfrac{k(1-\theta)}{\theta^2}$.

6.29 $f_2(x_2) = 2(1-x_2)$ for $0 < x_2 < 1$, $E(x_1) = \frac{1}{3}$, $E(x_2) = \frac{1}{3}$, $\rho(x_1, x_2) = -\frac{1}{2}$.

6.31 $f(x_1 \mid 0) = \begin{cases} \frac{1}{2} \text{ if } x_1 = 0, \\ \frac{1}{2} \text{ if } x_1 = 2, \end{cases}$ $E(x_1 \mid x_2 = 0) = 1$, Var $(x_1 \mid x_2 = 0) = 1$;

$f(x_1 \mid 1) = 1$ for $x_1 = 1$; $E(x_1 \mid x_2 = 1) = 1$, Var $(x_1 \mid x_2 = 1) = 1 - 1 = 0$.

$f(x_2 \mid 0) = 1$ for $x_2 = 0$; $E(x_2 \mid x_1 = 0) = 0$, Var $(x_2 \mid x_1 = 0) = 0$;

$f(x_2 \mid 1) = 1$ for $x_2 = 1$; $E(x_2 \mid x_1 = 1) = 1$, Var $(x_2 \mid x_1 = 1) = 0$;

$f(x_2 \mid 2) = 1$ for $x_2 = 0$; $E(x_2 \mid x_1 = 2) = 0$, Var $(x_2 \mid x_1 = 2) = 0$.

CHAPTER 7

7.1   (a) .3085   (b) .1587   (c) $A = 1829$ (d) $B = 392$.

7.3   (a) 2.275%   (b) .81746.

7.5   (a) .50135   (b) $\mu = .2500$; interval is [.2500 ± .0003]
(c) 13.362%.

7.9   (a) .4198   (b) $K = 3849$.

7.11   (a) .9857   (b) $\Phi\left(\dfrac{y - 200.5}{10}\right)$.

7.13   (a) .9780   (b) .01923.

7.15   (a) $E(x) = .250$ (b) Var $(x) = 59 \times 10^{-6}$, $\sigma(x) = .0077$
(c) .8061   (d) $K = .0127$.

7.17   (a) .9876   (b) 0.

CHAPTER 8

8.1   (a) $n = 22$, $K = 1528.1$

(b)

μ	1490	1500	1510	1520	1530	1540	1550	1560
$P(A \mid \mu)$	.01	.05	.14	.32	.54	.76	.90	.97

8.3 $K = 2.025$; $n \cong 174$.

8.5 $K = 400$; $n = 44$.

8.7 $K = 14925.6$; $n = 3$.

8.9   (a) $n = 11$; $d = .075$ (b) $P(A \mid \mu) = \Phi\left(\dfrac{12.225 - \mu}{.015}\right) - \Phi\left(\dfrac{12.075 - \mu}{.015}\right)$

8.11 $n = 7$, $d = 2$.

CHAPTER 9

9.1   [1331.6, 1488.4].

9.3 $n \geq \left(\dfrac{5.152\sigma}{L}\right)^2$.

9.5   [44.14, 265.86].

9.7   (a) [7.869, 7.971]   (b) [.010, .067].

9.9   (a) [−4.80, 5.30]   (b) [3.993, 12.923].

9.11   [43.257, 248.496].

9.13   [25.7, 72.8].

9.15   [−80.40, 353.80]. Yes, it does.

9.17   [−.0020, .0062].

9.19   [−3.959, 1.159].

9.21   [−.072, .712]. No, it doesn't.

9.23   [.22193, 3.5972].

9.25   For the data 9.15–9.18, the assumption that "true value of ratio of the variances is one" is borne out by the sample evidence.

9.27   [$113.06, $118.86].

9.29   [67.517, 68.583].

9.31   (a) [.422, .8668]   (b) [−.006, .806].

9.33   (a) [.261, .741]   (b) [−3.857, .143].

9.35   (a) [.4327, 1.1129];   (b) [−3019.98, −1194.02].

9.37   (a) [.4319, 1.7452]   (b) [1.6088, 2.0892].

9.39   [.358, .444].

9.41   [.24, .58].

9.43   Confidence interval is $\theta_1 < \theta < \theta_2$, where θ_1 and θ_2 are the smaller and larger of $\hat{\theta} \pm 1.96 \sqrt{\dfrac{\hat{\theta}(1 - \hat{\theta})}{n}}$. For the Example 9.8(a), this gives a confidence interval [.504, .696], and compares favourably with the interval [.50, .69] given in that example.

9.45   The 95% confidence interval for the difference in proportion of the population favouring Brand X before and after the campaign is [−.184, .004], which includes zero, i.e., the sample evidence indicates no significant change.

9.49   (b) Var $(t) = \dfrac{\theta^2}{n(n + 2)}$ (c) Var $(2\bar{x}) = \dfrac{\theta^2}{3n}$ (d) $\dfrac{V(t)}{V(2\bar{x})} = \dfrac{3}{n + 2}$, prefer the unbiased estimate $t = \dfrac{(n + 1)\hat{\theta}}{n}$.

9.51   Maximum likelihood estimator of σ^2 is $\hat{\sigma}^2 = \dfrac{\sum\limits_{1}^{n} (x_i - \mu_0)^2}{n}$;   maximum likelihood estimator is unbiased. The distribution of $\hat{\sigma}^2$ is $g(\hat{\sigma}^2) = \dfrac{n^{n/2}}{\Gamma\left(\dfrac{n}{2}\right)(2\sigma^2)^{n/2}} \times$

$(\hat{\sigma}^2)^{n/2-1}e^{-n\hat{\sigma}^2/2\sigma^2}$. That is, $\dfrac{n\hat{\sigma}^2}{\sigma^2} = \chi_n^2$. Hence $V(\hat{\sigma}^2) = \dfrac{(\sigma^2)^2}{n^2}$, $V(\chi_n^2) = \dfrac{2(\sigma^2)^2}{n}$.

ANSWERS TO SELECTED PROBLEMS

CHAPTER 9A

9A.1 $p(\theta = \theta_i \mid x_1, \ldots, x_n) = p_i \prod_{j=1}^{n} f(x_j \mid \theta_i) / \sum_{i=1}^{n} p_i \prod_{j=1}^{n} f(x_j \mid \theta_i)$.

9A.3 $N(404.2, 625)$; the Bayes estimate is 404.2, with variance 625; a 95% posterior interval is (355.2, 453.2).

9A.5 $N(51, 654.3; (521.4)^2)$; a 90% posterior interval is [50,796.6; 52,512.0].

9A.7 126.72; a 95% posterior interval is [18.55, 234.89].

9A.9 $N(-1.32; 3.56)$; a 98% posterior interval is [-5.716, 3.076].

9A.11 A 98% posterior interval is [-7.35, 3.55].

9A.15 $\mu = 8750 + \dfrac{175}{\sqrt{6}} t_5$; a 95% posterior interval for μ is [8566.4, 8933.6].
The sample evidence does not support the claim.

9A.17 $\sigma^2 = 5(175)^2/\chi_5^2$. The Bayes estimate is 51,041.7 and a 99% posterior interval for σ^2 is [9142; 371,897].

9A.19 $\mu = .0808 + \sqrt{\dfrac{.042735}{(16.4)(14)}}\, t_{14}$; a 95% posterior interval is [.05163, .10997].

9A.21 $\sigma^2 = .042735/\chi_{14}^2$; hence, the Bayes estimate is $.042735/12 = .0036$, and a 95% interval is [.0016, .0076].

9A.23 (a) $\mu_1 - \mu_2 = (24.7 - 22.1) + \sqrt{\dfrac{22.04}{38}\left(\dfrac{1}{20} + \dfrac{1}{20}\right)}\, t_{38}$; a 95% posterior interval is [2.11, 3.09] (b) $\sigma^2 = SS_w^2/\chi_{38}^2$; the Bayes estimate is $22.04/36 = .6122$, and a 95% posterior interval is [.3876, .9623].

9A.25 $\mu_1 - \mu_2 = (\bar{x}_1 - \bar{x}_2) + t'(6, 6; 58.8°)\sqrt{\dfrac{33 \times 10^{-6}}{7} + \dfrac{92 \times 10^{-6}}{7}}$;
$\phi = 58.8°$ is determined by letting $u = \sqrt{\dfrac{s_2^2}{n_2} / \dfrac{s_1^2}{n_1}} = 1.67 > 1$; so that letting $v = 1/u = .5988$, $\phi^* = 31.2°$ as found using (9A.3.63b), so that $\phi = 90 - 31.2 = 58.8°$; the Bayes estimate is $(\bar{x}_1 - \bar{x}_2) = .00027$, and a 95% posterior interval is $\left[.00027 \pm 2.435 \sqrt{\dfrac{125 \times 10^{-6}}{7}}\right] = [-.010061, .010601]$.

9A.27 Interval is $\left[\dfrac{s_1^2}{s_2^2} \dfrac{1}{F_{6,6;.025}}, \dfrac{s_1^2}{s_2^2} F_{6,6;.025}\right] = [.06088; 2.1133]$; hence sample evidence supports claim that $\sigma_1^2 = \sigma_2^2$, i.e., the claim $\sigma_1^2/\sigma_2^2 = 1$.

9A.29 [.899, 4.628].

9A.33 (a) $\mu_1 - \mu_2 = .4694 + (.4466)t'(13,11; 52.1°)$; the Bayes estimate is .4694 and a 95% posterior interval is [.4694 ± (.4466)(2.1661)] (b) $\sigma_1^2/\sigma_2^2 = .686791/F_{13,11}$ or $.686791 F_{11,13}$; the Bayes estimate is $(.68791) \times 13/11 = .811662$, and a 95% interval is [.2025, 2.1957].

9A.39   The Bayes estimate is $16/30 = .533$; a 98% posterior interval is [.3259, .7343].

9A.41   The Bayes estimate is $28/425 = .0659$, and a 95% posterior interval is [.0443, .0912].

9A.43   The likelihood is proportional to $e^{-n\tau}\tau^t$; the posterior of τ is such that $\tau = \chi^2_{2(t+1)}/2n$; the Bayes estimate is $(t+1)/n$, and a $100(1-\alpha)$% posterior interval is $[\chi^2_{2(t+1);1-\alpha/2}/2n; \ \chi^2_{2(t+1);\alpha/2}]$.

CHAPTER 10

10.1   (a) Observed $z = -3 < -2.326$, so reject H_0 (b) power $= \gamma(1400)$
$$Pr\left(z < \frac{1406.96 - 1400}{40}\right) = 0.5691.$$

10.3   (a) Critical region for $\bar{x}$ is: Reject if $\dfrac{(\bar{x} - .25)\sqrt{10}}{.0015} < -2.326$ or $\bar{x} <$ .2489; power $= \gamma(\mu_1 = .2490) = .4160$ (b) Use $n = 49$.

10.5   Reject H_0; power $= \gamma(\mu) =$
$$Pr\left(z < \frac{13{,}173.33 - \mu}{500/3}\right) + Pr\left(z > \frac{13{,}826.67 - \mu}{500/3}\right).$$

10.7   Critical region: $|\bar{x} - 40| > 0.9776$ or, equivalently, $\bar{x} < 39.0224$ or $\bar{x} > 40.9776$; power $= Pr(z < 2.635(39.0224 - \mu_1)) + Pr(z > 2.635(40.9776 - \mu_1))$.

10.9   Observed $t_9 = -.4102 > -t_{9;.05} = -1.833$, so we accept H_0.

10.11   Observed $t_{39} = -2.5298 < -t_{39;.01} = -2.4256$, so we reject H_0.

10.13   Observed $|t_{13}| = 1.394 < t_{13;.025}$, so we accept H_0.

10.15 $|t_8| = 1.786 \leq t_{8;.025} = 2.306$, so we accept H_0.

10.17   Observed $\chi^2_8 = 15.68 > \chi^2_{8;.05} = 15.51$ so we reject H_0.

10.19   No; the value of S is *not* significantly large. In fact, observed $\chi^2_8 = 14.182 < \chi^2_{8;.05} = 15.51$.

10.21   Accept H_0: $\mu_2 - \mu_1 = 0$ against H_1: $\mu_2 < \mu_1$ since observed $z = -1.6 > z_{.95} = -1.645$.

10.23   Observed $|z| = 2.74 > z_{0.025} = 1.96$, so we reject H_0.

$$\text{Power function} = \gamma(\delta) = Pr\left[\left|\frac{\bar{x}_A - \bar{x}_B}{56.56}\right| > 1.96 \,|\, \mu_A - \mu_B = \delta\right]$$
$$= 1 - \left[\Phi\left(1.96 - \frac{\delta}{56.56}\right) - \Phi\left(-1.96 - \frac{\delta}{56.56}\right)\right]$$
$$r(\delta = 100) = 1 - \left[\Phi\left(1.96 - \frac{100}{56.56}\right) - \Phi\left(-1.96 - \frac{100}{56.56}\right)\right]$$
$$= .4244 = \gamma(\delta = -100).$$

10.25 Observed $s_A^2/S_B^2 = .0022/.0035 = .629 > F_{7,7;.975} = 1/4.9949 = .2002$; hence, we accept H_0: $\sigma_A^2 = \sigma_B^2$ (b) Because of part (a), we reject if observed $|t_{14}| > t_{14;.025} = 2.145$; but observed $|t_{14}| = .139$, so that we accept H_0: $\mu_A = \mu_B$.

10.27 Observed $|t_{19}| = 3.297 > t_{19;.025} = 2.093$; we reject H_0: $\mu_A = \mu_B$.

10.29 (a) Accept $\sigma_A^2 = \sigma_B^2$, since observed $\dfrac{S_A^2}{S_B^2}$ lies between $F_{4.4;.975} = .104$ and $F_{4.4;.025} = 9.6045$ (b) Accept H_0: $\mu_A - \mu_B = 1\%$, since observed $t_8 = [(\bar{x}_A - \bar{x}_B) - 1]/\sqrt{(\frac{1}{5} + \frac{1}{3})S_w^2} = .976 < t_{8;.05} = 1.60$.

10.31 (a) Reject $\sigma_1^2 = \sigma_2^2$ (b) $C = .83$; $1/m = .0183$; $m = 54.64$; use $m = 54$; observed $t = (19.2 - 15.9)\sqrt{1.2288} = 2.976 > t_{54;.01} = 2.3973$, and so we reject $\mu_1 = \mu_2$.

10.35 (a) $\dfrac{S_1^2}{S_2^2} = 2.42 > F_{69.69;.025}$; we reject H_0: $\sigma_1^2 = \sigma_2^2$ (b) Observed $z = 2.514 > z_{.05}$.

10.37 Observed $|t_9| = .659 < t_{9;.025} = 2.262$, and so we accept $\mu_A = \mu_B$.

10.39 Observed $|t_9| = 2.206 < t_{9;.025} = 2.262$, i.e., there is no difference in treatments.

10.41 Test is: Accept if $F_{9.9;.975} = 1/4.026 <$ observed $F < F_{9.9;.025} = 4.026$; observed F is $.5006/.4332 = 1.156$, and so we accept.

10.43 Sampling continues as long as $-4.554 < .288t - .5m < 2.986$, where $t = \sum\limits_{i=1}^{n} x_i$. If $.288t - .5m < -4.554$, we stop sampling and accept H_0, while if $.288t - .5m > 2.986$, we stop and accept H_1.

10.45 Sampling continues as long as

$$-2.890 < m \log (\sigma_0/\sigma_1) + \frac{1}{2}\left(\frac{1}{\sigma_0^2} - \frac{1}{\sigma_1^2}\right) \sum_1^m (\chi_i - \mu^0)^2 < 2.251.$$

If $m \log (\sigma_0/\sigma_1) + \dfrac{1}{2}\left(\dfrac{1}{\sigma_0^2} - \dfrac{1}{\sigma_1^2}\right) \sum\limits_1^m (\chi_i - \mu^0)^2 < -2.890$, sampling stops and we accept H_0, etc.

10.47 Observed $|z| = \dfrac{.55 - .51}{\sqrt{\dfrac{(.51)(1 - .51)}{1000}}} = 2.532 < z_{.005} = 2.576$ so that he could be right, at the 1% level of significance.

10.49 (a) Reject if $\sqrt{\dfrac{n_1 n_2}{n_1 + n_2}}\,(\hat{\theta}_1 - \hat{\theta}_2)/\sqrt{\hat{\theta}_p(1 - \hat{\theta}_p)} > z_\alpha$ (b) Reject if

$$\sqrt{\frac{n_1 n_2}{n_1 + n_2}}\,|\hat{\theta}_1 - \hat{\theta}_2|/\sqrt{\hat{\theta}_p(1 - \hat{\theta}_p)} < z_{\alpha/2}.$$

CHAPTER 11

11.1 (a) $UCL = 107.805$, $LCL = 103.395$ (b) $UCL = 1.5234$, $LCL = 0$.

11.3 $UCL = 38.054$, $LCL = 10.946$.

11.5 $UCL = .659$, $LCL = .591$. P (at least one of the next two samples will have a T above the UCL) $= .031744$.

11.7 $UCL = 40$, $LCL = 10$, P (chart detects shift) $= .3707$, P (chart detects shift within 3 samples) $= 1 - (0.6293)^3 = 0.7508$·

11.9 (a) $UCL = .144616$, $LCL = .135384$ (b) $UCL = 0.01692$, $LCL = 0$.

11.11 For an empirical chart for $\bar{x}$, $UCL = 15.1$, $LCL = 6.5$. For an empirical chart for R, $UCL = 15.8625$, $LCL = 0$.

11.13 Control chart for $\bar{x}$ is: $UCL = 107.078$, $LCL = 36.156$.

CHAPTER 12

12.1 We reject the hypothesis of a fair coin; the observed $\chi_1^2 = 5.776 > \chi_{1;.05}^2 = 3.8416$.

12.3 (a) $\hat{\theta} = .567$ (b) The observed value of $\chi_3^2 = 8.4421 > \chi_{3;.05}^2 = 7.815$, so reject the hypothesis of "binomiality" of the data.

12.5 The observed value of $\chi_8^2 = 41.938 > \chi_{8;.05}^2 = 15.507$, so reject the hypothesis of independence.

12.7 The observed value of $\chi_1^2 = 30.356736 > \chi_{1;.05}^2 = 3.84146$, so reject the hypothesis of independence.

12.9 The observed value of $\chi_4^2 = 10.4699 > \chi_{4;.05}^2 = 9.48773$, so we reject the hypothesis of independence of range and deflection.

12.11 We reject the hypothesis that the data are Poisson.

12.13 The observed value of $\chi_3^2 = 3.111 < \chi_{3;.05}^2 = 7.81473$, so we accept the hypothesis of independence of the modes of classification.

CHAPTER 13

13.1 (a) $[1 - F(x)]^{100}$ (b) $100[1 - F(x)]^{99}f(x)$.

13.3 (a) $g(x_{(n)}) = nx_{(n)}^{n-1}, 0 < x_{(n)} < 1$
$= 0$ otherwise

(b) $g(x_{(1)}) = n(1 - x_{(1)})^{n-1}, 0 < x_{(1)} < 1$
$= 0$ otherwise

(c) $g(x_{(r)}) = \dfrac{n!}{(r-1)!(n-r)!} x_{(r)}^{r-1}[1 - x_{(r)}]^{n-r}, 0 < x_{(r)} < 1$
$= 0$ otherwise.

13.5 e^{-100cy}.

13.7 $n \simeq 4$.

13.9 $g(x_{(1)}) = ne^{-nx_{(1)}}, x_{(1)} > 0; = 0$ otherwise $E(x_{(1)}) = \dfrac{1}{n}$, Var $(x_{(1)}) = \dfrac{1}{n^2}$.

13.11 $g(v) = (n-2)(n-3)e^{-2v}(1-e^{-v})^{n-4}$ where $v = x_{(n-1)} - x_{(2)}$.

13.13 $\gamma = .95956$.

CHAPTER 14

14.1 $r = 15$, $n = 20$, significantly large at 5% level.

14.3 $W = 280$ is significantly small at the level of 1%.

14.5 Accept hypothesis of randomness.

14.7 Reject H_0, $u = 39$, significantly low at 5% level.

14.9 Accept the hypothesis of randomness of data.

14.11 $r = 2$, $n = 14$., significantly small at 1 or 5% level.

CHAPTER 15

15.1 $\hat{\beta} = \Sigma x_i y_i / \Sigma x_i^2$; $E(\hat{\beta}) = \beta$; $V(\hat{\beta}) = \sigma^2 / \Sigma x_i^2$; $\hat{y} = \hat{\beta}x$; $E(\hat{y}) = \beta x = \eta_x$; $100(1-\gamma)\%$ confidence interval for β is: $[\hat{\beta} \pm s_{ey}t_{n-1;\gamma/2}/\sqrt{\Sigma x_i^2}]$; $100(1-\gamma)\%$ confidence interval for $\eta_0 = \beta x_0$ is $[\hat{y}_{x_0} \pm x_0 s_{ey} t_{n-1;\gamma/2}/\sqrt{\Sigma x_i^2}]$. Here, $s_{ey}^2 = \Sigma(y_i - \hat{y}_i)^2 = \Sigma(y_i - \hat{\beta}x_i)^2 = \Sigma y_i^2 - \hat{\beta}^2 \Sigma x_i^2$, and $\hat{y}_{x_0} = \hat{\beta}x_0$.

15.3 $\eta = \hat{\alpha} + \hat{\beta}x = 48.02 - .32x$; 95% confidence interval: for α is $[40.00, 56.04]$; for β is $[-.442, -.192]$; for $\eta_0 = \alpha + \beta x_0$ is

$$\left[y_0 \pm 2.228 \sqrt{\left(\frac{1}{12} + \frac{(x_0 - 60)^2}{6000} \right) 18.97} \right].$$

15.5 $\hat{\eta} = \hat{\alpha} + \hat{\beta}x = 6.2842 + .183x$; 95% confidence interval: for α is $[6.1167, 6.4517]$; for β is $[.174, .192]$; for $\eta_0 = \alpha + \beta x_0$ is

$$\left[y_0 \pm 2.776 \sqrt{\left(\frac{1}{6} + \frac{(x_0 - 17.5)^2}{437.5} \right) .0042} \right].$$

15.7 $\hat{\eta} = 2.70 + .193x$; 95% confidence intervals are: for α: $[2.20, 3.25]$; for β: $[.129, .257]$; and for $\eta_0 = \alpha + \beta x_0$:

$$\left[y_0 \pm 2.776 \sqrt{\left(\frac{1}{16} + \frac{(x_0 - 7)^2}{70} \right) .0374} \right].$$

15.9 The analysis of variance table is

Source	Sum of Squares	Degrees of Freedom	Mean Square
Crude sum of squares	$\Sigma y_i^2 = 3{,}563.48$	10	
Due to $\delta = a + b\bar{x}$	$n\bar{y}^2 = 2{,}729.10$	1	
Due to $\hat{\beta} = b$	$bS(XY) = 724.51$	1	724.51
Residual sum of squares	109.87	8	13.73

Test rejects $\beta = 0$; $\hat{y} = -11.70 + .17x$; $\hat{y}_{145} = 12.95$; 95% confidence interval

for η_{145} is $\left[12.95 \pm 2.306 \sqrt{\left(\dfrac{1}{10} + \dfrac{(145 - 166)^2}{24{,}840} \right) 13.73} \right] = [10.02,\ 15.88]$.

15.11 $r = .862$; observed value of $r \dfrac{\sqrt{n - 2}}{\sqrt{1 - r^2}}$ is $6.131 > t_{13;.025} = 2.160$, so

that $H_0: \rho = 0$ is rejected. A 95% confidence interval for $\frac{1}{2} \log \dfrac{1 + \rho}{1 - \rho}$ is $[1.301 \pm$

$.566] = [.735,\ 1.867]$—see Table IV of Appendix VII. Inversely interpolating in this table, we find that the 95% confidence interval for ρ is $[.626,\ .959]$.

15.13 $\gamma = .996$; Reject $H_0: \rho = 0$; 95% confidence interval for ρ is $[.9825,\ 1]$.

15.15 $\hat{\alpha} = a = -63.5624$; $\hat{\beta}_1 = b_1 = 1.0742$; $\hat{\beta}_2 = b_2 = 1.0153$; $\hat{y} = 88.07$; $S_{ey}^2 = 970.9925/17 = 57.12$.(d) 95% confidence interval for $E(y \mid 75{,}70)$ is $[\hat{\eta}_{75.70} \pm t_{17;.025}\sqrt{AS_{ey}^2}] = [(88.07) \pm 2.110\sqrt{(.167308)(57.12)}] = [81.55,\ 94.59]$.

15.17 $\hat{y} = 18$ is the least squares estimate of $E[y \mid 0, 0]$. $b_1 = 1.4960$; $b_2 = .8774$; 95% confidence region for $E[y \mid 0, 0]$ is $[15.78,\ 20.22]$; a 99% (joint) confidence region for (β_1, β_2) is $2(11.56)(8.02) = 185.42 = 98(1.4960 - \beta_1)^2 + 32(1.4960 - \beta_1)(.8774 - \beta_2) + 65(.8774 - \beta_2)^2$.

15.19 $\hat{y} = -28.04 + 757.41x$.

15.21 $a = 57.10$; $b = 577.507$; observed value of $|t_{16}|$ is $34.68 > t_{16;.025} = 2.120$ and so hypothesis $H_0: \beta = 0$ is rejected.

15.23 $\hat{v} = 30.22t$.

CHAPTER 16

16.1 (a) ANOVA

Source	Degrees of Freedom	Sum of Squares	Mean Square	F-ratio
Preparations	3	877.2	292.4	9.766
Error	5	149.7	29.94	
Total	8	1026.9		

(b) We reject H_0, and estimate the effects to be: $\hat{\delta}_A = 10^{-6} \times (-11.39)$, $\hat{\delta}_B = 10^{-6} \times 6.11$, $\hat{\delta}_C = 10^{-6} \times (-6.22)$, $\hat{\delta}_D = 10^{-6} \times (14.61)$.

16.3 We reject H_0 and estimate the δ_i's as: $\hat{\delta}_1 = .0036$, $\hat{\delta}_2 = -.00844$, $\hat{\delta}_3 = .00256$.

16.5 (a) ANOVA

Source	Degrees of Freedom	Sum of Squares	Mean Square	F-ratio
Analysts	2	.031	.0155	2.0395
Batches	5	93.556	18.7112	2,462.0
Error	10	.076	.0076	
Total	17	93.663		

(b) We accept H_0 (c) We reject H_0' and estimate the effects to be: $\hat{\delta}_{.1} = 3.961$, $\hat{\delta}_{.2} = -1.339$, $\hat{\delta}_{.3} = -3.106$, $\hat{\delta}_{.4} = 1.661$, $\hat{\delta}_{.5} = -0.1039$, $\hat{\delta}_{.6} = -1.139$.

16.7 (a) ANOVA

Source	Degrees of Freedom	Sum of Squares	Mean Square	F-ratio
Machines	2	.371903	.1859515	4.984
Days	8	.365446	.0456808	1.224
Error	16	.596951	.0373094	
Total	26	1.334300		

(b) We accept the hypothesis of zero effects due to days   (c) We reject the hypothesis of zero effects due to machines, and estimate effects to be: $\hat{\delta}_{1.} = .142 = \hat{\delta}_A$, $\hat{\delta}_{2.} = -.145 = \hat{\delta}_B$, $\hat{\delta}_{3.} = .003 = \hat{\delta}_C$.

16.9 ANOVA

Source	Degrees of Freedom	Sum of Squares	Mean Square	F-ratio
Cloths	2	56.6742	28.3371	7.879
Machines	8	76.7803	9.5975	2.669
Interaction	16	69.8448	4.3653	1.214
Error	27	97.1017	3.5964	
Total	53	300.4010		

We accept the hypothesis of zero effects due to interaction. We reject the hypothesis of zero effects due to cloths and machines; we estimate the effects due to cloths as follows: $\hat{\delta}_{1.} = \hat{\delta}_A = 1.046$, $\hat{\delta}_{2.} = \hat{\delta}_B = -1.391$, $\hat{\delta}_{3.} = \hat{\delta}_C = .345$; we estimate the effects due to machines as: $\hat{\delta}_{.1} = -.5272$, $\hat{\delta}_{.2} = .4928$, $\hat{\delta}_{.3} = .3744$, $\hat{\delta}_{.4} = -1.9322$, $\hat{\delta}_{.5} = -.6589$, $\hat{\delta}_{.6} = -.2539$, $\hat{\delta}_{.7} = 1.2828$, $\hat{\delta}_{.8} = 2.2711$, $\hat{\delta}_{.9} = -1.0489$.

16.11   (a) ANOVA

Source	Degrees of Freedom	Sum of Squares	Mean Square	F-ratio
Rows	5	2.1897	.4379	2.511
Columns	5	2.5743	.5149	2.952
Humidity	5	23.5301	4.7060	2.6984
Error	20	3.4886	.1744	
Total	35	31.7827		

We *reject* H_0 "that effects due to humidity are zero," and estimate the effects to be: $\delta_A = 1.378$, $\delta_B = .460$, $\delta_C = .288$, $\delta_D = -.328$, $\delta_E = -.933$, $\delta_F = -.865$.

16.13   Reject H_0 if $F = \dfrac{(S_A + S_B)/(r + s - 2)}{S_E/(r - 1)(s - 1)} > F_{r+s-2,(r-1)(s-1);\alpha}$ and accept otherwise.

16.15   We accept the hypothesis of zero effects due to "Ages."

CHAPTER 17

17.1   (i) $y_{ij} = \mu + \delta_{i.} + \delta_{.j} + \epsilon_{ij}$, where y_{ij} are the observed penicillin yields, $i = 1, \ldots, 4$; $j = 1, \ldots, 5$; $\delta_{i.}$ are the treatment effects, $\delta_{.j}$ are the batch effects, and the ϵ_{ij} are $NID(0, \sigma^2)$. The corresponding analysis of variance table is:

	Σy^2	112480	20		
	CF	95772.8	1		
	$S(Y^2)$	16707.2	19		
Batch	SSq	1098.2	4	274.55	$F_{4,12} = 22.56$
Treatment	SSq	15463.6	3	5154.53	$F_{3,12} = 423.54$
Residual	SSq	145.4	12	$12.17 = s^2$	

(ii) The hypothesis that there are no treatment differences, that is, that the $\delta_{i.} = 0$, is rejected, since the observed ratio $F_{3,12} = 423.54 > F_{3,12,.05} = 3.49$.
(iii) The hypothesis that there are no differences between batches, that is, that the $\delta_{.j} = 0$ is rejected, since the observed ratio $F_{4,12} = 22.56 > F_{4,12,.05} = 3.26$.
(iv) x_1(pH effect): $\bar{y}_+ - \bar{y}_- = 5.18$
      x_2(Temperature effect): $\bar{y}_+ - \bar{y}_- = 20.2$

(v)

Source	SSq	Degrees of Freedom	Mean Square
x_1(pH effect)	13416.2	1	13416.2
x_2(temperature effect)	2040.2	1	2040.2
$x_1 x_2$(interaction)	7.2	1	7.2
Treatment SSq	15463.6	3	

(vi) $s^2 = (145.4)/12 = 12.17$ with $\nu = 12$ degrees of freedom; pH × temperature interaction effect: $(\bar{y}_+ - \bar{y}_-) = -1.2$; 95% confidence limits for the interaction effect:

$$-1.2 \pm 2.179 \sqrt{\frac{1}{(5)(2)}} \quad (12.17) \quad \text{or} \quad -1.2 \pm 2.4$$

		Penicillin Yields					Treatments Treatment	
Treatments	Temper-						Totals	Averages
pH	ature	Corn	Steep	Liquor	Batches			
−	−	40	35	28	27	33	163	32.6
+	−	95	80	94	76	83	428	85.6
−	+	66	50	48	45	51	270	54.0
+	+	124	98	105	96	100	523	104.6
Batch totals		325	263	275	244	277	1384	

17.3 (i)

x_1	x_2	x_3	1	2	3	Total	Yates	Algorithm		Estimates for
−	−	−	2.58	2.66	2.74	7.98	17.24	35.32	67.11	$2.80 = \bar{y}$: grand mean
+	−	−	3.04	2.96	3.26	9.26	18.08	31.79	3.63	$.3025 : x_1$ effect
−	+	−	2.81	2.63	3.07	8.51	15.67	2.34	1.29	$.1075 : x_2$ effect
+	+	−	3.12	3.17	3.28	9.57	16.12	1.29	0.09	$.0075 : x_1 x_2$ effect
−	−	+	2.45	2.49	2.65	7.59	1.28	.84	−3.53	$.2942 : x_3$ effect
+	−	+	2.65	2.62	2.81	8.08	1.06	.45	−1.05	$-.0875 : x_1 x_3$ effect
−	+	+	2.45	2.54	2.67	7.66	.49	−.22	−.39	$-.0325 : x_2 x_3$ effect
+	+	+	2.74	2.72	3.00	8.46	.80	.31	.53	$.0442 : x_1 x_2 x_3$ effect
Totals			21.84	21.79	23.48	67.11				

Replicate averages 2.7300 2.7238 2.9350 (ii) The analysis of variance table is:

		SSq	Degrees of Freedom		
	Σy^2	189.1471	24		
	CF	187.6563	1		
	$S(Y^2)$	1.4908	23		
Replicate	SSq	.2312	2	.1156	$F_{2,12} = 24.0833$
Treatment	SSq	1.2017	7	.1717	$F_{7,12} = 35.7708$
		.0577	12	$.0048 = s^2$	

(iii) Since the ratio $F_{2,12} = 24.088 > F_{2,12:.05} = 3.89$ the hypothesis that there are no differences between replicates is rejected.
(iv) Contrast: $\Sigma c_i \bar{y}_i = \frac{1}{2}(2.7300) + \frac{1}{2}(2.7238) - (2.9350) = -.2081$. Contrast sum of squares $= 8(-.2081)^2/1.5 = .2309$ with one degree of freedom. $F_{1,12} = .2309/.0048 = 48.12$. Since the ratio $F_{1,12} = 48.12 > F_{1,12:.05} = 4.75$ we reject the hypothesis that the constrast effect equals zero.
(v) The 95% confidence limits for the x_1 effect are given by

$$.3025 \pm 2.179 \sqrt{\frac{1}{3(4)}} \, (.0048) = .3025 \pm .0436$$

17.5 Treatments (i)

x_1	x_2	x_3			Totals		Yates Algorithm			Estimates are for
−	−	−	14.6	21.7	36.3	72.2	131.7	287.9		$17.99 = \bar{\bar{y}}$: grand mean
+	−	−	16.4	19.5	35.9	59.5	156.2	−14.3		-1.7875:x_1 effect
−	+	−	15.2	17.6	32.8	80.5	−6.5	−17.5		-2.1875:x_2 effect
+	+	−	10.2	16.5	26.7	75.7	−7.8	−12.1		-1.5125:x_1x_2 effect
−	−	+	19.3	21.3	40.6	−.4	−12.7	24.5		3.0625:x_3 effect
+	−	+	17.4	22.5	39.9	−6.1	−4.8	−1.3		$-.1625$:x_1x_3 effect
−	+	+	17.4	24.0	41.4	−.7	−5.7	7.9		$.9875$:x_2x_3 effect
+	+	+	16.0	18.3	34.3	−7.1	−6.4	−.7		$.0875$:Block $x_1x_2x_3$ effect

Block effect and three factor interaction effect are confounded.

Block	1	2	3	4
Totals	59.6	66.9	76.7	84.7
Averages	14.900	16.725	19.175	21.175

(ii) The analysis of variance table is:

	Σy^2	5355.1900	16	
	CF	5180.4006	1	
	$S(Y^2)$	174.7894	15	
Blocks	SSq	90.7869	3	
Treatment	SSq	82.6244	6	
Error	SSq	1.3781	6	$0.2297 = s^2$

(iii) Linear contrast between blocks is $-3(14.900) - 1(16.725) + 1(19.175) + 3(21.175) = 21.275$. Sum of squares for this contrast equals $4(21.275)^2/20 = 90.5251$ with one degree of freedom. Since $F_{1,6} = 90.5251/0.2297 = 394.1$ is greater than $F_{1,6;.05} = 5.99$ we reject the hypothesis that no linear trend exists amongst the four batches.

17.7 Data from Block 1 are:

x_1	x_2	x_3	x_4	Obs.		Yates Algorithm			Estimates are for
−	−	−	−	3.43	7.47	14.90	29.61		$3.70125 = \bar{\bar{y}}$:grand mean
+	−	−	+	4.04	7.43	14.71	1.13		$.2825$:$x_1 + x_2x_3x_4$ effect
−	+	−	+	3.57	7.36	.90	.05		$.0125$:$x_2 + x_1x_3x_4$ effect
+	+	−	−	3.86	7.35	.23	1.55		$.3875$:$x_1x_2 + x_3x_4$ effect
−	−	+	+	4.09	.61	−.04	−.19		$-.0475$:$x_3 + x_1x_2x_4$ effect
+	−	+	−	3.27	.29	−.01	−.67		$-.1675$:$x_1x_3 + x_2x_4$ effect
−	+	+	−	3.15	−.82	−.32	.03		$.0075$:$x_2x_3 + x_1x_4$ effect
+	+	+	+	4.20	1.05	1.87	2.19		$.5475$:$x_1x_2x_3 + x_4$ effect

Σy^2	110.7145	8
CF	109.5940	1
$S(Y^2)$	1.1205	7
x_1	.1596	1
x_2	.0003	1
x_3	.0045	1
x_4	.5995	1
$x_1 x_2 + x_3 x_4$	.3003	1
$x_1 x_3 + x_2 x_4$	.0561	1
$x_2 x_3 + x_1 x_4$	.0001	1
Residual	zero	0

The generator of this 2^{4-1} fractional factorial design is **1 2 3 4**

Three factor interactions are assumed equal to zero. No replication is available to provide a measure of σ^2. It is difficult to decide whether any of the estimated effects is really indistinguishable from errors alone. The largest effects: x_4, $x_1 x_2 + x_3 x_4$ and x_1 provide no easily understood explanation to the response changes.

17.9   First-order model: $y = \beta_0 + \beta_1 x_1 + \beta_2 x_2 + \epsilon$; $\epsilon \rightarrow NID(0, \sigma^2)$. Normal equations:

$$20b_0 + 0b_1 + 0b_2 = 1384$$
$$0b_0 + 20b_1 + 0b_2 = 518$$
$$0b_0 + 0b_1 + 20b_2 = 202$$

Thus $\hat{y} = 69.2 + 25.9 x_1 + 10.1 x_2$. Analysis of variance:

		Σy^2	112480	20	
SSq	b_0		95772.8	1	
SSq	b_1		13416.2	1	
SSq	b_2		2040.2	1	
Residual	SSq		1250.8	17	$73.576 = s^2$

Given the assumption the postulated model is appropriate to represent the response, $s^2 = 1250.8/17 = 73.576$. Since $F_{1,17,0.05} = 4.45$ is less than the observed values of $F_{1.17}$ appropriate to testing the separate hypothesis that $\beta_1 = 0$, and $\beta_2 = 0$ the fitted model is used to produce the following contours. *Example computation for the 50 response contour:* Set $\hat{y} = 50$ to give $0 = 19.2 + 25.9 x_1 + 10.1 x_2$. Points (x_1, x_2) on this straight line contour are $(0, -1.90)$; $(-.74, 0)$. Points on the 60 contour are $(0, -.91)$, $(-.36, 0)$. The contours are plotted in Figure 1.

17.11   (ii) Model: $y = \beta_0 + \beta_1 x_1 + \beta_2 x_2 + \epsilon$; ϵ are $NID(0, \sigma^2)$. Normal equations:

$$10b_0 + 0b_1 + 0b_2 = 42.3000$$
$$0b_0 + 3.0b_1 + 0b_2 = 9.3500$$
$$0b_0 + 0b_1 + 3b_2 = 3.2487$$

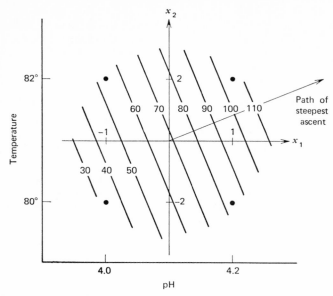

Figure 1. Contours of penicillin yields as a function of pH and temperature.

Solutions: $b_0 = 4.23$, $b_1 = 3.12$, $b_2 = 1.08$. Thus the fitted first-order model is $\hat{y} = 4.23 + 3.12x_1 + 1.08x_2$.

	Σy^2	212.03	10		
	$SSq\ b_0$	178.9290	1		
	$SSq\ b_1$	29.1408	1	29.1408	$F_{1,7} = 461.31$
	$SSq\ b_2$	3.5180	1	3.5180	$F_{1,7} = 55.69$
Residual	SSq	.4422	7	.06317	
Lack of Fit	SSq	.3147	4	.0787	$F_{4,3} = 1.85$
Error	SSq	.1275	3	.0425	

(iii) Since the observed lack of fit F ratio is smaller than $F_{4,3..05} = 9.12$ we accept the postulated model as being adequate to represent the data   (iv) Testing the hypothesis that $\beta_1 = 0$ yields an $F_{1,7} = 461.31$. Similarly the test that $\beta_2 = 0$ yields an $F_{1,7} = 55.69$. Since both these observed ratios are greater than $F_{1,7..05} = 5.59$, the individual hypotheses $\beta_1 = 0$ and $\beta_2 = 0$ are rejected. A test of the combined hypothesis that $\beta_1 = 0$ *and* $\beta_2 = 0$ is provided by $F_{2,7} = 16.3294/.06317 = 258.50$. Since $F_{2,7..05} = 4.74$ this joint hypothesis is also clearly rejected. Both x_1 and x_2 are required to explain the data   (v) The contours of the fitted response are given in Figure 2. The contour $y = 3$ is obtained from the fitted model, thus: $3.0 = 4.23 + 3.12x_1 + 1.08x_2$. Points lying on this contour line are $(0, -1.14)$; $(-.39, 0)$.

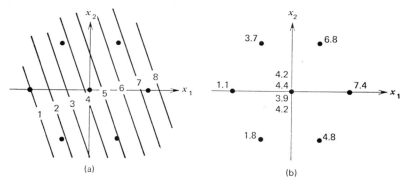

Figure 2. Response contours; $y = 4.23 + 3.12x_1 + 1.08x_2$.

17.13   (i)

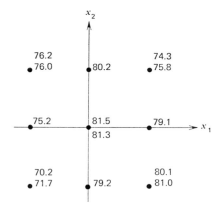

(ii) Second-order model: $y = \beta_0 + \beta_1 x_1 + \beta_2 x_2 + \beta_{11} x_1^2 + \beta_{22} x_2^2 + \beta_{12} x_1 x_2 + \epsilon$, $\epsilon \rightarrow NID(0, \sigma^2)$. Normal equations:

$$14b_0 + 0b_1 + 0b_2 + 10b_{11} + 10b_{22} + 0b_{12} = 1081.8$$
$$0b_0 + 10b_1 + 0b_2 + 0b_{11} + 0b_{22} + 0b_{12} = 21.0$$
$$0b_0 + 0b_1 + 10b_2 + 0b_{11} + 0b_{22} + 0b_{12} = .3$$
$$10b_0 + 0b_1 + 0b_2 + 10b_{11} + 8b_{22} + 0b_{12} = 759.60$$
$$10b_0 + 0b_1 + 0b_2 + 8b_{11} + 10b_{22} + 0b_{12} = 764.7$$
$$0b_0 + 0b_1 + 0b_2 + 0b_{11} + 0b_{22} + 8b_{12} = -21.3$$
$$b_1 = 21.0/10 = 2.10$$
$$b_2 = .3/10 = .03$$
$$b_{12} = -21.3/8 = -2.6625$$

Solving for b_0, b_1 and b_{22} given $b_0 = 81.336$, $b_{11} = -4.121$ and $b_{22} = -1.570$.

The fitted second-order model is: $\hat{y} = 81.34 + 2.10x_1 + 0.03x_2 - 4.12x_1^2 - 1.57x_2^2 - 2.66x_1x_2$ (iii) To determine whether the model is adequate to represent the data we set up the analysis of variance table and perform the lack of fit test.

Σy^2	83762.9800	14
$SSq\ b_0 = (1081.8)^2/14$	83592.23142	1
Corrected SSq	170.74858	13
SSq, first-order terms $= 2.10(21) + .03(.3) =$	44.10900	2
SSq, quadratic term $(-4.121)(759.6) + (-1.570)(764.7)$ $+ (81.336)(1081.8) - SSb_0 =$	66.16278	2
SSq, crossproduct term $(-2.6625)(-21.3)$	56.71125	1
Residual SSq	3.76555	1
	1.07055	3	.35685
			$F_{3.5} = 0.66 < F_{3.5,.05}$
Error $SSq = \Sigma d^2/2$	2.69500	5	.53900

The second-order model is adequate. *Note:* Rounding errors can be serious in these computations. One method for reducing their influence is to subtract a constant C from all the observations (for example, the average, or for this example $C = 70$ would be convenient). This will prevent the sum of squares for b_0 from dominating the computations of the sum of squares of the quadratic coefficients. Of course, the estimated b_0 obtained from the modified data will estimate $(\beta_0 - C)$.

Index